ADVANCES IN CHEMICAL PHYSICS

VOLUME LX

SERIES EDITORIAL BOARD

ADVANCES IN CHEMICAL PHYSICS—VOLUME LX

I. Prigogine and Stuart A. Rice—Editors

PHOTODISSOCIATION AND PHOTOIONIZATION

Edited by

K. P. LAWLEY

Department of Chemistry
Edinburgh University

A WILEY–INTERSCIENCE PUBLICATION

JOHN WILEY & SONS

CHICHESTER · NEW YORK · BRISBANE · TORONTO · SINGAPORE

Library of Congress Cataloging in Publication Data:
Main entry under title:

Photodissociation and photoionization.

(Advances in chemical physics; v. 60)
'A Wiley–Interscience publication.'
Includes index.
1. Photoionization—Addresses, essays, lectures.
2. Dissociation—Addresses, essays, lectures.
I. Lawley, K. P. II. Series.
QD453.A27 [QD561] 541.3'5 84-17333

ISBN 0 471 90211 X

British Library Cataloguing in Publication Data:

Lawley, K. P.
Photodissociation and photoionization.—
(Advances in chemical physics)
1. Photoionization 2. Dissociation
3. Molecules
I. Title II. Series
539'.6 QD561

ISBN 0 471 90211 X

Photoset at Thomson Press (India) Limited, New Delhi
Printed at The Bath Press, Avon

CONTRIBUTORS TO VOLUME LX

G. G. BALINT-KURTI, School of Chemistry, Bristol University, Bristol, BS8 1TS, UK

P. BRUMER, Department of Chemistry, University of Toronto, Toronto, Ontario, Canada

D. K. EVANS, Physical Chemistry Branch, Atomic Energy of Canada Limited, Chalk River Nuclear Laboratories, Chalk River, Ontario, KOJ 1JO, Canada

K. C. JANDA, Division of Chemistry and Chemical Engineering, California Institute of Technology, Pasadena, California 91125, USA

K. KIMURA, Institute for Molecular Science, Okazaki 444, Japan

L. V. LUKIN, Institute of Chemical Physics of USSR Academy of Sciences, Chernogolovka 142432, USSR

R. D. MCALPINE, Physical Chemistry Branch, Atomic Energy of Canada Limited, Chalk River Nuclear Laboratories, Chalk River, Ontario, KOJ 1JO, Canada

J. T. MOSELEY, Department of Physics and Chemical Physics Institute, University of Oregon, Eugene, Oregon 97403, USA

H. REISLER, Chemistry Department, University of Southern California, Los Angeles, California 90089-0484, USA

M. SHAPIRO, Department of Chemical Physics, Weizmann Institute of Science, Rehovet 766100, Israel

J. TELLINGHUISEN, Department of Chemistry, Vanderbilt University, Nashville, Tennessee 37235, USA

C. WITTIG, Chemistry Department, University of Southern California, Los Angeles, California 90089-0484, USA

B. S. YAKOVLEV, Institute of Chemical Physics of USSR, Academy of Sciences, Cherngolovka 142432, USSR

INTRODUCTION

Few of us can any longer keep up with the flood of scientific literature, even in specialized subfields. Any attempt to do more, and be broadly educated with respect to a large domain of science, has the appearance of tilting at windmills. Yet the synthesis of ideas drawn from different subjects into new, powerful, general concepts is as valuable as ever, and the desire to remain educated persists in all scientists. This series, *Advances in Chemical Physics*, is devoted to helping the reader obtain general information about a wide variety of topics in chemical physics, which field we interpret very broadly. Our intent is to have experts present comprehensive analyses of subjects of interest and to encourage the expression of individual points of view. We hope that this approach to the presentation of an overview of a subject will both stimulate new reasearch and serve as a personalized learning text for beginners in a field.

Ilya Prigogine

Stuart A. Rice

CONTENTS

Photodissociation and Photoionization
Edited by K. P. Lawley

MULTIPHOTON IONIZATION OF GASEOUS MOLECULES*

HANNA REISLER and CURT WITTIG

Chemistry Department,
University of Southern California,
Los Angeles, CA 90089-0484, USA

CONTENTS

I. INTRODUCTION

Multiphoton ionization (MPI) has been around for quite a long time and is too varied and complex a subject to review succinctly. We will therefore limit our comments to areas of interest to chemists, and in particular to the ionization of molecules and gaseous free radicals. We will not discuss the ionization of atoms, which has been studied thoroughly by the atomic physics community, even though a great deal of very thoughtful research has transpired in this area. For the reader who is interested in this subject, or who would like to see the theory developed in the most complete and elegant forms, we refer to recent reviews concerning the MPI of atoms.[1,2] Also, we will not discuss photoionization in condensed media,[3] although we will discuss collisional environments in the gas phase. In order to avoid confusion, let us decide to include sequential excitations through stationary intermediates (e.g. $A + h\nu \rightarrow A^*$; $A^* + h\nu \rightarrow A^+ + e$) under the umbrella of multiphoton ionization, even though many people would rather refer to such serial processes as 'multiple photon' ionization.

*Research supported by the U.S. National Science Foundation.

In Section II, several subjects of current interest will be discussed, and examples will be given in order to illustrate particular points. We do not intend to provide a historical summary of the development of MPI, since this would be terribly arduous and of questionable benefit in an area that is moving so quickly. Instead, we will focus on work which represents where the field is going and use examples from different research groups to punctuate our arguments. Every effort will be made to be critical and not simply act as a library retrieval service. In Section III, we will summarize what we believe are the most and least fruitful areas for future research, again trying to provide insight into a 'field' which is not really a field. In covering the literature as we have, it is inevitable that certain pieces of significant research are not referenced, when they should be. We extend in advance our apologies to parties who are thus slighted, and admit that our perspective reflects mainly our own interests, rather than a balanced overview of the whole subject.

II. RECENT RESULTS AND OBSERVATIONS

The pioneering work of Johnson and coworkers[4–9] and Dalby and coworkers[10–12] demonstrated conclusively that molecules could be efficiently ionized using MPI. This was a significant finding, since there was widespread fear at the time that molecules could not be pumped via multiphoton excitation to levels from which ionization could occur, because of access to

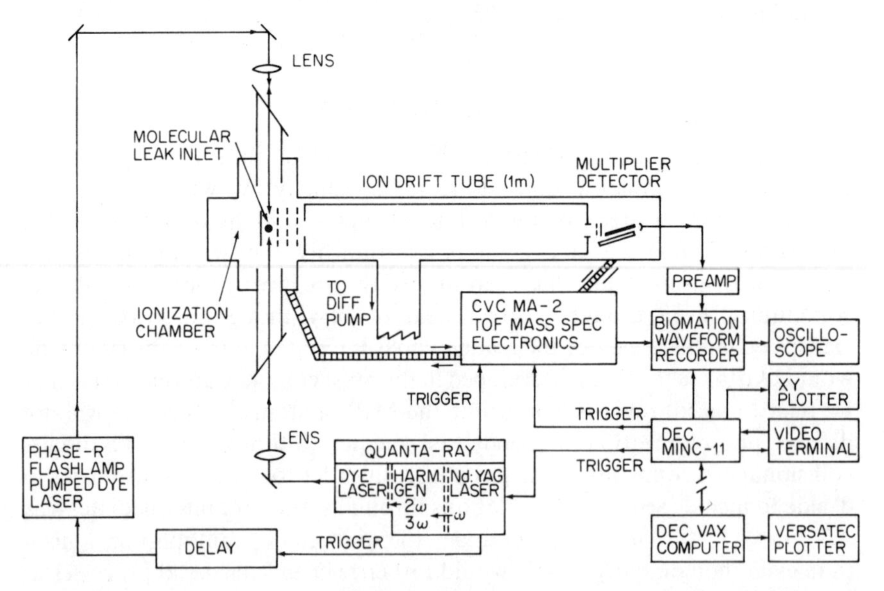

Fig. 1. Schematic diagram of the MPI apparatus used by Newton and Bernstein.[28] The phase-R dye laser pulse or electron ionization pulse could be triggered before or after the quanta-ray laser pulse.

dissociation channels at lower energies. Following these and other early contributions by several groups,[13–16] MPI developed momentum, receiving a major boost in the late 1970s and early 1980s due to experiments such as those from the research groups of Bernstein,[17–30] Smalley,[31–40] Schlag[41–48] and Zare.[49–53] By combining time-of-flight (TOF) mass analysis and computer-based data acquisition, they were able to accumulate data at an astounding rate. For groups which had previously worked at the technological frontiers of low S/N, it must have been gratifying to see entire mass spectra accumulate in < 1 s. Work in this area has continued at a very respectable pace until now, and it is appropriate at this time to take a hard look at MPI in order to soberly assess future prospects. The situation is similar in many ways to the 'discovery'

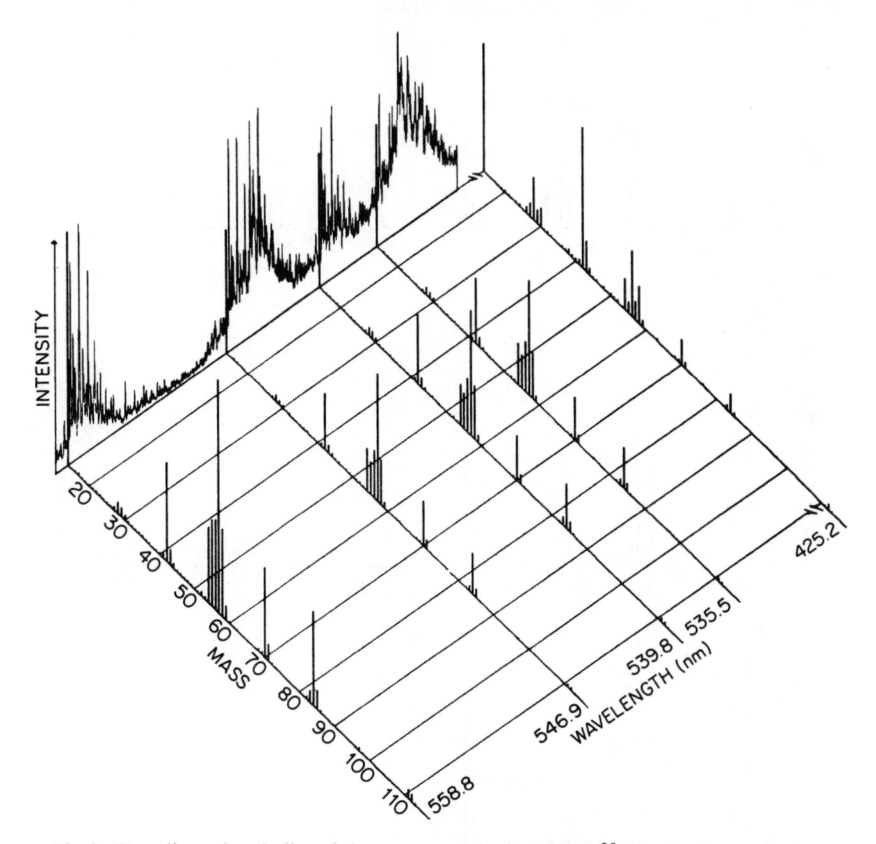

Fig. 2. Two-dimensional vibronic/mass spectrum of DABCO.[20] The 'back panel' gives essentially the total ion wavelength spectrum (actually the sum of the intensity of the principal ion 'clump' around $m/e = 57$); the 'stick diagrams' are digitized REMPI fragmentation patterns at the indicated laser wavelengths. The vertical line at the extreme left edge of each mass spectrum represents the sum of all the peaks shown in that spectrum, with a constant scaling factor throughout, except for that at 425.2 nm, where the entire spectrum has been scaled down by a factor of 10.

of infrared multiple photon excitation (IRMPE), in which a myriad of experimental observations were reported during a short period, but the more important research took some time to arrive.

Although it is possible in many instances to arrange conditions such that MPI leads predominantly to parent ions, this is not at all the rule. Fragmentation is often so severe that the ions produced do not resemble the starting material in any meaningful way. Using pulsed laser energies of several millijoules and focused laser beams, a complicated combination of excitation, fragmentation and ionization events lead to cracking patterns which depend

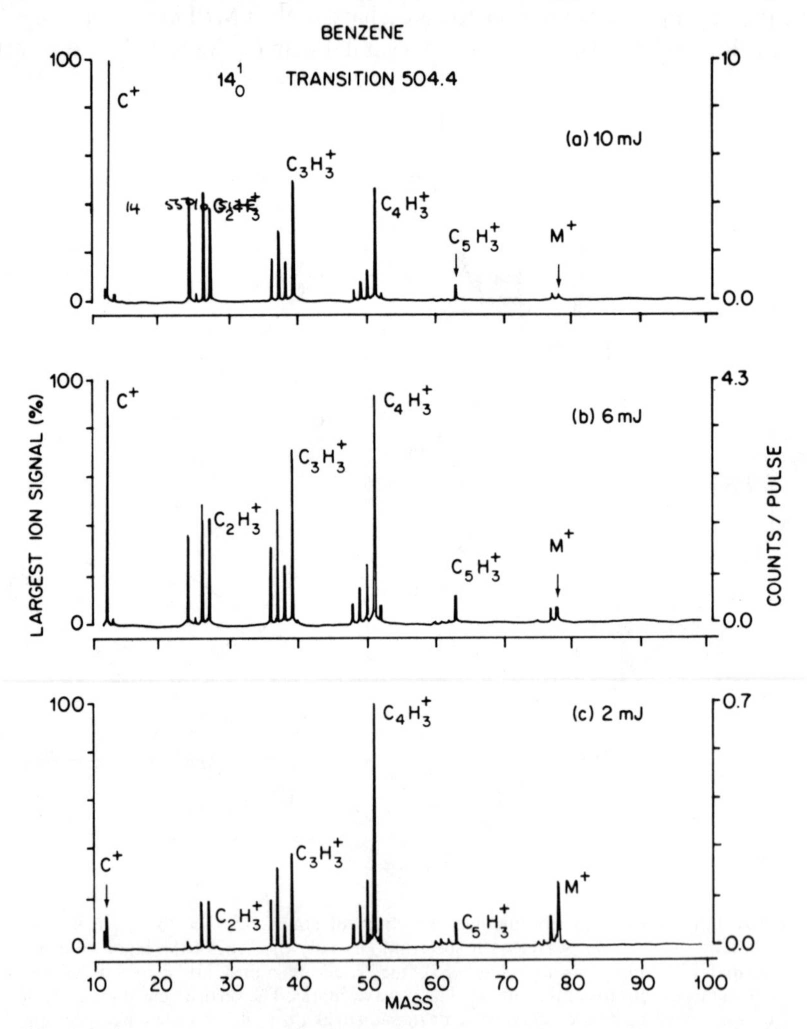

Fig. 3. The actual mass fragmentation patterns obtained for benzene of the 14^1 transition at the stated laser pulse energies.[29]

on both laser intensity and frequency. Several groups have tried to determine the pathways whereby ions are produced and the results are quite impressive for individual experiments.[17–30,54–60] However, there is little of a truly general nature which can be derived and one wonders if each case is not somewhat special. From conventional electron impact mass spectroscopy, we have learned that isolated electronically excited states of ions are the exception,[54–56] and therefore ions once formed will tend to fragment as per statistical theories.[61–65] The excitation/fragmentation of neutrals is more devious, and from the mass spectra alone it is hard to say whether the parent molecule dissociated and the fragments were then ionized or the ion was first produced and then fragmented.[66,67]

The work of Bernstein and coworkers[17–30] has provided massive amounts of information concerning intensity and wavelength-dependent fragmentation patterns of different types of molecules undergoing MPI. By developing a sophisticated computer controlled 'MPI machine', similar to that of Smalley's research group,[31–40] they were able to accumulate two-dimensional spectra (wavelength versus mass) at an enormous rate. The experimental arrangement is shown in Fig. 1 and typical data are shown in Fig. 2. One must bear in mind that data such as those shown in Fig. 2 were obtained with particular laser energies and that such two-dimensional spectra are therefore not unique. Examples of extreme cracking are shown in Figs. 3 to 5. In the case of benzene, C^+ is the dominant ion under conditions of high fluence, even though this species derives from the absorption of many photons, starting with C_6H_6. From the point of view of using MPI as a diagnostic tool, the intensity-dependent fragmentation is annoying, and the severe fragmentation at high fluence makes species identification impossible without a precise absorption signature.

The case of UF_6 is even more impressive.[68–70] In addition to U^+ being the dominant singly charged ion, the U^{2+} peak can be made as large as the U^+ peak[68] and U^{3+} is also detected (Fig. 6).[69] These highly charged species require that many photons are annihilated for their production, and Rhodes[71] has recently announced the MPI synthesis of highly charged uranium atoms which require ≥ 99 photons apiece for their manufacture. In sufficient concentration, these unique species could lead to coherent X-ray emission following charge transfer.

In scrutinizing the various cracking patterns, several groups[61–65] have pointed out that statistical treatments of the fragmentation which rely mainly on the information theoretic approach are in sensible accord with most observations. This is not surprising, in light of the tendency of ions to dissociate via ground electronic potential surfaces. Exceptions to the statistical distributions are not serious and are also of little significance because of the many experimental unknowns and the fact that there is no a priori reason to expect the distributions of fragments to be statistical. Thus, the information

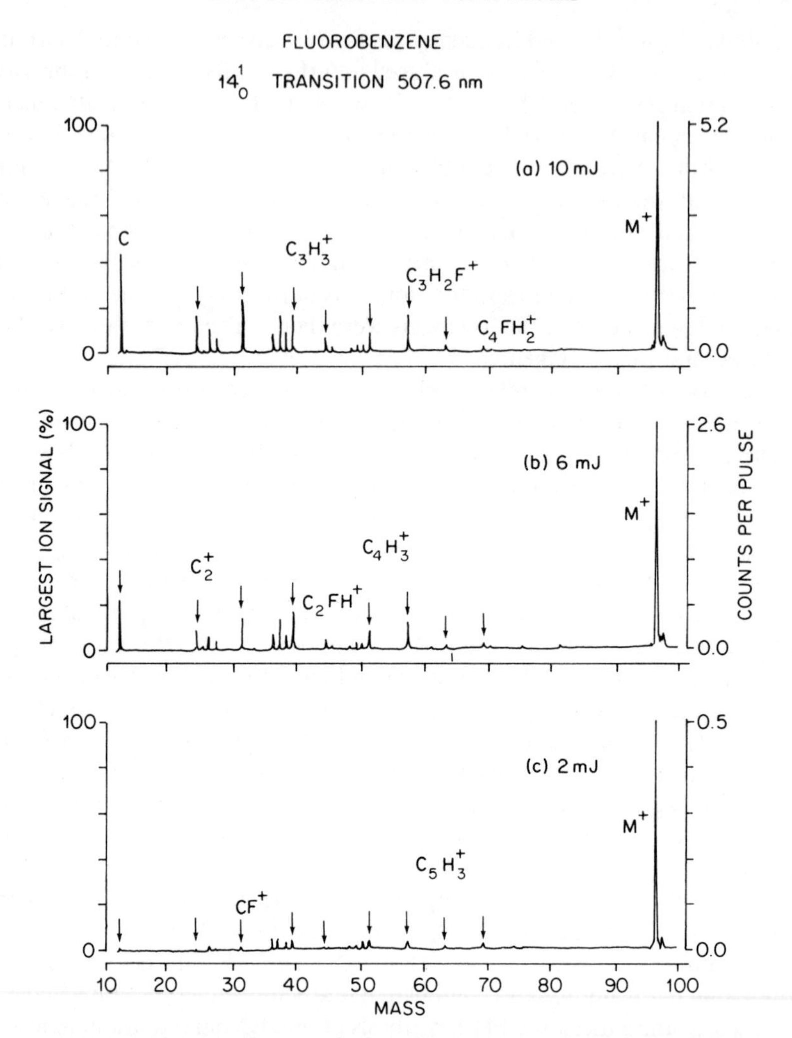

Fig. 4. Actual mass fragmentation patterns obtained for fluorobenzene on the 14^1 transition at 507.6 nm at the stated laser pulse energies.[29] These mass spectra were obtained with a 0.25-m focal length lens at nominal ion gauge pressures of 5.0×10^{-5} torr.

theoretic approach provides a convenient means of cataloguing data and making educated guesses.

Smalley and coworkers [31–40] have pioneered many of the most elegant applications of MPI, wherein the method is used to prepare or analyse unique species and to probe molecular systems with exquisite detail. They showed that nearly 100 per cent ionization efficiency could be achieved with pulsed laser intensities $\sim 10^7$ W cm^{-2},[31] and were first to introduce the computer-

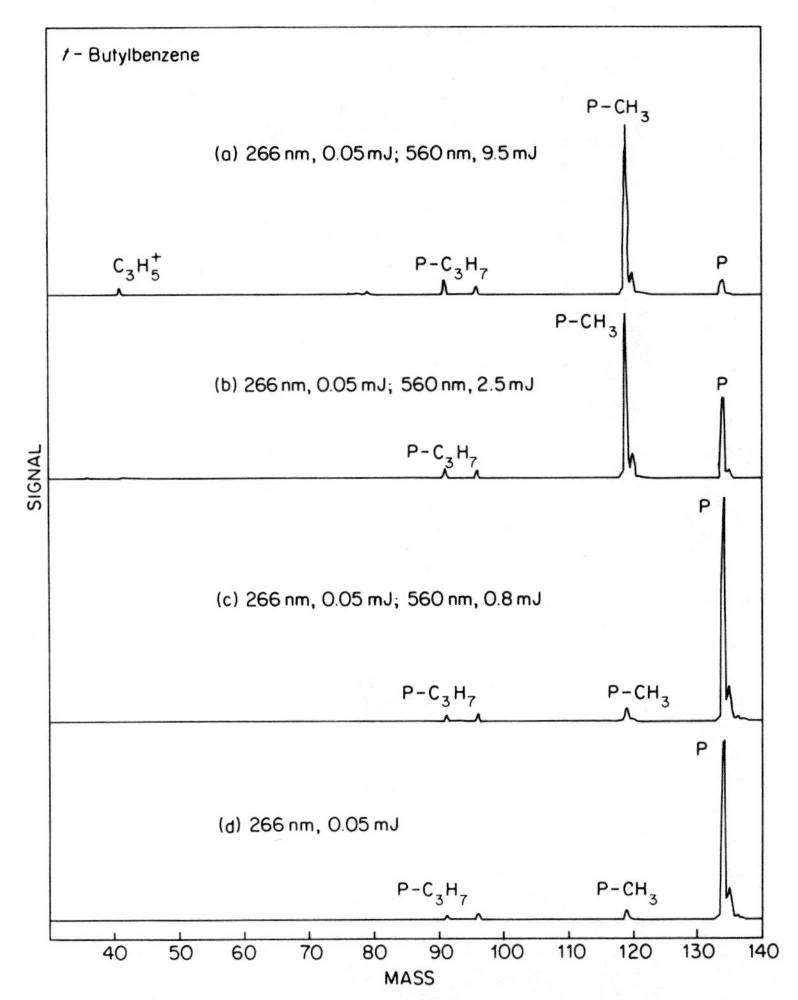

Fig. 5. Power dependence of two-colour fragmentation of tert-butylbenzene, showing the loss of a methyl group from the previously formed parent ion.[28] At the higher pulse energies several new fragments are seen to appear. Note the near complete destruction of the parent ion at the highest energy. The 'standard' MPI mass spectrum is shown in (d) for comparison.

controlled combination of TOF and free jet expansion technologies which has proven so invaluable in this area. They were also the first to use MPI for the detection of molecular triplets,[33,36,40] and by selectively exciting the $S_1(6^1)$ level of benzene, which undergoes intersystem crossing to the triplet, they were able to prepare vibrationally excited T_1, which subsequently decays to S_0 under completely collision-free conditions.[33] A partial energy level diagram indicating the relevant processes is shown in Fig. 7. Ionization from S_1 and/or

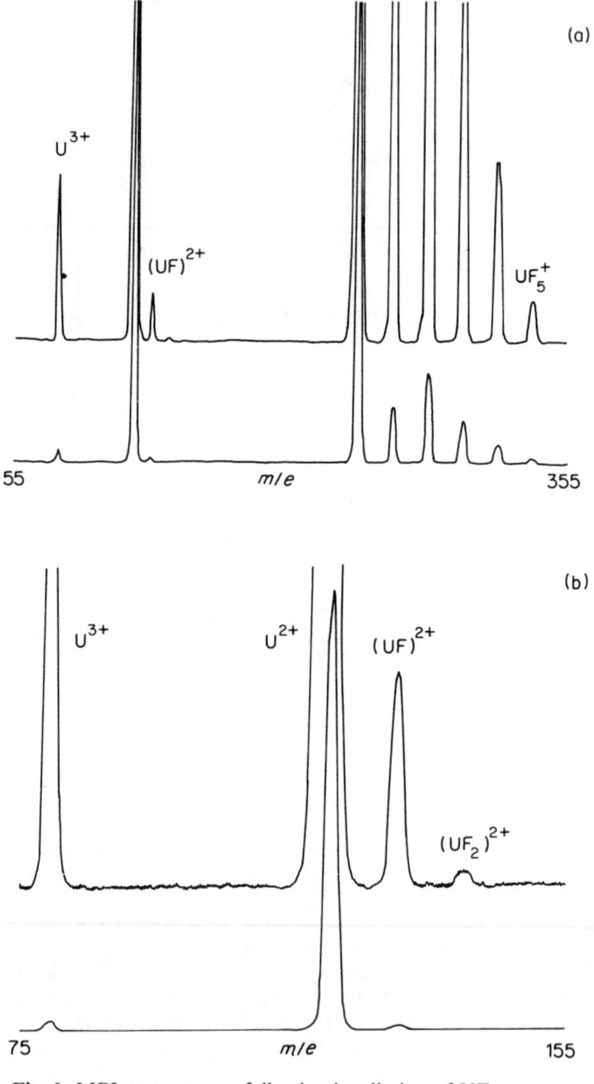

Fig. 6. MPI mass spectra following irradiation of UF_6 vapour with focused 266-nm radiation (10 mJ, 5 ns, $f = 200$ mm).[69] In (a), *m/e* in the range 55—355 is shown. The bottom scan shows the dominance of U^+ and U^{2+} in the mass spectrum. The upper scan ($\times$ 10 magnification) shows the smaller peaks more clearly. $UF_6{}^+$ was not detected. In (b), *m/e* in the range 75–155 is shown. Here, $(UF)^{2+}$ and $(UF_2)^{2+}$ are seen clearly, but there are no other detectable multiply charged molecular ions in the spectrum. U^{4+} is also absent.

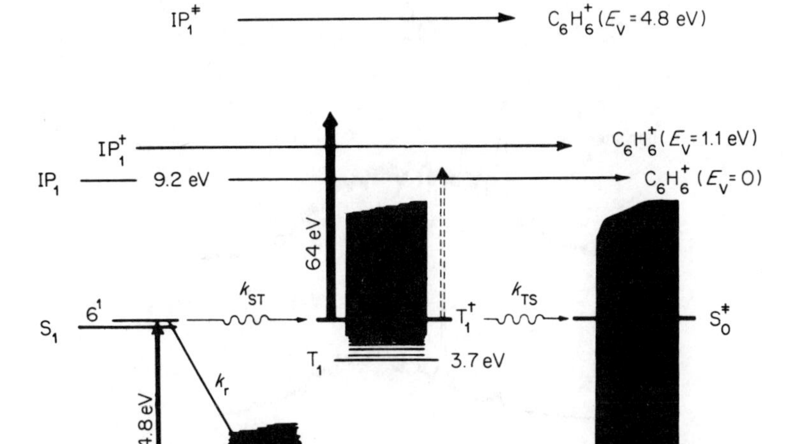

Fig. 7. Energy level diagram relevant to triplet detection by photoionization.[33] Dye laser radiation at 259 nm (4.8 eV) excites the cold benzene molecules to the 6^1 vibrational level of S_1 which then decays by fluorescence (at a rate k_r) and by intersystem crossing (at a rate k_{ST}) to isoenergetic levels of the triplet manifold. The resultant vibrationally hot triplet, $T_1^{\dagger}$, then decays (at a slower rate, k_{TS}) into highly vibrationally excited ground state levels, $S_0^{\dagger}$. Franck–Condon factors for photoionizing transitions in benzene are vanishingly small unless the resultant $C_6H_6^+$ ion has the same vibrational excitation as the molecular state to be ionized. Thus, for the $T_1^{\dagger}$ hot triplet, only the ArF excimer laser radiation (6.4 eV) can produce efficient ion formation by exciting above the first ionization threshold ($IP_1^{\dagger}$) which can produce $C_6H_6^+$ with 1.1 eV of vibrational energy.

T_1 is achieved with 193-nm radiation, and by scanning the delay between the excitation and probe lasers, the T_1 collision-free lifetime is obtained (470 $\pm$ 50 ns). Extensions to toluene,[36] pyrazine[40] and pyrimidine[40] underscore the generality of the method, and the utility of such precision measurements is evidenced in the discovery of the 'triplet lifetime mystery', in which triplets with more than a few thousand units of excess vibrational energy per centimetre decay many orders of magnitude faster than expected. In the threshold region, applied electric fields can influence ionization, just as for the case of atoms, and molecular Van der Waals clusters can be ionized softly enough to minimize fragmentation of the irradiated clusters (Figs. 8 and 9). Finally, Duncan, Dietz and Smalley,[35] discovered that metal clusters could be synthesized by irradiating microcrystals of metal carbonyls with 193 nm radiation (e.g. $[Fe(CO_5)]_n + h\nu(193\ \text{nm}) \rightarrow Fe_x, 1 \leq x \leq 30$), and they used the same radiation to ionize the metal clusters in a TOF mass spectrometer (Fig. 10). To date, this discovery has not been exploited, but it remains quite interesting nevertheless.

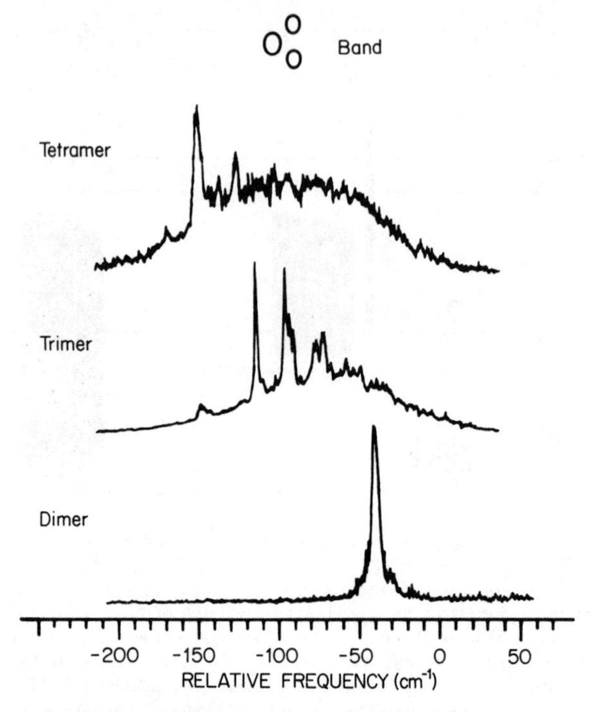

Fig. 8. Two-colour R2PI scan of the absorption bands of the benzene dimer, trimer and tetramer in the region of the $^1B_{2u}(\pi\pi^*) \leftarrow {}^1A_g$ origin of benzene monomer.[38] The intensity for each spectrum has been normalized here to a constant level. Actually, the two-colour observed peak photoion signals were in the ratio 1:0.05:0.03 for dimer:trimer:tetramer respectively. The zero of the relative frequency scale is set to the position of the forbidden origin of the benzene monomer at 38 086.1 cm^{-1}. The red shifts of the most prominent features from the monomer origin are —40, —115 and —149 cm^{-1} for the dimer, trimer and tetramer respectively. The strong features in the trimer spectrum centered 22 and 40 cm^{-1} to the blue of the origin are due to progression activity in the Van der Waals modes of the cluster. Similar features appear in the tetramer spectrum and constitute the main observed optical phonon modes in the bulk benzene crystal.

A. Photoelectron Spectra Following MPI

Photoelectron spectra can help elucidate molecular electronic structure, as well as the normal modes of molecular ions. Since the kinetic energies of the ejected electrons reflect the Franck–Condon factors for the transitions under consideration, data analyses are straightforward and complications arise primarly because of poor resolution ($\sim$ 15 meV) and the selection rules associated with the single step photoionization process. With MPI, resolution

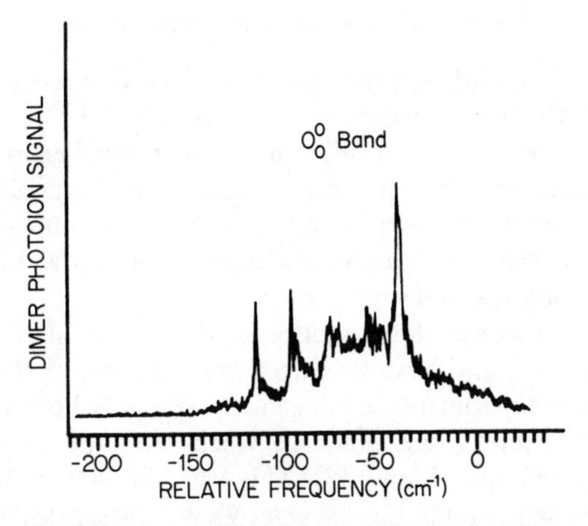

Fig. 9. One-colour R2PI spectrum of the benzene cluster beam monitoring the photoion signal in the benzene dimer channel.[38] The frequency scale is relative to the monomer origin at 38 086.1 cm^{-1}. Note by comparison to Fig. 8 that most spectral features in this scan result from fragmentation of the trimer into the dimer signal channel.

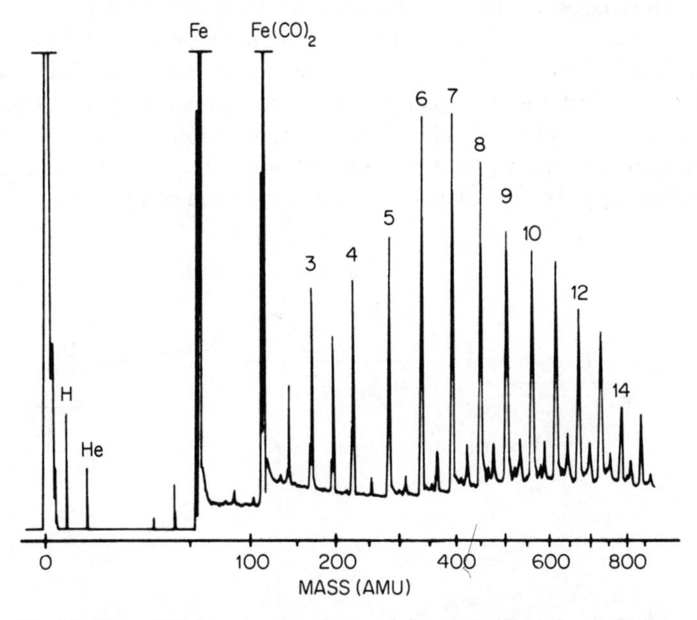

Fig. 10. Time-of-flight mass spectrum of ions produced by ArF laser excitation of a $Fe(CO)_5$ microcrystal beam.[35] Numbered peaks are predominantly due to bare iron clusters. Partially fragmented clusters of the type $Fe_x(CO)_y$, where y is odd, are seen as small peaks between the larger numbered features. Although these large peaks must contain some contribution from the even y clusters, there is no reason to expect a dominance of these even species in the partially fragmented distribution.

can be improved, and intermediate states can be used to make available many states which are inaccessible via one photon.[57,72–77] photoelectron spectra allow one to accurately determine the internal energy distributions of molecular ions, thereby allowing precision measurements such as unimolecular decomposition to be made, as described below. Thus, marked improvements and several new 'twists' are envisioned with MPI photoelectron spectroscopy of large molecules.

Reilly and coworkers have pioneered the use of MPI in photoelectron spectroscopy,[72–75] and have produced very impressive results using rather straightforward experimental arrangements (Fig. 11). For example, vibrationally resolved spectra were obtained when photoionizing toluene via the $S_1 \leftarrow S_0$ system (Figs. 12 and 13). This was the first high-resolution laser photoelectron spectrum published (7 meV) and the spectral features shown in Fig. 12 are better resolved, compared to one-photon photoelectron spectra,[78] than would be expected from improved resolution alone (7 versus 14 meV).

Detailed experiments by Durant and coworkers,[53] also using photoelectron spectroscopy, have endeavoured to prepare monoenergetic, or at least well-characterized, ensembles of ions which then undergo unimolecular decomposition, thus enabling one to test theories of unimolecular reactions. The experiments are profound and address the major issues in a forthright manner. Chlorobenzene is cooled in a free jet expansion, in order to minimize parent V, R excitation, and is ionized using two photons, as per the partial energy level diagram shown in Fig. 14. The chlorobenzene cation is then excited to states whose energies relative to the neutral chlorobenzene precursor are known rather precisely. The cation energies are determined by measuring the kinetic

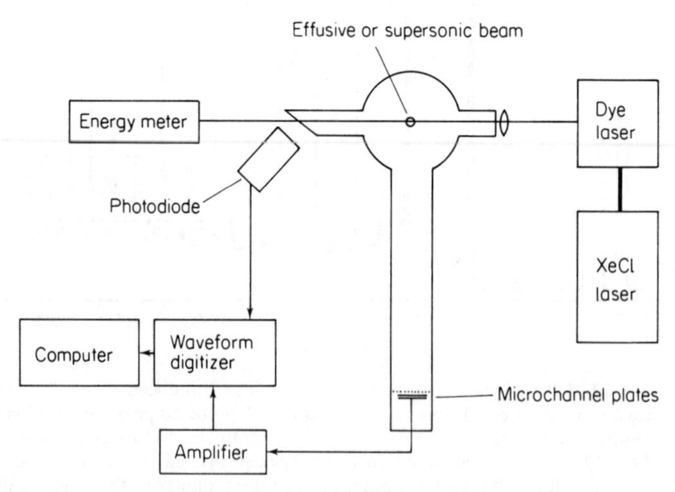

Fig. 11. Experimental arrangement used by Long, Meek and Reilly.[79] to study photoelectron spectra following MPI.

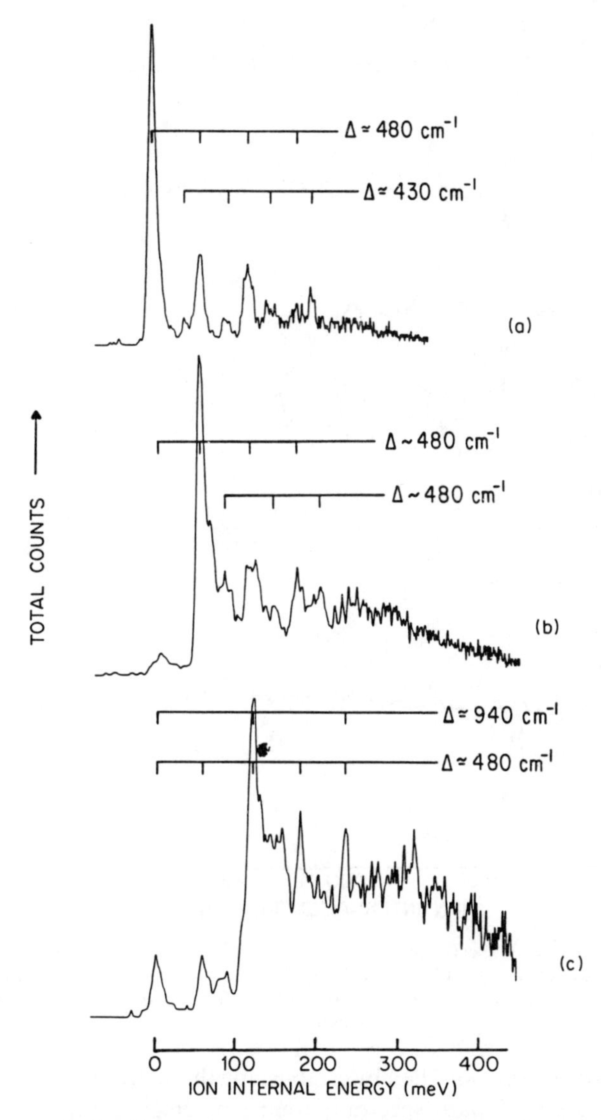

Fig. 12. Photoelectron spectra generated by two-photon ionization of gas-phase toluene.[74] Laser is resonant with (a) the $S_1 \leftarrow S_0$ origin, (b) $6b^1$ and (c) 12^1 transitions. The abscissa, ion internal energy, is the quantity $2h\nu - IP - KE$, where $h\nu$ is the laser photon energy and KE is the photoelectron kinetic energy.

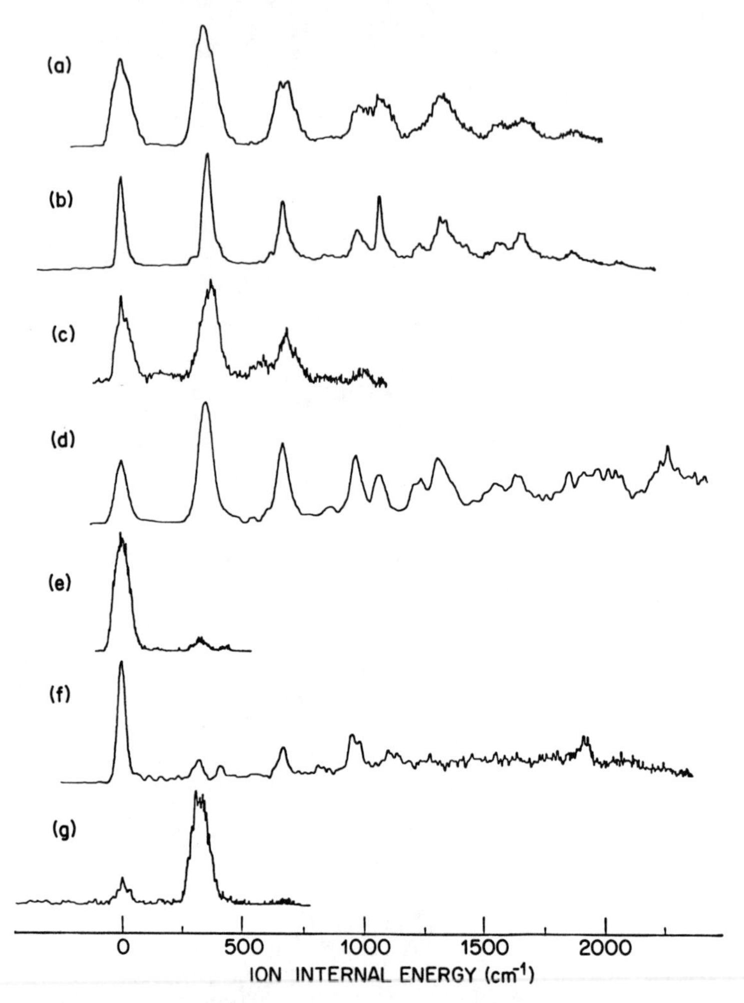

Fig. 13. Photoelectron spectra recorded by exciting the following transitions of C_6H_6:[79] (a) and (b): 6_0^1, (c) $6_0^1 1_1^0$, (d) $6_0^1 1_0^1$, (e) 6_1^0, (f) $6_1^0 1_0^1$, (g) $6_1^0 16_1^1$. In (a), (c), (e) and (g), an effusive molecular beam was used; in (b), (d) and (f), a pulsed supersonic beam was used.

energies of the photoelectrons, thereby establishing the distribution of internal energies of the cation. Typical spectra (parent ion versus wavelength, mass, photoelectron) are shown in Figs. 15 to 17. Unimolecular decay is monitored by careful analysis of the asymmetric, broadened peak in the TOF spectrum of the cation (Fig. 18). Not surprisingly, the results can be fitted using RRKM calculations, and the system behaves statistically, in agreement with the results from previous groups. Since both $S_1 \leftarrow S_0$ and the following ionization step are selective, the authors conclude that such selectivities remove the spectral

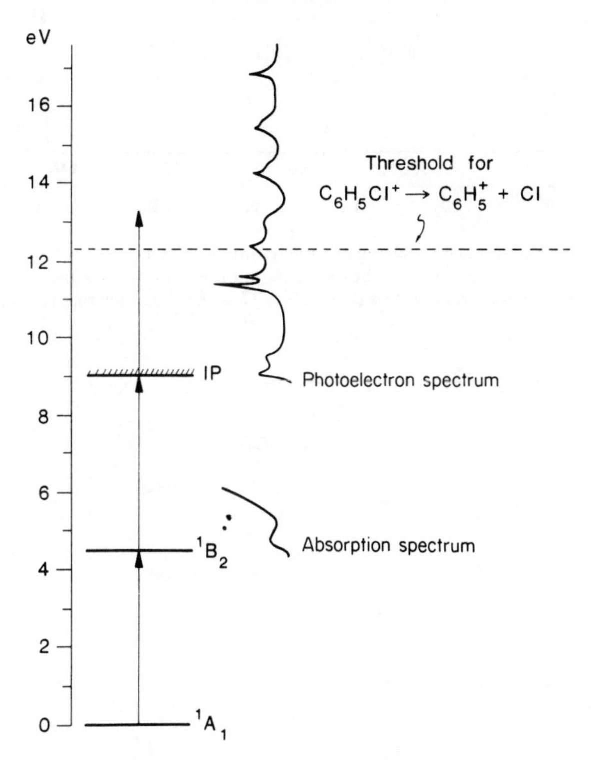

Fig. 14. Chlorobenzene energy level diagram showing the onset of absorption and photoionization.[53]

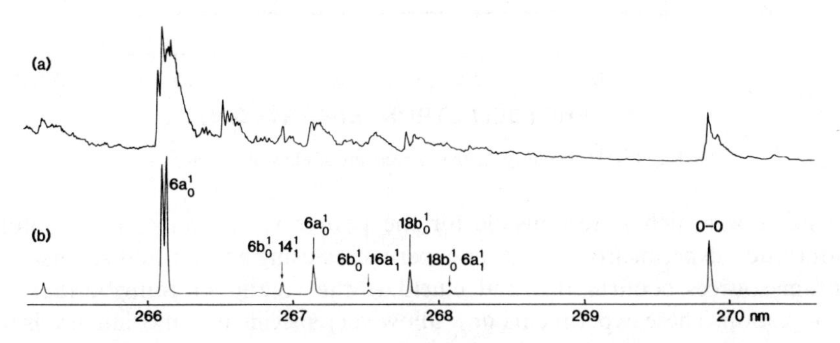

Fig. 15. Parent ion yield spectra of chlorobenzene: (a) room temperature and (b) jet-cooled.[53]

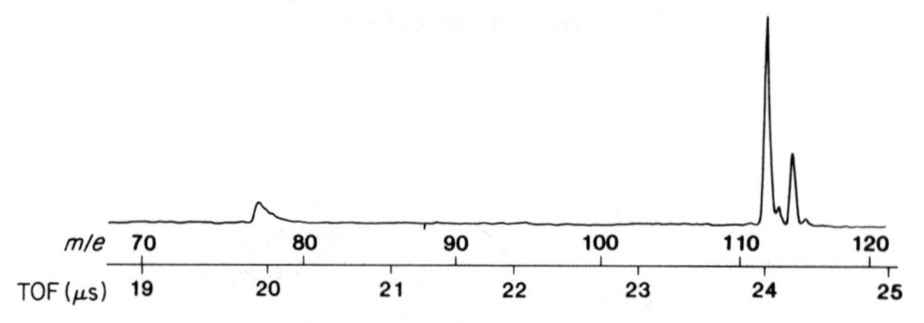

Fig. 16. MPI TOF mass spectrum of room temperature chlorobenzene at 266 nm.[53] The cluster at $m/e = 112$ to 115 shows the parent ion with its ^{37}Cl and ^{13}C isotopes. The asymmetric mass peak at $m/e = 77$ is the daughter phenyl ion.

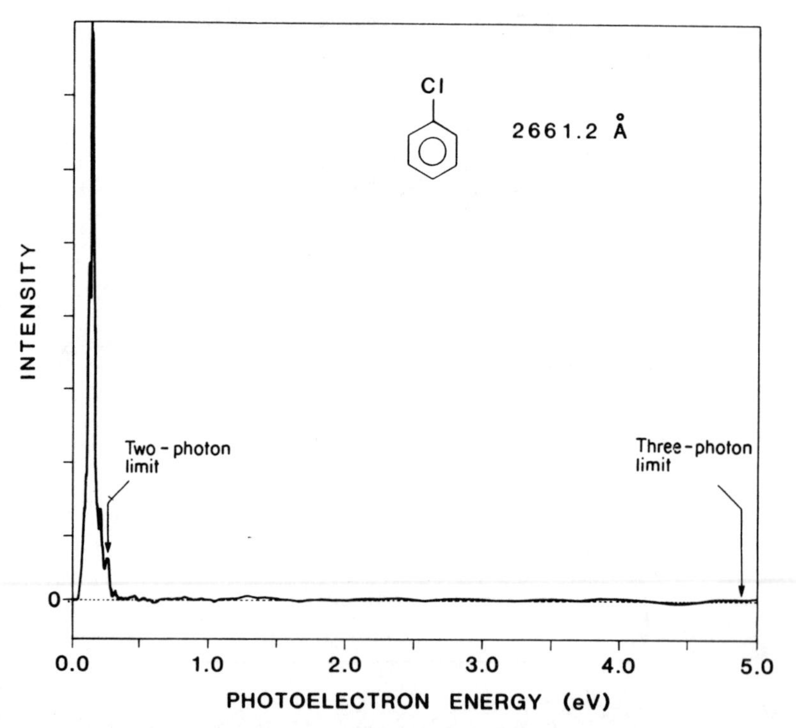

Fig. 17. TOF photoelectron spectrum of chlorobenzene.[53]

congestion which is responsible for the poorer resolution of one-photon ionization experiments.[78] Further experiments using effusive and supersonic beam sources confirm this and elucidate the rotational contributions to congestion. These experiments now allow very specific intermediate levels to be probed, and two-frequency ionization measurements in which both frequencies can be varied will allow even greater detail to be unveiled.

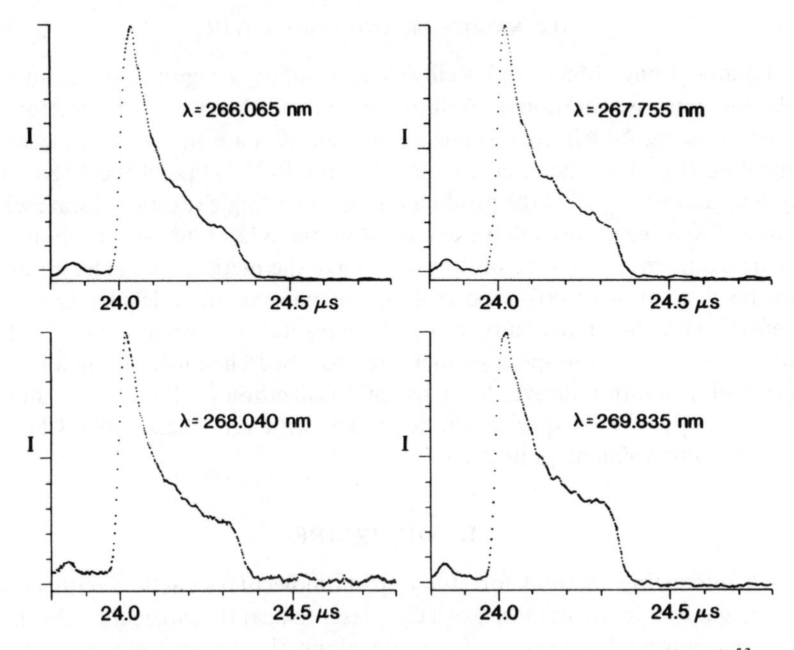

Fig. 18. Phenyl ion mass peak shape as a function of excitation wavelength.[53]

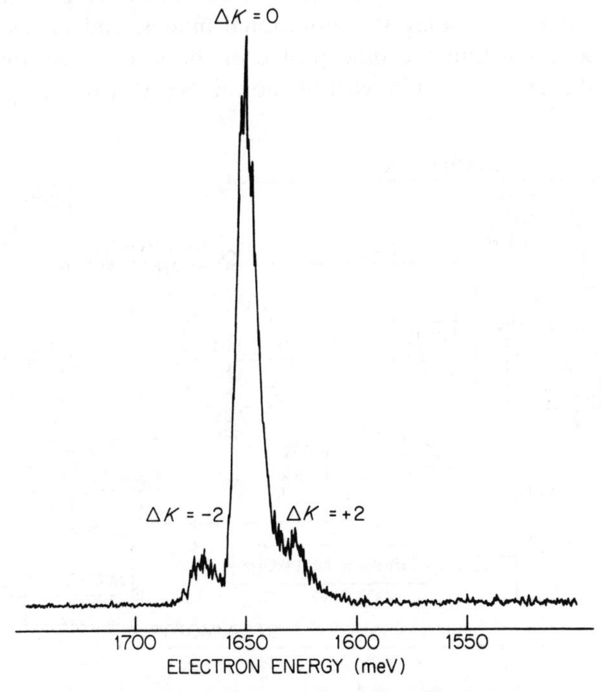

Fig. 19. Photoelectron spectrum following MPI of NO, showing rotational resolution. Nearly all of the NO^+ is formed with the same angular momentum as the $A^2\Sigma$ state which is ionized.

Finally, Long, Meek and Reilly[79] have made a significant advance by obtaining precise rotational resolution with their photoelectron spectrometer when ionizing NO from a single rovibronic K state in the $\mathbf{A}^2\Sigma$ electronic manifold (Fig. 19). The laser was tuned to the R(21.5) line of the $\mathbf{A}^2\Sigma \leftarrow \mathbf{X}^2\Pi$ system, and nearly all of the product ions are in a single rotational state whose value of K is the same as that excited within the $\mathbf{A}^2\Sigma$ state. At this point, only modest improvement in resolution will make the method viable for rotationally resolved photoelectron spectroscopy of many small molecules. Opsal and Reilly[80] have shown that surprisingly high resolution is obtained with a TOF mass filter by ionizing species which are absorbed/chemisorbed on a surface. The high resolution derives from spatial localization on the surface, and the spectroscopy of such species can be studies with very reasonable S/N since 'surface enhancement' is possible.[81]

B. MPI/IRMPE

An interesting method for studying unimolecular reaction pathways of molecular ions involves the use of a CO_2 laser to heat the molecular vibrations, thereby causing the ions to dissociate along the lowest energy electronic potential surfaces. Above reaction threshold, vibrational excitation is democratically allocated among the vibrational modes, and in instances where reaction occurs within the time profile of the laser pulse, the amount of vibrational energy present in reacting molecules is controlled by the approxi-

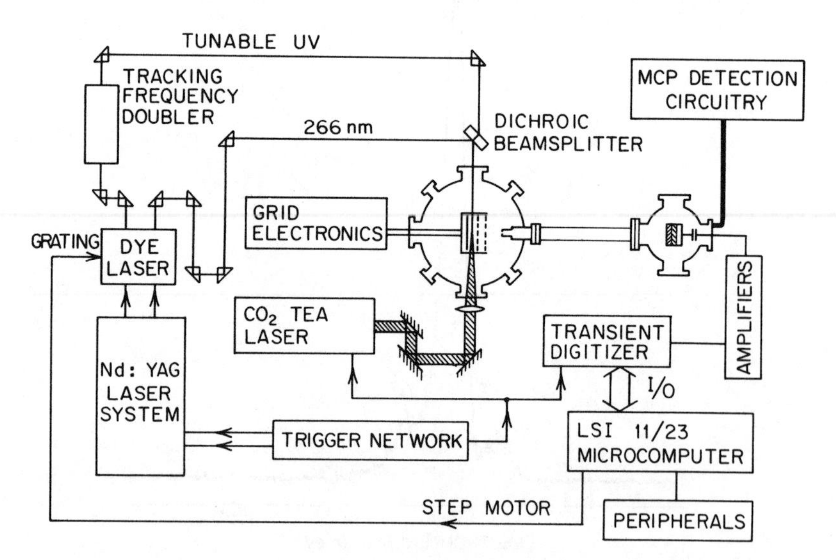

Fig. 20. Schematic drawing of an experimental arrangement for TOF photoionization detection.

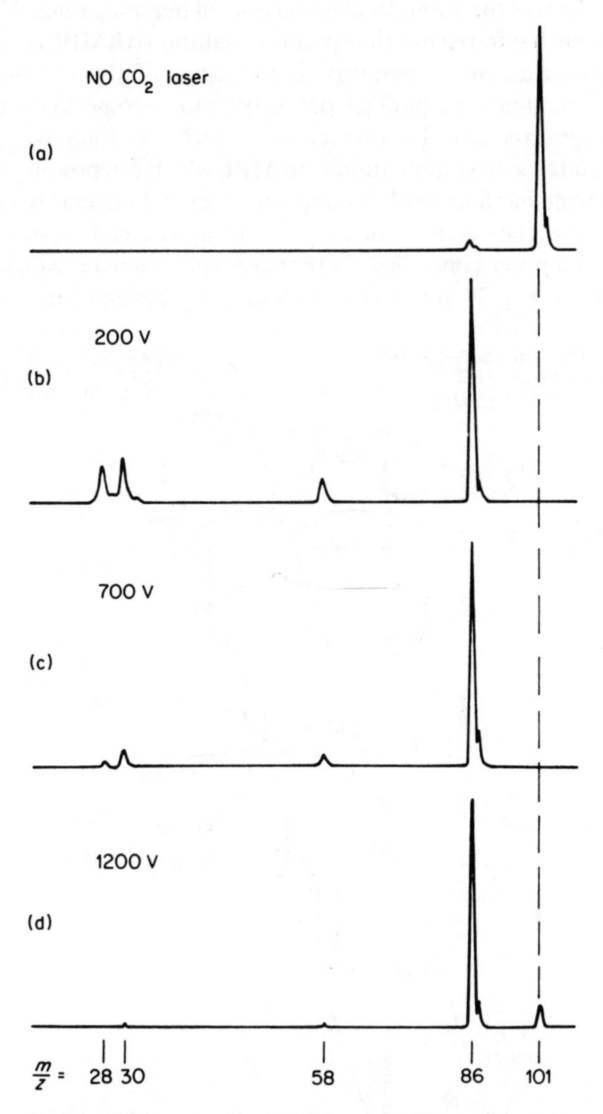

Fig. 21. TOF mass spectra of TEA, using MPI at 450 nm as a function of repeller voltage.[85] (a) The MPI process produces mainly the parent ion (TEA^+, $m/Z = 101$); the CO_2 laser is blocked. (b) to (d) The CO_2 laser (10.6 μm) onset is 50 ns before the dye laser output, while the repeller voltage is changed from 200 to 1200 V. The decreasing appearance of daughter ions with increasing repeller voltage reflects the change in the effective IR laser fluences experienced by the ions. Note the elimination of m/Z = 28, 30 and 58 as the repeller voltage is raised.

mate equivalence of the unimolecular and optical pumping rates. Thus, species energized by such infrared multiple photon excitation (IRMPE) can be trusted to favour the lowest energy pathways in the region of dissociation threshold. With high laser fluences, parallel pathways can become available and, in addition, fragments can also dissociate via IRMPE, thereby producing a myriad of products. In combination with MPI, which can produce unique and specific ions, the method has been applied to several systems which serve to illustrate most of the salient features.[82–86] The most useful apparatus for such work is a computer-controlled TOF mass spectrometer which is shown schematically in Fig. 20 and which includes many design features from other

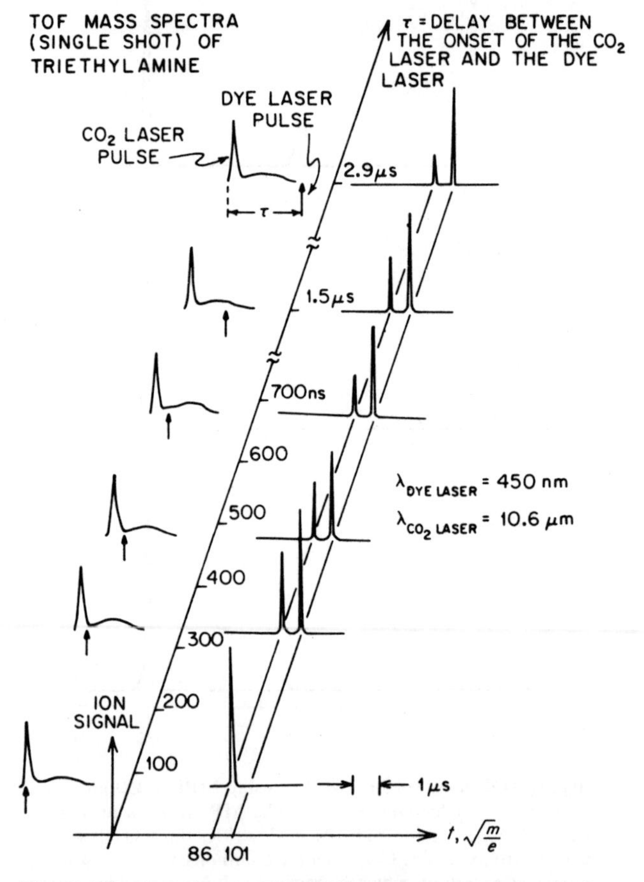

Fig. 22. The dependence of the TOF mass spectra on the relative timing between the MPI radiation and the IR laser radiation.[85] Dissociation of TEA^+ ($m/Z = 101$) declines as the dye laser is scanned through the IR laser profile. Timing between the lasers was monitored simultaneously with each TOF trace. No averaging was done. All spectra are from a signal firing of the lasers.

laboratories, as well as a few genuine innovations. Figures 21 and 22 show results for the case of triethylamine, where MPI is used to prepare one particular ion at a time, and IRMPD then establishes unambiguously the lowest energy reaction pathway for the ion of concern. A very decent map of the reaction pathways is obtained and the successive fragmentation of ions is finally terminated by the inability of small species to accumulate vibrational energy via IRMPE and/or by the termination of the CO_2 laser pulse. Note in particular that fragmentation via IRMPE can be very efficient, with almost complete dissociation of the irradiated ion. Similar results are obtained using an ion cyclotron resonance spectrometer to prepare and select ions which are then irradiated, and many very fine papers have evolved from the laboratories of Beauchamp[87,88] and Brauman.[89–92]

C. Taking the Non-linearities out of the Experiment: Coherent Tunable VUV Radiaton

From the point of view of detecting distinct chemical species, MPI has serious problems. The pronounced, intensity-dependent cracking which occurs makes species identification impractical, and the requirement that the analysis radiation be focused can cause the process of detection to also severely photolyse samples. Much of the analytical capability associated with making

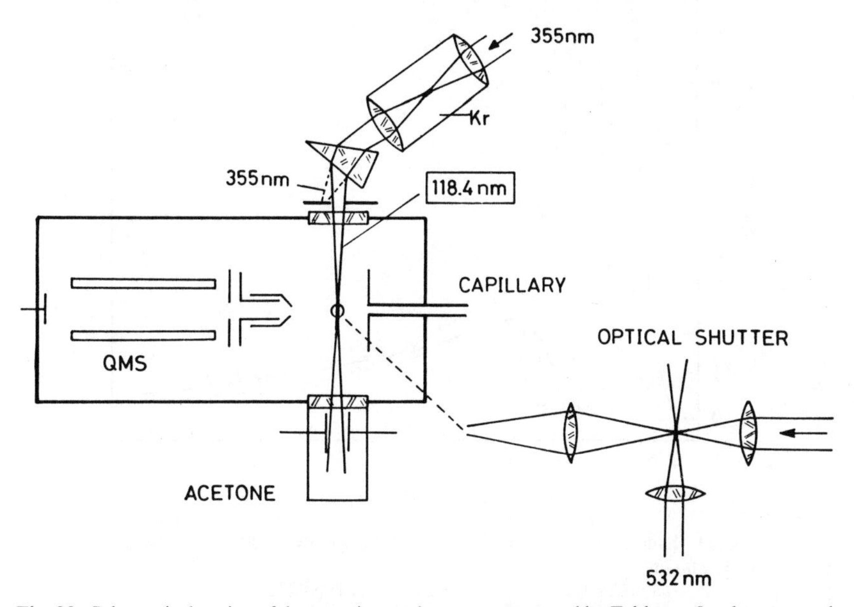

Fig. 23. Schematic drawing of the experimental arrangement used by Feldman, Laukemper and Welge[97] for the 118.4-nm one-photon photoionization detection of products from the IRMPE of acetic anhydride.

ions does not really concern the 'intermediate states' which spectroscopists tend to dwell on when extolling the virtues of MPI.

One way to enhance the generality, and perhaps ultimate utility, of laser ionization is to exploit non-linear effects before doing the real experiment, thereby producing coherent VUV radiation.[93] This can then be used for selective ionization of chemical species, including transients, with high sensitivity and excellent spatial and temporal resolution. Since ionization is via a one-photon process and frequency tuning is straightforward, it should be possible to minimize dissociation dramatically, thereby allowing minor constituents, including transients, to be detected easily. Soft ionization of this type could also be used to detect fragile entities such as Van der Waals clusters produced in free jet expansions, and in our laboratory we have had no difficulty in ionizing benzene clusters using 118.4 nm radiation—the 9th harmonic of an Nd:YAG laser.

These ideas are hardly new and a number of physicists have participated in

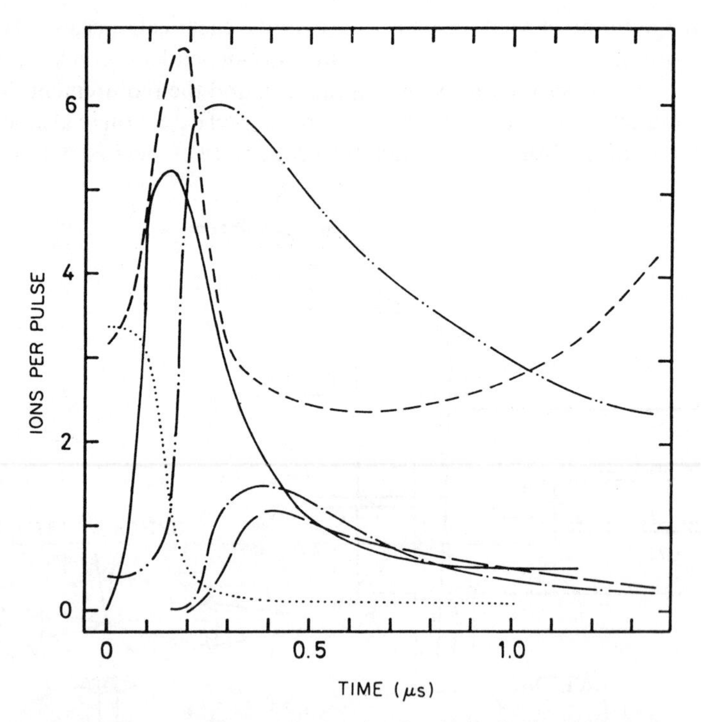

Fig. 24. Time dependence of the ion signals, obtained using the arrangement shown:[97]
..... $m^+ = 86$ $(CH_3CO)_2{}^+$; ——— $m^+ = 43$ (CH_3CO^+);
—..—..— $m^+ = 42$ (CH_2CO^+),—.—.$m^+ = 15$ $(CH_3{}^+)$;
—— $m^+ = 14$ $(CH_2{}^+)$; the solid line shows the intensity of the CO_2 laser pulse.

device development and early applications involving atoms and small molecules.[94–96] Recently, Feldmann, Laukemper and Welge[97] used 118.4-nm radiation to ionize species which are produced by radiating acetic anhydride with the focused output from a pulsed CO_2 laser. The experimental arrangement and essential results are shown in Figs. 23 and 24. Parent $(CH_3O)_2O$ is energized via IRMPE until dissociation occurs, and fragments are successively dissociated by the strong IR radiation field to varying degrees which depend on state densities and absorption cross-sections.

In these experiments, an optically triggered shutter blocks the CO_2 laser pulse after some specified time, thereby insuring that molecular ions are not dissociated by IRMPD. Nevertheless, the ionizing radiation interacts with a very complex ensemble of species, since IRMPE produces both parent and fragment molecules with broad vibrational distributions. Thus, these first experiments cannot be interpreted straightforwardly, insofar as the parentage of particular ions is concerned, and they serve primarily to alert and/or stimulate workers in this field. Given that the measurement techniques are straightforward and now well established, it seems inevitable that significant steps will be made towards using coherent VUV sources for single-photon photoionization detection of species in molecular and chemical physics.

III. DISCUSSION AND CONCLUSIONS

It is traditional to say that an enterprise such as MPI is in its 'infancy' and that marvellous discoveries and applications will be forthcoming from this day on. However, MPI is a method, not a field in itself, and will therefore be exploited in many different scientific environments. As a tool, its role will be secondary to the main scientific issues, as is proper. It is unlikely that the exploratory, survey-type research, which uses focused laser beams to produce effects which are then explained or rationalized, will remain a legitimate endeavour for much longer. Instead, we expect that MPI will be used to prepare and/or detect species as per the experimental requirements, and the uniqueness, sensitivity, spatial reslutions, etc., of the method will determine its use.

In the area of molecular dynamics, MPI is a viable detection method which can add a new dimension to existing experiments. It can be extremely sensitive, ionizing nearly all molecules in a specific rovibronic state, with excellent spatial and temporal resolution, and can discriminate against background as well as any method known. In the near future, maximum benefit will derive from studying a few systems with high precision, rather than trying every chemical on the shelf, and MPI has its niche here. State specificity has proven to be very important in molecular dynamics, and in a collision-free environment nascent rovibrational excitations are metastable and easily detected. Molecular beam scattering, elegant method that it is, will benefit greatly from improved sensitivity and/or the ability to detect quantum states of products. Beamists have tended to study only reactions with large cross-sections and have even suggested that translational energy is the only product degree of freedom worth considering. Kineticists and other chemists know better, but they must work within the confines imposed by thermal equilibrium and minimal control of important parameters. We predict that future experiments in which both reagents and products are state selected will require MPI detection and will establish new standards in reactive scattering experiments. Already, Rockney and Grant[98,99] have used MPI to detect angularly resolved unimolecular reaction products following the IRMPD of nitromethane under molecular beam conditions. They detect both CH_3 and NO_2 via MPI, and obtain information about internal as well as translational degrees of freedom. The experiment is quite impressive, since angularly resolved TOF spectra are obtained for specific product quantum states, and the products are not those species which are most easily detected via MPI. It is clear that further experiments using MPI to detect scattered products will lead to a level of refinement which has been hitherto unavailable. One should bear in mind, however, that tightly focused beams are often used for MPI detection, and thus undesirable effects, such as dissociation of the reactants or vibrationally excited precursors by the probe beam, may complicate interpretations.

MPI can be used in high-pressure situations (e.g. flames,[100] measuring rate coefficients, etc.) as well as the pristine environments of beam-scattering experiments. Here one measures the total current produced by MPI in the presence of a modest bias and thereby monitors and/or detects species via their spectroscopic signatures. Mass resolution is unavailable, so it is desirable that species with well-defined absorption features be perused one at a time. Obvious advantages over a method such as laser-induced fluorescence (LIF) are that (in principle) species which do not fluoresce can be detected, there is frequently little or no background and the sensitivity is quite high since all ions are detected. However, there are also serious drawbacks; these include the need for focused radiation, which can inadvertently perturb the environment, and unravelling complex spectra which often involve three or more photons. Since everyone cannot work on NO, the hydrogen atom of MPI, it is never obvious that MPI will be more sensitive that LIF until the experiment is actually tried. In our work, we find that sensitivity is often not so important, but turning spectra into state distributions is, and we must emphasize strongly that under high-pressure conditions obtaining such state distributions can be difficult.[101] Thus, we find that LIF is often the more desirable mode of detection and, if possible, we prefer to use LIF in high-pressure environments.

Several years ago, it was pointed out that MPI could be used to prepare high concentrations of interesting ions, thereby allowing these species to be studied in detail. One could imagine studying charge transfer, unimolecular decomposition, gas-phase ion chemistry, etc., with new vigour and direction. As it turned out, however, nothing of such lustre materialized and the results have tended to complement those obtained using more conventional methods. Our own work on the unimolecular decomposition of ions,[82–85] as well as those of Beauchamp and coworkers[87,88] and Brauman and coworkers,[89–92] although elegant and competently executed, have added little new knowledge. We believe that, in the future, there will be modest and sustained efforts in this area by users of MPI, but that the 'gold rush' is temporarily over.

MPI has been, and will continue to be, a powerful method for the spectroscopist.[102–124] The extremely high sensitivity alone opens new avenues for explorations (clusters, triplets, excited states, transients, etc.), which have been unavailable so far. It is hard to imagine that the precision work going on now in this area will not continue and even expand in the future. Spectroscopists also pave the way for people who use spectroscopy as an analytical tool; as they make information available, it will be consumed by other components of the scientific community. For example, it is now possible to detect H_2 and CO via MPI,[112,116,117,119,120] and these species are of fundamental importance in chemistry. Detecting H_2 and CO spectroscopically, under conditions where nascent excitations are preserved, will be of tremendous importance in the study of reactive collisions.

Finally, although not technically MPI, we wish to make the point that tunable VUV radiation, obtained via mixing or tripling, will be hard to beat for many diagnostic purposes, There are many advantages of a one-photon process and VUV photoionization is a proven and revered technique. State selection may be almost impossible, but a large number of species can be ionized efficiently, quantitatively and with some selectivity, thereby making the method attractive for a large number of experiments. Using TOF mass selection, data acquisition can be very rapid, and for purposes of identifying chemically distinct species this method could prove at least as effective as angularly resolved measurements using molecular beams. The ability to ionize species 'softly' makes the method agreeable for studying clusters, and we find that $(C_6H_6)_n^+$, $n \leq 8$, is produced efficiently via the 118.4-nm photoionization of an expanded He/C_6H_6 mixture. We also find that $Br^+ + Br^-$ can be produced via VUV photoionization of Br_2, and there is no fundamental reason why this route could not be followed for the production of negative ions and some of their interesting adducts (e.g. Br^-CH_3Cl, solvated electrons). The technology for producing coherent, tunable VUV radiation has been available for quire some time. However, chemical physicists do not usually pioneer both the development of a technology and its applications towards scientifically interesting problems. Nor does the applied physics community, which develops these marvellous gadgets and methods, always choose the most scientifically important applications for their inventions. Thus we believe that the future is very promising for experiments which use coherent, turnable

VUV radiation for one-photon photoionization. Although lacking the spectral signature of MPI via a resonant intermediate, there will be no shortage of important scientific applications.

Acknowledgements

The authors have benefitted from discussions with persons too numerous to list; we especially thank S. Novak and R. Senaha for the preparation of this manuscript.

References

1. Lambropoulos, P., *Adv. Atom. Mol. Phys.*, **12**, 87 (1976).
2. Chin, S. L., and Lambropoulos, P., *Multiphoton Ionization of Atoms*, Academic Press, 1984.
3. Siomos, K., and Christophorou, L. G., *Chem. Phys. Lett.*, **72**, 43 (1980).
4. Johnson, P. M., *Acc. Chem. Res.*, **13**, 20 (1980).
5. Johnson, P. M., Berman, M. R., and Zakheim, D., *J. Chem. Phys.*, **62**, 2500 (1975).
6. Johnson, P. M., *J. Chem. Phys.*, **62**, 4562 (1975),
7. Johnson, P. M., *J. Chem. Phys.*, **64**, 4143 (1976).
8. Johnson, P. M., *J. Chem. Phys.*, **64**, 4638 (1976).
9. Zakheim, D., and Johnson, P. M., *J. Chem. Phys.*, **68**, 3644 (1978).
10. Petty, G., Tai, C., and Dalby, F. W., *Phys. Rev. Lett.*, **34**, 1207 (1975).
11. Dalby, F. W., Petty-Sil, G., Pryce, M. H., and Tai, C., *Can. J. Phys.*, **55**, 1033 (1977).
12. Tai, C., and Dalby, F. W., *Can. J. Phys.*, **56**, 183 (1978).
13. Parker, D. H., Sheng, S. J., and El-Sayed, M. A., *J. Chem. Phys.*, **65**, 5534 (1976).
14. Nieman, G. C., and Colson, S. D., *J. Chem. Phys.*, **68**, 5656 (1978).
15. Robin, M. B., and Kuebler, N. A., *J. Chem. Phys.*, **69**, 806 (1978).
16. Lehmann, K. K., Smolarek, J., and Goodman, L., *J. Chem. Phys.*, **69**, 1569 (1978).
17. Zandee, L., Bernstein, R. B., and Lichtin, D. A., *J. Chem. Phys.*, **69**, 3427 (1978).
18. Zandee, L., and Bernstein, R. B., *J. Chem. Phys.*, **70**, 2574 (1979).
19. Zandee, L., and Bernstein, R. B., *J. Chem. Phys.*, **71**, 1359 (1979).
20. Lichtin, D. A., Datta-Ghosh, S., Newton, K. R., and Bernstein, R. B., *Chem. Phys. Lett.*, **75**, 214 (1980).
21. Parker, D. H., Bernstein, R. B., and Lichtin, D. A., *J. Chem. Phys.*, **75**, 2577 (1981).
22. Newton, K. R., Lichtin, D. A., and Bernstein, R. B., *J. Phys. Chem.*, **85**, 5728 (1981).
23. Lichtin, D. A., Bernstein, R. B., and Newton, K. R., *J. Chem. Phys.*, **75**, 5728 (1981).
24. Lichtin, D. A., Zandee, L., and Bernstein, R. B., Chap. 6 in *Laser in Chemical Analysis* (Eds. G. Hieftje, J. Travis and F. Lytle), The Humana Press, Clifton, New Jersey, 1981.
25. Parker, D. H., and Bernstein, R. B., *J. Phys. Chem.*, **86**, 60 (1982).
26. Bernstein, R. B., *J. Phys. Chem.*, **86**, 1178 (1982).
27. Lichtin, D. A., Bernstein, R. B., and Vaida, V., *J. Am. Chem. Soc.*, **104**, 1830 (1982).
28. Newton, K. R., and Bernstein, R. B., *J. Phys. Chem.*, **87**, 2246 (1983).
29. Squire, D. W., Barbalas, M. P., and Bernstein, R. B., *J. Phys. Chem.*, **87**, 1701 (1983).

30. Lichtin, D. A., Squire, D. W., Winnik, M. A., and Bernstein, R. B., *J. Am. Chem. Soc.*, **105**, 2109 (1983).
31. Dietz, T. G., Duncan, M. A., Liverman, M. G., and Smalley, R. E., *Chem. Phys. Lett.*, **70**, 246 (1980).
32. Dietz, T. G., Duncan, M. A., Liverman, M. G., and Smalley, R. E., *J. Chem. Phys.*, **73**, 4816 (1980).
33. Duncan, M. A., Dietz, T. G., Liverman, M. G., and Smalley, R. E., *J. Phys. Chem.*, **85**, 7 (1981).
34. Hopkins, J. B., Powers, D. E., and Smalley, R. E., *J. Phys. Chem.*, **85**, 3739 (1981).
35. Duncan, M. A., Dietz, T. G., and Smalley, R. E., *J. Am. Chem. Soc.*, **103**, 5245 (1981).
36. Dietz, T. G., Duncan, M. A., and Smalley, R. E., *J. Chem. Phys.*, **76**, 1227 (1982).
37. Duncan, M. A., Dietz, T. G., and Smalley, R. E., *J. Chem. Phys.*, **75**, 2118 (1981).
38. Smalley, R. E., *J. Chem. Ed.*, **59**, 934 (1982).
39. Dietz, T. G., Duncan, M. A., and Smalley, R. E., *J. Chem. Phys.*, **77**, 4417 (1982).
40. Dietz, T. G., Duncan, M. A., Puiu, A. C., and Smalley, R. E., *J. Phys. Chem.*, **86**, 4026 (1982).
41. Boesl, U., Neusser, H. J., Weinkauf, R., and Schlag, E. W., *J. Phys. Chem.*, **86**, 4857 (1982).
42. Boesl, U., Neusser, H. J., and Schlag, E. W., *Chem. Phys. Lett.*, **87**, 1 (1982).
43. Boesl, U., Neusser, H. J., and Schlag, E. W., in *Laser Induced Processess in Molecules* (Eds. K. L. Kompa and S. D. Smith), Vol. 6, Springer, Berlin, 1979.
44. Boesl, U., Neusser, H. J., and Schlag, E. W., *J. Am. Chem. Soc.*, **103**, 5058 (1981).
45. Boesl, U., Neusser, H. J., and Schlag, E. W., *Z. Naturforsch.*, **A33**, 1546 (1978).
46. Boesl, U., Neusser, H. J., and Schlag, E. W., *J. Chem. Phys.*, **72**, 3034 (1980).
47. Boesel, U., Neusser, H. J., and Schlag, E. W., *Chem. Phys.*, **55**, 193 (1981).
48. Boesl, U., Neusser, H. J., and Schlag, E. W., *J. Chem. Phys.*, **72**, 4327 (1980).
49. Lubman, D. M., Naaman, R., and Zare, R. N., *J. Chem. Phys.*, **72**, 3034 (1980).
50. Proch, D., Rider, D. M., and Zare, R. N., *Chem. Phys. Lett.*, **81**, 430 (1981).
51. Anderson, S. L., Rider, D. M., and Zare, R. N., *Chem. Phys. Lett.*, **93**, 11 (1982).
52. Feldman, D. L., Lengel, R. K., and Zare, R. N., *Chem. Phys. Lett.*, **52**, 413 (1977).
53. Durant, J. L., Rider, D. M., Anderson, S. L., Proch, F. D., and Zare, R. N., *J. Chem. Phys.*, **80**, 1817 (1984).
54. Carney, T., and Baer, T., *J. Chem. Phys.*, **75**, 477 (1981).
55. Carney, T. E., and Baer, T., *J. Chem. Phys.*, **76**, 5968 (1982).
56. Miller, J. C., Compton, R. N., Carney, T. E., and Baer T., *J. Chem. Phys.*, **76**, 11 (1982).
57. Compton, R. N., Miller, J. C., Carter, A. E., and Kruit, P., *Chem. Phys. Lett.*, **71**, 87 (1980).
58. Miller, J. C., and Compton, R. N., *J. Chem. Phys.*, **75**, 22 (1981).
59. Miller, J. C., and Compton, R. N., *J. Chem. Phys.*, **75**, 2020 (1981).
60. Baer, T., Peatman, W. B., and Schlag, E. W., *Chem. Phys. Lett.*, **4**, 243 (1969).
61. Rebentrost, F., and Ben-Shaul, A., *J. Chem. Phys.*, **74**, 3255 (1981).
62. Rebentrost, F., Kompa, K. L., and Ben-Shaul, A., *Chem. Phys. Lett.*, **77**, 394 (1981).
63. Silberstein, J., and Levine, R. D., *Chem. Phys. Lett.*, **74**, 6 (1980).
64. Silberstein, J., and Levine, R. D., *J. Chem. Phys.*, **75**, 5735 (1981).
65. Lubman, D. M., *J. Phys. Chem.*, **85**, 3752 (1981).
66. Pandolfi, R. S., Gobell, D. A., and El-Sayed, M. A., *J. Phys. Chem.*, **85**, 1779 (1981).
67. Hering, P., Maaswinkel, A. G. M., and Kompa, K. L., *Chem. Phys. Lett.*, **83**, 222 (1981).

68. Stuke, M., Reisler, H., and Wittig, C., *Appl. Phys. Lett.*, **39**, 201 (1981).
69. Stuke, M., and Wittig, C., *Chem. Phys. Lett.*, **81**, 168 (1981).
70. Chou, J. S., Sumida, D., Stuke, M., and Wittig, C., *Laser Chem.*, **1**, 1 (1982).
71. Rhodes, C. K., unpublished.
72. Meek, J. T., Jones, R. K., and Reilly, J. P., *J. Chem. Phys.*, **73**, 3503 (1980).
73. Long, S. R., Meek, J. T., and Reilly, J. P., *J. Chem. Phys.*, **79**, 3206 (1983).
74. Meek, J. T., Long, S. R., and Reilly, J. P., *J. Phys. Chem.*, **86**, 2809 (1982).
75. Long, S. R., Meek, J. T., Harrington, P. J., and Reilly, J. P., *J. Chem. Phys.*, **78**, 3341 (1983).
76. Hepburn, J. W., Trevor, D. J., Pollard, J. E., Shirley, D. A., and Lee, Y. T., *J. Chem. Phys.*, **76**, 4287 (1982).
77. Miller, J. C., Compton, R. N., Carney, T. E., and Baer, T., *J. Chem Phys.*, **76**, 5648 (1982).
78. Rabalais, J. W., *J. Chem. Phys.*, **57**, 960 (1972).
79. Long, S. R., Meek, J. T., and Reilly, J. P., *J. Chem. Phys.*, **79**, 3206 (1983).
80. Opsal, R. B., and Reilly, J. P., *Chem. Phys. Lett.*, **99**, 461 (1983).
81. Lubman, D. M., and Naaman, R., *Chem. Phys. Lett.*, **95**, 325 (1983).
82. Stuke, M., Sumida, D., and Wittig, C., *J. Phys. Chem.*, **86**, 438 (1982).
83. Hass, Y., Reisler, H., and Wittig, C., *J. Chem. Phys.*, **77**, 257 (1982).
84. Catanzarite, J. H., Hass, Y. Reisler, H., and Wittig, C., *J. Chem. Phys.*, **78**, 5506 (1983).
85. Chou, J. S., Sumida, D., and Wittig, C., *Chem. Phys. Lett.*, **100**, 209 (1983).
86. Glownia, J. H., Romero, R. J., and Sander, R. K., *Chem. Phys. Lett.*, **88**, 292 (1982).
87. Woodin, R. L., Bomse, D. S., and Beauchamp, J. L., in *Chemical and Biochemical Applications of Lasers* (Ed. C. B. Moore), Vol. IV, Academic Press, New York, 1979.
88. Thorne, L. R., and Beauchamp, J. L., *J. Chem. Phys.*, **74**, 5100 (1981).
89. Rosenfeld, R. N., Jasinski, J. M., and Brauman, J. I., *Chem. Phys. Lett.*, **71**, 400 (1980).
90. Rosenfeld, R. N., Jasinski, J. M., and Brauman, J. I., *J. Am. Chem. Soc.*, **101**, 3999 (1979).
91. Jasinski, J. M., Rosenfeld, R. N., Meyer, F. K., and Brauman, J. I., *J. Am. Chem. Soc.*, **104**, 652 (1982).
92. Rosenfeld, R. N., Jasinski, J. M., and Brauman, J. I., *J. Am. Chem. Soc.*, **104**, 658 (1982).
93. Miller, J. C., Compton, R. N., and Cooper, C. D., *J. Chem. Phys.*, **76**, 3967 (1982).
94. McIlrath, T. J., and Freeman, R. R. (Eds.), *Laser Techniques for Extreme Ultraviolet Spectroscopy*, Am. Inst. Phys., New York, 1982.
95. Harris, S. E., *Appl. Phys. Lett.*, **31**, 498 (1977).
96. New, G. H. C., and Ward, J. F., *Phys. Rev. Lett.*, **19**, 556 (1967).
97. Feldmann, D., Laukemper, J., and Welge, K. H., *J. Chem. Phys.*, **79**, 278 (1983).
98. Rockney, B. H., and Grant, E. R., *J. Chem. Phys.*, **77**, 4257 (1982).
99. Rockney, B. H., and Grant, E. R., (a) *Chem. Phys. Lett.*, **79**, 15 (1981); (b) *J. Chem. Phys.*, **79**, 708 (1983).
100. Goldsmith, J. E. M., *J. Chem. Phys.*, **78**, 1610 (1983).
101. Chou, J. S., Sumida, D., and Wittig, C., *Chem. Phys. Lett.*, **100**, 397 (1983).
102. Duignan, M. T., Hudgens, J. W., and Wyatt, J. R., *J. Phys. Chem.*, **86**, 4156 (1982).
103. DiGiuseppe, T. G., Hudgens, J. W., and Lin, M. C., *Chem. Phys. Lett.*, **82**, 267 (1981).
104. Hudgens, J. W., DiGiuseppe, T. G., and Lin, M. C., *J. Chem. Phys.*, **79**, 571 (1983).
105. Dulcey, C. S., and Hudgens, J. W., *J. Phys. Chem.*, **87**, 2296 (1983).

106. Murakami, J., Kaya, K., and Ito, M., *Chem. Phys. Lett.*, **91**, 401 (1982).
107. Murakami, J., Ito, M., and Kaya, K., *Chem. Phys. Lett.*, **80**, 203 (1981).
108. Murakami, J., Kaya, K., and Ito, M., *J. Chem. Phys.*, **72**, 3263 (1980).
109. Rianda, R., Moll, D. J., and Kuppermann, A., *Chem. Phys. Lett.*, **73**, 469 (1980).
110. McDiarmid, R., and Auerbach, A., *Chem. Phys. Lett.*, **76**, 520 (1980).
111. Even, U., Amirov, A., Leutwyler, S., Ondrechen, M. J., Berkovitch-Yellin, Z., and Jortner, J., *Faraday Disc. Chem. Soc.* **73**, 153 (1982)
112. Rottke, H., and Welge, K. H., *Chem. Phys. Lett.*, **99**, 456 (1983).
113. Shinohara, H., and Nishi, N., *Chem. Phys. Lett.*, **87**, 561 (1982).
114. Morrison, R. J. S., Rockney, B. H., and Grant, E. R., *J. Chem. Phys.*, **75**, 2643 (1981).
115. Kliger, D. J., Bokor, J., and Rhodes, C. K. *Phys. Rev.*, **A21**, 607 (1980).
116. Marinero, E. E., Rettner, C. T., and Zare, R. N., *Phys. Rev. Lett.*, **48**, 1323 (1982).
117. Marinero, E. E., Vasudev, R., and Zare, R. N., *J. Chem. Phys.*, **78**, 692 (1983).
118. Cooper, C. D., Williamson, A. D., Miller, J. C., and Compton, R. N., *J. Chem. Phys.*, **73**, 1527 (1980).
119. Jones, R. W., Sivakumar, N., Rockney, B. H., Houston, P. L., and Grant, E. R., *Chem. Phys. Lett.*, **91**, 271 (1982).
120. Pratt, S. T., Dehmer, P. M., and Dehmer, J. L., *J. Chem. Phys.*, **79**, 3234 (1983).
121. Gerrity, D. P., Rothberg, L. J., and Vaida, V., *Chem. Phys. Lett.*, **74**, 1 (1980).
122. Rothberg, L. J., Gerrity, D. P., and Vaida, V., *J. Chem. Phys.*, **75**, 4403 (1981).
123. Rothberg, L. J., Gerrity, D. P., and Vaida, V., *J. Chem. Phys.*, **73**, 5508 (1980).
124. Leutwyler, S., Even, U., and Jortner, J., (a) *Chem. Phys.*, **79**, 5769 (1983); (b) *Chem. Phys. Lett.*, **86**, 439 (1982).

Photodissociation and Photoionization
Edited by K. P. Lawley

LASER ISOTOPE SEPARATION BY THE SELECTIVE MULTIPHOTON DECOMPOSITION PROCESS*

ROBERT D. McALPINE and D. K. EVANS

Physical Chemistry Branch, Atomic Energy of Canada Limited Research Company, Chalk River Nuclear Laboratories, Chalk River, Ontario, KOJ 1JO, Canada

CONTENTS

*Issued as AECL Number 8233.

List of Symbols

C_{op}	Operating cost for a cascade
C_F	Unit cost of feed material
C_E	Unit cost of enrichment
D_c	Separative duty per unit time
$D(1)$	Number of molecules decomposed in one laser pulse
E_I to E_{III}	Regions of vibration excitation for a ground electronic state
f	Decomposition Probability
J	Total flow rate in a cascade
$k_n^{(i)}$	The *n*th order rate constant for *i*th isotopic species
$K(m)$	Measured enrichment after *m* laser pulses
L	Cascade interstage flow rate
$M_j^{(i)}(m)$	Concentration of the *i*th isotopic species of the *j*th product after *m* laser pulses
N	Number density of molecules undergoing MPD
N_F, N_W, N_P	Quantity of feed, waste and product for a cascade
$\langle n \rangle$	Average number of photons absorbed per molecule
P	Pressure
Q_L	Laser related energy required for unit quantity of an enriched isotope

R_F, R_W, R_P	Flow rate of feed, waste and product for a cascade
R_i	Fractional recovery of the *i*th isotopic species
$S(\Phi)$	Absorption cross-section ratio for fluence Φ
T	Tails enrichment (or depletion) factor
W_c	Separative work
x	Atom fraction of the desired isotope
X_E, X_W, X_P	Composition of feed, waste and product for a cascade
y	Atom fraction of the undesired isotope
α	Stage separation factor
β	Heads enrichment factor
i_Γ	Decomposition probability of *i*th isotopic species
δ	Fraction of the molecules containing a specific isotope in an equivalent position
ε_l	Efficiency of the *l*th part of a process
$\bar{\nu}_L$	Wavenumber of laser line
ξ	Abundance ratio
η	Stoichiometric ratio
ρ_{VIB}	Density of vibrational levels
σ	Absorption cross-section
$\Delta\tau$	Laser pulse width
ψ	Separation potential
Φ	Local fluence

I. INTRODUCTION

As early as 1920 Merton and Hartley (1920) considered a photochemical technique for the separation of chlorine isotopes. Since ^{35}Cl and ^{37}Cl occur in a 3:1 ratio (Aston, 1920), a natural sample of Cl_2 will contain isotopic species in the ratio $^{35}Cl_2$:$^{35}Cl^{37}Cl_2$:$^{37}Cl_2 = 9:6:1$. If a beam of white light is passed through a filter of natural abundance Cl_2 such that the wavelengths absorbed by $^{37}Cl_2$ are attenuated by a factor of 10, the corresponding wavelengths absorbed by $^{35}Cl_2$ and $^{35}Cl^{37}Cl$ will be decreased by 10^9 and 10^6 respectively. This filtered light source will now be selectively absorbed by $^{37}Cl_2$ in a natual abundance chlorine sample placed after the filter. If the $^{37}Cl_2$ is thus selectively photodissociated, the resulting ^{37}Cl atom will react with H_2 to form $H^{37}Cl$ and isotopically enriched $^{37}Cl_2$ could be obtained from the $H^{37}Cl$ sample. This scheme, of course, failed (Hartley *et al.*, 1922) due to free radical scrambling reactions which were not known in 1922.

In 1930, a photochemical scheme to separate mercury isotopes was suggested (Mrozowski, 1930). A filter consisting of a column of mercury vapour in a magnetic field was constructed which transmitted only the ^{200}Hg and ^{202}Hg components of natural abundance mercury selectively. The electronically excited Hg, but not ground state Hg, reacts with O_2 to give HgO enriched in ^{200}Hg and ^{202}Hg (Zuber, 1935a, 1935b, 1936). Later workers constructed mercury resonance lamps from single isotopes of mercury and used these lamps to induce isotopically selective photochemical reactions of mercury with air (McDonald and Gunning, 1952), H_2O (Billings, Hitchcock and Zelikoff, 1953; Gunning, 1958; Pertel and Gunning, 1959; Zelikoff, Aschenbrand and Wyckoff, 1953), HCl (McDonald and Gunning, 1952; McDonald, McDowell and Gunning, 1959) and organic molecules (Gunning and Strausz, 1963; Osborne and Gunning, 1959; Sherwood and Gunning, 1960). If there were sufficient demand for enriched mercury isotopes, the photochemical enrichment scheme could likely be developed on a commercial scale (Voiland, 1959).

In 1943, a team of distinguished scientists working at Columbia University under the direction of H. C. Urey considered photochemical means to enrich uranium in ^{235}U (Urey, 1943). Spectra of many gas-phase uranium molecules were studied. Conventional light sources were inadequate to achieve photochemical or photophysical uranium enrichment. However, with the advent of high-powered, spectrally sharp lasers in the late 1960s, the potential for uranium enrichment using lasers was considered at Avco Everett Laboratories, Los Alamos, Livermore Laboratories (Snavely, 1974) and elsewhere (Smith, 1972). In the early 1970s laser isotope separation became a topic of widespread interest throughout the world (Letokhov and Moore, 1977). The discovery of the multiphoton decomposition (MPD) processes (Isenor and Richardson, 1971) and the fact that these could be isotopically selective (Ambartzumian *et al.*, 1974b) has led to a great deal of research into the understanding of infrared multiphoton processes and the potential for their application to isotope enrichment of many elements.

The scope of the present review is limited to a discussion of selective MPD among laser isotope separation (LIS) processes. In Section II is a discussion of the theory of isotope separation processes and the construction of a cascade. These concepts are extended to LIS processes and the theory of the selective MPD process is outlined. Laser sources and their relationships to LIS processes are described in Section III. Subsequent sections are devoted to a discussion of selective MPD processes for individual isotopes of various elements. This review ends with a brief summary of the state of laser isotope separation by the MPD process and the research topics most needed to fully understand these processes.

II. THE PRINCIPLES OF ISOTOPIC ENRICHMENT

A. The Theory of an Ideal Cascade

Isotopic enrichment is achieved through the exploitation of a mass-dependent physical or chemical property such as vapour pressure, chemical exchange, chemical decomposition rate, diffusion, centrifugation and frequencies for the absorption of light. The mass-dependent property is used to divide the feed stream containing an isotopic mixture into two streams: the heads stream (which is enriched in the desired isotope) and the tails stream (which is depleted in the desired isotope). If the degree of enrichment attained is not sufficient, the heads stream is again subjected to further enrichments in higher 'stages' of a 'cascade'. The tails stream can also be subjected to further depletion, thus 'stripping' out more of the desired isotope from the feed.

For simplicity, let us assume a mixture of two isotopes only. The atom fraction of the desired isotope is labelled x and that of the other isotope $y = 1 - x$. The isotopic abundance ratio ξ is given by

$$\xi = \frac{x}{1 - x} \tag{1}$$

Consider one stage of enrichment and label properties of the feed, heads and tails streams (x, ξ, L), (x', ξ', L') and (x'', ξ'', L'') respectively where L denotes the flow rate. The single stage of enrichment may be achieved by one or more units operating in parallel, each having the same ξ, ξ' and ξ''. The degree of isotopic enrichment is defined by the separation factor α and the heads separation factor β:

$$\alpha = \frac{\xi'}{\xi''} \tag{2}$$

$$\beta = \frac{\xi'}{\xi} \tag{3}$$

If α is not sufficiently large to achieve the desired enrichment in one stage, a cascade is constructed in which the heads of one stage become the feed for a higher stage and the tails may become the feed for a lower stage. Although many different types of cascades are possible, isotopic enrichment is achieved most efficiently in an arrangement of individual stages fulfilling the conditions of an 'ideal cascade' (Benedict and Pigford, 1957; Cohen, 1951).

The ideal cascade is constructed from a series of n stages. Abundance ratios and gas flows for the ith stage are labelled ξ_i, ξ_i', ξ_i'', L_i, L_i', L_i'', corresponding to the feed, heads and tails streams of each stage. The stages are interconnected in such a way that

$$\xi_{i-1}' = \xi_i = \xi_{i+1}'' \tag{4}$$

These conditions imply that (Benedict and Pigford, 1957)

$$\alpha = \beta^2 \tag{5}$$

and

$$\xi_i' = \beta \xi_i \tag{6}$$

$$\xi_i'' = \beta^{-1} \xi_i \tag{7}$$

The feed is brought in at some point in the cascade (see, for example, the ideal cascade shown in Fig. 1). The n_s stages downstream (in terms of the heads flow) of this point are known as the stripping stages and the n_e stages upstream are termed the enriching stages. A quantity N_F of feed material, of composition X_F, is introduced into the cascade at a rate of R_F. At the top of the cascade a quantity N_P of product (composition X_P) is obtained at a rate of R_p. An example of an ideal cascade with $\beta = 20$, $n_e = 4$ and $n_s = 2$ is shown in Fig. 1.

For an ideal cascade, the enrichment in the product and depeletion in the waste can be calculated from

$$\xi'_n = \beta^{n_e} \xi_{n_s + 1} \tag{8}$$

$$\xi''_1 = \beta^{-(n_s + 1)} \xi_{n_s + 1} \tag{9}$$

where $\xi_{n_s + 1}$ can be calculated from X_F and ξ'_n and ξ''_1 can be used to obtain X_P and X_W respectively using Eq. (1).

For each stage in the cascade, the following mass balance relationships hold:

$$L_i = L'_i + L''_i \tag{10}$$

$$x_i L_i = x'_i L'_i + x''_i + L''_i \tag{11}$$

$$R_F = \frac{R_P(X_P - X_W)}{X_F - X_W} \tag{12}$$

$$R_W = \frac{R_P(X_P - X_F)}{X_F - X_W} \tag{13}$$

Gas flows within the cascade can be calculated within the enriching section from:

$$L'_i = R_P + L''_{i+1} \tag{14}$$

$$L''_{i+1} = \frac{R_P}{(\beta - 1)} [X_P(1 - \beta^{i-n}) + (1 - X_P)\beta(\beta^{n-i} - 1)] \qquad n_s < i \leq n \tag{15}$$

and within the stripping section from:

$$L'_j = \frac{R_W}{(\beta - 1)} [X_W \beta(\beta^j - 1) + (1 - X_W)(1 - \beta^{-j})] \qquad 0 < j \leq n_s \tag{16}$$

Many of the capital and operating costs of enriching isotopes are proportional to the magnitude of the gas flows. Hence, several useful functions, obtained from considerations of gas flows, are defined for a particular cascade. The total gas flow rate in a cascade, J, is given by

$$J = \frac{\beta + 1}{(\beta - 1)\ln \beta} D_c \tag{17}$$

where

$$D_c = R_P \psi_P + R_W \psi_W - R_F \psi_F \tag{18}$$

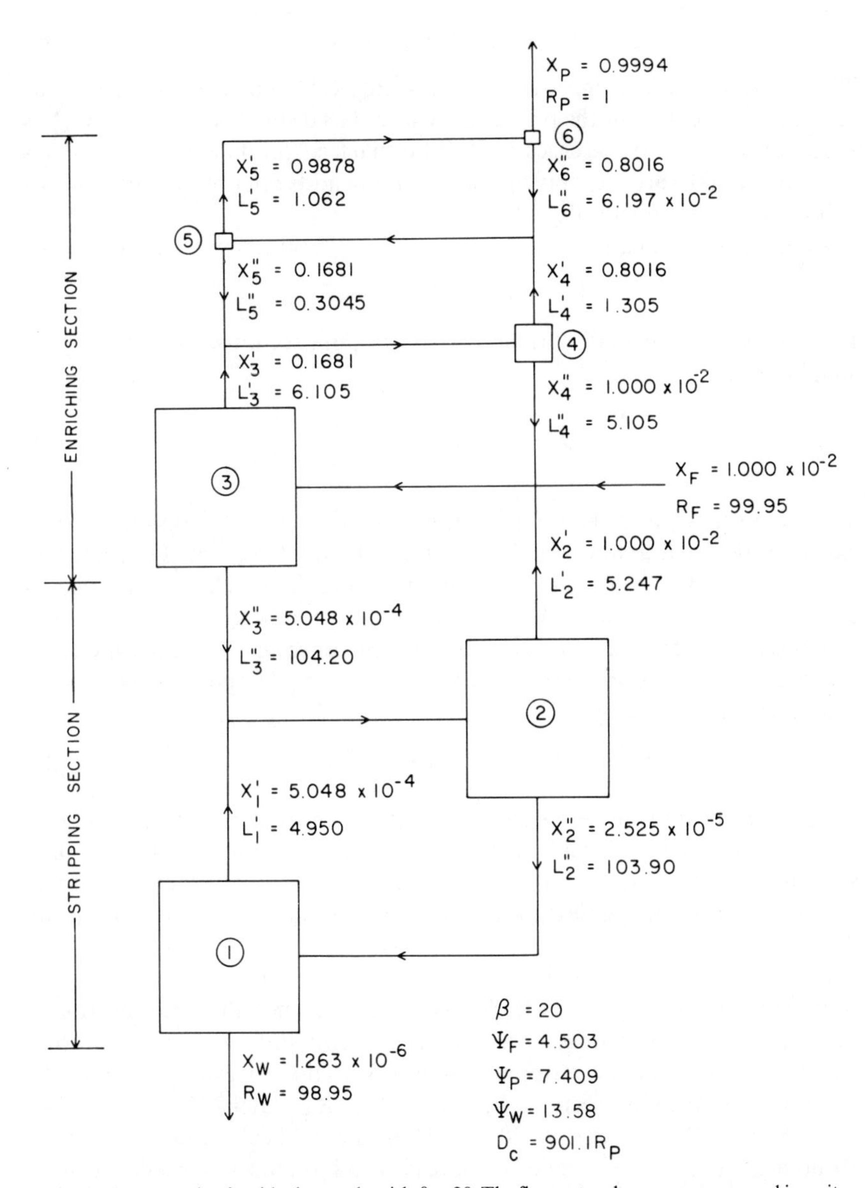

Fig. 1. An example of an ideal cascade with $\beta = 20$. The flow rates shown are expressed in units of $R_p = 1$ and the areas of each stage are scaled to the gas flow through that stage. Important parameters of the cascade are shown in the right-hand corner and for each stage beside the lines indicating heads or tails flow streams.

and

$$\psi_i = (2x_i - 1)\ln \xi_i \tag{19}$$

The functions ψ_P, ψ_W and ψ_F, defined in Eqs. (18) and (19), are termed the separation potential for the product, waste and feed streams respectively. D_c is the separative duty of the cascade. If additional feed, product or waste streams are added to the cascade, appropriate terms are added (if product or waste) or subtracted (if feed) from Eq. (18).

Another useful quantity, closely related to D_c is the separative work, W_c:

$$W_c = N_P\psi_P + N_W\psi_W - N_F\psi_F \tag{20}$$

For a cascade with both enriching and stripping sections, Eq. (20) may be simplified to

$$\frac{W_c}{N_P} = \psi_P - \psi_F - (X_P - X_F)\frac{\psi_P - \psi_W}{X_F - X_W} \tag{21}$$

The quantities D_c and W_c are quoted in units of R_P and N_P respectively. For example, W_c for uranium enrichment is usually quoted in units of kilogram swu. D_c and W_c are useful in calculating costs of enrichment, change in enrichment costs associated with changing the cascade and in making comparisons of enrichment costs for different processes. The capital cost of building a cascade is usually proportional to D_c, while operating costs, C_{op}, will be given by

$$C_{op} = N_F C_F + W_c C_E \tag{22}$$

where C_E is the unit cost of enrichment and C_F is the unit cost of feed material. The relative value of C_F and C_E will largely determine the size of a stripping section, and if C_F is sufficiently small so that $N_F C_F \ll W_c C_E$, then a stripping section is not justified unless there is a significant market for the isotopically depleted waste.

For the cascade shown in Fig. 1, the area of the box representing each stage is proportional to the gas flow through that stage. Hence, the major portion of the capital and operating costs associated with this cascade is the first enriching stage and the stripping section. For this cascade $D_c = 901.1R_P$. Without the stripping section, $R_F = 105.20R_P$, $R_W = 104.20R_P$, $\psi_W = 7.583$, $X_W = 5.045 \times 10^{-4}$ and $D_c = 320.0R_P$. Thus, the smaller cascade without the stripping section uses ~ 5 per cent. more feed and produces ~ 5 per cent. more waste but at a reduction of $580.2R_P$ in separative duty. Clearly, unless there were a significant market for the depleted waste (which is highly enriched to 99.9999 per cent. in the second isotope) the cascade shown in Fig. 1 would not be constructed with the stripping section.

B. The Fundamentals of Laser Isotope Separation (LIS)

Here we will briefly outline the selective multiphoton decomposition process and show how the basic principles of isotopic enrichment relate to this process. We will also outline the general morphology of all LIS processes.

1. The Selective Multiphoton Absorption (MPA) and Decomposition (MPD) Process

The current status of theoretical understanding of infrared laser-induced MPA and MPD has been extensively reviewed elsewhere (Aldridge *et al.*, 1976; Ambartzumian and Letokhov, 1977; King, 1982; Quack, 1982; Wittig, 1985). Only a brief outline is given here.

The frequency $\bar{\nu}_L$ of a pulsed IR laser is selected to be in resonance with only one isotopic species in a mixture. Provided the laser fluence is sufficiently large, this species will absorb sufficient photons to decompose. If the laser pulse is of sufficiently short duration that collisions do not occur during the laser pulse and MPD occurs quickly, then selective decomposition of the species initially excited is possible (Marling, Herman and Thomas, 1980) and the product of MPD will be enriched in the isotope selected.

As the molecule absorbs photons, it is characterized as moving through three distinct regions of excitation, termed E_I to E_{III}.

For the region of lowest energy, E_I, the density of vibrational states, ρ_{VIB}, is small and radiationless processes between the pumped mode(s) and other degenerate levels (known as 'background' levels) are slower than the pumping process (Black *et al.*, 1979; Mukamel and Jortner, 1976). The molecule is usually considered to be coherently pumped (Goodman, Seliger and Minkowski, 1970; Stone, Thiele and Goodman, 1973) through this region up a ladder of overtone levels of the mode initially in resonance. As the molecule moves through E_I, anharmonicity in the pumped mode decreases the intervals between successive overtone levels, meaning that the laser is increasingly off resonance with these levels. This causes an anharmonic 'bottlenecking' in the pumping process (Ambartzumian *et al.*, 1976a; Bagratashvili *et al.*, 1976; Letokhov, Ryabov and Tumanov, 1972). Factors which have been suggested to reduce the severity of this bottleneck include: power broadening (Bloembergen, 1975), anharmonic splittings (Cantrell and Galbraith, 1976; Jensen *et al.*, 1977), rotational selection rules (Ambartzumian and Letokhov, 1977) and collisions (Evans, McAlpine and Adams, 1982). For small molecules (such as alcohols) pumped with a laser which is initially in resonance with a highly anharmonic mode (such as an HF laser), either collisions or very high laser intensities may be essential to overcome anharmonic bottlenecking (McAlpine, Evans and McClusky, 1980).

For region E_{II}, often termed the quasi-continuum region, ρ_{VIB} is sufficiently large that radiationless processes between the pumped and background levels are sufficiently fast to compete with laser pumping. At this point in the excitation, the laser has access to broadened levels of mixed character. The harmonically allowed intensity associated with single-photon excitations in the pumped mode is distributed over the broadened levels, allowing the absorption of photons not in resonance with the zeroth-order intervals of the pumped mode. Structure within the quasi-continuum is complex (von Puttkamer, Dubal and Quack, 1983). This structure can contribute to bottlenecking; however, lasers of many different wavelengths will be absorbed by the quasi-continuum, allowing a weak resonant laser to be used to pump to E_{II} followed by a strong laser of a different colour to pump the molecule the rest of the way to dissociation (Ambartzumian and Letokhov, 1972). This two-colour approach reduces power broadening in E_I, allowing selective excitation of specific isotopes in molecules such as OsO_4 or UF_6 having small isotopic shifts (Ambartzumian *et al.*, 1978b; Rabinowitz *et al.*, 1982).

Region E_{III} lies above the threshold for the lowest decomposition channel. In this region, decomposition, described by the statistical RRKM theory (Forst, 1973), competes with the up-pumping process.

The MPD probability, f, is in general a function of the incident fluence Φ, the laser pulse width, $\Delta\tau$, and the pressure P. For many MPD applications, focused beam geometries (approximately described as a dog-bone geometry—see Hallsworth and Isenor, 1973) are used and the number of molecules decomposed in one pulse, $D(1)$, will be:

$$D(1) = N \int_V f(\Phi, P, \Delta\tau)\, \mathrm{d}V \tag{23}$$

where N is the number density of absorber molecules and V specifies the photolysis geometry.

For fixed values of P and $\Delta\tau$, three suggested approximate forms of $f_i(\Phi)$ are listed in Table I. The HF laser-induced MPD of 2, 2, 2-trifluoroethanol was measured in an approximately cylindrical geometry (Anderson *et al.*, 1981). These results were fitted to the various forms of $f_i(\Phi)$ given in Table I and Eq. (23) was evaluated for two different focused beam geometries. The values of $D(1)$ thus calculated agreed crudely with the experimental results to within a factor of 3 or 4 and not one of these three forms of $f_i(\Phi)$ could be judged to give results superior to the others (Evans, McAlpine and Goodale, 1983; McAlpine, Evans and Goodale, 1983).

For MPD, the reactant depletion is given as a function of the number of laser pulses, m, by

$$\frac{\mathrm{d}N^{(i)}}{\mathrm{d}m} = -k_n^{(i)}[N^{(i)}]^n \tag{24}$$

where $k_n^{(i)}$ a signifies the nth-order rate constant for MPD of the ith isotopic

TABLE I

Several suggested approximations for the decomposition probability, $f(\Phi)$, for fixed values of pressure and laser pulse width.

Functional form	Parameters	References
1. $f_1(\Phi) = A_1 \exp\left(\frac{-\Phi_1}{\Phi}\right)$	A_1, Φ_1 are adjustable	Fuss (1979)
2. $f_2(\Phi) = A_2\left(\frac{\Phi_2}{\Phi}\right)^{n_2}$ $\quad \Phi < \Phi_2$	A_2, Φ_2, n_2 are adjustable	Marling, Herman and Thomas (1980); Speiser and Jortner (1976)
$= A_2 \quad \Phi > \Phi_2$		
3. $f_3(\Phi) = \frac{1}{\sqrt{2\pi}} \int_{-\infty}^{(\log_e \Phi - \mu_3)/\sigma_3} \exp(-t^2/2)\, dt$	μ_3 is the centre of a cumulative log-normal distribution and σ_3 is the standard deviation	Baldwin and Barker (1981a, 1981b); Barker (1980)

species. (The superscript (i) is used to denote quantities which are specific to one isotopic species.) For collisionless MPD, $n=1$, but $n=2$ is often encountered at high pressure where collisionally assisted MPD can occur. For MPD of the reactant N to give the products M_1, M_2, etc.,

$$q_0 N \rightarrow q_1 M_1 + q_2 M_2 + \cdots \tag{25}$$

the product concentrations are

$$M_j^{(i)}(m) = \eta_j N^{(i)}(m) \tag{26}$$

where the stoichiometric ratio ($\eta_j = q_j/q_0$) can be pressure dependent in some cases (Anderson *et al.*, 1981; McAlpine, Evans and Adams, 1983).

If $f(\Phi, P, \Delta\tau)$ is not a function of pressure, then for $n=1$:

$$k_1^{(i)} = -\ln\left[1 - \frac{1}{V_0}\int_V f(\Phi, \Delta\tau)\,\mathrm{d}V\right] \tag{27a}$$

where V_0 is the volume of the entire cell in which the photolysis volume V is contained. For low conversions Eq. (27a) becomes

$$k_1^{(i)} \approx \frac{1}{V_0}\int_V f(\Phi, \Delta\tau)\,\mathrm{d}V \tag{27b}$$

For $n=2$:

$$k_2^{(i)} = \frac{1}{N(0)(V-I)} \tag{28}$$

where

$$I = \int_V f(\Phi, \rho, \Delta\tau)\,\mathrm{d}V \tag{29}$$

Now consider selective MPD of isotope 2 in a binary mixture of two isotopes specified by $i=1$ and 2. Further, assume first-order kinetics since this is most likely to occur under the conditions of few kinetic collisions during the laser pulse which are necessary for high isotopic enrichments. After m pulses, the abundance ratio in the product M extracted will be, by analogy with Eq. (3):

$$\xi' = \beta_2(m)\xi \tag{30}$$

where $\beta_2(m)$ is the heads enrichment factor given by

$$\beta_2(m) = \frac{N^{(1)}(0)M^{(2)}(m)}{N^{(2)}(0)M^{(1)}(m)} \tag{31}$$

$$\beta_2(m) = \frac{1-\exp(-k_1^{(2)}m)}{1-\exp(-k_1^{(1)}m)} \tag{32}$$

For $m = 1$, we define β by

$$\beta = \beta_2(1) \approx \frac{k_1^{(2)}}{k_1^{(1)}} \tag{33}$$

In the tails:

$$\xi'' = \frac{\xi}{T_2(m)} \tag{34}$$

where $T_2(m)$ is the tails depletion factor of isotope 2 (or the enrichment factor of isotope 1 in the tails) given by

$$T_2(m) = \frac{N^{(2)}(0)N^{(1)}(m)}{N^{(1)}(0)N^{(2)}(m)} \tag{35}$$

$$T_2(m) = \exp[(k_1^{(2)} - k_1^{(1)})m] \tag{36}$$

The separation factor $\alpha(m)$ is defined by

$$\xi'(m) = \alpha(m)\xi''(m) \tag{37}$$

where

$$\alpha(m) = \frac{N^{(1)}(m)M^{(2)}(m)}{N^{(2)}(m)M^{(1)}(m)} \tag{38}$$

and hence

$$\alpha(m) = \beta_2(m)T_2(m) \tag{39}$$

This condition is necessary for an ideal cascade to be constructed, and Eqs. (5) to (7) require

$$\beta_2(m) = T_2(m) \tag{40}$$

For various values of β, the values of $k_1^{(2)}m$ which satisfy Eq. (40) are plotted in Fig. 2. Hence, the number of pulses which should be delivered to the sample before the products and depleted reactants pass to higher and lower stages respectively can be determined, for an ideal cascade, from Eq. (40). The rate constant $k_1^{(2)}$ can be varied to some degree by varying the photolysis geometry according to Eq. (27). The fractional recovery, R_2, of isotope 2 in the photolysis volume V_0 is

$$R_2 = 1 - \exp(-k_1^{(2)}m) \tag{41}$$

An important feature of LIS is that often large separation factors are possible. This minimizes the necessity to use a cascade to recover products which are highly enriched. Indeed, in the most favourable cases, sufficient enrichment may be obtained in a single stage, eliminating the need to use a

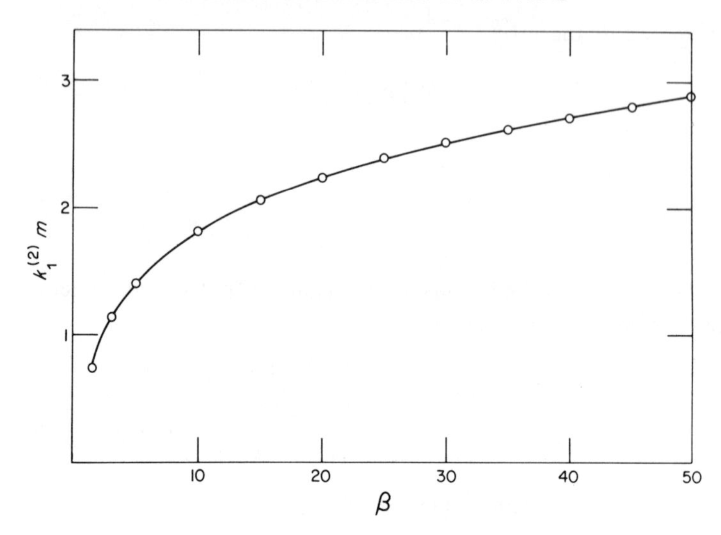

Fig. 2. The value of $k_1^{(2)}m$ necessary to construct an ideal LIS cascade as a function of β. The rate constant $k_1^{(2)}$ may, to some extent, be varied by the choice of photolysis geometry (see text).

cascade. When a cascade is required, the fact that the products are of different chemical species than the reactants introduces additional complications in the design of a cascade. The product must be separated from the reactants and the next stage must either perform selective MPD on these products (Abdushelishvili *et al.*, 1979) or convert them back to the initial reactant species before being passed to the next stage. Kojima and coworkers (1983) have considered this problem as applied to the LIS of ^{13}C.

Experiments performed mainly on SF_6 have led to the conclusion that $f(\Phi, P, \Delta\tau)$ is virtually independent of $\Delta\tau$ (Kolodner, Winterfeld and Yablonovitch, 1977). However, an intensity ($\Phi/\Delta\tau$) dependence has been observed for the MPD of CH_3NH_2 (Ashfold, Hancock and Ketley, 1979; Evans *et al.*, 1983; Hancock, Hennessy and Villis, 1978), C_2H_3CN (Ashfold, Hancock and Hardaker, 1980; McKillop, Gordon and Zare, 1982; Miller and Zare, 1980; Renlund, Reisler and Wittig, 1981; Yu, Levy and Wittig, 1980), $(CH_3CO)_2O$ and CH_3COOH (Grimley and Stephenson, 1981). Recent reports suggest that there may be some intensity dependence in the MPA (Hancock and MacRobert, 1983) and MPD (Apatin and Makarov, 1983) of SF_6.

The pressure dependence of $f(\Phi, P, \Delta\tau)$ is of great interest for LIS processes. For most molecules, there is initially a rise in MPD yield with increasing pressure to some plateau value followed by an exponential decrease for further increase of pressure (Kuzmin and Stuchebrukhov, 1984). Two effects seem to be important in the low-pressure region of initial rise. For small molecules in which anharmonicity bottlenecking is important (e.g. CH_3OH pumped by an

HF laser), collisions assist the MPD process by reducing these bottlenecks (McAlpine, Evans and McClusky, 1980). The second effect which contributes to the initial rise of MPD yields with pressure is termed 'rotational hole filling'. Initially the laser is in resonance with a restricted set of ground state rotational levels and MPD is restricted to these molecules. As P increases, rotational relaxation allows more of the molecules to come into resonance with the laser during the laser pulse (the hole burned by the laser in the rotational distribution is refilled by collisional rotational relaxation) and consequently MPD will occur for a larger fraction of the molecules (Ambartzumian and Letokhov, 1977). For the molecules CDF_3 (McAlpine, Evans and Adams, 1983) an CH_3OH (Evans, McAlpine and Adams, 1984) the MPA cross-section σ, which depends only on Φ and the number of collisions $P\Delta\tau$, is given by

$$\sigma(\Phi, P\Delta\tau) = \sigma(\Phi, 0) + [\sigma(\Phi, 0) - \sigma(\Phi, \infty)]\left[1 - \exp\left(\frac{-P\Delta\tau}{\tau_0}\right)\right] \quad (42)$$

where $\sigma(\Phi, 0)$ and $\sigma(\Phi, \infty)$ are respectively the 'zero collision' and plateau values of σ, and τ_0 is the macroscopic rotational hole-filling characteristic time.

The plateau level for MPD is reached when there are sufficient collisions that the laser has access to all of the rotational distribution that will participate in MPD. Eventually, a further increase in the number of collisions leads to V–V and V–T transfer processes and some of the excitation which previously went into MPD is degraded, leading to an exponential fall of MPD yield with further increases in pressure. For some molecules, the laser has access to the complete rotational distribution, even at 'zero' pressure, and for these molecules there is no initial rise of MPD yield.

The measured separation factor $\beta_2(m)$ is dependent on the number of collisions during the laser pulse. Vibrational energy transfer from the selected isotopic species to other isotopic species opens up decomposition channels of the non-selected species and often $\beta_2(m)$ falls off as P^{-1} (Bittenson and Houston, 1977; Evans, McAlpine and McClusky, 1979; Gauthier *et al.*, 1978; Ishikawa *et al.*, 1980). For some molecules, $\beta_2(m)$ initially increases before falling off as P^{-1} (Gauthier *et al.*, 1980). These effects will be discussed further as they apply to LIS of individual elements.

2. *General Morphology of LIS Processes*

To provide a basis of comparison of different LIS processes and to set realistic goals for their economic exploitation, a general morphology of these processes (McAlpine, 1975) is shown in Fig. 3. The overall separation process, from a convenient, cheap source of the desired isotope to the final isotopically enriched product, can be divided into five steps which may be accomplished by

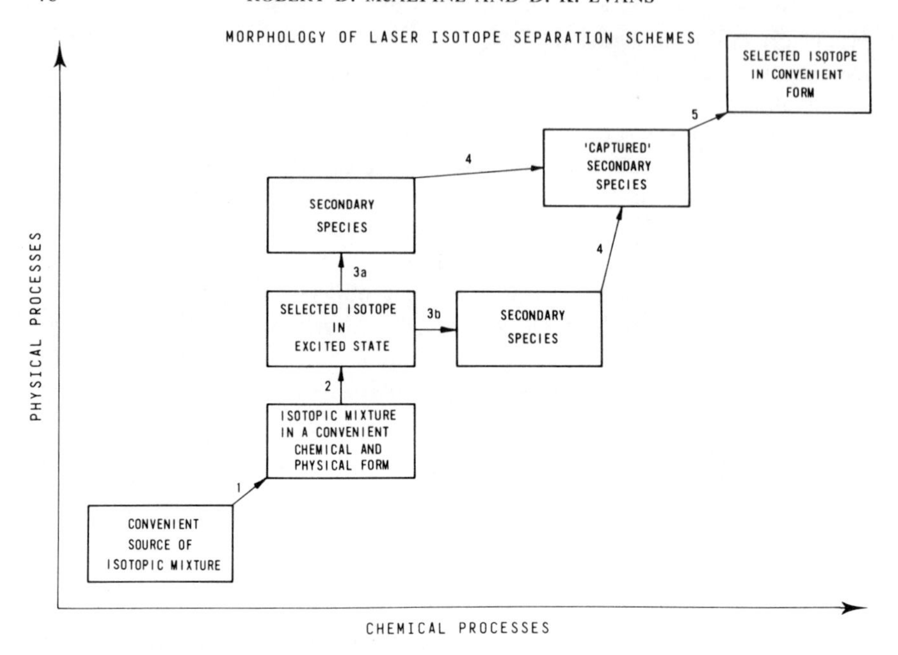

Fig. 3. General morphology of LIS schemes. Chemical processes are those involving change to a new chemical entity and physical processes are those involving only a change in physical state (such as absorption of light, vaporization, distillation, etc.).

chemical (involving the formation of new chemical entities) or physical (absorption of light, vaporization, distillation, etc.) processes.

a. Isotopic exchange with bulk feed source (1)

If, for some reason, LIS cannot be performed directly on an abundant cheap source of the desired isotope (such as water or methane in the case of deuterium) then the desired isotope must be chemically exchanged into a second molecule (called the 'working gas' or 'working molecule') and LIS is performed on this second molecule. This step is perceived to be necessary for the production of the very large quantities of heavy water which are required of any practical heavy-water process and more will be said about this in Section IV. It may also be necessary to convert this working molecule from a liquid to a vapour. Although step 1 involves conventional and well-understood processes such as chemical exchange and vaporization, the cost associated with this step may dominate the overall LIS process.

b. Selective excitation of the working molecule (2)

A laser is tuned to a frequency which will selectively excite the deuterium-containing species with an optical selectivity, $S(\Phi)$, which is given by (Marling,

Herman and Thomas, 1980)

$$S(\Phi) = \frac{\sigma_2(\Phi)}{\sigma_1(\Phi)} \tag{43}$$

where $\sigma_1(\Phi)$ and $\sigma_2(\Phi)$ are respectively the absorption cross-section, at fluence Φ, for the undesired and desired isotopic species in the working gas. The photon efficiency $\varepsilon_p(\Phi)$, the fraction of the absorbed photons giving rise to an excited species containing isotope 2, is given by

$$\varepsilon_p(\Phi) = \frac{\sigma_2(\Phi)N_2}{\sigma_1(\Phi)N_1 + \sigma_2(\Phi)N_2} \tag{44a}$$

$$\approx \left[1 + \frac{1}{\delta S(\Phi)}\right]^{-1} \tag{44b}$$

where N_1 and N_2 are respectively the number of illuminated molecules containing isotopes 1 and 2 and $\delta = N_2/N_1$. The effect of S and δ on ε_p is shown in Fig. 4 for values of δ and ε_p which are typical of deuterium LIS. If the exchange step is operated to give enrichment of isotope 2 in the working gas, then ε_p will be improved. A choice of a working molecule containing more than one equivalent atom of isotope 2 (which will double, triple, etc., δ) will also increase ε_p for the same value of S.

Figure 4 demonstrates that values of $S \gtrsim 10^3$ are required before more than

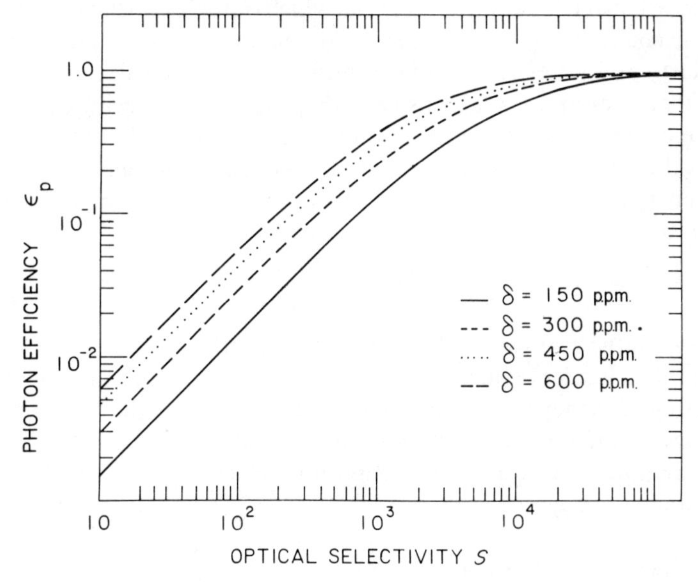

Fig. 4. The effect of parameters S (the optical selectivity) and δ (the fraction of the working molecules containing the desired isotope in a symmetry equivalent position) on the photon efficiency ε_p.

10 per cent of the photons are used to excite the species containing isotope 2. The fraction $(1 - \xi_p)$ of the absorbed photons will have been absorbed by the other species. This results in three undesired effects: (1) energy costs are increased due to the loss of the photons to the species containing isotope 1, (2) overall enrichment factors are reduced since the excited molecules containing species 1 will be 'collected' and, finally, (3) the subsequent destruction of isotope 1 containing working molecules contributes their value to the cost of isotope separation (Marling, Herman and Thomas, 1980).

c. Creation of secondary species (3a and b)

Following the laser-induced selective activation, excited state molecules are converted to secondary species, which are new chemical entities such as new stable or reactive molecules, free radicals or ions, all of which are enriched in the isotope desired. The conversion may be spontaneous (e.g. selective predissociation) or it may be accomplished by physical processes (such as the further absorption of photons) or chemical processes (such as the addition of a molecule which will react with excited, but not ground state, working gas molecules).

d. Capture of the secondary species (4)

Having created the secondary species, they must be 'captured' by chemical or physical conversions into a form from which isotopic enrichment will not be lost by exchange with, or back reaction to, the working gas. In practice, if the secondary species are highly reactive entities such as radicals, this step may be difficult or prohibitively expensive. LIS processes producing stable, non-reactive (or non-exchangeable) secondary species need no capture step and these will have a much greater chance of being economical.

The energy requirements for steps 2 to 4, per kilogram of enriched product, is given by

$$Q_L(\Phi) = \frac{\langle n \rangle hc\bar{\nu}_L N_k}{\varepsilon_e \varepsilon_l \varepsilon_c \varepsilon_p \varepsilon_d \varepsilon_x} \tag{45}$$

where $\langle n \rangle$ is the average number of photons absorbed per molecule, N_k is the number of molecules required to give a kilogram of enriched product, ε_e is the conversion efficiency for thermal heat to electricity, ε_l is the 'wall plug' laser efficiency, ε_c is the photon utilization (fraction of the total photons available which are absorbed), ε_d is the decomposition efficiency and ε_x is the product extraction efficiency.

e. Conversion of captured species (5)

The captured secondary species may be required to be converted to a more useful chemical or physical form. This is a process which works on a greatly

TABLE II
Convention for the classification of LIS experiments.

I	Photolysis of actual isotopic mixtures
II	Photolysis of the individual isotopes unmixed
A	Determination of isotopic ratios from actually separated or easily separable stable products
B	Determination of isotopic ratios from a transient intermediate which may later exchange or further react
C	Determination of isotopic ratios from unreacted initial molecules following photolysis without physical separation of products
D	Inference of isotopic ratios from an analysis of products not containing the desired isotope

reduced volume of the initial feed material, and consequently somewhat increased unit energy costs are tolerable. The requirements for this step will vary according to the particular LIS process being considered.

3. *Assessment of LIS Experiments as a Basis for a Process*

In Sections IV to X describing experimental studies of LIS, we have used the convenient classification scheme outlined in Table II. For example, an experiment which achieves steps 2, 3 and 4 of Fig. 3 (i.e. for which selective MPD is done on an actual isotopic mixture and for which $\beta_2(m)$ measurements are made on 'capture' products) is classified as IA.

III. LASER SOURCES FOR LIS

A key component of any LIS process is the laser source used to carry out selective excitation, and also separation in some cases. Economic considerations and requirements of scale impose quite severe limitations on such laser parameters as average power, efficiency, wavelength, tunability, spectral purity, temporal pulse width, pulse repetition rate and maintenance requirements, which will vary from process to process.

Two approaches to LIS are taken. In the first, a molecule is sought which will undergo selective MPD with an existing, convenient scalable laser source. Alternatively, a molecule may be selected for its chemical or exchange properties and then one attempts to find or develop a laser source which is suitable for selective MPD of the chosen molecule. Often a laser source which is appropriate only for research is used for the fundamental photochemical and photophysical phase of the LIS studies of a particular molecule, and a second source must be developed if this molecule is to be used in a practical LIS process. In this section, we will discuss the developments of several lasers which may prove useful for LIS studies, processes or both.

A. The CO_2 Laser

The pulsed TEA CO_2 laser, operating on the 9- or 10-μm bands, is a convenient, efficient and scalable source and, consequently, since the first MPD experiments were performed on SiF_4 (Isenor and Richardson, 1971), it has become the 'work horse' of the field. In fact, many molecules have been selected for MPD studies principally because they possessed a resonance with one of the CO_2 laser 9- or 10-μm lines. Isotopic variants of CO_2 can provide lines which are shifted in frequency from those available with the dominant isotopic form, $^{12}C^{16}O_2$, and these valuable isotopic variants of CO_2 can be conserved by the use of a gas recycling system (Norris and Smith, 1979; Willis, Hackett and Parsons, 1979; Willis and Purdon, 1979). Laser transitions between a number of different sets of levels are possible in CO_2. A number of these possibilities (some of which have only so far been observed in CW operation) are shown in Fig. 5 (Beck, Englisch and Gurs, 1980; Herzberg, 1945; Pressley, 1971). Transition 1 corresponds to the 10-μm bands having lines from 10.0 to 11.2 μm, of which the most intense occur in the vicinity of 10.2 and 10.6 μm (Beck, Englisch and Gurs, 1980). Transition 2 gives the 9-μm lines from 9.1 to 10.0 μm, with the most intense occurring near 9.3 and 9.6 μm. The sequence band transitions 3, 4 and 9, 10 occur in the vicinity of 1 and 2, as does transition 8. The sequence bands 3 and 4 can be obtained by placing a cell of

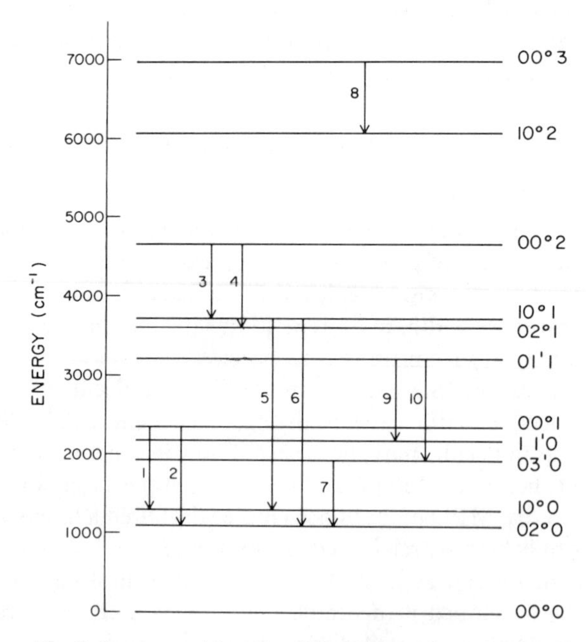

Fig. 5. Laser transitions for CO_2. The numbers signifying each transition are used in the text.

hot CO_2 in the cavity Brimacombe, Reid and Znotins, 1981; Reid, Shewchun and Garside, 1978; Reid and Siemsen, 1976, 1977, 1978; Znotins, Reid and Brimacombe, 1982). High-pressure CO_2 lasers can be used to broaden transitions 1, 2, 3, 4, 9 and 10 to give regions of continuous tunability in the 9- and 10-μm bands (Alcock, Fedosejevs and Walker, 1975; Alcock, Leopold and Richardson, 1973; Bagratashvili, Knyazev and Letokhov, 1971; Basov *et al.*, 1971a, 1971b; Chang and Wood, 1977; O'Neil and Whitney, 1977; Reid and Siemsen, 1978). Transition 7, which gives lines near 16 μm was obtained by collisional pumping of CO_2 with CO (Buckwald *et al.*, 1976). Finally, the cascade transitions 5 and 6 give lines in the vicinity of 4.3 μm (Znotins *et al.*, 1979, 1981).

B. Hydrogen Halide Chemical Lasers

A population inversion of rotational/vibrational levels can be achieved by an exothermic reaction allowing a chemical laser to be constructed (Kasper and Pimental, 1965; Polanyi, 1961). An HF laser is an example of such a chemical laser (Gross and Bott, 1976). Two reactants, e.g. SF_6 and H_2, are subjected to an electrical discharge giving

$$SF_6 + H_2 \rightarrow HF^* + SF_5 + H \tag{46}$$

where HF* denotes HF in excited rotational/vibrational levels. Since this laser is of such high gain, it is difficult to get single-line operation. However, Greiner (1980) has described a device which will allow a single line or simultaneously several lines to be selected. For selective MPD of the small molecule formaldehyde, using several DF lines rather than a single line greatly enhanced the MPD yield (Evans, McAlpine and McClusky, 1979).

The following hydrogen halide lasers and their wavelength ranges (Beck, Englisch and Gurs, 1980) have been constructed: HF, 2.4 to 3.3 μm and 10.2 to 126.5 μm; DF, 3.5 to 4.2 μm and 1.8 μm (Suchard and Pimental, 1971); HCl, 3.6 to 4.1 μm and 13.9 to 24.6 μm; DCl, 5.0 to 5.6 μm; HBr, 4.0 to 4.6 μm and 19.3 to 29.8 μm; DBr, 5.8 to 6.3 μm. Arrowsmith and coworkers (1983) demonstrated laser action for HCl and HBr of 120 and 35 mJ respectively for multiple lines and 55 and 6 mJ respectively for single lines.

C. Optically Pumped Lasers

This type of laser is used to convert a readily available laser frequency (such as a CO_2 frequency) to a new lower frequency, by creating an inversion within the rotational/vibrational levels of a lasing medium through optical pumping. There are many such possible lasers, two of which, the NH_3 and the CF_4 laser, have been useful for MPD studies.

The CO_2 pumped NH_3 laser was first constructed by Chang and McGee

(1976). For pumping with the CO_2 9R(30) line, MPA occurs in the NH_3, and geometrical factors are important in the design of a laser (Waller, Evans and McAlpine, 1983). A very large number of transitions are available (Beck, Englisch and Gurs, 1980), with those in the 12-μm region being of particular interest for LIS of tritium (see Section IV). In addition, this laser was used for selective MPD studies of CCl_4 (Ambartzumian *et al.*, 1978a) and UF_6 (Vasil'ev, Dyad'kin and Sukhanov, 1980).

The optically pumped CF_4 laser, which produces lines from 15.3 to 16.9 μm (Beck, Englisch and Gurs, 1980), was first constructed by Tiee and Wittig (1977) and then by Rabinowitz, Stein and Kaldor (1978) for use in MPD studies of UF_6 (Rabinowitz, Stein and Kaldor, 1978; Tiee and Wittig, 1978b, 1978c).

D. Raman Shifted Lasers

The stimulated Raman effect is used to shift available laser radiation to longer wavelengths in an efficient manner. Interest in 16-μm radiation for MPD studies of UF_6 spurred development of the Raman shifted lasers. Using *para*-H_2 at 77 K to shift CO_2 radiation, photon conversion and energy conversion efficiencies of 85 and 55 per cent respectively have been obtained (Rabinowitz *et al.*, 1978). Radiation of wavelength 16 μm, but at low power and efficiency, was obtained using a triply Raman shifted iodine laser (Jetter, Jill and Volk, 1980). If the pump laser is tunable (as is the case for a high-pressure CO_2 laser or an iodine laser tuned by a magnetic field) this tunability will be preserved in the Raman shifted frequencies.

E. Conclusions

The list of lasers discussed in this section is not complete and it is expected that continuing research in laser physics will provide new sources for both MPD studies and processes. There has been interplay between LIS studies and research in laser physics. On the one hand, the need for sources to study LIS has stimulated the development of new lasers and, on the other, the development of new laser sources has allowed MPD studies to be extended to more systems. This fruitful interplay between the two fields is likely to continue.

IV. SELECTIVE MPD OF HYDROGEN ISOTOPES

A. Market for Hydrogen Isotope Enrichment

The current and foreseeable future market for deuterium enrichment is dominated by the production of heavy water for use in CANDU (Canada

Deuterium Uranium) nuclear power reactors (at a requirement of about 0.8 Mg of heavy water of 99.75 per cent deuterium concentration per megawatt of installed capacity). Since losses of heavy water for a CANDU reactor are small (< 1 per cent of inventory per year), heavy water is a mainly capital item whose production rate is tied to the expected long-term growth rate for CANDU generating capacity. Current versions of CANDU operate on natural uranium fuel and consequently no uranium isotope enrichment is required.

Shown in Fig. 6 is the cumulative Canadian heavy water production, from 1970 to the end of 1982. Also shown are the US and worldwide (excluding USSR and China) cumulative total heavy water production to 1975. Since that date, about 80 to 90 per cent of Western world heavy water production capacity has been in Canada meaning that by 1982 roughly half the western world's total heavy water was produced in Canada.

The growth of the domestic CANDU grid is shown in Fig. 7. This capacity has grown from 230 MW_e in 1970 to 7 GW_e in 1983. Plants currently under

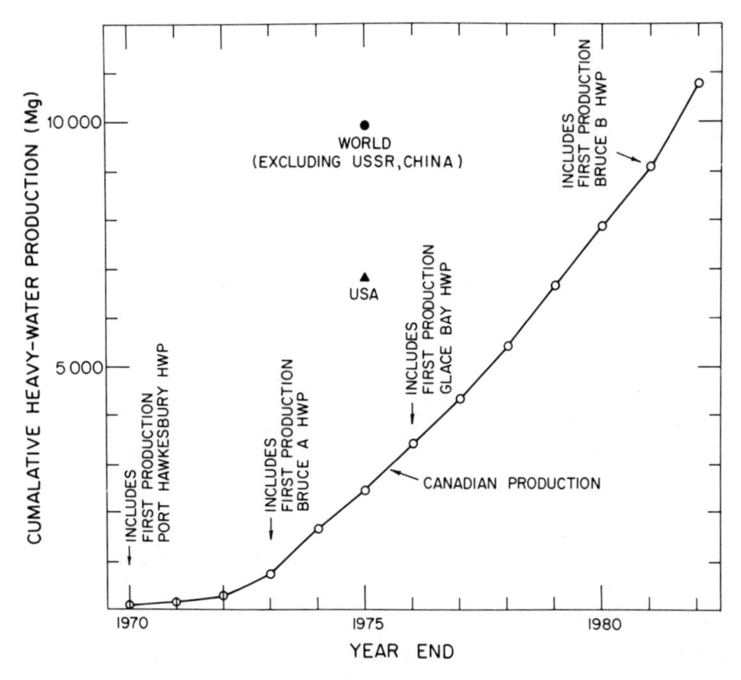

Fig. 6. Cumulative Canadian heavy-water production from 1970 to 1982. The cumulative total production at the end of each year is plotted as a single point. Also shown are the cumulative US and Western world productions to the end of 1975. Since this time, most of the Western world's heavy-water production has been in Canada. First production for each of Canada's new heavy-water plants brought into production during the 12 years is also indicated.

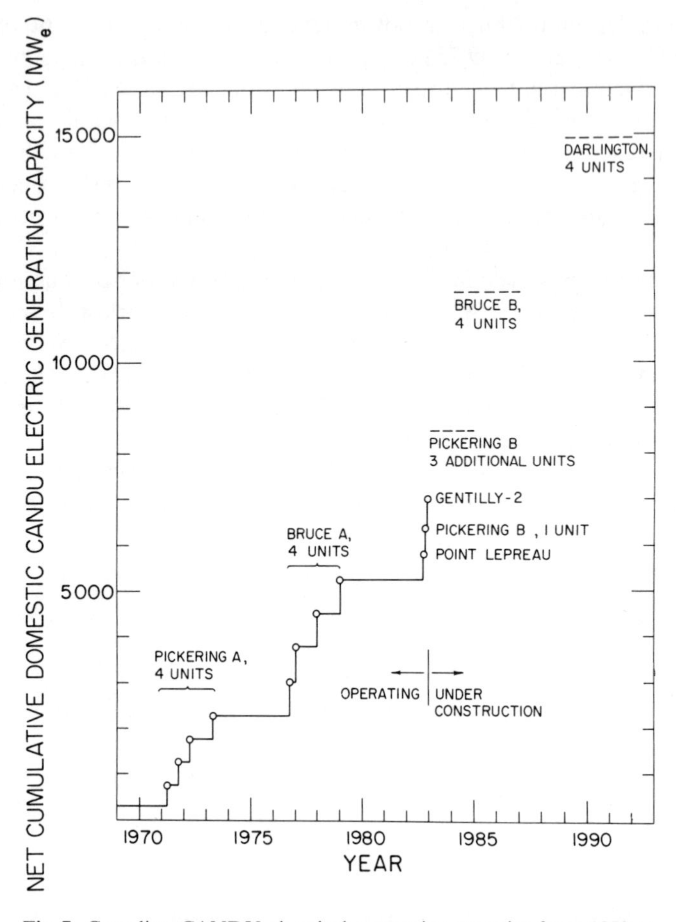

Fig. 7. Canadian CANDU electrical generating capacity from 1970 to 1990. The approximate time that each new plant first contributed electricity to the grid is shown. Plants under construction on 1 Jan. 1983 are shown (dashed lines) at the approximate time they are scheduled to contribute to the grid.

construction will increase this to $\sim 15\,GW_e$ in the early 1990s. The export market for CANDU reactors is difficult to predict due to the current worldwide recession; however, there will likely be between 2 and $5\,GW_e$ of installed capacity outside Canada by the early 1990s. Hence, a cumulative production of 16 Gg of heavy water will be required by the early 1990s. Clearly the currently installed Canadian heavy water capacity is sufficient to service this market and any significant commitment to new enrichment capacity is unlikely to occur until the 1990s or later.

The recovery of tritium from either deuterium or hydrogen is also an area of interest to the nuclear industry. In the moderator (and to a lesser extent in the coolant) of a CANDU reactor, tritium is formed mainly according to (Mughabghab, Divadeenam and Holden, 1981)

$$^{2}H + n \rightarrow {}^{3}H \qquad \sigma = 0.519 \pm 0.007\,\text{mb} \tag{47}$$

and decays according to (Lederer and Shirley, 1978)

$$^{3}H \rightarrow \beta^{-} + {}^{3}He \qquad \tau_{1/2} = 12.33\,\text{yr} \tag{48}$$

Tritium builds up in the moderator at a rate of ~ 0.1 kg per year per 500 MW_e reactor, and would, if permitted, build up to a theoretical equilibrium value $\sim 60\,\text{Ci}\,\text{kg}^{-1}$ of D_2O ($1\,\text{Ci}\,\text{kg}^{-1}$ of T in D_2O is 0.35 p.p.m. T/D and $1\,\text{Ci}\,\text{kg}^{-1}$ of T in H_2O is 0.31 p.p.m. T/H). Current tritium levels are a long way from this theoretical equilibrium. For example, in 1978, those for the moderator of the Pickering reactors were $\sim 16\,\text{Ci}\,\text{kg}^{-1}$ of D_2O. Tritium is also encountered in light water, during fuel reprocessing for light water reactors, at levels $\sim 0.2\,\text{Ci}\,\text{kg}^{-1}$ of H_2O.

Recovery of tritium reduces employee and environmental exposures and also provides a potential revenue source. Currently tritium is used in emergency signs and in watches, and is predicted to be required for fusion test facilities, between about 1985 and 1995, in very large quantities.

B. Conventional Isotope Enrichment

1. *Chemical Exchange Processes*

Although many different chemical exchange isotope enrichment processes were considered for heavy-water production (Rae, 1978) only four have proved suitable for commercial production. These are H_2S/HDO (GS), H_2/HDO, NH_3/HD and CH_3NH_2/HD. Deuterium occurs in nature in a ratio of D/H ~ 150 p.p.m.; consequently very large volumes of feed material must be processed to extract megagram quantities of heavy water. For example, the GS process requires a feed of more than 500 Mg of water per kg of D_2O produced. Currently only water or natural gas are sufficiently abundant to provide feed for a heavy water process; consequently the GS and the H_2/HDO processes have the advantage that large plants can be built. If hydrogen streams become more abundant in the future, the NH_3/HD and CH_3NH_2/HD processes may share this advantage and become viable as a stand-alone process rather than as a parasitic process linked to an ammonia plant (Holtslander and Lockerby, 1978; Nitschke, Ilgner and Walter, 1978; Wynn, 1978). In addition, a version of the H_2O/HD process using hydrogen rather than water feed is an attractive alternative heavy-water process.

The principle of the chemical exchange process can be demonstrated by considering the bithermal GS process which currently accounts for over 90 per cent of the Western world's production capacity. The GS exchange reaction between liquid water and gaseous H_2S is given by

$$H_2O + HDS \rightleftharpoons HDO + H_2S \quad (49)$$

with the distribution of deuterium given by the temperature-dependent equilibrium constant:

$$\alpha(T) = \frac{(D/H)_{LIQ}}{(D/H)_{GAS}} \quad (50)$$

By operating a bithermal exchange, H_2S at the top of the hot section of the first tower is enriched to $x' = 600$ p.p.m. This enriched gas is fed to further, smaller stages for enrichment to 18 to 20 per cent. Final enrichment to 99.75 per cent is by electrolysis or water distillation. In such a process only a small fraction of the deuterium originally in the feed water may be recovered; typically, only about 18 per cent is obtained.

Table III compares parameters of the various important chemical exchange processes. In terms of energy requirements and recovery, the GS process is at a significant disadvantage compared to the other processes. However, since the GS process is based on a water feed, and requires no catalyst, it has been the process chosen for large-scale production. Developments of suitable catalysts for H_2/HDO exchange or wide-scale implementation of the hydrogen economy could lead to other chemical exchange processes displacing the GS processes (Butler, 1980; Butler, Rolston and Stevens, 1978). The HD/H_2O process could also operate on a hydrogen feed, and this would significantly reduce the energy requirement even further. Many economic considerations of laser-based processes have compared laser-related energy costs with the energy requirements for the GS process. The figures in Table III clearly demonstrate that this is an overly simplified comparison for a process.

TABLE III

Comparison of bithermal chemical exchange heavy water production processes.

	H_2O/H_2S	H_2O/H_2	NH_3/H_2	CH_3NH_2/H_2
Deuterium source	H_2O	H_2O	H_2[a]	H_2
Temperature (°C) Cold	30	25	−40	−50
Hot	130	150	60	40
α_c	2.33	3.8	6.0	7.9
α_H	1.82	2.2	3.0	3.6
Recovery	0.22	0.42	0.50	0.55
Energy (GJ kg^{-1} of D_2O)	30	9	7	11
Catalyst required	No	Yes	Yes	Yes

[a]Can use a water feed by putting an H_2/H_2O exchange step before the first stage.

2. Other Conventional Processes

Many enrichment processes for deuterium have been suggested and most have been rejected for practical application on the basis of energy cost, scale, throughput or for some other reason. For example, both hydrogen distillation and electrolysis have been used in the past but are not competitive in the current market. Hydrogen distillation will probably be important for tritium recovery and a combined electrolysis catalytic exchange (CECE) process (Hammerli, Stevens and Butler, 1978) is likely to become more attractive if hydrogen production becomes important. With an abundant supply of hydrogen, hydrogen distillation might also be considered for heavy water production. A liquid vapour phase exchange process has been used in Switzerland for a tritium recovery process (Pautrot and Damiani, 1978).

C. Demonstration of Selective MPD

Selective MPD of water or methane directly does not appear to be likely to form the basis of a practical selective MPD process. This means that LIS must

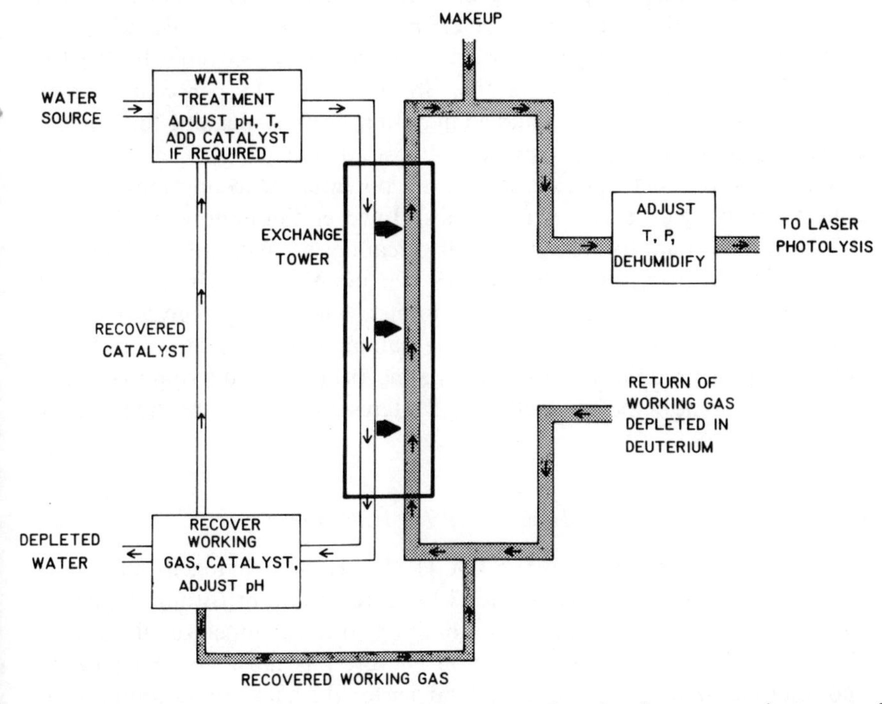

Fig. 8. The exchange step which is necessary to incorporate deuterium from a convenient source of deuterium (e.g. water) into a suitable molecule on which LIS may be performed. This redeuteration, which corresponds to step 1 of Fig. 3, can be quite expensive unless a large fraction of the deuterium is recovered from the working gas.

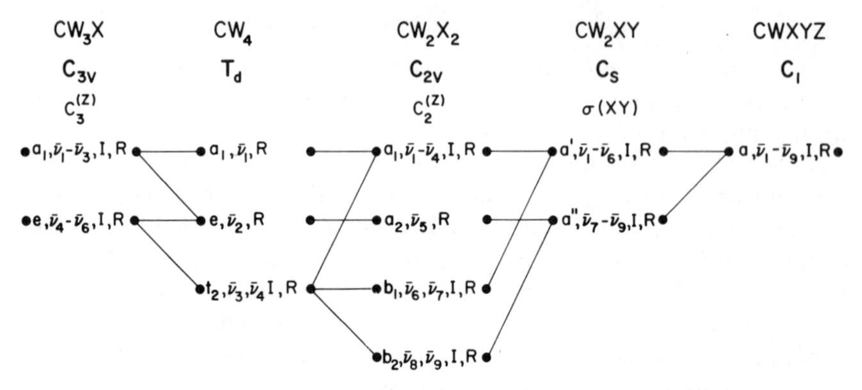

Fig. 9. The point group, symmetry species, numbering conventions and selection rules for substituted methanes. The symbols I and R are used to indicate the allowed infrared or Raman respectively and the symmetry element which defines the axis convention is stated in line 3.

be performed on a working molecule and the unreacted working molecule following photolyis must be redeuterated by exchange with either water or methane according to step 1 in Fig. 3. The exchange redeuteration (see Fig. 8) step is similar to the first stage of the GS process. Unless a significantly larger fraction of the deuterium in the working molecule is recovered by the LIS process than that recovered in the GS process (~ 18 per cent), similar quantities of water would be handled in both processes, meaning that the LIS process is unlikely to have a significant cost advantage.

Where convenient, the discussion of potential working molecules is collected into logical chemical groupings. However, this approach, while easy to follow, does tend to obscure the historical development of the field.

For both hydrogen and carbon LIS (Section V) halocarbons have been extensively studied. The symmetry species, numbering convention and selection rules for normal modes of the halomethanes are shown in Fig. 9. Selected spectroscopic and thermodynamic properties are summarized in Table IV for some of the halomethanes of interest for LIS of either hydrogen or carbon isotopes.

1. Methane and the Haloforms

Several members of the series CX_3H ($X = H$ or a halogen) have been considered as working molecules for LIS of deuterium or tritium. Figure 10 shows the endothermicity for various decomposition channels of this series. If the reverse channel has an activation energy, this must be added to the endothermicity to obtain the energy barrier for the haloform decomposition channel.

For fluoroform and chloroform, the channel of lowest endothermicity is the elimination of a hydrogen halide to form CX_2. For hydrogen LIS, this is an

TABLE IV

Vibrational frequencies (cm^{-1}) and heats of formation ($kJ\,mol^{-1}$) for various halomethanes of interest for hydrogen or carbon LIS. For symmetries and selection rules see Fig. 9. Values shown in brackets are calculated.

Formula	Freon Number	$\bar{\nu}_1$	$\bar{\nu}_2$	$\bar{\nu}_3$	$\bar{\nu}_4$	$\bar{\nu}_5$	$\bar{\nu}_6$	$\bar{\nu}_7$	$\bar{\nu}_8$	$\bar{\nu}_9$	$\Delta H^\circ f(298\,K)$ ($kJ\,mol^{-1}$)	References
CH_4	50	2917	1534	3039	1306						−74.873	a, b
CF_4	14	909	435	1281	632						−933.12	a, c
CCl_4	10	459	217	776	314						−95.81	a, c
CBr_4	10B4	267	122	672	182						−22.22	a, d
CDH_3		2945	2200	1300	3017	1471	1155					a
CHF_3	23	3036	1117	700	1372	1152	507				−693.29	a, b
CDF_3		2261	1111	694	1202	975	502					a
CTF_3		1930	1077	687	1200	835	496					e
$CHCl_3$	20	3034	680	363	1220	774	261				−102.93	c
$CDCl_3$		2266	659	369	914	749	262					a
$CTCl_3$		1932	633	361	835	673	259					f
$CHBr_3$	20B3	3042	541	222	1149	669	155				16.74	a, b
$CDBr_3$		2251	521	222	850	632	153					a, b
$CTBr_3$		(1898)	(504)	(232)	(704)	(607)	(153)					g
$CClF_3$	13	1105	781	476	1212	563	350				−707.97	a, b
$CBrF_3$	13Bl	1089	760	349	1210	547	306				−649.78	g
CIF_3		1080	742	286	1187	537	260				−589.53	g
CH_2F_2	32	2949	1508	1116	529	1262	3012	1176	1437	1090	−452.71	c
$CHDF_2$		(2986)	(2202)	(1374)	(1116)	(990)	(523)	1367	1103	952		j
CD_2F_2		2129	1165	1027	522	907	2284	962	1158	1002		j
$CHTF_2$		(1936)	(1303)	(1143)	(574)	(815)	(3041)	(856)	(1383)	(1109)		i
CCl_2F_2		1101	667	458	362	322	1159	446	902	437	−493.29	a, g
CH_2Cl_2	30	2999	1467	717	282	1153	3040	898	1268	758	−95.40	a, c
$CHDCl_2$		3024	2249	1282	778	692	283	1223	890	738		a
CD_2Cl_2		2205	1052	687	282	826	2304	712	957	727		a
$CHTCl_2$		(1904)	(1230)	(632)	(299)	(770)	(3036)	(753)	(1221)	(687)		h

TABLE IV (Contd.)

Formula	Freon number	$\bar{\nu}_1$	$\bar{\nu}_2$	$\bar{\nu}_3$	$\bar{\nu}_4$	$\bar{\nu}_5$	$\bar{\nu}_6$	$\bar{\nu}_7$	$\bar{\nu}_8$	$\bar{\nu}_9$	ΔH°f(298 K) (kJ mol^{-1})	References
CH_2Br_2	30B2	3009	1382	588	169	1095	3073	812	1195	653	−14.77	a, d
$CHDBr_2$		3040	2249	1220	701	565	172	1154	838	632		a
$CHTBr_2$		(1898)	(1201)	(522)	(186)	(616)	(3035)	(655)	(1137)	(597)		h
$CHClF_2$	22	3010	1310	1100	802	589	410	1349	1118	367	−483.67	j
$CDClF_2$			1013	1104	750			970	1162			j
$CTClF_2$		(1930)	(848)	(1095)	(692)	(586)	(425)	(834)	(1122)	(393)		h
$CHFCl_2$	21										−281.17	b
$CTFCl_2$		(1915)	(847)	(1103)	(668)	(450)	(283)	(804)	(700)	(392)		h
$CHFBr_2$	21B2	3015	1295	1063	620	358	171	1170	704	295	−191.63	g
$CTFBr_2$		(1908)	(821)	(1103)	(561)	(365)	(176)	(726)	(623)	(323)		h
$CHBrCl_2$	20B1	3028	1177	734	597	330	220	1270	773	215	−48.83	g
$CTBrCCl_2$		(1900)	(732)	(670)	(558)	(325)	(229)	(774)	(695)	(219)		h
$CHClBr_2$	20B2	3034	1191	756	576	279	168	1149	669	201	−8.95	g
$CTClBr_2$		(1898)	(750)	(684)	(528)	(280)	(169)	(706)	(609)	(204)		h

References:
(a) Shimanouchi (1972, 1977).
(b) Stull and Prophet, (1971 and annual supplement).
(c) Rodgers *et al.* (1974).
(d) Kudchadker and Kudchadker (1975).
(e) Herman and Marling (1981), Magnotta, Herman and Aldridge (1982).
(f) Herman *et al.* (1983).
(g) Kudchadker and Kudchadker (1978).
(h) Ishikawa, Arai and Nakane (1980).
(i) Suzuki and Shimanouchi (1973).
(j) McLaughlin, Poliakoff and Turner (1982).

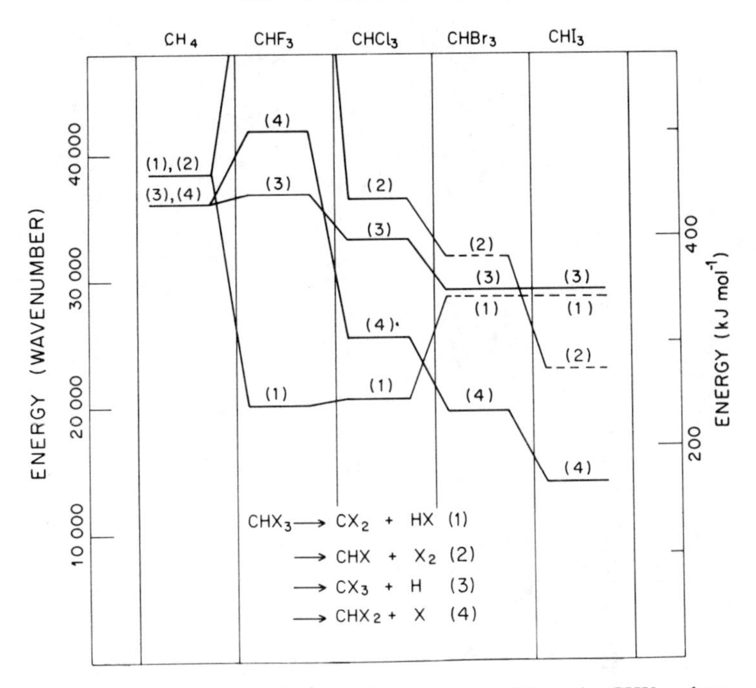

Fig. 10. Endothermicity for decomposition channels of the series CHX_3, where X = H, F, Cl, Br, I. Values shown in solid lines were calculated from measured heats of formation and those shown as dashed lines were estimated. Data for this and for Figs. 12 to 14 were obtained from the following sources: Baulch *et al.* (1980); Benson and Buss (1958); Eggers and Cocks (1973); Kudchadker and Kudchadker (1975, 1978); Levy, Taft and Hehre (1977); Pamidimukkala, Rogers and Skinner (1982); Rodgers *et al.* (1974); Stull and Prophet (1971); Weissman and Benson (1983).

attractive possibility since the desired isotope is 'captured' in a stable, albeit reactive, species. For bromoform and iodoform, the lowest decomposition channel is likely elimination of a halogen atom, leaving the desired hydrogen isotope in the CDX_2 fragment which must now be 'captured' in some way. In addition, if the halogen atom abstracts from CHX_3 to form X_2, the CHX_2 also formed will further reduce the selectivity inherent in the 'captured' product.

a. Methane

Since methane (as natural gas) is so abundant, it is of obvious interest as a feed material for D-LIS. However, other properties of methane make it a poor choice for efficient MPD for the following reasons:

1. The threshold for MPD via the lowest channel is about twice that of other members of the series (see Fig. 10).

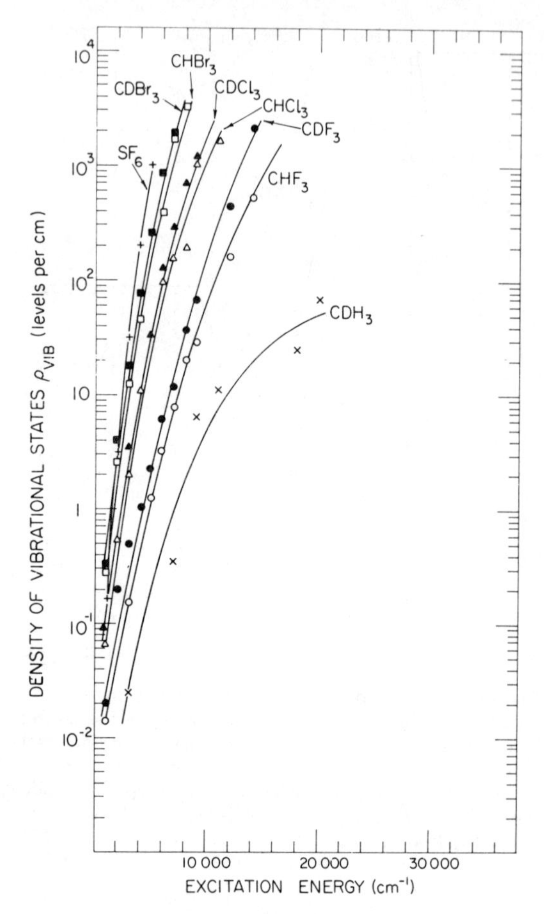

Fig. 11. The density of vibrational states as a function of excitation energy for various protonated and deuterated members of the series CHX_3 and CDX_3 and for SF_6 (for comparison). Since the curves for CDX_3 rise a little more quickly than for CHX_3, the deuterated molecules should undergo MPD slightly easier than their protium analogue.

2. The density of states versus the number of absorbed CO_2 photons is less steeply rising for CH_4 than for the other members of the series (see Fig. 11).
3. Absorptions of CDH_3 occur in a spectral region which is not attainable using currently available high-power IR laser such as CO_2 (cf. Table IV).
4. The selective MPD of CDH_3 is likely to produce D in CH_2D, making 'capture' difficult.

As a result, little work has been done on the selective MPD of CDH_3 in CH_4.

b. Fluoroform

The $\bar{\nu}_5$ bands (Table IV) of CHF_3, CDF_3 and CTF_3 are spectroscopically well separated, making selective excitation of CDF_3 and CTF_3 possible. For CDF_3 in CHF_3 (at natural abundance), $S(0) \geq 2000$ in the region of the CO_2 laser 10R(26) and 10R(28) lines, which selectively excite the R-branch ($\sim 980\,cm^{-1}$) of the CDF_3 $\bar{\nu}_5$ mode (Herman and Marling, 1979), and S(0) ~ 6300 for the 10R(12) and 10R(14) lines which selectively excite the CDF_3 $\bar{\nu}_5$ P-branch ($\sim 970\,cm^{-1}$). However, for values of Φ sufficiently large to give MPD, $S(\Phi)$ is reduced; e.g. $S(25) \sim 900$ for the 10R(26) and 10R(28) lines (Herman and Marling, 1980). Selective excitation of CTF_3 in CHF_3 can be achieved by the CO_2 laser lines in the vicinity of $1075\,cm^{-1}$ (Makide *et al.*, 1980, 1981b; Takeuchi *et al.*, 1982). Selective excitation of CTF_3 may also be possible with a CO or DCl laser (Ishikawa *et al.*, 1980) and selective excitation of CDF_3 in the presence of both CHF_3 and CTF_3 occurs with the CO_2 10R(12) line (Makide *et al.*, 1981a).

Selective MPD of CDF_3 in CHF_3 appears to be primarily via the lowest channel in Fig. 10:

$$\left\{\begin{matrix} CDF_3 \\ CHF_3 \end{matrix}\right\} \rightarrow \left\{\begin{matrix} DF \\ HF \end{matrix}\right\} + CF_2 \tag{51}$$

$$2CF_2 \rightarrow C_2F_4 \tag{52}$$

In glass cells, the DF/HF mixture further reacts according to

$$4\left\{\begin{matrix} DF \\ HF \end{matrix}\right\} + SiO_2 \rightarrow SiF_4 + 2\left\{\begin{matrix} D_2O \\ HDO \\ H_2O \end{matrix}\right\} \tag{53}$$

Using ^{13}C-labelled fluoroform, Herman and Marling (1980) demonstrated that the following equation is unimportant:

$$CF_2 + \left\{\begin{matrix} CDF_3 \\ CHF_3 \end{matrix}\right\} \rightarrow \left\{\begin{matrix} DF \\ HF \end{matrix}\right\} + C_2F_4 \tag{54}$$

At quite low fluences ($\sim 1\,J\,cm^{-2}$), the CO_2 laser radiation field reportedly stimulated an exchange reaction between DF and CHF_3 (Gauthier *et al.*, 1979; Parthasarathy *et al.*, 1982). However, highly deuterium enriched water samples, from Eq. (53) (reduced to $D_2/HD/H_2$ mixtures with uranium), were recovered from selective MPD of natural abundance CDF_3 in CHF_3 at $\sim 20\,J\,cm^{-1}$ (Evans, McAlpine and Adams, 1982).

Collisions play an important, but not essential, role in the MPD of CDF_3. The addition of CHF_3 or argon initially increases and then decreases the MPD yield (Gauthier *et al.*, 1979; Herman and Marling, 1979, 1980; Marling, Herman and Thomas, 1980). The initial rise is mirrored in the MPA cross-sections (Evans, McAlpine and Adams, 1982; McAlpine, Evans and Adams, 1983) and is thought to be due to rotational relaxation processes overcoming rotational hole-burning saturation effects (Ambartzumian and Letokhov, 1977). At higher collision rates V–V transfer processes compete with MPA or MPD. The fluence dependence of the MPD yield may be described by form 2 of Table I (Marling, Herman and Thomas, 1980).

Recently Gozel, Braichotte and van den Bergh (1983) have demonstrated that static electric (up to 4.3 kV cm^{-1}) and magnetic (up to 17.3 kG) fields can induce increases in the collisionless MPD yield of CDF_3 up to a factor of 5.

Of systems investigated to date for deuterium LIS, fluoroform has the highest selectivity. Using class IB, IC and ID experiments (see Table II), Tuccio and Hartford (1979) concluded that $5000 \leq \beta \leq 19\,000$ for CDF_3/CHF_3 mixtures. By class ID experiments, Herman and Marling (1980) estimated that $\beta > 20\,000$. Marling, Herman and Thomas (1980) demonstrated that a 2-ns CO_2 laser pulse could be used to carry out selective MPD of natural abundance ($CDF_3/CHF_3 \sim 150$ p.p.m.) fluoroform at a pressure of 13.3 kPa with $\beta \sim 11\,000$ (class ID). This latter result was an important development, since it demonstrated that high selectivities are possible at the high pressures which are essential for the engineering of a practical isotope separation process. Also using a 6-ns laser pulse and a natural abundance fluoroform mixture at a pressure of 13.3 kPa, Evans, McAlpine and Adams (1982) demonstrated that an enriched product (a $D_2/HD/H_2$ mixture obtained by reduction of the $D_2O/HDO/H_2O$ mixtures from Eq. (53) with metallic uranium) could be physically separated from the fluoroform mixture, a class IA experiment, with $\beta \geq 2000$.

Both the MPD (Marling, Herman and Thomas, 1980) and MPA (Evans, McAlpine and Adams, 1982) 'spectrums' of CDF_3 show broadening, but no significant red shift, even for $\Phi = 20$ J cm^{-2}. This indicates that there is no significant anharmonicity bottlenecking in this molecule, an observation which is consistent with a low anharmonicity constant (X_{55}) for the $\bar{\nu}_5$ mode. Kirk and Wilt (1975) estimate that $X_{55} = -0.25$. Two colour MPD studies of CDF_3 are also consistent with small anharmonicity bottlenecking (Herman, 1983).

Selective MPD of CTF_3 in CTF_3/CHF_3 mixtures (~ 0.2 to 6 p.p.m.) gives β values as high as 580 for class IC experiments (Takeuchi, Kurihara and Nakane, 1981; Takeuchi *et al.*, 1981, 1982). The presence of CDF_3 in these mixtures interferes with selective MPD of the CTF_3 component for the 9R(14) CO_2 laser line.

In order to use fluoroform for a working molecule for either deuterium or tritium LIS, the following exchange reaction is important:

$$CHF_3 + HDO \rightleftharpoons CDF_3 + H_2O \tag{55}$$

For this exchange, Symons and Bonnett (1984) measure

$$\ln \alpha_{\mathrm{I}} = -0.3929 + \frac{263.5}{T} - \frac{21\,200}{T^2} \tag{56}$$

for $343\ \mathrm{K} \leq T \leq 403\ \mathrm{K}$ and where

$$\alpha_{\mathrm{I}} = 2K_{\mathrm{eq}} = \frac{(\mathrm{D/H})_{\mathrm{fluoroform}}}{(\mathrm{D/H})_{\mathrm{water}}} \tag{57}$$

Hence, the fluoroform is significantly enriched in deuterium by Eq. (5) for this temperature range. Without catalyst, Eq. (55) is too slow to be practical. Both hydroxide ion (Symons and Clermont, 1981) and hydroxide in the presence of dimethyl sulphoxide (Symons, Clermont and Coderre, 1981) were considered as catalysts. Both have the disadvantage of being expensive to remove before process water is returned to the environment. The use of supported OH^--type anion exchange resins was also considered. These eliminate the need to remove the catalyst from process water; however, the resins considered lacked sufficient thermal stability (Symons *et al.*, 1983). Base-catalysed hydrolysis of CHF_3 is too slow to compete with Eq. (55). Bigeleisen, Hammond and Tuccio, (1983) considered several feed cycles of fluoroform exchange based on water, hydrogen and ammonia feed stocks. The latter two feed stocks currently imply a parasitic rather than a stand-alone heavy-water process.

Schematics of a heavy-water process based on selective MPD of fluoroform

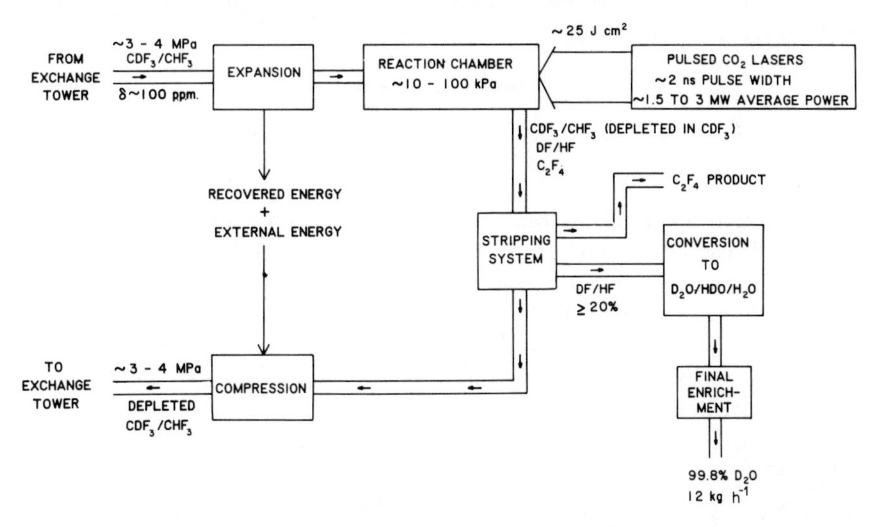

Fig. 12. A deuterium enrichment scheme using the selective MPD of fluoroform to produce $\sim 100\ \mathrm{Mg\ yr^{-1}}$ of 99.8 per cent. D_2O. Laser powers, pressures, etc., shown are only estimates based on the current understanding of fluoroform MPD. These estimates could be significantly modified by further experimentation.

are shown in Fig. 12. This molecule currently appears to be the best candidate for deuterium LIS. Marling, Herman and Thomas, (1980) estimate the laser photolysis part of the energy requirements at $Q_L < 0.6\,GJ_t mol^{-1}$ or $< 30\,GJ_t kg^{-1}$ of D_2O (in terms of thermal energy equivalent), which, at the upper limit, is high in terms of competitive processes. Also this does not include the energy requirements for the exchange step. In addition, the fluences and pulse widths required for selective MPD of fluoroform exceed the damage threshold values for windows, complicating the design of a photolysis cell. A preliminary study of the two-colour selective MPD of the CDF_3 indicates that this approach may somewhat reduce both Q_L and the necessary high fluence (Herman, 1983).

c. Chloroform

Both $CDCl_3$ and $CTCl_3$ can be selectively excited in the $\bar{v}_4$ mode using a CO_2 laser operating on the 10P(50) line for $CDCl_3$ (Herman *et al.*, 1983) or the 828 cm^{-1} NH_3 laser line for $CTCl_3$ (Magnotta, Herman and Aldridge, 1982). For $CTCl_3$ in $CDCl_3$, $S(0) = 12000^{+\infty}_{-3000}$ at 835 cm^{-1} and $S(0) = 6500^{+\infty}_{-1900}$ at 828 cm^{-1}.

From MPD studies of $CDCl_3$ in a molecular beam, Herman and coworkers (1983) conclude that >99.1 per cent. of the $CDCl_3$ that decomposed did so by elimination of DCl (channel 1 of Fig. 10). The energy barrier for the reverse reaction was estimated to be small ($\sim 29\,kJ\,mol^{-1}$). The products, C_2Cl_4, C_2Cl_2, result from

$$2CCl_2 \rightarrow C_2Cl_4^* \tag{58}$$

$$C_2Cl_4^* \rightarrow C_2Cl_4 \tag{59}$$

$$\rightarrow C_2Cl_2 + Cl_2 \tag{60}$$

The source of a small amount of CCl_4 as a product has not been adequately explained. The saturation fluence (Φ_2 in functional form 2 of Table I) for 0.027 Pa of $CDCl_3$ is 23 $J\,cm^{-2}$ (Magnotta, Herman and Aldridge, 1982) compared to 37 $J\,cm^{-2}$ for CDF_3 at a pressure of 0.267 kPa (Marling, Herman and Thomas, 1980). Based on a comparison of ρ_{VIB} versus excitation energy for CHF_3 and $CHCl_3$ (see Fig. 11), it would be expected that $CDCl_3$ would undergo MPD more easily than CDF_3.

Selective MPD experiments on a 200 p.p.m. mixture of $CTCl_3$ in $CDCl_3$ (class IC) gives $\beta \geq 165$ with indications that β may be higher than 2200. Extrapolating their results to mixtures of 5 p.p.m., Magnotta, Herman and Aldridge (1982) suggest that collisional enhancement of the selective MPD of $CTCl_3$ in $CDCl_3$ could give $\beta \sim 6400$.

Using two NH_3 lasers, Herman, Magnotta and Aldridge, (1984) measured the two-colour MPD probability for $CTCl_3$ in $CDCl_3$ (20 to 200 p.p.m.). In the most favourable case, the first pulse enhanced the dissociation probability, relative to just a one-laser photolysis, by 250 per cent. However, Herman, Magnotta and Aldridge (1984) conclude that further studies will be required to determine if there is a net decrease in fluence or electrical energy requirements for a two-colour compared to a single-laser MPD of $CTCl_3$.

The results to date on chloroform, though not extensive, suggest that this molecule might be a better choice as a working molecule for tritium and deuterium LIS than fluoroform.

2. *The Difluorohalomethanes*

The endothermicities for members of this series, which have the formula $CHXF_2$ (X = H, Cl, Br, I), are shown in Fig. 13. Thc climination of IIX, in each case, is the channel having the lowest endothermicity. Vibrational frequencies and heats of formation for members of this series are given in Table IV.

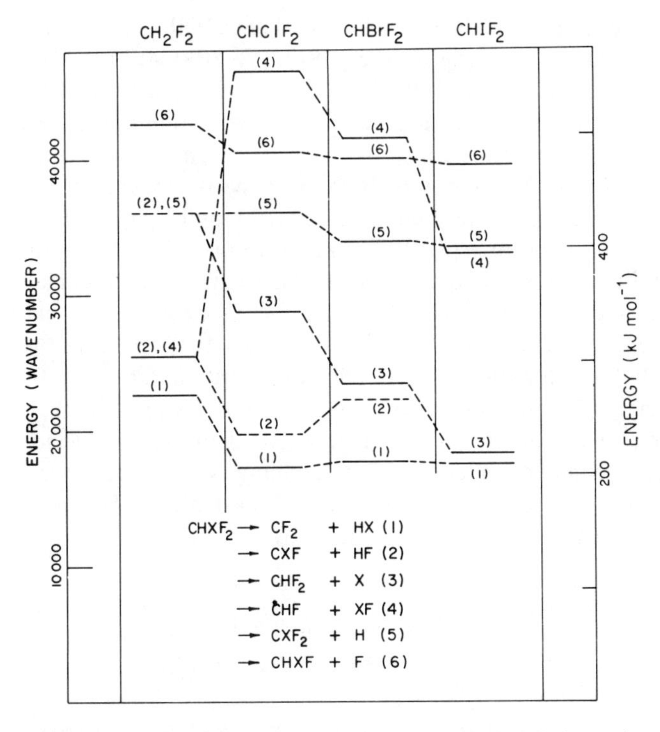

Fig. 13. Endothermicity for decomposition channels of the series $CHXF_2$, where X = H, Cl, Br, I. For X = F, see Fig. 10. Values shown as solid lines were calcualted from measured heats of formation and those shown as dashed lines were estimated. The data sources are listed in the caption to Fig. 10.

a. Difluoromethane

Selective excitation of $CHDF_2$ in CH_2F_2 can be achieved with various CO_2 laser lines in the vicinity of 954 cm^{-1} with $S(0.2)$ as high as 1800 ± 300. For fluence at which significant MPD occurs, $S(23) \sim 350$. Since CH_2F_2 has two equivalent hydrogens, $\delta = 300$ p.p.m. and $\varepsilon_p(23) \sim 0.1$, giving an estimate of the energy requirement for the laser part of a heavy-water process of ~ 51 GJ_t kg^{-1} of D_2O. (Marling, Herman and Thomas, 1980).

Channel 2 of Fig. 13, which is not the channel having the lowest endothermicity, is the predominant MPD channel, and the deuterium occurs in both major products: (HF, DF) and (FCCH, FCCD) (Marling, Herman and Thomas, 1980). Selective MPD experiments (class ID) of $CHDF_2/CH_2F_2$ mixtures from 350 p.p.m. (natural abundance) to 1 per cent. gives $1800 \leq \beta \leq 3500$ (Marling, Herman and Thomas, 1980).

b. Chlorodifluoromethane

Selective excitation of the $\bar{v}_2$ fundamental of $CDClF_2$ in $CHClF_2$ is possible with a CO_2 laser operating on the 10R(28) to 10R(34) lines (the vicinity of 985 cm^{-1}) (Hason Gozel and van den Bergh, 1983; Moser *et al.*, 1983) or on the 9P(30) to 9P(42) lines (the vicinity of 1030 cm^{-1}) (Kutschke, Gauthier and Hackett, 1983). The best selectivity, however, would probably be obtained with excitation in the vicinity of 1015 cm^{-1} (Moser *et al.*, 1983).

Channel 1 of Fig. 13 is the predominant MPD route for $CHClF_2$ (Stephenson and King, 1978; Sudbo *et al.*, 1978) with HCl and C_2F_4 being the principal stable products (Gauthier *et al.*, 1982; Kutschke, Gauthier and Hackett, 1983; Martinez and Herron, 1981). Selective MPD experiments (class IC) for excitation in the region of 985 cm^{-1} suggest $\beta > 10^4$ is possible (Moser *et al.*, 1983) with $\beta > 35$ for selective MPD of a 1 per cent mixture (class IC) using the CO_2 laser 9P(42) line (Kutschke, Gauthier and Hackett, 1983).

Moser and coworkers (1983) suggest that deuterium-depleted chlorodifluoromethane may be redeuterated by exchange with a large excess of HCl at temperatures above 425 °C. The deuterium-depleted HCl can then be redeuterated by exchanges with water.

3. Other Halocarbons

a. Dichloromethane

The CO_2 laser 10P(38) line was used to carry out selective MPD of $CDTCl_2$ in CD_2Cl_2 (4 μCi ml^{-1}), giving $\beta = 29^{+156}_{-15}$ at a pressure of 0.4 kPa (Yokoyama *et al.*, 1983).

b. Dichlorofluoromethane

A preliminary investigation of the selective MPD of $CDFCl_2$ in $CHFCl_2$ (1 per cent mixture) was reported by Hason, Gozel and Van den Bergh (1982) using a CO_2 laser operating in the 920 to 960 cm^{-1} region. The selectivities are reportedly $\sim 10^4$ when Br_2 is added to prevent reaction between CFCl and $CHFCl_2$. The isotopic exchange between $CDFCl_2$ and water (catalysed by an NaOH/dimethyl sulphoxide mixture) is fast and quantitative near 100 °C. Arisawa and coworkers (1982) have studied selective MPD of the $CDFCl_2/CHFCl_2$ system measuring both $\beta_2(m)$ and $T_2(m)$. Their results fit well to a model including collisional processes as well as MPD.

Sugita, Ishikawa and Aral (1983) have studied the effect of the partial pressure of $CHFCl_2$, $CDFCl_2$ or Ar as well as of fluence, temperature and laser pulse width on β_2 and MPD yield. Sudbo and coworkers (1978) conclude that the primary MPD channel for $CHCFCl_2$ is elimination of HCl.

c. 2, 2-dichloro-1, 1, 1-trifluoroethane (freon 123)

CF_3CCl_2H was identified as a potential working molecule for deuterium LIS by a systematic survey of over 200 compounds (Marling and Herman, 1979). Near 944 cm^{-1}, $S(0) \sim 100$ for natural abundance CF_3CCl_2D in CF_3CCl_2H. For a fluence of 10 J cm^{-2}, $S(10) \sim 40$, giving a larger laser energy-related requirement of 1 $TJ_t\, kg^{-1}$ of D_2O and making this molecule unattractive for a heavy-water process (Marling, Herman and Thomas, 1980).

d. 2-chloro, 1, 1, 1, 2-tetrafluoroethane (freon 124)

CF_3CTClF was studied by Kurihara and coworkers (1983) with a view to tritium LIS. The tritiated version has a lower critical fluence (Φ_2 of Table I) than CTF_3, making the former molecule potentially more attractive than CTF_3 for scale-up to practical throughputs.

e. Pentafluoroethane (freon 125)

Selective MPD of C_2F_5T in C_2F_5H (0.2-p.p.m.) were performed using the 10P(20) line of a CO_2 laser. Under conditions of soft focus with a long (52 cm) focal length lens, $\beta > 500$ (classes IA and IC) were measured. The MPD of C_2F_5T occurs via two channels and the tritium is distributed among several free radical products (Hackett *et al.*, 1980; Makide *et al.*, 1982). Selective MPD of C_2F_5T in C_2F_5D was accomplished for 0.2-p.p.m. mixtures at 1.3 to 4 kPa pressure and a temperature of -78 °C. Values of β up to 3000 (class IC) were obtained using the CO_2 laser 10P(34) line (Makide *et al.*, 1983).

4. *Hydroxyl-Containing Molecules*

Hydroxyl-containing molecules, such as alcohols or acids, exchange rapidly with water and the OD stretch frequency (which is often in near resonance with the lines of a DF laser) is usually well separated from absorptions of the OH-containing species. The MPA and selective MPD of formic acid (Evans, McAlpine and McClusky, 1978) and other hydroxyl-containing molecules were studied with either an HF or a DF laser tuned to the OH or OD stretch mode respectively. The smaller molecules, such as methanol, show large anharmonic bottlenecking for HF/DF laser-induced MPD under collisionless conditions (Bhatnagar, Dyer and Oldershaw, 1979; Bialkowski and Guillory, 1977; Chin *et al.*, 1979; McAlpine *et al.*, 1979; McAlpine, Evans and McClusky, 1979 and 1980), meaning that very large fluences are required for MPD. Collisions greatly reduce this bottleneck, however; collisions also reduce selectivity. Collisionless MPD of alcohols can be obtained with a CO_2 laser at lower fluences than an HF/DF laser (Evans, McAlpine and Adams, 1984) but isotopic selectivities are much lower with a CO_2 laser compared to an HF/DF laser. Larger alcohols, such as ethanol or 2, 2, 2-trifluoroethanol do not show large anharmonic bottlenecking for either collisionless or collisional MPD induced by an HF/DF laser (Anderson *et al.*, 1981; McAlpine, Evans and Goodale, 1983; McAlpine, Evans and McClusky, 1980; Selwyn, Back and Willis, 1978). Consequently, collisionless selective MPD can be carried out at lower fluences than the small hydroxyl-containing molecule. However, the decomposition mechanism of 2, 2, 2-trifluoroethanol is complex and free-radical in nature, probably making practical LIS with this molecule difficult.

D. Conclusions

The conventional isotope separation processes are very efficient for hydrogen isotopes, making LIS of this element a particularly challenging problem. The working molecule which is suitable for LIS of hydrogen atoms must fulfil the following specifications:

1. Highly selective excitation must be possible with a laser which is efficient, cheap, reliable and can be operated with pulse widths ≤ 2 ns and which may be scaled up to the size necessary for a production of at least 100 Mg per year of D_2O (this may be achieved with several parallel lasers in the plant). For tritium LIS, this scale-up need not be as large.
2. Exchange of the working molecule with either water or methane must be cheap and rapid and without significant decomposition of the working molecule.
3. MPD of the working molecule should occur at fluences which are sufficiently low to allow efficient design of a photolysis cell.

4. The working molecule should be cheap and preferably non-polluting, non-toxic and non-explosive.

To date, no molecule which fulfils all of these specifications has been identified. In addition, most work has been done using a CO_2 laser, since this currently the laser most likely to meet the requirements of condition 1. However, optical selectivities in the spectral region corresponding to OD or CD sketching modes are higher than those observed in the 9- and 10-μm regions. Consequently, future developments in laser sources may make possible a broader range of possible molecules from which to choose a suitable working molecule for LIS of hydrogen.

V. SELECTIVE MPD OF CARBON ISOTOPES

A. Market

The market for enriched carbon isotopes is very small when compared to that of deuterium or uranium. The current production of 90 per cent. ^{13}C is only a few tens of kilograms per year at a 1983 price of US \$130 per gram of ^{13}C (Monsanto Research Corporation, 24 Jan. 1983 price list for commercial users). From 1966 to 1972, this price fell from US \$3850 to US \$80 per gram of ^{13}C which was accompanied by an increase in sales from 2.4 to 695 g yr^{-1} (Eck, 1973). Recently, Marling (1979) estimated that a hundredfold increase in market would arise from a ten-fold reduction in price. He further estimated that selective MPD of $^{13}CF_2Cl_2$ would produce 90 per cent. ^{13}C at a price of US \$20 per gram using a 50-W (average power) CO_2 laser and of US \$2 per gram using a 10-kW (average power) CO_2 laser.

LIS of carbon isotopes is a subject that has aroused considerable interest and 'pilot plants' with a scale of kilograms per year have been under development in Canada (Gauthier, Hackett and Willis, 1983) and the USSR (Abdushelishvili *et al.*, 1982). The major uses of ^{13}C are currently mainly in research applications, though there is a widely held belief that new uses will develop when cheap plentiful supplies are available. Enrichment of ^{14}C has application for improved ^{14}C-dating techniques (Fuss and Schmid, 1983; Hedges, Ho and Moore, 1980).

B. Demonstration of Selective MPD

1. The Trifluoromethylhalides

The series CXF_3 (where X = F, Cl, Br, I) has been extensively considered for LIS of ^{13}C. The decomposition channel of lowest enthalpy is, in each case, elimination of X (see Fig. 14) and the enthalpy threshold decreases across the series. Since this threshold is so high for CF_4, it is less likely than other

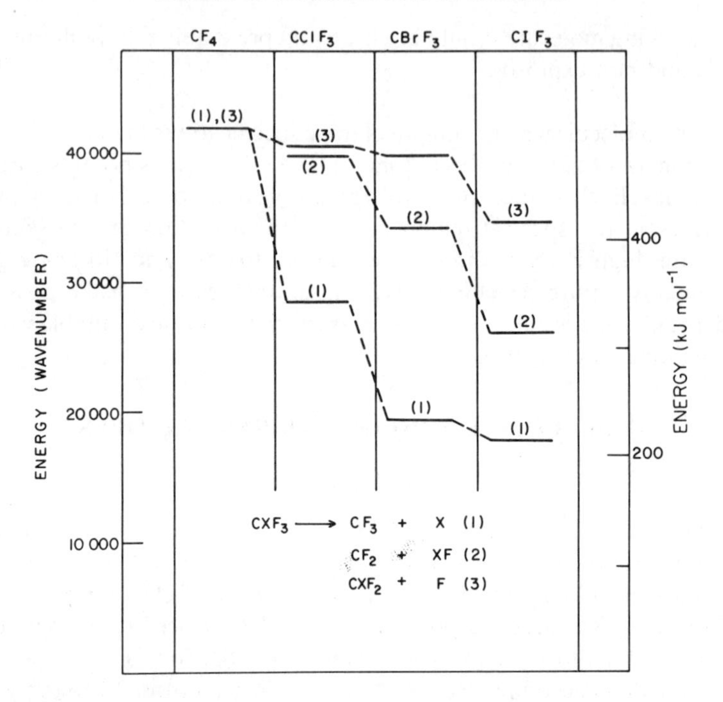

Fig. 14. Endothermicity for decomposition channels of the series CXF_3, where X = F, Cl, Br, I. For X = H, see Fig. 10. Values shown as solid lines were calculated from measured heats of formation. The data sources are listed in the caption to Fig. 10.

members of the series to be of interest for LIS applications. The other three members have been the object of much research and further discussion will be limited to them.

The enthalpy thresholds for channel 1 (Fig. 14) are well removed from other channels making single-channel MPD likely (Sudbo *et al.*, 1978, 1979). However, secondary MPD of CF_3 to $CF_2 + F$ was observed by Würtzberg, Kovalenko and Houston (1978) and by Horowitz and coworkers (1981).

The secondary reactions of CF_3 and X are as follows:

$$2CF_3 \rightarrow C_2F_6 \quad (61)$$

$$2X \rightarrow X_2 \quad (62)$$

$$CF_3 + X \rightarrow CXF_3 \quad (63)$$

$$CF_3 + X_2 \rightarrow CXF_3 + X \quad (64)$$

$$^{13}CF_3 + {}^{12}CXF_3 \rightarrow {}^{13}CXF_3 + {}^{12}CF_3 \quad (65)$$

Reaction (65), which is slower for X = Cl and Br than for I, can be reduced by cooling the cells (Gauthier *et al.*, 1978). Reactions (63) or (64), both of which reduce the net yield of C_2F_6, are important for the series CXF_3. Up to 40 per cent. of the $^{12}CF_3$ from selective MPD of $^{12}CIF_3$ at room temperature is recombined (Bagratashvili *et al.*, 1978) and for $^{12}CBrF_3$ this fraction recombined is higher (Abdushelishvili *et al.*, 1979). The competition between reaction (61) versus (63) or (64) is temperature dependent (Abdushelishvili *et al.*, 1982; Bagratashvili *et al.*, 1978; Bittenson and Houston, 1977). In addition, both reactions (63) and (64) can be prevented by the addition of scavengers (which react with CF_3) such as Br_2 (Abdushelishvili *et al.*, 1979), O_2 (Nève de Mévergnies and del Marmol, 1980) or NO (Abdushelishvili *et al.*, 1982; Quack and Seyfang, 1982).

The $\bar{\nu}_1$ modes of $^{12,13}CIF_3$, $^{12,13}CBrF_3$ and $^{13}CClF_3$ occur near CO_2 laser lines. The ^{13}C isotope shift for this mode is $\sim 26\,cm^{-1}$, meaning that CO_2 laser lines 30 to $70\,cm^{-1}$ to the red of the $^{12}CXF_3\bar{\nu}_1$ mode should be ideal for selective MPD of the $^{13}CXF_3$ component ($X_F = 1.08$ per cent). Selective excitation of the indicated isotopic species have been achieved by the following CO_2 laser lines: $^{12}CIF_3$—9R(8) to 9R(14) (Abdushelishvili *et al.*, 1979, 1982; Bagratashvili *et al.*, 1978; Bittenson and Houston, 1977; Fuss and Schmid, 1983; Kudryavtsev, 1980; Quack and Seyfang, 1982), $^{13}CIF_3$—9P(20) to 9P(32) (Abdushelishvili *et al.*, 1982; Bagratashvili *et al.*, 1978; Gauthier, Hackett and Willis, 1980; Quack and Seyfang, 1982), $^{12}CBrF_3$—9R(26) to 9R(40) (Abdushelishvili *et al.*, 1979; Horowitz *et al.*, 1981; Würtzberg, Kovalenko and Houston, 1978), $^{13}CBrF_3$—9P(12) to 9P(32) (Abdushelishvili *et al.*, 1982, Doljikov, Kolomisky and Ryabov, 1981; Gauthier, Hackett and Willis, 1980; Hackett *et al.*, 1981; Nève de Mévergnies, 1982) and $^{13}CClF_3$—9R(14) to 9R(18) (Gauthier, Hackett and Willis, 1980; Gauthier, Willis and Hackett, 1980; Nève de Mévergnies, 1982).

Both the dependence of yield and selectivity is a complex function of both total pressure and of the partial pressure of the selectively excited species when this is dilute (Abdushelishvili *et al.*, 1982; Gauthier *et al.*, 1980). In order of increasing pressure, effects associated with R–R, V–V and V–T energy transfer processes dominate (Abdushelishvili *et al.*, 1982; Doljikov, Kolomisky and Ryabov, 1981; Gauthier *et al.*, 1980; Gauthier, Willis and Hackett, 1980; Nève de Mévergnies, 1982). A proper choice of conditions is essential to optimize the yield of selectivity for applications to LIS (Abdushelishvili *et al.*, 1982). The use of focused beams can create regions of the gas for which pressure-dependent processes dominate, thus further complicating the design of an efficient LIS process (Gauthier *et al.*, 1980; Nève de Mévergnies, 1982). One advantage of CIF_3 as a working molecule is that focused beams are not necessary for high-yield selective MPD (Abdushelishvili *et al.*, 1982).

Abdushelishvili and Coworkers (1982) constructed a pilot plant which can produce up to $2.2\,g\,h^{-1}$ of highly enriched ^{13}C (> 99 per cent) from CIF_3 in two stages with a 1-kW (average power) CO_2 laser constructed for this application. The energy cost was $Q_L = 120\,GJ_t\,kg^{-1}$ of ^{13}C; however, it could probably be improved by the use of scavengers for CF_3. $CBrF_3$ was also considered to be suitable for large-scale ^{13}C LIS, and would have a commercial advantage over CIF_3 since, to produce a few kilograms per year of

^{13}C requires ~ 10 Mg of raw material of which only 2 per cent is consumed. The remaining 98 per cent could be sold. The construction of this pilot plant clearly demonstrates that LIS can be competitive with conventional isotope separation processes (McInteer, 1980) in both cost and scale of production.

2. *Dichlorodifluoromethane*

The two lowest decomposition channels and their enthalpies for CCl_2F_2 are

$$CCl_2F_2 \rightarrow CClF_2 + Cl \qquad \Delta H = 322\ \text{kJ mol}^{-1} \tag{66}$$

$$CCl_2F_2 \rightarrow CF_2 + Cl_2 \qquad \Delta H = 308\ \text{kJ mol}^{-1} \tag{67}$$

There exists an activation energy for the reverse of reaction (67) which gives (66) as the lowest MPD channel of CCl_2F_2 (Krajnovich *et al.*, 1982). However, reaction (67) is also observed (King and Stevenson, 1977; Morrison and Grant, 1979; Morrison *et al.*, 1981; Stevenson and King, 1978), with a maximum yield of 10 per cent of reaction (66) (Morrison *et al.*, 1981). Rayner, Kimel and Hackett (1983) concluded that secondary MPD of $CClF_2$ to $CF_2 + Cl$ also occurs. The major products of CCl_2F_2 are CF_3Cl, C_2F_4, Cl_2, C_3F_6 and $C_2F_4Cl_2$ (Chou and Grant, 1981; Fettweis and Nève de Mévergnies, 1978; Freeman, Travis and Goodman, 1974; Lyman and Rockwood, 1976). The free radical photochemistry products can be simplified by the addition of scavengers such as: O_2 to give COF_2 (which will hydrolyse) and Cl_2 (Fettweis and Nève de Mévergnies, 1978; Nève de Mévergnies, 1979; Nip, Hackett and Willis, 1981; Ritter and Freund, 1976); H_2 to give C_2F_4, Cl_2 and CH_2F_2 (Fettweis and Nève de Mévergnies, 1978); H_2O to give C_2F_4, Cl_2 and COF_2 (Fettweis and Nève de Mévergnies, 1978); Br_2 to give $CBrClF_2$ and CBr_2F_2 (Morrison and Grant, 1979; Morrison *et al.*, 1981). Other radical scavengers such as NO, HCl (Ritter and Freund, 1976) and various olefins (Ritter, 1978) were also studied.

CCl_2F_2 was one of the first molecules to be considered for ^{13}C LIS. Selective MPD of $^{12}CCl_2F_2$ can be achieved with the 10P(20) CO_2 laser line for which $T_2(m) = 1.7$ (class IC) (^{13}C enrichment in the tails) was measured (Lyman and Rockwood, 1976). Selective MPD of $^{13}CCl_2F_2$ is achieved with the 9P(10) to 9P(34) CO_2 laser lines (King and Stevenson, 1978; Nip, Hackett and Willis, 1981). Enrichment factors for ^{13}C comparable to those obtained with the series CXF_3 have been seen with CCl_2F_2 (Chou and Grant, 1981; King and Stevenson, 1978; Nip, Hackett and Willis, 1981) but no attempt has yet been made to design an LIS scheme using this molecule.

3. *Chlorodifluoromethane*

The selective MPD of this molecule was discussed in Section IV. The molecule gives very efficient selective MPD of $^{13}CHClF_2$ for the CO_2 laser

9P(12) to 9P(32) lines. The laser-related energy requirements and the pressures and fluences required for efficient LIS are quite competitive for those obtained with ClF_3 (Gauthier *et al.*, 1982) and this molecule was selected for use in a 'pilot plant' scale-up (Gauthier, Hackett and Willis, 1983).

4. *Hexafluoroacetone*

The 10R(12) CO_2 laser line is absorbed into the $\bar{\nu}_{15}$ mode of $(CF_3)_2CO$ to give efficient MPD at fluences of only a few joules per square centimetre (Hackett *et al.*, 1978, 1979; Hackett, Willis and Gauthier, 1979). The products at low pressures are CO and C_2F_6, while at high pressures small quantities of CF_4 and C_2F_4 are also observed. The 10R(36) line is selectively absorbed by the $(CF_3)_2{}^{13}CO$ isotopic species, resulting in the product CO enriched in ^{13}C by as much as a thousandfold over natural abundance. Enrichment in the C_2F_6 product is considerably smaller than that in CO (Hackett *et al.*, 1979; Hackett, Willis and Gauthier, 1979). Two-colour LIS experiments have also been performed (Hackett *et al.*, 1981), but these show little advantage over one-colour selective MPD. The energy cost for ^{13}C enrichment to 50 per cent is estimated to be 98 $GJ_t\,kg^{-1}\,{}^{13}C$.

C. Conclusions

Practical enrichment of ^{13}C by selective MPD of ClF_3, $CBrF_3$ $CHClF_2$ or CCl_2F_2 appears to be competitive with conventional techniques and promises to reduce significantly the cost of enriched ^{13}C. Research is continuing and other better molecules for ^{13}C LIS may be identified.

VI. SELECTIVE MPD OF BORON ISOTOPES

Isotopically selective MPD of boron was reported in 1974 by Ambartzumian and coworkers (1974b) and shortly after this by Freund and Ritter (1975) and Lyman and Rockwood (1976). These experiments were the first examples of selective MPD.

A. MPD of BCl_3

The strong absorption of CO_2 laser radiation by BCl_3 was well established in studies using it as a saturable absorber to Q-switch a CO_2 laser (Karlov *et al.*, 1968b). The observation that $^{11}BCl_3$ absorption at 956 cm^{-1} was in the range of the 10-μm P-branch transitions and the $^{10}BCl_3$ absorption at 995 cm^{-1} in the R-branch region was applied in using BCl_3 as a tuning device for a CO_2 laser (Karlov, Petrov and Stel'makh, 1968) and in an attempt to construct a BCl_3 laser (Karlov *et al.*, 1968a). At pressures higher than that required for Q-switching they were able to tune from 944 to 1085 cm^{-1}. Based on these observations Karlov and coworkers (1970) studied the recombination radiation after the MPD of BCl_3. They used very high pressures of BCl_3 (20 to 90 kPa) and over this range they found that the power (P_L) or fluence dependence of the intensity of the recombination radiation (I_F) decreased from $I_F \propto P_L^{2.5}$ at 20 kPa to $I_F \propto P_L^2$ at 90 kPa. The onset of the glow was determined to be delayed after the laser pulse by a time which did not depend on the irradiation intensity but which decreased as pressure increased. They also observed infrared luminescence and used this to determine the spectroscopic parameters of $^{11}BCl_3$ given in Table V, refining the observations of Scruby, Lacher and Park (1951). They were unable to observe infrared fluorescence from $^{10}BCl_3$, which indicated that the excitation is selective. Possibly, at low pressures, the MPD of BCl_3 would also be selective.

The possibility of scavenging the radicals formed was recognized in a study of the CO_2 laser initiated chain reaction of H_2 and BCl_3 (Karlov *et al.*, 1971). The growth of the reaction zone was monitored by a fast motion picture camera (Karlov *et al.*, 1971). Houston, Nowak and Steinfeld (1973) used infrared double-resonance spectroscopy to measure V–V relaxation times by CO_2 laser pumping of $^{11}BCl_3$ and to monitor $^{10}BCl_3$. They determined that intermolecular V–V transfer was almost as fast as intramolecular V–V relaxation. Ambartzumian and coworkers (1974b) found visible luminescence stimulated by a pulsed CO_2 laser to be composed of two temporal components. The first of these was an 'instantaneous' luminescence which followed the CO_2 laser pulse with a pressure-independent delay of ≤ 20 ns for pressures from 6.7 Pa to 2.6 kPa. The addition up to 5 kPa of helium does not reduce the intensity of this component. Since, at pressures ≤ 67 Pa, the time between BCl_3 collisions is less than 20 ns, the process responsible for this luminescence is clearly collisionless. The second temporal component shows a broad peak which is delayed from the CO_2 laser pulse by a delay time which decreases as the BCl_3 pressure increases until at a pressure of 0.8 kPa the delayed luminescence overlaps the instantaneous luminescence. The addition of helium reduces the intensity of the delayed luminescence. Spatially, the

TABLE V
Spectroscopic parameters for BCl_3.

Normal mode	$^{10}BCl_3$[a]	$^{11}BCl_3$[b]	Anharmonicity for $^{11}BCl_3$[b]	Dissociation onset (955 cm^{-1} photons)
$\bar{\nu}_1$	471	472.05	−0.570	102
$\bar{\nu}_2$	480	466.68	−1.405	41
$\bar{\nu}_3$	995	957.67	−1.648	146
$\bar{\nu}_4$	244	242.23	−6.213	72

[a]Scruby, Lacher and Park (1951).
[b]Karlov *et al.* (1970).

instantaneous luminescence predominates in the focal region where laser intensity is highest and the delayed luminescence predominates in the conical regions where intensity is lower. The behaviour of the delayed luminescence probably indicates that the process leading to the luminescent species involves V–V collisional up-pumping processes, now often called energy pooling (Ambartzumian *et al.*, 1975a). The observation that the instantaneous luminescence was clearly a 'collisionless' phenomenon was one of the important results establishing that MPD was a new phenomenon yet to be understood.

Very quickly Ambartzumian and coworkers (1974b) established that it was possible to separate boron isotopes using the selective MPD process. A natural abundance mixture of 0.5 kPa of BCl_3 and either 2.6 kPa of He or $(O_2 + N_2)$ was photolysed with a CO_2 laser tuned to either the $^{10}BCl_3$ 10P(26) line or $^{11}BCl_3$ 10R(24) line component; isotopically specific fluorescence from either ^{10}BO or ^{11}BO respectively for $(O_2 + N_2)$ as the added gas clearly demonstrates (class IB) isotopically selective MPD of BCl_3 (Ambartzumian *et al.*, 1974b). A value of $\beta(m) = 24$ (class IB) for selective MPD of $^{10}BCl_3$ using the 10R(24) line was measured (Ambartzumian *et al.*, 1975a).

MPD of BCl_3 appears to result in recombination of the fragments to BCl_3 with no net production of a new chemical species. Hence, to achieve LIS, scavengers (such as O_2) must be added to capture the isotopically enriched fragments. Freund and Ritter (1975) added either H_2S or D_2S to the BCl_3. Using a CO_2 laser, tuned to either $^{10}BCl_3$ 10R(30) line or $^{11}BCl_3$ 10P(16) line, consistent isotopic enrichment of physically separated, unreacted BCl_3 (tails) could be measured following long photolyses (class IC). The optimum ratio of H_2S to BCl_3 differed, depending on the BCl_3 component selectively photolysed. In addition, a reversal of enrichment from that expected for a BCl_3/D_2S mixture led Freund and Ritter to conclude that different reactions were responsible for the MPD induced by the two different CO_2 lines. Takeuchi, Kurihara and Nakane (1980, 1981) showed that H_2S does not scavenge fragments of BCl_3 MPD, but rather that H_2S reacts with vibrationally excited BCl_3, formed by MPA with insufficient vibrational energy to dissociate. Lyman and Rockwood (1976) used H_2 as a scavenger for BCl_3 MPD products to give exclusively $BHCl_2$ and HCl. Small enrichments ($T_2(m) \sim 1.7$) were observed (class IC) which went through a maximum as the number of pulses increased. This was thought to be due to isotopic scrambling from the reverse reaction occurring on the walls. Lin, Atvars and Pessine (1977) irradiated similar H_2/BCl_3 mixtures in the presence of titanium powder, which, they concluded, either acts as a catalyst for selective MPD or reduces back-reaction of HCl and $BHCl_2$.

B. Selective MPD of Other Boron-Containing Molecules

There have been several studies of the MPD of other boron-containing molecules. *trans*-2-Chloroethenyldichloroborane was synthesized and found to decompose at relatively low fluence (Jensen *et al.*, 1980). However, the selectivity obtained was low. This was explained by suggesting either a non-specific thermal reaction of some of the molecules, a low initial optical selectivity or scrambling due to a chain reaction. The complex $BCl_3 \cdot CH_3SH$ was also found to decompose at low fluence (Ishikawa *et al.*, 1981). However, high selectivity was not found in the products even though pure isotopic samples showed that the dissociation rate of pure $^{10}BCl_3 \cdot CH_3SH$ was ten times that of $^{11}BCl_3 \cdot CH_3SH$ for irradiation with the 10P(20) line.

C. Conclusions

The high initial selectivity for absorption of CO_2 lasers by boron compounds such as BCl_3 has not yet been preserved in physically separated products. The molecules selected for study display various isotopic scrambling mechanisms which have frustrated overall LIS. However, LIS of boron has not been extensively studied and the discovery of new scavengers or new molecules on which to perform LIS may reveal more practical approaches to boron LIS.

VII. SELECTIVE MPD OF SULPHUR ISOTOPES

A. MPD of SF_6

The MPA and MPD of SF_6 have been studied extensively as a standard for a molecule interacting with an intense CO_2 laser, and many studies of this molecule have appeared in the literature. Indeed, SF_6 has been referred to as the hydrogen atom of MPD.

The four isotopes of sulphur span a large range of natural abundance and give large isotope shifts in the part of the infrared spectrum of SF_6 which strongly absorbs 10P-branch CO_2 laser lines. Some of the properties of SF_6 important to MPD are listed in Table VI. In principle, LIS may be studied for a variety of different material conditions and in this review we have attempted to limit our consideration of the MPD of SF_6 to only those reports relevant to LIS.

The first reports of selective MPD of sulphur in SF_6/H_2 mixtures were published in 1975 almost simultaneously by Ambartzumian and coworkers (1975b, 1975c) and by Lyman and coworkers (1975). Selective MPD of the following isotopic species were obtained using the given CO_2 laser line: $^{32}SF_6$—10P(12), 10P(16), 10P(20); $^{34}SF_6$—10P(32), 10P(40). The species undergoing selective MPD is 'burned out' of the mixture leaving the unreacted SF_6 almost arbitrarily highly enriched in the other isotopes depending on the number of laser pulses and the fraction of SF_6 consumed. The decomposition process and resulting enrichment can be interpreted in terms of competing first-order processes with different rate constants for each isotope. This means

TABLE VI
Properties of SF_6.

Isotopic distribution[a] $^{32}S:^{33}S:^{34}S:^{36}S = 95.0:0.76:4.22:0.014$

Vibrational frequencies[b]

$\bar{\nu}_1(a_{1g}) = 773.6 \pm 0.5\ cm^{-1}$	$\omega_1 = 782 \pm 3\ cm^{-1}$
$\bar{\nu}_2(e_g) = 642.1 \pm 0.5\ cm^{-1}$	$\omega_2 = 649 \pm 2\ cm^{-1}$
$\bar{\nu}_3(t_{1u}) = 947.968 \pm 0.001\ cm^{-1}$	$\omega_3 = 966 \pm 3\ cm^{-1}$
$\bar{\nu}_4(t_{1u}) = 615.03 \pm 0.02\ cm^{-1}$	$\omega_4 = 620 \pm 5\ cm^{-1}$
$\bar{\nu}_5(t_{2q}) = 522.9 \pm 0.5\ cm^{-1}$	$\omega_5 = 528 \pm 5\ cm^{-1}$
$\bar{\nu}_6(t_{2u}) = 346 \pm 1\ cm^{-1}$	$\omega_6 = 352 \pm 2\ cm^{-1}$

Anharmonicity constants[b] X_{ij} (see ref. b)

Isotope shifts $^{32}SF_6 - {}^{34}SF_6$	$\bar{\nu}_3$	$17.4 \pm 0.3\ cm^{-1}$
	$\bar{\nu}_4$	$3.3 \pm 0.4\ cm^{-1}$

Density of vibrational states (levels cm^{-1})[c]

after absorption of 3–4 CO_2 laser photons ~ 100
10 CO_2 laser photons ~ 1.7×10^8
20 CO_2 laser photons ~ 3.1×10^{11}

V–V relaxation time[d] $P_{\Delta\tau} = 0.2\ \mu s\ kPa$
V–T relaxation time[e] $P_{\Delta\tau} = 16.3\ \mu s\ kPa$

[a] Weast (1972).
[b] McDowell, Aldridge and Holland (1976).
[c] Lyman and Rockwood (1976).
[d] Bates *et al.*(1971).
[e] Steinfeld *et al.* (1970).

that, if one can collect the products of selective MPD, high enrichments, $\beta(m)$, may be obtained with only small fractional consumption; however, for enrichment in the residual SF_6, enrichment factors, $T(m)$, will continue to increase only as larger fractions of the SF_6 are 'burned out'. Clearly, collection of the enriched products is the most efficient option, depending critically on an understanding of the chemistry of SF_6 MPD. For the same fractional conversion of SF_6, $T(m)$ decreases exponentially to a minimum value of 1 as pressure increases (Ambartzumian *et al.*, 1975b).

Early workers made the following important observations concerning selective MPD of SF_6:

1. $\langle n \rangle$ is large, indicating MPA. When consideration is taken of the effective laser line width (Rabi line width) relative to the ground state rotational distribution, it is evident that only a fraction of the entire rotational

distribution can couple to the laser. This means that the values of $\langle n \rangle$ for this fraction are very high (Ambartzumian *et al.*, 1975b).
2. The MPA cross-section (which is smaller than the small-signal cross-section) decreases as intensity (or fluence since this was always proportional to intensity in the early experiments) increases (Ambartzumian *et al.*, 1975b).
3. Power broadening is not sufficiently large at the intensities used (1 to 70 MW cm^{-2}) to overcome the non-resonance between the laser and the high intervals of the anharmonic ladder of the $\bar{\nu}_3$ mode (Ambartzumian *et al.*, 1975b).

These observations were an important consideration for theoretical efforts to understand MPD. Ambartzumian and coworkers (1975b, 1976a, 1976b) suggested that MPA could occur up to 3 to 4 levels of $\bar{\nu}_3$ by a P–Q–R rotational compensation model. Letokhov and Makarov (1976) suggested that, for levels above 3 or 4 CO_2 laser quanta, 'leakage' from the ladder of $\bar{\nu}_3$ levels to a 'quasi-continuum' due to other modes could play an important role in MPA. Cantrell and Galbraith (1976) suggested that anharmonic splitting of the degenerate excited $\bar{\nu}_3$ levels could provide convenient resonances for MPA of 5 to 10 CO_2 laser photons. At about the same time, Bloembergen (1975) pointed out that after absorption of about 3 CO_2 photons, SF_6 would display quasi-continuum absorption similar to the spectral broadening obtained by heating. This would also shift the absorption maximum to the red of that observed for room temperature SF_6. Such a red shift in the spectrum of MPD probability versus wavelength of SF_6 was observed by Ambartzumian *et al.* (1976a, 1976b).

The dependence of decomposition probability and tails enrichment factor (class IC) on pressure, laser pulse number, wavelength, energy, pulse duration and photolysis geometries were described, using a model of competing first-order reactions, by Lyman and Rockwood (1976), Hancock, Campbell and Welge (1976), Campbell, Hancock and Welge (1976), Gower and Billman (1977), Lyman, Rockwood and Freund (1977) and Fuss and Cotter (1977). The enrichment of $^{34}SF_6$ in the tails (or depletion of $^{32}SF_6$) is defined as

$$K(m) = \left[\frac{1 - {}^{34}\Gamma}{1 - {}^{32}\Gamma}\right]^m \tag{68}$$

where ${}^{i}\Gamma$ represents the decomposition probability of the ith isotope. Gower and Billman (1977) expressed $K(m)$ in terms of a cross-section which is independent of photolysis geometry. The wavelength and fluence dependence of MPD (Ambartzumian *et al.*, 1976a, 1976b; Gower and Billman, 1977; Schulz *et al.*, 1980) and of MPA (Alimpiev *et al.*, 1978; Ham and Rothschild, 1977) shows a shift to the red (by 7 to 10 cm^{-1}) and a broadening, both of which increase with increasing fluence. This effect was seen to be equivalent to heating the SF_6 (Bloembergen, 1975; Schulz *et al.*, 1980).

Intramolecular and intermolecular energy transfer and the effect of these on selectivity and probability of decomposition have been considered in many reports (Gower and Billman, 1977; Hancock, Campbell and Welge, 1976; Lin, Lee and Ronn, 1978; Lyman and Rockwood, 1976; Lyman, Rockwood and Freund, 1977. These studies included varying the pressure of SF_6 with various additives. In a neat SF_6 experiment an optimum pressure for selectivity of about 6 Pa was observed (Gower and Billman, 1977). At both high and lower pressures the selectivity decreases and at higher pressure the probability of decomposition decreases. A similar trend was seen by Fuss and Cotter (1977), but they did not look at pressures as low as those studied by Gower and Billman (1977) and did not report the decrease at very low pressures. In Fig. 15, plots of decomposition probability for two sulphur isotopes and enrichment (as defined in Eq. 68) are shown. The data are taken from Fuss and Cotter (1977). Because the values of pressure at which various phenomena occur depend on the pulse length of the laser used (Bagratashvili *et al.*, 1977) no numerical values are given on the axes. The trends shown can be interpreted as follows. At low pressures, $^{32}\Gamma$ and $^{34}\Gamma$ are constant. In this pressure region, the

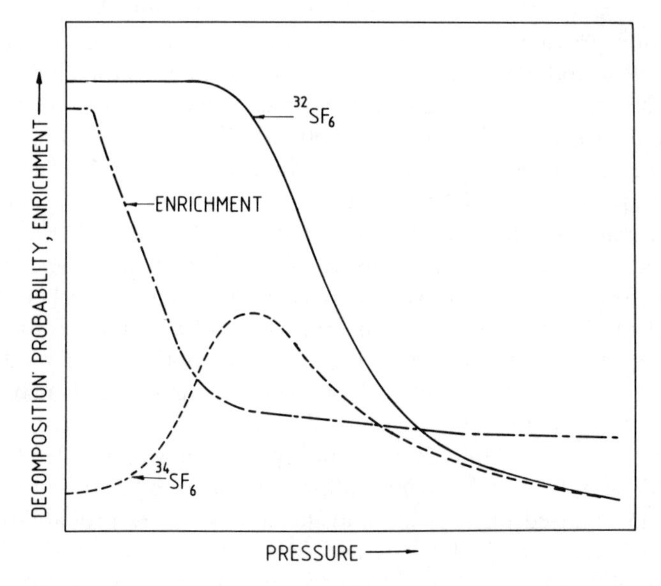

Fig. 15. The effect of pressure on decomposition probability and selectivity for the MPD of SF_6 by a CO_2 laser. The collisional transfer of energy to the transparent isotopic species ($^{34}SF_6$) can be seen by its rising decomposition probability, which occurs at the expense of selectivity. At higher pressures, both decomposition and selectivity decrease. (The plot is based on the results of Fuss and Cotter, 1977.)

time between collisions is long with respect to the laser pulse length. As pressure increases, V–V transfer from $^{32}SF_6$ to $^{34}SF_6$ causes the latter to absorb the last part of the laser pulse, exciting it to dissociation. The fact that $^{32}\Gamma$ does not decrease suggests this mechanism rather than any energy pooling after the pulse. As pressure continues to increase, the two species share energy and $^{32}\Gamma$ and $^{34}\Gamma$ tend to become equal. However, there are now enough collisions during the pulse to transfer energy to more molecules in the system and we tend towards a gentle warming of the gas to a temperature too low for significant decomposition.

An interesting fundamental question for an understanding of MPA and MPD of SF_6 is the parametrical importance of laser intensity versus fluence, as this can be determined by studying the dependence of MPA or MPD on laser pulse profile. To the extent that $P\Delta\tau$ (where P is SF_6 pressure and $\Delta\tau$ is the laser pulse width) is sufficiently small that collisional relaxation and energy transfer processes are unimportant, the statistical heating model of MPD predicts that fluence rather than intensity determines the degree of MPA (Black *et al.*, 1979; Bloembergen, 1975; Kolodner, Winterfeld and Yablonovitch, 1977). Two approaches to this question have been taken: (1) comparing MPA or MPD using both single-mode and mode-locked pulses; (2) comparing MPA or MPD for different values of $\Delta\tau$. Since the intensity variation during a pulse is complex for a mode-locked pulse, method one can only provide qualitative information. The interconnection among the variables P, Φ and $\Delta\tau$ for studies in which Φ and $\Delta\tau$ are varied independently means that the effects of collisions (which vary as $P\Delta\tau$) must also be considered in method 2, and preferably all of the variables P, Φ and $\Delta\tau$ should be varied systematically to distinguish between fluence, intensity ($\Phi/\Delta\tau$) and collisional ($P\Delta\tau$) dependences (McAlpine, Evans and Adams, 1983). Another potential problem in interpreting MPA and MPD studies of SF_6 is the observation of self-focusing of intense CO_2 laser beams by long paths of SF_6 (Nowak and Ham, 1981). Lyman and coworkers (Lyman, Feldman and Fischer, 1978; Lyman, Hudson and Freund, 1977) showed that only an 8 per cent. increase in MPD yield was obtained for mode-locked compared to single-mode pulses. Kolodner, Winterfeld and Yablonovitch (1977) studied a change in $\Delta\tau$ from 0.5 to 200 ns and observed a decrease in MPD probability of only 30 per cent. Apatin and Makarov (1983) studied the MPD of 2 Pa of SF_6 using multimode pulses of $\Delta\tau$ of either 100 or 900 ns and reported up to a ten fold increase in MPD probability for the shorter pulse. Recently, Hancock and MacRobert (1983) observed $\langle n \rangle$ to be $\sim$twofold larger for $\Delta\tau = 50$ ns compared to $\Delta\tau = 200$ ns for the same fluence; however, the ambiguity of the potential participation of collisional processes was not eliminated from these results. The results appear to be contradictory. The early results indicated that MPA and MPD of SF_6 depends principally on fluence with only a very weak dependence on intensity.

However, more recent results (Apatin and Makarov, 1983; Hancock and MacRobert, 1983) indicate that the intensity dependence may be stronger. Further careful studies are required to determine the source of the disagreement.

Sulphur LIS was accomplished for SF_6 by 'burning out' one component (class IC) and collecting the tails. Also of practical interest is the collection of the enriched products (class IA), and, for this, the MPD chemistry of SF_6 must be understood. Diebold and coworkers (1977), Grant and coworkers (1977) and Schulz and coworkers (1980) have studied the MPD of SF_6 in a molecular beam, observing the primary process to be

$$SF_6 \rightarrow SF_5 + F \tag{69}$$

At higher laser fluences, the SF_5 can undergo further MPD induced by the same laser pulse to give

$$SF_5 \rightarrow SF_4 + F \tag{70}$$

B. MPD of Other Sulphur-Containing Molecules

Several studies have been carried out on the molecules SF_5X where X = Cl, NH_2 and SF_6. In all cases, the S—X bond is the weakest bond and is the one that breaks, consistent with the statistical nature of MPD (Karl and Lyman, 1978; Leary *et al.*, 1978; Lyman and Leary, 1978; Lyman *et al.*, 1979). All of the SF_5X series undergo MPD at fluences lower than SF_6, thus demonstrating that increasing the density of states by an appropriate substitution will greatly increase the ease of MPA. The molecules S_2F_{10} and SF_5NH_2 are sufficiently large that the entire ground state rotational distribution is accessible to the effective laser line width (Judd, 1979).

C. Conclusions

Although there is little demand for sulphur isotope enrichment, more theoretical and experimental studies of SF_6 MPD have been performed than for any other molecule. SF_6 is expected to resemble UF_6, which is important for uranium isotope enrichment, and SF_6 has proven to be particularly tractable for both experimental and theoretical studies. Most of our current understanding of selective MPD was obtained from consideration of experiments on SF_6.

Demonstrations of LIS for SF_6 were usually restricted to class IC experiments and little attention was given to class IA experiments which would likely lead to more economical schemes for large-scale production. The preliminary results obtained on SF_5Cl, SF_5NH_2 and S_2F_{10} indicate that these molecules may be preferable to SF_6 for practical applications.

VIII. SELECTIVE MPD OF OSMIUM TETROXIDE

There is, at present, no large market for isotopically enriched Os; however, since OsO_4 is a convenient example of a volatile compound of a heavy element a number of interesting MPA and MPD studies have been performed on OsO_4. As expected for a heavy element, the isotopic shift is small ($\Delta\bar{v}_3 \sim 0.26\,cm^{-1}$ per unit mass difference; see McDowell and Goldblatt, 1971) compared to a width of 3 to $4\,cm^{-1}$ for the Q-branch. Kompanets and coworkers (1975) studied the non-linear absorption of CO_2 laser radiation by monoisotopic forms of OsO_4 observing selective absorption of 10R(6), 10R(8) and 10P(8) by $^{189}OsO_4$ and 10R(26) by $^{187}OsO_4$, as well as predominant absorption of 10R(24) by $^{192}OsO_4$ and 10P(12) and 10P(16) by $^{187}OsO_4$.

Isotopic enrichment of ^{192}Os relative to ^{187}Os (class IC) was observed by Ambartzumian and Coworkers (1975d) using OCS as a radical scavenger to prevent back-reactions. Visible luminescence, which is proportional to the MPD yield and which was used in several studies as a measure of MPD yield (Ambartzumian *et al.*, 1977a, 1977b, 1978b; Ambartzumian, Makarov and Puretzky, 1978a, 1978c), was attributed to dissociation products. Later studies concluded that this luminescence was from electronically excited OsO_4 prepared by ground state MPA (Ambartzumian *et al.*, 1979; Ambartzumian, Makarov and Puretzky, 1980). The observation that this luminescence is proportional to the MPD yield will be valid over all conditions only if the proportion of excited molecules which decompose compared to those which luminesce is constant. Both MPA and MPD spectra, as a function of fluence and wavelength, show a shift to longer wavelength (red shift) from the small-signal spectrum along with structure which was attributed to different one-photon resonances in the $\bar{v}_3$ ladder (Ambartzumian *et al.*, 1978b; Ambartzumian, Makarov and Puretzky, 1978a). The major technical improvement in these papers was the use of two separately tunable lasers to probe the MPA characteristics at different levels of excitation. The advantages of two compared to one laser is clearly demonstrated by the fact that comparable MPD could be obtained by two lasers with the sum of their fluences considerably lower than the fluence required for one laser. This is particularly important for situations of small isotope shift where the fluence of the first laser must be kept low to avoid power broadening. Lower fluences also greatly facilitate the design of reaction vessels to give uniform photolysis of large gas volumes. The wavelength of the second laser is also important since the MPD probability was enhanced by more than a factor of 1000 when the second laser was tuned to the red of the first laser.

Studies on OsO_4 have estimated the fraction of molecules which are in resonance with the beam and also the average level acquired by an individual molecule (Ambartzumian, Makarov and Puretzky, 1978a, 1978b). The luminescence attributed to excited OsO_4 is thought to be phosphorescence from a metastable state formed by an inverse radiationless transition (Ambartzumian *et al.*, 1979; Ambartzumian, Makarov and Puretzky, 1978c, 1980; Markarov *et al.*, 1980). Ashfold, Atkins and Hancock (1981) showed that the phosphorescence was intensity as well as fluence dependent.

Studies of OsO_4 have demonstrated that selective MPD is achievable, using a two-colour approach, even for a molecule having a small isotope shift. The observation of an inverse radiationless transition for OsO_4 is quite interesting since theories of radiationless processes derived from the approach of Robinson and Frosch (1962) predict that radiationless transitions from an excited to a ground state should have a much higher probability than the proposed inverse radiationless transition.

IX. SELECTIVE MPD OF URANIUM ISOTOPES

A. Introduction

Uranium enrichment by MPD is probably the most studied and least publicly reported elemental enrichment. Both industrial security and the possible potential for nuclear proliferation (Krass, 1977) have contributed to information being classified. Recently the choice has been made in the United States to pursue the Atomic Vapor LIS (AVLIS) process which was invented by Levy and Janes and pursued at Lawrence Livermore National Laboratory (Bethe, 1982). This might mean that more of the fundamental results obtained in the US MPD LIS programmes will be released. A recent article by Jensen, Judd and Sullivan (1982) has given a broad overview of the laser isotope separation research programme at Los Alamos. Becker and Kompa (1982) have also recently reviewed the LIS of uranium.

B. MPD of UF_6

Uranium hexafluoride was chosen by many researchers as the molecule on which to base an LIS program. Since fluorine is monoisotopic all isotopic shifts will be due to uranium. UF_6 is well characterized chemically and is used in diffusion enrichment plants. There is a well-developed technology for handling this material and for practical use it has a relatively high vapour pressure at room temperature. Unfortunately, there is no strong fundamental absorption among the wavelengths of the CO_2 laser lines. Eerkens (1976) calculated intensities that one might expect for $^{235}UF_6$ and $^{236}UF_6$ in various combination bands and predicted that even for weak overlapping bands selective excitation would be possible.

Optically pumped CF_4 lasers were developed to be absorbed by $\bar{\nu}_3$ of UF_6 (see Section III) and were used by Tiee and Wittig (1978c), and by Rabinowitz, Stein and Kaldor (1978) and by Alimpiev and coworkers (1979) to study the MPD of UF_6. The first two groups used IR fluorescence from HF to detect UF_6 decomposition in the presence of hydrogen as follows:

$$UF_6 \xrightarrow{nh\nu} UF_5 + F \tag{71}$$

$$F + H_2 \rightarrow HF^* + H \tag{72}$$

$$HF^* \rightarrow HF + h\nu \tag{73}$$

The observation of 'laser snow' was also used by Rabinowitz *et al.* (1978). Tiee and Wittig (1978c) performed a two-colour photolysis of UF_6 using a CF_4 laser tuned to the $\bar{\nu}_3$ mode of UF_6 and a synchronously fired CO_2 laser which was not absorbed by unexcited UF_6. When both lasers were used there was a large increase in MPD of UF_6 compared to either laser only. This they interpreted as due to a shift of the 1160 cm^{-1} UF_6 absorption to lower energy as the molecule was vibrationally heated. Tiee and Wittig (1978c) concluded that it is 'trivial to photodissociate UF_6' with a CF_4 laser and that an appropriately tuned CO_2 laser can be used to significantly enhance the rate of photodissociation. Rabinowitz, Stein and Kaldor (1978) used only a CF_4 laser to carry out their MPD study of UF_6. From RRKM calculations, they determined that about 15 photons above the dissociation threshold were required to ensure decomposition in a time shorter than the mean collisional period.

Horsley and coworkers (1980a, 1980b) used 16-μm radiation from an optical parametric oscillator to study MPA of UF_6. The weak combination band absorptions accessible to a CO_2 laser were also measured to be much smaller than those estimated by Eerkens (1976). They also studied the reactions of different species with vibrationally excited UF_6. Stuke, Reisler and Wittig (1981) studied the UV photodissociation of UF_6 monitored by multiphoton ionization.

Isotopically selective MPD was reported by Rabinowitz and coworkers (1982) using two 16-μm sources produced by Raman shifting CO_2 lines with a multipass H_2 cell. The low fluence (selective) source operated in the region 613 to 629 cm^{-1} and the high fluence source at 596 to 598 cm^{-1}. Depletion of $^{235}UF_6$ compared to $^{238}UF_6$ was measured as a function of the low fluence source wavelength and was found to be 1.20 for $\bar{\nu}_L = 627.8$ cm^{-1}. Maximum selectivity occurred to lower frequencies than the maximum of the ratio of the small signal cross-sections of two isotopic species. In these experiments, both lasers pumped different regions of the ladder of the same mode in contrast to the experiment of Tiee and Wittig (1978c) in which the two lasers pumped different modes. While the selectivities reported by Rabinowitz and coworkers (1982) not negligible, neither were they of the magnitude reported for selective MPD of some other elements. Investigations using cooled UF_6 such as those carried out at Los Alamos (Jensen, Judd and Sullivan, 1982) might give higher selectivities.

C. MPD of Other Uranium-Containing Molecules

An alternative to developing a laser suitable for selective MPD of UF_6 is to identify a volatile, uranium-containing molecule for which selective MPD may be achieved using an existing, efficient source such as a CO_2 laser. To date, the molecules thus considered have been volatile, large molecules with a uranium atom surrounded by organic ligands and with absorption bands near 10 μm which show a modest U isotopic shift.

Selective MPD of $U(OCH_3)_6$ both for natural uranium and for a sample enriched to 1.5 per cent in ^{235}U gave a small enrichment factor ($\beta(m) \sim 1.034$) for CO_2 10P(38) line irradiation at a fluence of 3.2 J cm^{-2} (Cuellar *et al.*, 1983; Miller *et al.*, 1979). Such a small enrichment factor, while much larger than the β for gaseous diffusion, is too small to allow economic staging that would be required to produce fuel for light-water reactors.

The selective MPD of several volatile uranyl complexes were studied by groups at Exxon Research and Engineering Company in New Jersey and at the Australian Atomic Energy Commision at Lucas Heights, New South Wales. The molecules $UO_2(hfacac)_2 \cdot B$ (where hfacac is $C_5F_6O_2H$ and B is a Lewis base) are interesting since the asymmetric stretch mode of UO_2 lies near 950 cm^{-1} and is easily excited by a CO_2 laser to give reversible detachment of B at low fluences (≤ 200 mJ cm^{-2}):

$$UO_2(hfacac)_2 \cdot B \rightleftharpoons UO_2(hfacac)_2 + B \tag{74}$$

For higher fluences (≥ 600 mJ cm^{-2}) $UO_2(hfacac)_2$ undergoes irreversible decomposition to give UO_2F_2 and a cyclic furanone (Eberhardt *et al.*, 1983; Eberhardt, Knott and Pryor, 1983). Kaldor and coworkers (1979) noted that for such large molecules the density of vibrational states was so large that they were essentially occupying region E_{II} before the absorption of photons at room temperature. The isotopic shift is small (~ 0.7 cm^{-1}) compared to the absorption profile and consequently quite small enrichment factors can be anticipated. The vibrational energy content at room temperature is $\geq 10^4$ cm^{-1} and the high-energy tail of the energy distribution of molecules is very important in limiting the measured selectivity since a significant fraction of the molecules can dissociate after absorption of only one photon. Techniques which increase the gap between the high-energy tail of the distribution and the energy required for dissociation are important to the optimization of selectivity. Compounding of selectivity is possible if the number of photons which must be absorbed to give MPD is increased. One way that this can be achieved is by cooling in bulk or supercooling in a beam (Bray *et al.*, 1983; Cox *et al.*, 1979; Cox and Horsley, 1980; Cox and Maas, 1980; Dietz *et al.*, 1982; Kaldor and Woodin, 1982). Published separation factors for the uranyl complexes are not large ($\beta(m) < 2$), though selective MPD can be achieved with quite low fluences.

D. Conclusions

Since security classification is widespread in studies of uranium LIS it is not possible to assess the possibilities of applying these studies to a practical situation. It appears that the separation factors for either UF_6 or uranyl complexes will be small and staging will be required. In addition, it is likely that cooling techniques such as supersonic expansion will be necessary. The United States has chosen to terminate studies on the selective MPD process for uranium LIS and to take to pilot plant stage the competing AVLIS process. This latter process is claimed to have the potential for a very significant reduction in the cost of uranium enrichment compared to gaseous diffusion and advanced gaseous centrifugation (Davis, 1982, 1983).

X. SELECTIVE MPD OF ISOTOPES OF OTHER ELEMENTS

The isotopes of these elements have been separated in MPD experiments using only a very small effort.

A. Chlorine

Ambartzumian and coworkers (1975e, 1976c) decomposed CCl_4 selectively. They investigated the dependence of enrichment on laser frequency, finding a maximum enrichment in the decomposition products of about 15 per cent for ^{37}Cl for MPD with the CO_2 10R(14) line.

B. Molybdenum

Freund and Lyman (1978) looked at the MPD of MoF_6 as an extension of SF_6 MPD studies to heavier elements. Here only relatively weak combination bands ($\bar{\nu}_3 + \bar{\nu}_5$ and $\bar{\nu}_2 + \bar{\nu}_4$; see McDowell *et al.*, 1975) are in the region of the CO_2 laser. The isotope masses range from 92 to 100 and each species has an overlapping absorption. By irradiating on the high-frequency side of the absorption band with the 9P(10) line they could selectively deplete the lower mass isotopes and by switching to the low-frequency side with the 9P(16) line they depleted the heavier isotopes selectively. Irradiation with an intermediate frequency showed no selectivity. The enrichment was small but significant, indicating that for a very valuable isotope, a LIS process might be based on a poorly resolved, weak transition; however, staging would be required.

C. Selenium

Tiee and Wittig (1978a) photolysed SeF_6 with an NH_3 laser with $\bar{\nu}_L =$ 7805 cm^{-1}) as an example of a molecule having small isotopic shifts and a number of abundant isotopes. The NH_3 laser frequency chosen lies between absorption peaks due to $^{80}SeF_6$ and $^{78}SeF_6$ and, arguing by analogy with results for SF_6, they expected the peak of the decomposition probability versus frequency to be shifted to lower frequencies than the small signal absorption spectrum for each isotopic species. As expected, they observed that the heavier isotopes were dissociated less efficiently than the lighter ones by the NH_3 laser.

They concluded from their study that: (1) in general terms one can predict the selectivity of MPD from the low signal spectrum, (2) that even heavy molecules with small isotope shifts can show significant enrichment in MPD, (3) that cooling the gas is important for separation of isotopes of heavy elements and (4) that efficient conversion of CO_2 frequencies to other convenient frequencies increases the number of molecules which can be potential bases for LIS.

XI. CONCLUDING REMARKS

Research on photochemical isotopic enrichment schemes has been carried out sporadically over the last 64 years. The results of the first 47 years indicated that a practical photochemical enrichment scheme for mercury isotopes is possible, though there has not been sufficient commercial interest for this to be pursued. During the last 14 years, an explosion of interest in photochemical and photophysical isotopic enrichment schemes, initiated by the widespread availability of appropriate laser sources, has occurred. In this period, the selective MPD process has developed from a puzzling curiosity to a field of study which, though still not well understood, is yielding important new fundamental principles and interesting potential applications. Yet much remains to be done and the next decade promises to be as fruitful and as interesting as the last.

For many elements large, single-step enrichments have been demonstrated for empirically selected molecules without assurance that these molecules represent the best choice for a practical LIS process. For practical processes such 'mundane' factors as a too-slow exchange rate (see step 1 of Fig. 3) can relegate an otherwise interesting molecule to commercial non-viability. Selective MPD studies of ^{13}C and sulphur isotopes indicate that this process is competitive with conventional enrichment schemes for these isotopes. The results to date of tritium LIS studies are also encouraging. Selective MPD of ^2H has been demonstrated though it is still too early to determine if this process will be competitive with the very cheap conventional processes such as H_2S/HDO or H_2/HDO exchange. For uranium, studies on the selective MPD processes were discontinued in favour of taking the AVLIS process to pilot plant stage.

The MPA and MPD process is not sufficiently well understood to predict how changes in such parameters as molecular structure or laser beam characteristics can be exploited to optimize the efficient conversion of photons to reaction energy. Multiple laser experiments have demonstrated that this approach can increase the efficiency of MPD. To take full advantage of this technique requires improved understanding of the spectroscopy of excited molecules and the detailed vibrational/rotational level structure of ground electronic states. As has been the case in the past decade, future research on MPA and MPD will likely contribute valuable new fundamental knowledge to the understanding of reactive processes and the interaction of molecules with lasers.

References

Abdushelishvili, G. I., Avatkov, O. N., Andryushchenko, V. I., Bagratashvili, V. N., Bakhtadze, A. B., Vetsko, V. M., Doljikov, V. S., Esadze, G. G., Letokhov, V. S., Ryabov, E. A., and Tkeshelashivil, G. I. (1979). *Pis'ma. Zh. Tekh. Fiz.*, **5**, 849 (*Sov. Tech. Phys. Lett.*, **5**, 350).

Abdushelishvili, G. I., Avatkov, O. N., Bagratashvili, V. N., Baranov, V. Yu., Bakhtadze, A. B., Velikhov, E. P., Vetsko, V. M., Gverdtsiteli, I. G., Doljikov, V. S., Esadze, G. G., Kazakov, S. A., Kolomiiski, Yu. R., Letokhov, V. S., Pigul'skii, S. V., Pis'mennyi, V. D., Ryabov, E. A., and Tkeshelashvili, G. I. (1982). *Kvant. Elektron.*, **9**, 743 (*Sov. J. Quantum Electron.*, **12**, 459).

Alcock, A. J., Fedosejevs, R., and Walker, A. C. (1975). *IEEE J. Quantum Electron*, **QE-11**, 767.

Alcock, A. J., Leopold, K., and Richardson, M. C. (1973). *Appl. Phys. Lett.*, **23**, 562.

Aldridge, J. P., Birely, C. D., Cantrell, C. D., and Cartwright, D. C. (1976). In *Physics of Quantum Electronics*, Vol. 4, *Laser Photochemistry and Tunable Lasers* (Eds. S. F. Jacobs, M. Sargent, M. O. Scully and C. T. Walker), p. 57, Addison-Wesley, Reading.

Alimpiev, S. S., Babichev, A. P., Baranov, G. S., Karlov, N. V., Karchevskii, A. I., Kulikov, S. Yu., Martsynk'yan, V. L., Nabiev, Sh. Sh., Nikiforov, S. M., Prokhorov, A. M., Sartakov, B. G., Skvortsova, E. P. and Khokhlov, E. M. (1979). *Kvant. Electron.*, **6**, 2155 (*Sov. J. Quantum Electron.*, **9**, 1263).

Alimpiev, S. S., Karlov, N. V., Sartakov, B. G., and Khokhlov, E. M. (1978). *Opt. Commun.*, **26**, 45.

Ambartzumian, R. V., and Letokhov, V. S. (1972). *Appl. Optics*, **11**, 354.

Ambartzumian, R. V., and Letokhov, V. S. (1977). In *Chemical and Biochemical Applications of Lasers* (Ed. C. B. Moore), Vol. 3, p. 167, Academic Press, New York.

Ambartzumian, R. V., Chekalin, N. V., Doljikov, V. S., Letokhov, V. S., and Ryabov, E. A. (1974a). *Chem. Phys. Lett.*, **25**, 515.

Ambartzumian, R. V., Letokhov, V. S., Ryabov, E. A., and Chekalin, N. V. (1974b). *ZhETF Pis. Red.*, **20**, 597 (*JETP Lett.*, **20**, 273).

Ambartzumian, R. V., Doljikov, V. S., Letokhov, V. S., Ryabov, E. A., and Chekalin, N. V. (1975a). *Zh. Eksp. Teor. Fiz.*, **69**, 72 (*Sov. Phys. JETP*, **42**, 36).

Ambartzumian, R. V., Gorokhov, Yu. A., Letokhov, V. S., and Makarov, G. N. (1975b). *Zh. Eksp. Theor. Fiz.*, **69**, 1956 (*Sov. Phys. JETP*, **42**, 993).

Ambartzumian, R. V., Gorokhov, Yu. A., Letokhov, V. S., and Makarov, G. N. (1975c). *ZhETF Pis. Red.*, **21**, 375 (*JETP Lett.*, **21** 171).

Ambartzumian, R. V., Gorokhov, Yu. A., Letokhov, V. S., and Makarov, G. N. (1975d). *Pis'ma Zh. Eksp. Teor. Fiz.*, **22**, 96 (*JETP Lett.*, **22**, 43).
Ambartzumian, R. V., Gorokhov, Yu. A., Letokhov, V. S., and Puretzky, A. A. (1975e). *Pis'ma Zh. Eksp. Teor. Fiz.*, **22**, 374 (*JETP Lett.*, **22**, 177).
Ambartzumian, R. V., Gorokhov, Yu. A., Letokhov, V. S., Makarov, G. N., and Puretzky, A. A. (1976a). *Zh. Eksp. Teor. Fiz.*, **71**, 440 (*Sov. Phys. JETP*, **44**, 231).
Ambartzumian, R. V., Gorokov, Yu. A., Letokhov, V. S., Makarov, G. N., and Puretzky, A. A. (1976b). *Pis'ma Zh. Eksp. Teor. Fiz.*, **23**, 26 (*JETP Lett.*, **23**, 22).
Ambartzumian, R. V., Gorokhov, Yu. A., Letokhov, V. S., Makarov, G. N., and Puretzky, A. A. (1976c). *Phys. Lett.*, **56A**, 183.
Ambartzumian, R. V., Furzikov, N. P., Gorokhov, Yu. A., Letokhov, V. S., Makarov, G. N., and Puretzky, A. A. (1977a). *Opt. Lett.*, **1**, 22.
Ambartzumian, R. V., Gorokhov, Yu. A., Makarov, G. N., Puretzky, A. A., and Furzikov, N. P. (1977b). *Chem. Phys. Lett.*, **45**, 231.
Ambartzumian, R. V., Furzikov, N. P., Letokhov, V. S., Dyad'kin, A. P., Grasyuk, A. Z., and Vasil'yev, B. I. (1978a). *App. Phys.*, **15**, 27.
Ambartzumian, R. V., Letokhov, V. S., Makarov, G. N., and Puretsky, A. A. (1978b). *Opt. Commun.* **25**, 69.
Ambartzumian, R. V., Makarov, G. N., and Puretzky, A. A. (1978a). *Opt. Commun.*, **27**, 79.
Ambartzumian, R. V., Makarov, G. N., and Puretzky, A. A. (1978b). *Pis'ma Zh. Eksp. Teor. Fiz.*, **28**, 246 (*JETP Lett.*, **28**, 228).
Ambartzumian, R. V., Makarov, G. N., and Puretzky, A. A. (1978c). *Pis'ma Zh. Eksp. Teor. Fiz.*, **28**, 696 (*JETP Lett.*, **28**, 647).
Ambartzumian, R. V., Knyazev, I. N., Lobko, V. V., Makarov, G. N., and Puretsky, A. A. (1979). *Appl. Phys.*, **19**, 75.
Ambartzumian, R. V., Makarov, G. N., and Puretzky, A. A. (1980). *Appl. Phys.*, **22**, 77.
Anderson, D., McAlpine, R. D., Evans, D. K., and Adams, H. M. (1981). *Chem. Phys. Lett.*, **79**, 337.
Apatin, V. M., and Makarov, G. N. (1983). *Appl. Phys.*, **B30**, 207.
Arisawa, T., Kato, M., Maruyama, Y., and Shiba, K. (1982). *J. Phys. B: At. Mol. Phys.*, **15**, 1671.
Arrowsmith, P., Polanyi, J. C., Telle, H., and B. Yang (1983). *Appl. Opt.*, **22**, 2716.
Ashfold, M. N. R., Atkins, C. G., and Hancock, G. (1981). *Chem. Phys. Lett.*, **80**, 1.
Ashfold, M. N. R., Hancock, G., and Hardaker, M. L. (1980). *J. Photochem.*, **14**, 85.
Ashfold, M. N. R., Hancock, G., and Ketley, G. (1979). *Faraday Discuss. Chem. Soc.*, **67**, 204.
Aston, F. W. (1920). *Phil. Mag.*, **39**, 611.
Bagratashvili, V. N., Doljikov, V. S., Letokhov, V. S., Makarov, A. A., Ryabov, E. A., and Tyakht, V. V. (1979). *Zh. Eksp. Teor. Fiz.*, **77**, 2238 (*Sov. Phys. JETP*, **50**, 1075).
Bagratashvili, V. N., Doljikov, V. S., Letokhov, V. S., and Ryabov, E. A. (1978). *Pis'ma Tekh. Fiz.*, **4**, 1181 (*Sov. Tech. Phys. Lett.*, **4**, 475).
Bagratashvili, V. N., Knyazev, I. N., and Letokhov, V. S. (1971). *Opt. Commun.*, **4**, 154.
Bagratashvili, V. N., Knyazev, I. N., Letokhov, V. S., and Lobko, V. V. (1976). *Opt. Commun.*, **18**, 525.
Bagratashvili, V. N., Kolomisky, Yu. R., Letokhov, V. S., Ryabov, E. A., Baranov, V. Yu., Kazakov, S. A., Nizjev, V. G., Pismenny, V. D., Starodubtsev, A. I., and Velikhov, E. P. (1977). *Appl. Phys.*, **14**, 217.
Baldwin, A. C., and Barker, J. R. (1981a). *J. Chem. Phys.*, **74**, 3813.
Baldwin, A. C., and Barker, J. R. (1981b). *J. Chem. Phys.*, **74**, 3823.
Barker, J. R. (1980). *J. Chem. Phys.*, **72**, 3686.

Basov, N. G., Belenov, E. M., Danilychev, V. A. Kerimov, O. M., Kovsh, I. B., and Suchkov, A. F. (1971a). *Zh. ETF Pis. Red.*, **14**, 421 (*JETP Lett.*, **14**, 285).

Basov, N. G., Belenov, E. M., Danilychev, V. A., and Suchkov, A. F. (1971b). *Zh. ETF Pis. Red.*, **14**, 545 (*JETP Lett.*, **14**, 375).

Bates, R. D., Knudtson, J. T., Flynn, G. W., and Ronn, A. M. (1971). *Chem. Phys. Lett.*, **8**, 103.

Baulch, D. L., Cox, R. A., Hampson, R. F., Kerr, J. A., Troe, J., and Watson, R. T. (1980). *J. Phys. Chem. Ref. Data*, **9**, 295.

Beck, R., Englisch, W., and Gurs, K. (1980). *Table of Laser Lines in Gases and Vapors*, Springer-Verlag, Berlin.

Becker, F. S., and Kompa, K. L. (1982). *Nucl. Technol.*, **58**, 329.

Benedict, M., and Pigford, T. H. (1957). *Nuclear Chemical Engineering*, McGraw-Hill, New York.

Benson, S. W., and Buss, J. H. (1958). *J. Chem. Phys.*, **29**, 546.

Bethe, H. A. (1982). *Science*, **217**, 398.

Bhatnager, R., Dyer, P. E., and Oldershaw, G. A. (1979). *Chem. Phys. Lett.*, **61**, 339.

Bialkowski, S., and Guillory, W. A. (1977). *J. Chem. Phys.*, **67**, 2061.

Bigeleisen, J., Hammond, W. B., and Tuccio, S. (1983). *Nucl. Sci. and Eng.*, **83**, 473.

Billings, B. H., Hitchcock, W. J., and Zelikoff, M. (1953). *J. Chem. Phys.*, **21**, 1762.

Bittenson, S., and Houston, P. L. (1977). *J. Chem. Phys.*, **67**, 4819.

Black, J. G., Yablonovitch, E., Bloembergen, N., and Mukamel, S. (1977). *Phys. Rev. Lett.*, **38**, 1131.

Black, J.G., Kolodner, P., Schultz, M. J., Yablonovitch, E., and Bloembergen, N. (1979). *Phys. Rev.*, **A19**, 704.

Bloembergen, N. (1975). *Opt. Commun.*, **15**, 416.

Bray, R. G., Cox, D. M., Hall, R. B., Horsley, J. A., Kaldor, A., Kramer, G. M., Levy, M. R., and Priestley, E. B. (1983). *J. Phys. Chem.*, **87**, 429.

Brimacombe, R. K., Reid, J., and Znotins, T. A. (1981). *Appl. Phys. Lett.*, **39**, 302.

Buckwald, M. I., Jones, C. R., Fetterman, H. R., and Schlossberg, H. R. (1976). *Appl. Phys. Lett.*, **29**, 300.

Butler, J. P. (1980). *Sep. Sci. and Tech.*, **15**, 371.

Butler, J. P., Rolston, J. R., and Stevens, W. H. (1978). In *Separation of Hydrogen Isotopes* (Ed. H. K. Rae), p. 93, ACS Symp. Number 68, Washington.

Campbell, J. D., Hancock, G., and Welge, K. H. (1976). *Chem. Phys. Lett.*, **43**, 581.

Cantrell, C. D., and Galbraith, H. W. (1976). *Opt. Commun.*, **18**, 513.

Cantrell, C. D., and Galbraith, H. W. (1977). *Opt. Commun.*, **21**, 374.

Casassa, M. P., Bomse, D. S., and Janda, K. C. (1981). *J. Phys. Chem.*, **85**, 2623.

Chang, T. Y., and McGee, J. D. (1976). *App. Phys. Lett.*, **28**, 526.

Chang, T. Y., and Wood, O. R. (1977). *IEEE J. Quantum Electron*, **QE-13**, 907.

Chin, S. L., Evans, D. K., McAlpine, R. D., McClusky, F. K., and Selkirk, E. B. (1979). *Opt. Commun.*, **31**, 235.

Chou, J-S. J., and Grant, E. R. (1981). *J. Chem. Phys.*, **74**, 5679.

Cohen, K. (1951). *The Theory of Isotope Separation*, McGraw-Hill, New York.

Cox, D. M., Hall, R. B., Horsley, J. A., Kramer, G. M., Rabinowitz, P., and Kaldor, A. (1979). *Science*, **205**, 390.

Cox, D. M., and Horsley, J. A. (1980). *J. Chem. Phys.*, **72**, 864.

Cox, D. M., and Maas, E. T. (1980). *Chem. Phys. Lett.*, **71**, 330.

Cuellar, E. A., Miller, S. S., Marks, T. J., and Weitz, E. (1983). *J. Am. Chem. Soc.*, **105**, 4580.

Davis, J. I. (1982). *Lasers in Chemical Processing*, Invited talk: Am. Inst. of Chem. Engineers 1982 Winter meeting, Orlando, Fla., Lawrence Livermore National Laboratory Report Number UCRL-53276.

Davis, J. I. (1983). *Status and Prospects for Large-Scale Laser Isotope Separation*, 1983 Conf. of Canadian Nuclear Soc., Montreal, P.Q.
Diebold, G. J., Engelke, F., Lubman, D. M., Whitehead, J. C., and Zare, R. N. (1977). *J. Chem. Phys.*, **67**, 5407.
Dietz, T. G., Duncan, M. A., Smalley, R. E., Cox, D. M., Horsley, J. A., and Kaldor, A. (1982). *J. Chem. Phys.*, **77**, 4417.
Doljikov, V. S., Kolomisky, Yu. R., and Ryabov, E. A. (1981). *Chem. Phys. Lett.*, **80**, 433.
Eberhardt, J. E., Hoarde, I. E., Johnson, D. A., Knott, R. B., Pryor, A. W., and Waugh, A. B. (1983). *Chem. Phys.*, **72**, 41.
Eberhardt, J. E., Knott, R. B., and Pryor, A. W. (1983). *Chem. Phys.*, **72**, 51.
Eck, C. F. (1973). *Research/Development*, August issue, p. 32.
Eerkens, J. (1976). *Appl. Phys.*, **10**, 15.
Eggers, K. W., and Cocks, A. T. (1973). *Helv. Chem. Acta*, **56**, 1516.
Evans, D. K., McAlpine, R. D., and Adams, H. M. (1982). *J. Chem. Phys.* **77**, 3551.
Evans, D. K., McAlpine, R. D., and Adams, H. M. (1984). *Israel J. Chem.*, **24**, 187.
Evans, D. K., McAlpine, R. D., Adams, H. M., and Creagh, A. L. (1983). *Chem. Phys.*, **80**, 379.
Evans, D. K., McAlpine, R. D., and Goodale, J. W. (1983). In *Quantum Electronics and Electro-Optics* (Ed. P. L. Knight, p. 179), Wiley, Chichester.
Evans, D. K., McAlpine, R. D. and McClusky, F. K. (1978). *Chem. Phys.*, **32**, 81.
Evans, D. K., McAlpine, R. D., and McClusky, F. K. (1979). *Chem. Phys. Lett.*, **65**, 226.
Fettweis, P., and Nève de Mévergnies, M. (1978). *J. Appl. Phys.*, **49**, 5699.
Filseth, S. V., Danon, J., Feldman, D., Campbell, J. D., and Welge, K. H. (1979). *Chem. Phys. Lett.*, **63**, 615.
Forst, W. (1973). *Theory of Unimolecular Reactions*, Academic Press, New York.
Freeman, M. P., Travis, D. N., and Goodman, M. F. (1974). *J. Chem. Phys.*, **60**, 231.
Freund, S. M., and Lyman, J. L. (1978). *Chem. Phys. Lett.*, **55**, 435.
Freund, S. M., and Ritter , J. J. (1975). *Chem. Phys. Lett.*, **32**, 255.
Fuss, W., and Cotter, T. P., (1977). *Appl. Phys.*, **12**, 265.
Fuss, W., and Hartmann, J. (1979). *J. Chem. Phys.*, **70**, 5468.
Fuss, W., and Schmid, W. E. (1983). *Ber. Bunsenges. Phys. Chem.*, **83**, 1148.
Gauthier, M., Cureton, C. G., Hackett, P. A., and Willis, C. (1982). *Appl. Phys.*, **B28**, 43.
Gauthier, M., Hackett, P. A., Drouin, M., Pilon, R., and Willis, C. (1978). *Can. J. Chem.*, **56**, 2227.
Gauthier, M., Hackett, P. A., and Willis, C. (1980). *Chem. Phys.*, **45**, 39.
Gauthier, M., Hackett, P. A., and Willis, C. (1983). In *Synthesis and Applications of Isotopically Labeled Compounds* (Eds. W. P. Duncan and A. B. Susan), p. 413, Elsevier, Amsterdam.
Gauthier, M., Nip, W. S., Hackett, P. A., and Willis, C. (1980). *Chem. Phys. Lett.*, **69**, 372.
Gauthier, M., Pilon, R., Hackett, P. A., and Willis, C. (1979). *Can. J. Chem.*, **57**, 3173.
Gauthier, M., Willis, C., and Hackett, P. A. (1980). *Can. J. Chem.*, **58**, 913.
Goodman, M. F., Seliger, H. H., Minkowski, J. M. (1970). *Photochem. and Photobiol.*, **12**, 355.
Gower, M. C., and Billman, K. W. (1977). *Opt. Commun.*, **20**, 123.
Gozel, P., Braichotte, D., and van den Bergh, H. (1983). *J. Chem. Phys.*, **79**, 4924.
Grant, E. R., Coggiola, M. J., Lee, Y. T., Schulz, P. A., Sudbo, Aa. S., and Shen, Y. R. (1977). *Chem. Phys. Lett.*, **52**, 595.
Greiner, N. R. (1980). *Rev. Sci. Instrum.*, **51**, 392.
Grimley, A. J., and Stephenson, J. C. (1981). *J. Chem. Phys.*, **74**, 447.
Gross, R. W. F., and Bott, J. F. (1976). *Handbook of Chemical Lasers*, Wiley, New York.

Gunning, H. E. (1958). *Can. J. Chem.*, **36**, 89.
Gunning, H. E., and Strausz, O. P. (1963). In *Advances in Photochemistry* (Eds. W. A. Noyes, Jr., G. S. Hammond and J. N. Pitts, Jr.), Vol. I, pp. 209–274, Interscience, New York.
Hackett, P. A., Gauthier, M., and Willis, C. (1978). *J. Chem. Phys.*, **69**, 7924.
Hackett, P. A., Gauthier, M., Willis, C., and Pilon, R. (1979). *J. Chem. Phys.*, **71**, 546.
Hackett, P. A., Malatesta, V., Nip, W. S., Willis, C., and Corkum, P. B. (1981). *J. Phys. Chem.*, **85**, 1152.
Hackett, P. A., Willis, C., and Gauthier, M. (1979). *J. Chem. Phys.*, **71**, 2682.
Hackett, P. A., Willis, C., Drouin, M., and Weinberg, E. (1980). *J. Phys. Chem.*, **84**, 1873.
Hallsworth, R. S., and Isenor, N. R. (1973). *Chem. Phys. Lett.*, **22**, 283.
Ham, D. O., and Rothschild, M. (1977). *Opt. Lett.*, **1**, 28.
Hammerli, M., Stevens, W. H., and Butler, J. P. (1978). In *Separation of Hydrogen Isotopes* (Ed. H. K. Rae), p. 110, ACS Symposium Number 68, Washington.
Hancock, G., Campbell, J. D., and Welge, K. H. (1976). *Opt. Commun.*, **16**, 177.
Hancock, G., Hennessy, R. J., and Villis, T. (1978). *J. Photochem.*, **9**, 197.
Hancock, G., and MacRobert, . . (1983). *Chem. Phys. Lett.*, **101**, 312.
Hartley, H., Ponder, A. O., Bowen, E. J., and Merton, T. R. (1922). *Phil. Mag.*, **43**, 430.
Hason, A., Gozel, P., Duperrex, R., and van den Bergh, H. (1982). *Appl. Phys.*, **B29**, 188.
Hason, A., Gozel, P., and van den Bergh, H. (1982). *Helv. Phys. Acta*, **55**, 187.
Hedges, R. E. M., Ho, P., and Moore, C. B. (1980). *Appl. Phys.*, **23**, 25.
Herman, I. P. (1983). *Chem. Phys.*, **75**, 121.
Herman, I. P., Magnotta, F., and Aldridge, F. T. (1984). *Israel J. Chem.*, **24**, 192.
Herman, I. P., Magnotta, F., Buss, R. J., and Lee, Y. T. (1983). *J. Chem. Phys.*, **79**, 1789.
Herman, I. P., and Marling, J. B. (1979). *Chem. Phys. Lett.*, **64**, 75.
Herman, I. P., and Marling, J. B. (1980). *J. Chem. Phys.*, **72**, 516.
Herman, I. P., and Marling, J. B. (1981). *J. Phys. Chem.*, **85**, 493.
Herzberg, G. (1945). *Molecular Spectra and Molecular Structure*, Vol. 2, *Infrared and Raman Spectra of Polyatomic Molecules*, Van Nostrand Reinhold, New York.
Holtslander, W. J., and Lockerby, W. E. (1978). In *Separation of Hydrogen Isotopes* (Ed. H. K. Rae), p. 40, ACS Symposium Number 68, Washington.
Horowitz, A. B., Preses, J. M., Weston, R. E., and Flynn, G. W. (1981). *J. Chem. Phys.*, **74**, 5008.
Horsley, J. A., Cox, D. M., Hall, R. B., Kaldor, A., Mass, E. T., Priestley, E. B., and Kramer, G. M. (1980). *J. Chem. Phys.*, **73**, 3660.
Horsley, J. A., Rabinowitz, P., Stein, A., Cox, D. M., Brickman, R. O., and Kaldor, A. (1980). *IEEE J. Quant. Elec.*, **QE-16**, 412.
Houston, P. L., Nowak, A. V., and Steinfeld, J. I. (1973). *J. Chem. Phys.*, **58**, 3373.
Isenor, N. R., and Richardson, M. C. (1971). *Appl. Phys. Lett.*, **18**, 224.
Ishikawa, Y., Arai, S., and Nakane, R. (1980). *J. Nucl. Sci. Technol.*, **17**, 275.
Ishikawa, Y., Kurihara, O., Arai, S., and Nakane, R. (1981). *J. Phys. Chem.*, **85**, 3817.
Ishikawa, Y., Kurihara, O., Nakane, R., and Arai, S. (1980). *Chem. Phys.*, **52**, 143.
Jensen, C. C., Person, W. B., Krohn, B. J., and Overend, J. (1977). *Opt. Commun.*, **20**, 275.
Jensen, R. J., Hayes, J. K., Cluff, C. L., and Thorne, J. M. (1980). *IEEE J. Quant. Elec.*, **QE-16**, 1352.
Jensen, R. J., Judd, O. P., and Sullivan, J. A. (1982). *Los Alamos Science*, **3**, 2.
Jetter, H. L., Fill, E. E., and Volk, R. (1980). German Patent Document 2808955/B/1.
Judd, O. P. (1979). *J. Chem. Phys.*, **71**, 4515.
Kaldor, A., Hall, R. B., Cox, D. M., Horsley, J. A., Rabinowitz, P., and Kramer, G. M. (1979). *J. Am. Chem. Soc.*, **101**, 4465.
Kaldor, A., and Woodin, R. L. (1982). *Proc. IEEE*, **70**, 565.

Karl, R. R., and Lyman, J. L. (1978). *J. Chem. Phys.*, **69**, 1196.
Karlov, N. V., Karpov, N. A., Petrov, Yu. N., Prokhorov, A. M., and Stel'makh, O. M. (1971). *Zh. ETF Pis. Red.*, **14**, 214 (*JETP Lett.*, **14**, 140).
Karlov, N. V., Konev, Yu. B., Petrov, Yu. N., Prokhorov, A. M., and Stel'makh, O. M. (1968a). *Zh. ETF Pis. Red.*, **8**, 22 (*JETP Lett.*, **8**, 12).
Karlov, N. V., Kuz'min, G. P., Petrov, Yu. N., and Prokhorov, A. M. (1968b). *Zh. ETF Pis'ma*, **7**, 174 (*JETP Lett.*, **7**, 134).
Karlov, N. V., Petrov, Yu. N., Prokhorov, A. M., and Stel'makh, O. M. (1970). *Zh. ETF Pis. Red.*, **11**, 220 (*JETP Lett.*, **11**, 135).
Karlov, N. V., Petrov, Yu. N., and Stel'makh, O. M. (1968). *Zh. ETF Pis. Red.*, **8**, 363 (*JETP Lett.*, **8**, 224).
Kasper, J. V. V., and Pimental, G. C. (1965). *Phys. Rev. Lett.*, **14**, 352.
King, D. S. (1982). *Adv. Chem. Phys.*, **50**, 105.
King, D. S., and Stephenson, J. C. (1977). *Chem. Phys. Lett.*, **51**, 48.
King, D. S., and Stephenson, J. C. (1978). *J. Am. Chem. Soc.*, **100**, 7151.
Kirk, R. W., and Wilt, P. M. (1975). *J. Mol. Spectrosc.*, **58**, 102.
Kojima, H., Fukumi, T., Nakajima, S., Maruyoma, Y., and Kosasa, K. (1983). *Appl. Phys.*, **B30**, 143.
Kolodner, P., Winterfeld, C., and Yablonovitch, E. (1977). *Opt. Commun.*, **20**, 119.
Kompanets, O. N., Kukudzhanov, A. R., Letokhov, V. S., Minogin, V. G., and Mikhailov, E. L. (1975). *Zh. Eksp. Teor. Fiz.*, **69**, 32 (*Sov. Phys. JETP*, **42**, 15).
Krajnovich, D., Huisken, F., Zhang, Z., Shen, Y. R., and Lee, Y. T. (1982). *J. Chem. Phys.*, **77**, 5977.
Krass, A. S. (1977). *Science*, **196**, 721.
Kudchadker, S. A., and Kudchadker, A. P. (1975). *J. Phys. Chem. Ref. Data*, **4**, 457.
Kudchadker, S. A., and Kudchadker, A. P. (1978). *J. Phys. Chem. Ref. Data*, **7**, 1285.
Kudryavtsev, Yu. A. (1980). *Kvant. Elektron*, **7**, 1985 (*Sov. J. Quantum Electron*, **10**, 1143).
Kurihara, O., Takeuchi, K., Satooka, S., and Makide, Y. (1983). *J. Nucl. Sci. and Tech.*, **20**, 617.
Kutschke, K. O., Gauthier, M., and Hackett, P. A. (1983). *Chem. Phys.*, **78**, 323.
Kuzmin, M. V., and Stuchebrukhov, A. A. (1984). *Chem. Phys.*, **83**, 115.
Leary, K. M., Lyman, J. L., Asprey, L. B., and Freund, S. M. (1978). *J. Chem. Phys.*, **68**, 1671.
Lederer, C. M., and Shirley, V. S. (1978). *Table of Isotopes*, 7th ed., John Wiley and Sons, New York.
Letokhov, V. S., and Makarov, A. A. (1976). *Opt. Commun.*, **17**, 250.
Letokhov, V. S., and Moore, C. B. (1977). In *Chemical and Biochemical Applications of Lasers* (Ed. C. B. Moore), Vol. 3, p. 1, Academic Press, New York.
Letokhov, V. S., Ryabov, E. A., and Tumanov, O. A. (1972). *Zh. Eksp. Teor. Fiz.*, **63**, 2025 (*Sov. Phys. JETP*, **36**, 1069).
Levi, B. A., Taft, R. W., and Hehre, W. J. (1977). *J. Am. Chem. Soc.*, **99**, 8454.
Lin, C. T., Atvars, T. D. Z., and Pessine, F. B. T. (1977). *J. App. Phys.*, **48**, 1720.
Lin, S. T., Lee, S. M., and Ronn, A. M. (1978). *Chem. Phys. Lett.*, **53**, 260.
Lyman, J. L., Danen, W. C., Nilsson, A. C., and Nowak, A. V. (1979). *J. Chem. Phys.*, **71**, 1206.
Lyman, J. L., Feldman, B. J., and Fischer, R. A. (1978). *Opt. Commun.*, **25**, 391.
Lyman, J. L., Hudson, J. W., and Freund, S. M. (1977). *Opt. Commun.*, **21**, 112.
Lyman, J. L., Jensen, R. J., Rink, J., Robinson, C. P., and Rockwood, S. D. (1975). *Appl. Phys. Lett.*. **27**, 87.
Lyman, J. L., and Leary, K. M. (1978). *J. Chem. Phys.*, **69**, 1858.
Lyman, J. L., and Rockwood, S. D. (1976). *J. Appl. Phys.*, **47**, 595.

Lyman, J. L., Rockwood, S. D., and Freund, S. M. (1977). *J. Chem. Phys.*, **67**, 4545.
McAlpine, R. D. (1975). *The Use of Lasers for the Separation of Isotopes*, p. 179, Proc. 1974 Conf. of the Nuclear Target Development Soc. AECL Report-5503.
McAlpine, R. D., Evans, D. K., and Adams, H. M. (1983). *J. Chem. Phys.*, **78**, 5990.
McAlpine, R. D., Evans, D. K., and Goodale, J. W. (1983). *Can. J. Chem.*, **61**, 1481.
McAlpine, R. D., Evans, D. K. and McClusky, F. K. *Chem. Phys.*, **39**, 263.
McAlpine, R. D., Evans, D. K., and McClusky, F. K. (1980). *J. Chem. Phys.*, **73**, 1153.
McDonald, C. C., and Gunning, H. E. (1952). *J. Chem. Phys.*, **20**, 1817.
McDonald, C. C., McDowell, J. R., and Gunning, H. E. (1959). *Can. J. Chem.*, **37**, 930.
McDowell, R. S., Aldridge, J. P., and Holland, R. F. (1976). *J. Phys. Chem.*, **80**, 1203.
McDowell, R. S., and Goldblatt, H. (1971). *Inorg. Chem.*, **10**, 635.
McDowell, R. S., Sherman, R. J., Asprey, L. B., and Kennedy, R. C. (1975). *J. Chem. Phys.*, **62**, 3974.
McInteer, B. B. (1980). *Sep. Sci. and Tech.*, **15**, 491.
McKillop, J. S., Gordon, R. J., and Zare, R. N. (1982). *J. Chem. Phys.*, **77**, 2895.
McLaughlin, J. G., Poliakoff, M., and Turner, J. J. (1982). *J. Mol. Struct.*, **82**, 51.
Magnotta, F., Herman, I. P., and Aldridge, F. T. (1982). *Chem. Phys. Lett.*, **92**, 600.
Makarov, A. A., Makarov, G. N., Puretsky, A. A., and Tyakht, V. V. (1980). *Appl. Phys.*, **23**, 391.
Makide, Y., Hagiwara, S., Kurihara, O., Takeuchi, K., Ishikawa, Y., Arai, S, Tominaga, T., Inoue, I., and Nakane, R. (1980). *J. Nucl. Sci. and Technol.*, **17**, 645.
Makide, Y., Hagiwara, S., Tominaga, T., Kurihara, O., and Nakane, R. (1981a). *Int. J. Appl. Rad. Isotopes*, **32**, 885.
Makide, Y., Hagiwara, S., Tominaga, T., Takeuchi, K., and Nakane, R. (1981b). *Chem. Phys. Lett.*, **82**, 18.
Makide, Y., Kato, S., Tominaga, T., and Takeuchi, K. (1982). *Appl. Phys.*, **28**, 34.
Makide, Y., Kato, S., Tominaga, T., and Takeuchi, K. (1983). *Appl. Phys.*, **B32**, 33.
Marling, J. B. (1979). *Proposal to Realize a Cost Breakthrough in Carbon-13 Production by Photochemical Separation*, Lawrence Livermore National Laboratory Report Number VCID-18604.
Marling, J. B., and Herman, I. P. (1979). *Appl. Phys. Lett.*, **34**, 439.
Marling, J. B., Herman, I. P., and Thomas, S. J. (1980). *J. Chem. Phys.*, **72**, 5603.
Martinez, R. I., and Herron, J. T. (1981). *Chem. Phys. Lett.*, **84**, 180.
Merton, T. R., and Hartley, H. (1920). *Nature*, **105**, 104.
Meyer, C. F., Woodin, R. L., and Kaldor, A. (1981). *Chem. Phys. Lett.*, **83**, 26.
Miller, C. M., and Zare, R. N. (1980). *Chem. Phys. Lett.*, **71**, 376.
Miller, S. S., DeFord, D. D., Marks, T. J., and Weitz, E. (1979). *J. Am. Chem. Soc.*, **101**, 1036.
Morrison, R. J. S., and Grant, E. R. (1979). *J. Chem. Phys.*, **71**, 3537.
Morrison, R. J. S., Loring, R. F., Farley, R. L., and Grant, E. R. (1981). *J. Chem. Phys.*, **75**, 148.
Moser, J., Morand, P., Duperrex, R., and van den Bergh, H. (1983). *Chem. Phys.*, **79**, 277.
Mrozowski, S. (1930). *Bull. Acad. Pol.*, **A**, 464.
Mrozowski, S. (1932). *Z. Physik*, **73**, 826.
Mughabghab, S. F., Divadeenam, M., and Holden, N. E. (1981). *Neutron Cross Sections*, Vol. 1, Academic Press, New York.
Mukamel, S., and Jortner, J. (1976). *Chem. Phys. Lett.*, **40**, 150.
Nève de Mévergnies, M. (1979). *Appl. Phys. Lett.*, **34**, 853.
Nève de Mévergnies, M. (1982). *App. Phys.*, **B29**, 125.
Nève de Mévergnies, M., and del Marmol, P. (1980). *J. Chem. Phys.*, **73**, 3011.
Nip, W. S., Hackett, P. A., and Willis, C. (1981). *Can. J. Chem.*, **59**, 2703.
Nitschke, E., Ilgner, H., and Walter, S. (1978). In *Separation of Hydrogen Isotope* (Ed. H. K. Rae), p. 177, ACS Symposium Number 68, Washington.
Norris, B., and Smith, A. L. S. (1979). *Appl. Phys. Lett.*, **34**, 385.

Nowak, A. V., and Ham, D. O. (1981). *Opt. Lett.*, **6**, 185.
O'Neill, F., and Whitney, W. T. (1977). *App. Phys. Lett.*, **31**, 270–272.
Osborne, K. R., and Gunning, H. E. (1959). *Can. J. Chem.*, **37**, 1315.
Pamidimukkala, K. M., Rogers, D., and Skinner, G. B. (1982). *J. Phys. Chem. Ref. Data*, **11**, 83.
Papulov, Yu. G., Chulkova, L. V., Levin, V. P., and Stepan'yan, (1972). *Z. Struk. Khim.*, **13**, 956 (*J. Struct. Chem.*, **13**, 895).
Parthasarathy, V., Sarkar, S. K., Singhal, V. P., Pandey, A., Rama Rao, K. V. S., and Mittal, J. P. (1982). *Chem. Phys. Lett.*, **86**, 259.
Pautrot, Ph., and Damiani, M. (1978). In *Separation of Hydrogen Isotope* (Ed. H. K. Rae), p. 163, ACS Symposium Number 68, Washington.
Pertel, R., and Gunning, H. E. (1959). *Can. J. Chem.*, **37**, 35.
Polanyi, J. C. (1961). *J. Chem. Phys.*, **34**, 347.
Pressley, R. J. (1971). *Handbook of Lasers*, Chemical Rubber Co., Cleveland.
Quack, M. (1982). *Adv. Chem. Phys.*, **50**, 395.
Quack, M., and Seyfang, G. (1982). *Chem. Phys. Lett.*, **93**, 442.
Rabinowitz, P., Kaldor, A., Gnauck, A., Woodin, R. L., and Gethner, J. S. (1982). *Opt. Lett.*, **7**, 212.
Rabinowitz, P., Stein, A., Brickman, R., and Kaldor, A. (1979). *Appl. Phys. Lett.*, **35**, 739.
Rabinowitz, P., Stein, A., and Kaldor, A. (1978). *Opt. Commun.*, **27**, 381.
Rae, H. K. (1978). In *Separation of Hydrogen Isotopes* (Ed. H. K. Rae), p. 1, ACS Symposium Number 68, Washington.
Rayner, D. M., Kimel, S., and Hackett, P. A. (1983). *Chem. Phys. Lett.*, **96**, 678.
Reid, J., Shewchun, J., and Garside, B. K. (1978). *Appl. Phys.*, **17**, 349.
Reid, J., and Siemsen, K. (1976). *Appl. Phys. Lett.*, **29**, 250.
Reid, J., and Siemsen, K. (1977). *J. Appl. Phys.*, **48**, 2712.
Reid, J., and Siemsen, K. (1978). *IEEE J. Quant. Electron*, **QE14**, 217.
Renlund, A. M., Reisler, H., and Wittig, C. (1981). *Chem. Phys. Lett.*, **78**, 40.
Ritter, J. J. (1978). *J. Am. Chem. Soc.*, **100**, 2441.
Ritter, J. J., and Freund, S. M. (1976). *J. Chem. Soc. Chem. Comm.*, 811.
Robinson, G. W., and Frosch, R. P. (1962). *J. Chem. Phys.*, **37**, 1962.
Rodgers, A. S., Chao, J., Wilhoit, R. C., and Zwolinski, R. J. (1974). *J. Phys. Chem. Ref. Data*, **3**, 117.
Salvetat, G., and Bourene, M. (1980). *Chem. Phys. Lett.*, **72**, 348.
Schulz, P. A., Sudbo, Aa. S., Grant, E. R., Shen, Y. R., and Lee, Y. T. (1980). *J. Chem. Phys.*, **72**, 4985.
Scruby, R. E., Lacher, J. R., and Park, J. D. (1951). *J. Chem. Phys.*, **19**, 386.
Selwyn, L., Back, R. A., and Willis, C. (1978). *Chem. Phys.*, **32**, 323.
Sherwood, A. G., and Gunning, H. E. (1960). *Can. J. Chem.*, **38**, 466.
Shimanouchi, T. (1972). *Tables of Molecular Vibrational Frequencies, Consolidated Volumes I*, NBS Publication Number NSADS-NBS 39.
Shimanouchi, T. (1977). *J. Phys. Chem. Ref. Data*, **6**, 993.
Smith, A. C. (1972). *Photoseparation of Isotopes*, Report Number RM-5825-1-PR.
Snavely, B. B. (1974). *Separation of Uranium Isotopes by Laser Photochemistry*, Paper G-9, VIII International Quantum Electronics Conf., San Francisco.
Steinfeld, J. I., Burak, I., Sutton, D. G., and Nowak, A. V. (1970). *J. Chem Phys.*, **52**, 5421.
Stephenson, J. C., and King, D. S. (1978). *J. Chem. Phys.*, **69**, 1485.
Stone, J., and Goodman, M. F. (1976). *Chem. Phys. Lett.*, **44**, 411.
Stone, J., Thiele, E., and Goodman, M. F. (1973). *J. Chem. Phys.*, **59**, 2909.
Stuke, M., Reisler, H., and Wittig, C. (1981). *Appl. Phys. Lett.*, **39**, 201.
Stull, D. R., and Prophet, H. (1971). *JANAF Thermochemical Tables*, 2nd ed., NBS Publication Number NSRDS-NAS 37.
Sudbo, Aa. S., Schulz, P. A., Grant, E R., and Lee, Y. T. (1979). *J. Chem. Phys.*, **70**, 912.

Sudho, Aa. S., Schulz, P. A., Grant, E. R., Shen, Y. R., and Lee, Y. T. (1978). *J. Chem. Phys.*, **68**, 1306.
Sugita, K., Ishikawa, Y., and Aral, S. (1983). *J. Phys. Chem.*, **87**, 3469.
Suzuki, I., and Shimanouchi, T. (1973). *J. Mol. Spectrosc.*, **46**, 130.
Symons, E. A., and Bonnett, J. D. (1984). *J. Phys. Chem.*, **84**, 866.
Symons, E. A., and Clermont, M. J. (1981). *J. Am Chem. Soc.*, **103**, 3127.
Symons, E. A., Clermont, M. J., and Coderre, L. A. (1981). *J. Am. Chem. Soc.*, **103**, 3131.
Symons, E. A., Rolston, J. R., Baldisera, L. A., Drover, J. C. G., and Bonnett, J. D. (1983). *Can. J. Chem.*, **61**, 1301.
Takeuchi, K., Inoue, I., Nakane, R., Makida, Y., Kato, S., and Tominaga, T. (1982). *J. Chem. Phys.*, **76**, 398.
Takeuchi, K., Kurihara, O., and Nakane, R. (1980). *J. Chem. Eng. Jap.*, **13**, 246.
Takeuchi, K., Kurihara, O., and Nakane, R. (1981). *Chem. Phys.*, **54**, 383.
Takeuchi, K., Kurihara, O., Satooka, S., Makide, Y., Inoue, I., and Nakane, R. (1981). *J. Nucl. Sci and Tech.*, **18**, 972.
Tiee, J. J., and Wittig, C. (1977). *Appl. Phys. Lett.*, **30**, 420.
Tiee, J. J., and Wittig, C. (1978a). *Appl. Phys. Lett.*, **32**, 236.
Tiee, J. J., and Wittig, C. (1978b). *J. Chem Phys.*, **69**, 4756.
Tiee, J. J., and Wittig, C. (1978c). *Opt. Commun.*, **27**, 377.
Tuccio, S. A., and Hartford, A. (1979). *Chem. Phys. Lett.*, **65**, 234.
Urey, H. C. (1943). *Investigation of the Photochemical Methods for Uranium Isotope Separation*, Columbia University Report Number A-750.
Vasil'ev, .B. I., Dyad'kin, A. P., and Sukhanov, A. N. (1980). *Pis'ma Zh. Tekh. Fiz.*, **6**, 311–313 (*Sov. Tech. Phys. L ett.*, **6**, 135).
Voiland, E E. (1959). *Separation of Mercury Isotopes by Selective Photoexcitation*, Hanford Laboratory Report Number HW-59329.
von Puttkammer, K., Dubal, H-., and Quack, M. (1983). *Faraday Disc.*, **75**, 197.
Waller, I., Evans, D. K., and McAlpine, R. D. (1983). *Appl. Phys.*, **B32**, 75.
Weast, R. C. (1972). *Handbook of Physics and Chemistry*, 53rd ed. Chemical Rubber Co., Ohio.
Weissman, M., and Benson, S. W. (1983). *J. Phys. Chem.*, **87**, 243.
Willis, C., Dosi, M., and James, D. J. (1979). *Rev. Sci. Instrum.*, **50**, 622.
Willis, C., Hackett, P. A., and Parsons, J. M. (1979). *Rev. Sci. Instrum.*, **50**, 1141.
Willis, C., and Purdon, J. G. (1979) *J. Appl. Phys.*, **50**, 2539.
Wittig, C. (1985). *Adv. Chem. Phys.*, this volume.
Würtzberg, E., Kovalenko, L. J., and Houston, P. L. (1978). *Chem. Phys.*, **35**, 317.
Wynn, N. P. (1978). In *Separation of Hydrogen Isotopes* (ed. H. K. Rae), p. 53, ACS Symposium Number 68, Washington.
Yokoyama, A., Suzuki, K., Fujisawa, G., Ishikawa, N., and Iwasaki, M. (1983). *Chem. Phys. Lett.*, **99**, 221.
Yu, M. H., Levy, M. R., and Wittig, C. (1980). *J. Chem. Phys.*, **72**, 3789.
Zelikoff, M., Aschenbrand, L. M., and Wyckoff, P. H. (1953). *J. Chem. Phys.*, **21**, 376.
Znotins, T. A., Reid, J., and Brimacombe, R. K. (1982). *J. Appl. Phys.*, **53**, 2843.
Znotins, T. A., Reid, J., Garside, B. K., and Ballik, E. A. (1981). *Appl. Phys. Lett.*, **39**, 199.
Znotins, T. A., Reid, J., Garside, B. K., and Ballik, E. A. (1979). *Opt. Lett.*, **4**, 253.
Zuber, K. (1935a). *Helv. Phys. Acta*, **8**, 370.
Zuber, K. (1935b). *Nature*, **136**, 796.
Zuber, K. (1936). *Helv. Phys. Acta*, **9**, 285.

Photodissociation and Photoionization
Edited by K. P. Lawley

PHOTOIONIZATION IN NON-POLAR LIQUIDS

B. S. YAKOVLEV and L. V. LUKIN

Institute of Chemical Physics of USSR Academy of Sciences, Chernogolovka, Moscow Region, 142432, USSR

CONTENTS

I. INTRODUCTION

During the past decade many experimental and theoretical works have been devoted to photoionization processes in non-polar molecular liquids.

On the one hand, they were initiated by the great interest in both intra- and intermolecular photon-induced charge separation in condensed media. On the other hand, the understanding of the mechanism of the processes provides a challenging problem in the general area of the electronic structure and electron behaviour in disordered materials. Recently, the behaviour of electrons in non-polar liquids, especially in liquid hydrocarbons, has received great attention among photo and radiation chemists, mainly with respect to electron transport and electron reactivities. It appears that the behaviour of electrons in non-polar liquids is similar to that for early stages of ionization in polar media, where full dielectric relaxation requires nuclear rearrangements and takes on the order of picoseconds or more. Therefore, it is probable that the data obtained for non-polar molecular liquids give an insight into very rapid ($<10^{-12}$ s) photoionization processes in more polar systems and more complicated micellar and biological systems.

The picture of ionization processes in molecular condensed media is essentially different from that for gases at low pressures. An important feature of ionization in the condensed medium consists in rapid dissipation of the excess kinetic energy of an electron (or a hole) ejected in the medium. Because of that, even if the excitation energy is several times more than the ionization potential, the majority of electrons produced by ionization are thermalized within several tens of angstroms from their partner cation. The distance, especially for non-polar media, is not large enough to neglect the Coulomb interaction between charged particles. (For example, at room temperature in a liquid of dielectric constant $\varepsilon = 2$, the Coulomb energy equals the thermal energy $k_B T$ at a distance of 290Å.) Therefore the thermalized electron and cation then move in their mutual Coulomb field either escaping each other (to produce charged particles with negligible low interaction, usually called for liquids as free ions) or, with much larger probability, geminately recombining.

The mechanism of ionization was first proposed[1-4] for non-polar hydrocarbon liquids owing to radiation investigations. Extensive data obtained during the last two decades on free ion radiation yields, steady state and pulse radiolysis scavenging measurements, and an optical study of electrons and positive ions in pulse-irradiated liquids have provided conclusive evidence for the creation under radiation of geminate ion pairs and their rapid recombination or dissociation to free ions. The evidence of this feature of ionization in condensed media is also obtained for different molecular matrix: glassy hydrocarbons,[5] molecular crystals[6,7] and polymers.[8,9]

According to current thinking about the ionization mechanism in the condensed phase, the photoionization process in non-polar molecular liquids can be divided into two stages:

$$\mathrm{M} \underset{\substack{\text{geminate}\\ \text{recombination}}}{\overset{h\nu}{\rightleftarrows}} \text{ion pair} \rightarrow \text{free ions} \tag{1}$$

where ion pairs are pairs of oppositely charged particles (ions and electrons) produced at the same time; these remain close enough together for the interaction between them not to be negligible compared to $k_B T$.

The existence of a 'dark' stage in photogeneration of quite separated charged particles (i.e. free ions in scheme 1) makes the definition of an ionization rather ambiguous. Since the free ion photogeneration efficiency in non-polar liquids is usually small and strongly dependent on temperature and the external electric field, it is convenient for practical purposes to include the ion pair under the definition of ionization. The definition of ionization—'an event during which an electron separates from an ion or molecule, and after which the electron may be considered to have a separate existence for a finite time'—has been accepted[10] at the Conference on Electrons in Fluids at Banff in 1976 and will be used by us below. In the case, however, an ionization is not always readily distinguishable from an excitation of neutral states.

Dissociation of ion pairs on free ions is now the most experimentally investigated stage of photoionization processes in non-polar liquids, mainly owing to photoconductivity measurements. The measurements are very suitable for this purpose because of low 'dark' conductivity of non-polar liquids, and have been used in numerous works[11–24] to obtain data on free ion quantum yield, its dependence on temperature, electric field, excitation energy and the nature of solutes and solvents.

Less is known about two other processes in scheme (1): the creation of ion pairs and geminate recombination. Some traits of the processes are elucidated by means of various methods developed lately: studying the effects of an electric field[25–31] and electron-attaching solutes[32–38] on the fluorescence yield, laser kinetics spectroscopy [38–42], photoassisted dissociation of ion pairs,[43–46] multiphoton ionization[23,47–49] and transient d*c* photoconductivity.[50–54] The experimental works have been mainly devoted to solutions of molecules with low gas-phase ionization potentials such as tetramethyl-*p*-phenylenediamine (TMPD). For these compounds the ionization threshold in non-polar electron non-attaching liquids (like hydrocarbons) is above 200 nm and, therefore, conventional light sources may be utilized.

In the present paper we will focus almost entirely on ionization processes in these systems. Particular attention will be given to those aspects which are conceptually far removed from the normal intermolecular charge transfer processes by including the stage of thermalization of quasi-free electrons to interpret the correlation observed between free ion generation efficiency and properties of excess electrons in the liquids.

II. FREE ION GENERATION

A. The Onsager Model

The recent treatment of photoionization in non-polar liquids is based on Onsager's theory,[55] which allows the rigorous analysis of the diffusive motion of a pair of oppositely charged particles in the presence of an applied electric field. Use of Onsager's theory implies that production of free ions is a two-stage process. In the first stage, a photoelectron scatters away from its parent positive ion until it is thermalized. In the second stage, the charge pair (called below the ion pair) undergoes mutual diffusion. The usual application of the theory of free ion generation involves the following assumptions: (1) the isolated ion pairs are taken to be formed in thermal equilibrium with the medium; (2) the initial distribution function $f(r)$ of distances r between ions of a pair is spherically symmetric and, as well as the quantum yield of ion pairs φ_0, independent of applied field strength, E; (3) ions remain in thermal equilibrium with the medium so that their mobility and diffusion constant are related by the Nernst–Einstein relationship.

In this model the theory allows calculation of the average probability $P(r, E)$ that an ion pair with a separation r will escape geminate recombination:[56]

$$P(r, E) = \mathrm{e}^{-r_c/r}\left[1 + \mathrm{e}^{-\xi} \sum_{n=1}^{\infty} \frac{\xi^n}{(n+1)!} \sum_{m=1}^{n} \frac{n-m+1}{m!}\left(\frac{r_c}{r}\right)^m\right] \tag{2}$$

where $r_c = e^2/\varepsilon k_B T$ (the Onsager radius) is the distance from an ion at which the coulombic potential energy is equal to $k_B T$, k_B is the Boltzman constant, T is the absolute temperature, e is the electron charge and $\xi = eEr/2k_B T$. Using $P(r, E)$ one can find the quantum yield of free ions φ_{fi}:

$$\varphi_{fi} = \varphi_0 \int_0^{\infty} 4\pi r^2 f(r) P(r, E)\, \mathrm{d}r \tag{3}$$

As first pointed out by Onsager, the model predicts that the field and temperature dependence of $P(r, E)$ is very simple in the low-field limit:

$$P(r, E) = \mathrm{e}^{-r_c/r}\left(1 + \frac{e r_c E}{2k_B T}\right) \tag{4}$$

and gives, together with Eq. (3),

$$\frac{\varphi_{fi}(E)}{\varphi_{fi}(0)} = 1 + \frac{e r_c E}{2k_B T} \tag{5}$$

Although the Onsager model is strictly applicable to the classical particles, the model has been surprisingly successful in allowing interpretation of various data on ionization in both liquids[2-4] and solids, including organic single crystals,[6,7] a polymeric photoconductor[8] and amorphous selenium.[57,58] A major concern of such works has been to verify Onsager's theory predictions and to derive electron thermalization lengths and their distribution.

B. Low-Field Photoconductivity

Measurements of photocurrent in TMPD solutions in various non-polar solvents[17,18,20,22] have allowed verification of the Onsager model prediction concerning the dependence of φ_{fi} on E in the low-field limit (Eq. 5). For this purpose the results obtained at conditions where bimolecular charge recombination can be negligible are useful. In this case, the photocurrent I is proportional to the quantum yield φ_{fi} and, according to Eq. (5), must increase linearly with the applied field:

$$I(E) = I(0)[1 + (s/i)E] \tag{6}$$

where s/i is the slope to intercept ratio.

Figure 1 demonstrates a good linear dependence of I on E for TMPD solution in 2, 2, 4-trimethylpentane,[22] except as very low fields. As the field increases more and more ions drift to the electrodes; this accounts for the rapid rise in observed current at very low fields. At higher fields the current rises less rapidly; here the volume recombination is less important.

It was found in some early studies of X-ray ionization of organic liquids[59,60]

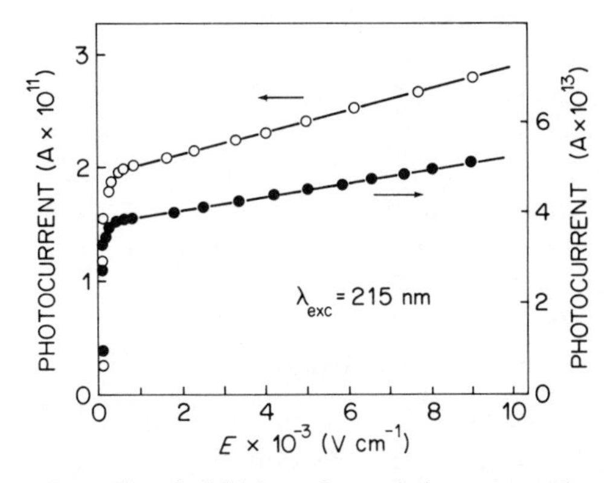

Fig. 1. Electric field dependence of photocurrent due to single-photon ionization of TMPD in 2,2,4-trimethylpentane at the excitation wavelength indicated and room temperature. The lower curve (solid points) was measured with roughly a fiftyfold reduction in light intensity as compared with the intensity of 1×10^{10} photons cm^{-2} used for the upper curve. Approximately the same slope to intercept ratio for both curves (5.32×10^{-5} and 5.51×10^{-5} cm V^{-1} for the upper and lower curves) indicates that the linear increase of the photocurrent at $E > 1 \times 10^3$ V cm^{-1} does not result from the bimolecular volume recombination. This recombination is evident below about 10^3 V cm^{-1}. (From Ref. 22. Reproduced by permission of the American Institute of Physics.)

that the slope to intercept ratio s/i for photoionized non-polar liquids conformed to Onsager's predictions: $s/i = er_c/2k_BT$. Holroyd and Russel[17] observed that, because of bimolecular recombination, their s/i values decreased sharply as the incident intensity decreased. Their lowest intensity values of s/i agreed quite well (within 20 per cent) with theory for TMPD photocurrents in n-hexane, n-pentane, n-butane, 2, 2,4-trimethylpentane and tetramethylsilane. According to data of Choi, Sethi and Braun,[22] the differences between experimental and theoretical s/i values were -0.23 per cent. for n-hexane, -2.7 per cent. for 2,2,4-trimethylpentane and -22.5 per cent. for tetramethylsilane.

Because of undoubted simplicity of the Onsager model there are a few reasons for discrepancy between low-field Onsager theory and experiment. (The possible origins of this discrepancy, discussed for example by Choi, Sethi and Braun,[22] are, in particular, the following: (1) the electron mean free path is so long as to invalidate the assumption about diffusive motion of a geminate pair or (2) the effect of field-dependent mobility on escape probability.) Therefore, more interesting in the authors opinion is the observed quite good agreement of s/i values with theory.

Does this result show the validity of the fundamental assumption that diffusive motion governs the fate of the majority of ion pairs? It seems the answer is 'no', at least for broad thermalization length distribution usually assumed because, as discussed in Section IV.E, the large majority of ion pairs with relatively small separation distances weakly contributes to the photocurrent. Moreover, as will be shown in Section III.C, the s/i value can conform to Onsager's predictions even though the motion of geminate ions becomes diffusive only for times $t > t^*$, where t^* is near the time taken to become completely separated ions.

C. The Temperature Effect

Studies of the temperature effect on photocurrent in non-polar solutions[17] show that, in general, the lower the temperature, the lower is the quantum yield of free ions at all wavelengths. For hydrocarbon liquids investigated, a marked temperature effect on φ_{fi} is observed in n-alkanes while in branched hydrocarbons it is less pronounced.[17] The temperature effect on the free ion quantum yield well conforms with the assumption that there is a 'dark' stage in photogeneration of free ions.

In the zero field the Onsager model predicts a linear dependence of $\ln[\varphi_{fi}(0)]$ on $(\varepsilon T)^{-1}$ if all pairs have the same initial separation distance r_0 which is independent of T, and if the initial quantum yield of ion pairs φ_0 is also independent of T. In this case Eqs. (3) and (4) give

$$-\frac{\mathrm{d}}{\mathrm{d}(\varepsilon T)^{-1}} \ln[\varphi_{fi}(0)] = \frac{e^2}{k_B r_o} \tag{7}$$

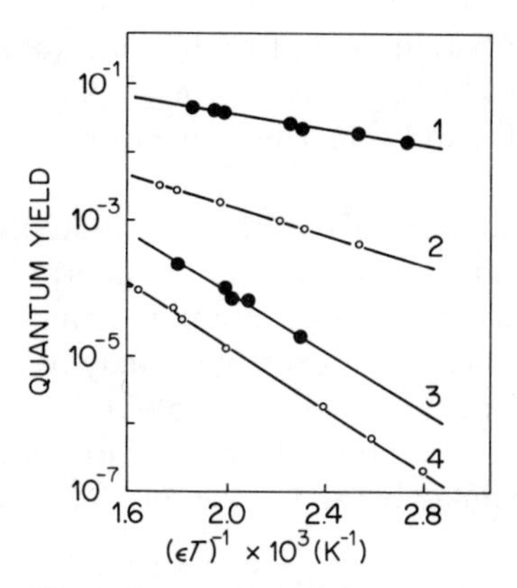

Fig. 2. Free carrier quantum yield for TMPD photoionization (at $E=0$) versus $(\varepsilon T)^{-1}$, $\varphi_{fi}(E)/(1+r_c Ee/2k_B T)$: (1) tetramethylsilane, (2) 2,2,4-trimethylpentane, (3) *n*-pentane, (4) *n*-hexane. The φ_{fi} values were measured at photon energies of $h\nu \approx 5.7$ eV, corresponding to the maximum of the spectrum $\varphi_{fi}(h\nu)$. (Reprinted with permission from Ref. 17. Copyright © 1974 American Chemical Society.)

The linear dependence of $\ln[\varphi_{fi}(0)]$ on $(\varepsilon T)^{-1}$ has in fact been observed[17] in various non-polar liquids (see, for example, Fig. 2). Within the model above different slopes observed for the dependence indicate that the separation distances r_0 are different in each liquid. The values of r_0 derived from the slopes by using Eq. (7) range from 31 Å for *n*-hexane to 106 Å for tetramethylsilane.[17] With r_0 evaluated, Holroyd and Russel[17] have calculated the escape probability $P(r_0, E)$ to compare with the experimental free ion quantum yield. The near agreement of these two values allows them to conclude that the quantum yield of ion pair formation φ_0 is approximately equal to unity for ionization of TMPD at $h\nu \approx 6$ eV in various non-polar liquids.

The temperature effect observed definitely indicates the good qualitative agreement of Onsager's theory with experiment. It should be noted, however, that the average separation distance of ion pairs and their quantum yield values derived from comparing the theory with experimental data are very sensitive to assumptions used for application of the theory. In particular, in the case above, r_0 and φ_0 values can greatly change if the possibility of a weak temperature dependence of r_0 is taken into account. In this case, the right hand part of Eq. (7) will contain the additional term, including the temperature derivations of ε and r_0:

$$\frac{1}{k_B}\left(\frac{e}{r_0}\right)^2 \frac{dr_0}{dT}\left(\frac{1}{T}+\frac{1}{\varepsilon}\frac{d\varepsilon}{dT}\right)^{-1}$$

For example, $r_0 = 40$ Å is required at room temperature to fit the n-hexane data[17] if r decreases only by 10 per cent., with temperature decreasing from 300 down to 200 K, and if $d\varepsilon/dT = 0$. This value of r_0 (instead of 31Å) leads to $\varphi_0 \simeq 0.1$.

Moreover, in the framework of the Onsager model, the delta distribution of geminate pairs assumed[17] is not unique for interpretation of data on the temperature effect. Some broad distribution functions also allow reasonable fits to the experimental data. For example, it may be reached for the usually assumed exponential function $f(r) \sim r^{-2}\exp(-r/B)$ (see Section II.E). In this case, it is helpful to use the empirical equation[61] for the average probability of escape, which is approximately valid in the zero field for reasonable values of B:

$$P(\mathrm{EXP}) \simeq b e^{-2b} \tag{8}$$

where $b = \sqrt{r_c/B}$.

Provided that φ_0 is independent of temperature, Eq. (8) predicts a very weak temperature dependence of the slope of the $\ln[\varphi_{fi}(0)]$ versus $(\varepsilon T)^{-1}$ curve:

$$-\frac{d\ln[\varphi_{fi}(0)]}{d(\varepsilon T)^{-1}} \simeq (b - \tfrac{1}{2})\left[\varepsilon T - \frac{1}{B}\frac{dB}{d(\varepsilon T)^{-1}}\right] \tag{9}$$

This gives good agreement with experimental data and, in particular, with data shown in Fig. 2 for n-hexane. It is of interest to note that, if the parameter B is independent of T, the best-fit value of B is too small (3 Å) and in order to obtain the experimental value of φ_{fi} with this parameter $\varphi_0 \approx 10^4$ should be assumed. This, of course, is physically impossible. However, the φ_0 value becomes approximately equal to unity (and $B = 10$ Å at $T = 296$ K) if the B value decreases by only 30 per cent (or less if $d\varepsilon/dT < 0$) with decreasing T from 300 down to 200 K.

Apart from the temperature dependence of liquid density, a possible reason for a decrease of the B value with cooling is expected to be the experimentally observed increase of ionization potential of solute molecules in non-polar liquids as the temperature decreases. According to the data,[17] the photoionization threshold energy for TMPD in liquid n-hexane increases by ≈ 0.2 eV when the temperature decreases from 300 down to 200 K. This appears to be quite enough to result in the change of B by ≈ 10 per cent. (see Section II.E).

D. The Wavelength Dependence in the Threshold Region

The usual Onsager description implies a ballistic model, in which the excess energy of excitation of an electron-hole pair is dissipated during the initial translational motion of the particles through the medium.[7] A prediction of this model is that the free ion quantum yield is strongly dependent on the incident photon energy, since a highly excited electron-hole pair will require a larger separation distance for thermalization. The prediction conforms in general with experimental data on wavelength dependence of photocurrent in non-polar liquids. For example, in the most investigated systems, TMPD + hydrocarbon, it has been found[17] that above onset the photocurrent rises rapidly with increasing photon energy (by a factor of 30 for each 0.1 eV at photon energies near onset). The increase appears to be due mainly to an increase in the effective thermalization length (see Section II.E).

Study of the wavelength dependence of photocurrent is a widespread method used to estimate the ionization threshold, I_L, in non-polar liquids. It has long been known that the ionization potential of a solute molecule in a condensed phase is lowered relative to the value measured in the gas phase, I_G. This has been experimentally checked in numerous works[11-13,16-17,62-65] by measuring the photocurrent threshold, I_{th}, for some solute molecules, especially for TMPD, in various non-polar liquids. Of course, since I_{th} values are dependent on the detection technique, identification of the I_{th} values with the photoionization threshold in liquids is rather relative. Nevertheless, the I_{th} values obtained by different authors are near each other (within ~ 0.2 eV) because the photocurrent is strongly dependent on the photon energy in the threshold region.

The unexpected result of photoconductivity measurements was that I_{th} values could differ considerably, even for liquids with very similar chemical and physical properties such as saturated hydrocarbon liquids. This fact was very striking for the early intrepretation of Lyons and Mackie[66,67] who have emphasized the importance of polarization of the medium around a cation E_p and an ejected electron. In the theoretical model developed more recently by Springett, Jortner and Cohen[68] for liquid rare gases and modified by Davis, Schmidt and Minday[69] for molecular liquids, the importance of the energy V_0 of an excess electron in the lowest quasi-free level (conduction band) has been demonstrated for energetics of the ejected electron in non-polar liquids. The energy V_0 enters the equation giving the photoionization threshold if the ionization results in formation of a solute cation and a quasi-free electron removed far away from the cation:[70]

$$I_L(\text{q.f.}) = I_G + E_p + V_0 \tag{10}$$

It is customary to assume that $I_L(\text{q.f.}) = I_{th}$. In this case, the difference of I_{th} values measured for the same solute in various saturated hydrocarbon liquids, where E_p values appear to be approximately equal, could reasonably be explained by the difference of V_0. Figure 3 shows that I_{th} values obtained[17] for TMPD solution in hydrocarbon liquids are in fact well correlated with V_0 values determined[71] by a photoelectric injection method, and the dependence $I_{th}(V_0)$ is linear as Eq. (10) predicts ($I_G = 6.2$ eV[72] for the TMPD molecule). This approach also accounts for the temperature dependence of I_{th} values obtained[17,65] for TMPD solute in some hydrocarbon solvents. It was found

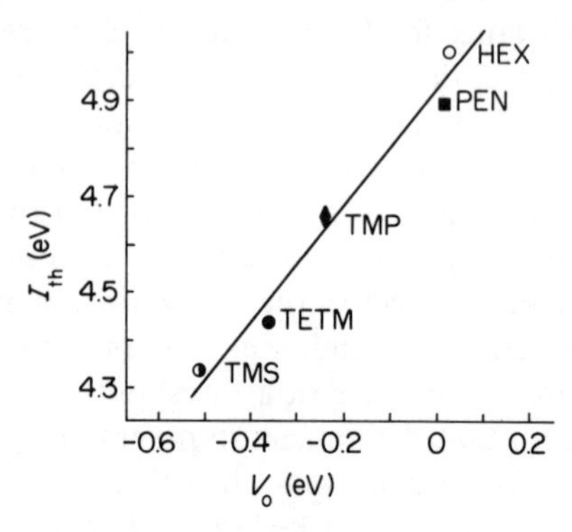

Fig. 3. The photocurrent threshold I_{th} for a single-photon TMPD ionization versus the level V_0 of the conduction band bottom in *n*-hexane (HEX), *n*-pentane (PEN), 2,2,4-trimethylpentane (TMP), 2,2,4,4-tetramethylpentane (TETM) and tetramethylsilane (TMS) at 296 K. The values of I_{th} were taken from Ref. 17, the V_0 values from Ref. 71. (Reprinted with permission from Ref. 17. Copyright © 1974 American Chemical Society.)

that I_{th} increases when the temperature decreases. According to Eq. (10), the result suggests that V_0 increases with decreasing temperature. Such temperature behaviour of V_0 is in fact observed[71] in hydrocarbon liquids and is in accord with the SJC model if the temperature dependence of the free volume in the liquid is taken into consideration.[71]

It is of interest to compare experimental values of the photocurrent threshold I_{th} with I_L (q.f.) values calculated from Eq. (10) using the experimental V_0 values and E_p estimated from the Born formula:[73] $E_p = -(e^2/2a)(1 - 1/\varepsilon)$, where a is the cation radius calculated from the molecular volume. The comparison has shown[24] that I_{th} coincides with I_L (q.f.) within 0.3 eV for some aromatic solute in hydrocarbon liquids. Aside from calculation of E_p, the accuracy of equality of the I_L and I_{th} values is limited by an unclear definition of what are the photoionization and photocurrent thresholds in condensed media.

In the Onsager model the photocurrent threshold corresponds to the lowest photon energy I_{th} (i.p.) needed for the ion pair creation. Evidently, this energy depends on the nature of the charge particles of an ion pair and on the separation distance between them. If an excess electron of an ion pair is formed in a localized state with an energy E_{tr}, the I_{th}(i.p.) value may be proposed as

$$I_{th}(\text{i.p.}) = I_L(\text{q.f.}) + E_c - (V_0 - E_{tr}) \tag{11}$$

where E_c is the Coulomb energy. The minimum value of E_c is approximately equal to $-e^2/(\varepsilon R_a) \approx -1$ eV, where $R_a \approx 10$ Å is the neutralization radius. However, the probability P that a pair with this radius escapes geminate recombination in a non-polar liquid is too low ($\sim 10^{-13}$) to be detected by means of the conductivity method. In the liquid, the detection sensitivity is limited by the dark current value ($\sim 10^{-13}$ A at the room temperature) and by the necessity of using low light intensity to avoid a contribution of the high-order photonic processes. The minimum detectable value of P is usually $\gtrsim 10^{-5}$ at the room temperature and corresponds to $|E_c| \lesssim 0.3$ eV.

Little is known about excess electron states in non-polar molecular liquids. For low mobility hydrocarbon liquids the electron localized states have been observed by optical absorption measurements.[74–77] A crude estimate of the energy of these states using the excess electron mobility activation energy[78] leads to $V_0 - E_{tr} \lesssim 0.2$ eV. Thus, it should be expected in the Onsager model that the I_{th} experimental values can be lower by $\lesssim 0.5$ eV than the I_L (q.f.) value calculated from Eq. (10). The estimate agrees with the experimental data discussed above.

Recent photoionization studies have shown that photocurrent near the threshold can be represented by a power law of the form:[24,79–81]

$$I(h\nu) = C(h\nu - I_{th})^m \tag{12}$$

For ionization of neutral molecules in non-polar solvents the value of $m = 5/2$ gives the best fit[24,81] (see, for example, Fig. 4). Although the power dependence

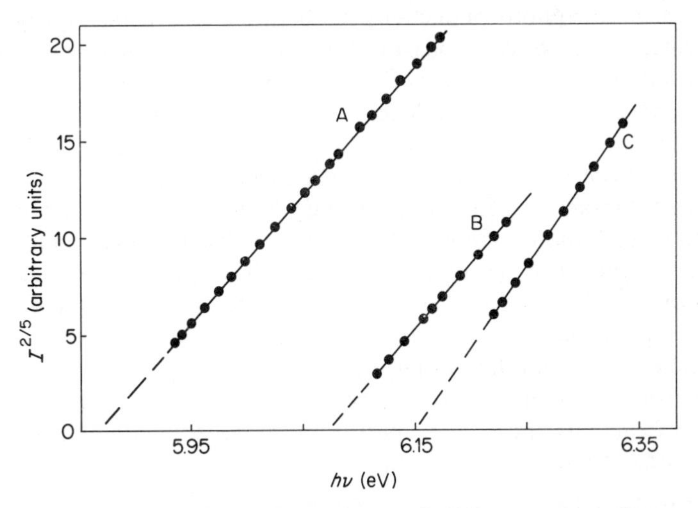

Fig. 4. Plot of normalized photocurrent to the 2/5 power versus photon energy for a single-photon induced conductivity of anthracene in tetramethylsilane (A), 2,2,4,4-tetramethylpentane (B) and 2,2,4-trimethylpentane (C). For simplicity, only some experimental points obtained for every curve are shown (From Ref. 24 Reproduced by permission of the American Institute of Physics.)

of 5/2 is an empirical fit, there is a theoretical support for an equation of this form.[82] If Eq. (12) is valid, the intercept of $I^{2/5}$ with the $h\nu$ axis gives the photocurrent threshold I_{th} which is independent of the detection sensitivity.

Holroyd, Preses and Zevos[24] have found that I_{th} values obtained by means of this method for various aromatic solutes in some hydrocarbon liquids is lower by $\lesssim 0.3$ eV than I_L values estimated from Eq. (10), with E_p calculated from the Born equation. The difference (≈ 0.3 eV) was also obtained[24] for anthracene in liquid 2, 2, 4-trimethylpentane by using the experimental value of the anthracene anion polarization energy[80] instead of E_p (Holroyd and coworkers [24] argued the anion polarization energy $|E_p|$ to be less than that of cation. If it is the case, the difference $I_L - I_{th}$ is with more certainty less than 0.3 eV.)

According to Eq. (11), the value of I_L(q.f.) $- I_{th}$ gives the Coulomb energy to be $|E_c| \lesssim 0.3$ eV. Though the accuracy of such estimate is not high enough, it is interesting to speculate about a threshold value of E_c. If, in fact, $|E_c| \lesssim 0.3$ eV at the photoionization threshold, the value of an initial ion-pair separation distance is $r_{th} > 30$ Å. This means that there is a threshold value for the ion pair separation distance which is larger than an ion recombination radius which is, as expected, on the order of the molecular diameter. It can be understood if we assume, as Wu and Lipsky,[32] that a large-orbit, excited state is a precursor of the geminate pairs.

E. The Electric Field Effect

Apart from the photoconductivity method, the electric field quenching of recombination luminescence has been found to be a useful probe of free charge generation for non-polar liquids.[25-31] These two rather disparate methods have given the consistent dependences $\varphi_{fi}(E)$ of a free ion quantum yield on the electric field strength. A major concern of the study of the electric field effect on φ_{fi} has been to derive electron thermalization lengths and their distribution on the basis of Onsager's theory. For this purpose it is commonly assumed that the initial distribution function of thermalization distances of geminate pairs is spherically symmetric and independent of E. (The validity of these assumptions is discussed in Section III.C.) In this case, the problem is to find the form of the one-parameter function, $f(r)$, for the pair distribution which would best fit to the experimental data using Eq. (3).

Numerous functional forms have been tested against experimental data, but the most frequently chosen are, apart from the delta function, the three trial functions:

1. Narrow Gaussian (N-GAUS): $f(r) \sim \exp[-16(r-M)^2/M^2]$
2. Gaussian (GAUS): $f(r) \sim \exp(-r^2/G^2)$
3. Exponential (EXP): $f(r) \sim r^{-2}\exp(-r/B)$

These functions with a short-range cutoff or neutralization radius R_a of ~ 10 Å have been used to interpret photoconductivity and radiation chemistry data for molecular solids and liquids. N-GAUS has theoretical justification in the work by Sano and Mozumder,[83] who have considered, using the Fokker–Planck equation, the velocity evolution for an electron ejected from a donor with some excess kinetic energy. EXP has been proposed by Abell and Funabashi[84] on the basis of a quantum mechanical consideration of electron scattering. Concerning experimental studies, most radiation chemistry data have been found to be consistent with broad distributions, while most solid photoconductivity studies have assumed a spherical delta function.

1. *High-Field Photoconductivity*

Under experimental conditions, provided that the free ion bimolecular recombination is insignificant, the values of the photo-current I are proportional to the free ion quantum yield and may be directly connected with the average escape probability, $P(\mathrm{E})$, for an initial thermalized ion pair, which is determined by the integral of Eq. (3):

$$\frac{I(\mathrm{E})}{I(\mathrm{O})}=\frac{P(E)}{P(\mathrm{O})} \tag{13}$$

Using the highest field reached in liquid-phase photoconductivity studies, Choi, Sethi and Brown[20,22] have recently measured the photo-current induced by single-photon ionization of TMPD in some hydrocarbon liquids. Figure 5 shows, for example, the electric field dependence of $I(E)/I(\mathrm{O})$ obtained[22] for *n*-hexane, where the current ratio exhibits the largest change. It was found for this liquid that all three functional forms and the delta function allow reasonable fits to the data but, in general, the fits by EXP is better. For 2, 2, 4-trimethylpentane and tetramethylsilane, the GAUS curves fit the data fairly well, the EXP curves fitting them almost perfectly. The best-fit delta and N-GAUS curves exhibit the downward curvature which is far too pronounced, although the deviation from the experimental data is not too large (<10 per cent.).

It is difficult to say now whether the differences which have been obtained between the experimental and theoretical $P(\mathrm{E})$ curves are enough to discriminate any trial function forms of $f(r)$. The point is that the usual application of Onsager's theory to field dependence experiments involves many assumptions. Apart from assumptions concerning the diffusive description of a geminate pair motion, the two above-mentioned assumptions (φ_0 and $f(r)$ are independent of E) are important for application of the theory. It would be surprising if the thermalized pair distribution is actually independent of applied fields. A crude estimate made in Section III.C shows that even displacing hot electrons in the external electric field may markedly distort the just-thermalized ion pair distribution away from spherical symmetry. Unfor-

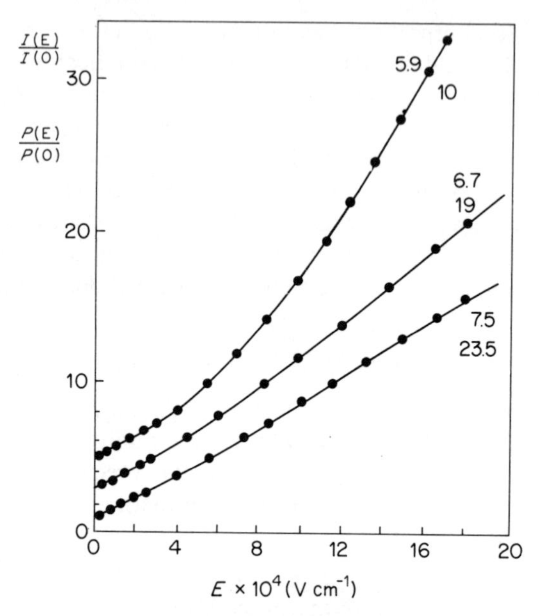

Fig. 5. Electric field dependence of photocurrent ratio $I(E)/I(0)$ at various excitation energies for TMPD hexane solutions (points) compared with best-fit theoretical $P(E)/P(0)$ values for EXP geminate pair distribution function (lines). All curves actually intercept at unity on the Y axis but have been displaced from one another by two units for clarity. The upper number near each curve is the excitation energy (eV) and the lower number is the thermalization length parameter B(Å). Only some experimental points are shown for simplicity (From Ref. 22. Reproduced by permission of the American Institute of Physics.)

tunately, there is no available theory that would allow correct predictions of the effect.

The more reliable information derived at this stage from field experiments seems to concern the dependence of electron thermalization lengths on the wavelength and the nature of a liquid. A marked effect of excitation energy and solvent on the functional form of electric field dependence of the photocurrent has been recently found for TMPD in various solvents.[20,22] Figure 5 demonstrates the effect for the photocurrent observed in the TMPD–n-hexane solution. In the conventional Onsager model, this data evidently indicates an alteration of the thermalized pair distribution function. Figure 6 summarizes B-parameter values obtained for the theoretical curves $P(E)/P(O)$ which fit to the $I(E)$ experimental data at assumption of the exponential distribution function. (For this function the B value equals the average distance between ions in an ion pair.)

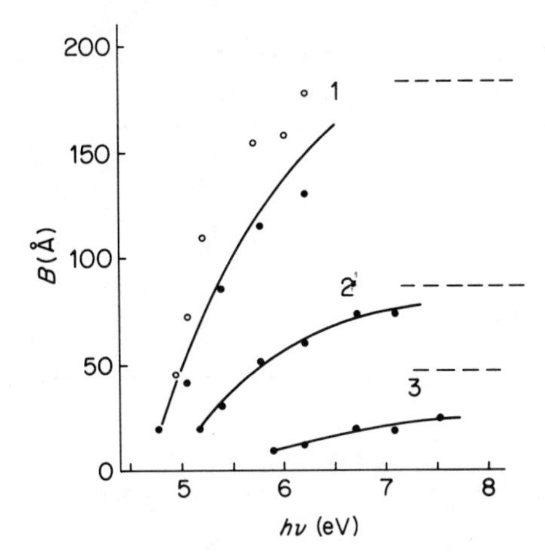

Fig. 6. Photoelectron thermalization length parameter *B* versus photon energy for a single-photon ionization of TMPD in different solvents: tetramethylsilane (1), 2,2,4-trimethylpentane (2), *n*-hexane (3). The *B* values are taken from Ref. 22 (●) and Ref. 29 (○). The dashed lines correspond to the electron ranges derived from radiation data;[85,86] in the case of tetramethylsilane (1) the *B* parameter was calculated to give experimental radiation yields.[85] (Reproduced by permission of the American Institute of Physics.)

In general, the thermalization lengths have been found to increase monotonically with photon energy. At any fixed excitation energy, they vary with the solvent, with the importance of the degree of sphericity of the solvent molecules being clearly seen. As mentioned, the increase in thermalization length with photon energy compares well with the 'ballistic' model of thermalization processes. Moreover, in accordance with a simple variant of the model, both the wavelength dependence of free ion quantum yield and the huge yield differences accompanying a change of the solvent appear to be almost completely the result of alterations of ion pair thermalization lengths rather than that of the quantum yields of ion pairs φ_0.[22]

It should also be emphasized that thermalization lengths for photoelectrons derived[17,22] from the photoconductivity data do not differ markedly from those for the secondary electrons produced by γ- and X-ray absorption in non-polar liquids[85,86] (see Fig. 6). Differences in the thermalization lengths might be expected to result from much larger average kinetic energies of electrons ejected by high-energy radiation. The fact that they are comparable has not surprised radiation chemists. In their interpretation of radiation data Mozumder and Magee[87] argued relatively long ago that the distance which an

TABLE I
Some excess electron properties and TMPD photoionization parameters in liquids at room temperature.

Liquid	μ_e[a] ($cm^2 V^{-1} s^{-1}$)	V_0[b] (eV)	I_{th}[c] (eV)	B(Å)[d] $h\nu - I_{th} \approx 1$ eV
n-Hexane	0.09	+0.02	4.92	10
Iso-octane (2, 2, 4-trimethylpentane)	7	−0.24	4.63	42
Tetramethylsilane	90	−0.51	4.29	75

[a]Ref. 78.
[b]The conduction band energy; Ref. 71.
[c]The photoionization threshold of TMPD; Ref. 13.
[d]The ion pair exponential distribution parameter derived from data of Ref. 22.

electron travels from its sibling positive ion before thermalization is controlled mainly by energy loss processes in the < 1 eV and lower range where the electron energy was relatively slowly degraded by excitation of vibrational and librational modes of molecules.

In addition, the thermalization lengths obtained for radiation-produced electrons as well as for photoelectrons are well correlated with the drift mobility of excess electrons in non-polar liquids: the more the mobility, the more the length at any fixed energy above the photoionization threshold (see also Section III. A and Table I). The nature of the correlation has been discussed for radiation data[83,88,89] on the basis of the supposition that the efficiency of electron-scattering processes determines both the electron motion during thermalization and its motion in the applied electric field.

These correlations between photo- and radiation-produced thermalized ion pairs give good support to the model in which formation of the pairs by photons proceeds via hot electron ejection into a liquid, since the mechanism is generally accepted for radiation conditions.

2. *Electric Field Fluorescence Quenching*

Bullot, Cordier and Gauthier[25,26] were the first to apply the method for studying photoionization in non-polar liquids. They found that, if an external electric field of sufficiently high strength is applied, a quenching of the fluorescence emitted by a dilute solute of a photoionization molecule is observed. This effect was observed earlier in organic solids[90] and has been mainly studied with TMPD in some hydrocarbon solvents.[25–29]

The suggested scheme of processes for describing the field effect is shown in Fig. 7. A primary excited state A_n^* either produces an ion pair with the probability φ_0 or, with the probability $1 - \varphi_0$, converts into the lowest

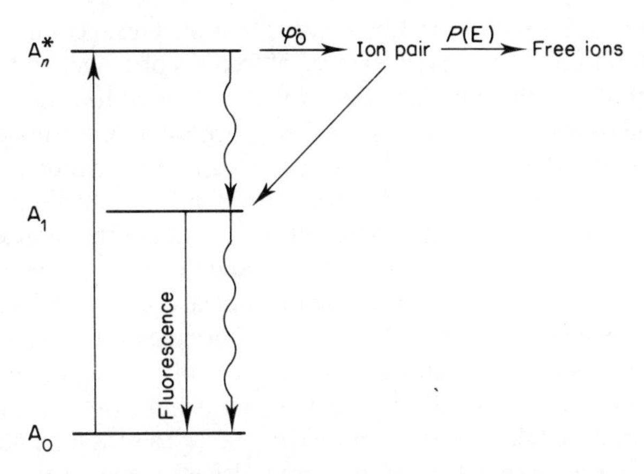

Fig. 7. Simple scheme for illustration of recombination fluorescence quenching by an applied electric field. The field raises the escape probability P leading to reduction of the fluorescence intensity. It is assumed [25–30] that electrons and positive ions which escape geminate recombination do not form singlets A_1 upon bimolecular volume recombination.

singlet excited state A_1 responsible for fluorescence. Practical fluorescence quantum yield independence of photon energy for TMPD solutions[32,34] and the theoretical analysis[91,92] allow us to conclude that, in this case, geminate recombination results in an A_1 state with probability equal to unity. In the model the quenching of the fluoresence intensity F by the applied field E is given by the equation[28]:

$$\frac{\Delta F}{F} = \frac{\varphi_{\text{fi}}(0)}{1-\varphi_{\text{fi}}(0)}\left[\frac{P(E)}{P(0)} - 1\right] = \frac{\varphi_{\text{fi}}(0)}{1-\varphi_{\text{fi}}(0)}\left[\frac{I(E)}{I(0)} - 1\right] \tag{14}$$

which connects the variation ΔF of the fluorescence intensity with that of the free ion quantum yield $\Delta\varphi_{\text{fi}}$. The linear relationship between the quantities predicted by Eq. (14) has in fact been observed[26,28] for all investigated TMPD solutions (tetramethylsilane; 2,2,4-trimethylpentane, dimethylpentane, cyclopentane). From the slope to intercept ratio of the $\Delta F/F$ versus $I(E)$ experimental curve, one can find the zero-field quantum yield $\varphi_{\text{fi}}(0)$ and, by using the experimental data on $\Delta F/F(E)$, determine the field dependence $P(E)/P(0)$.

It was on the basis of this method that Bullot, Cordier and Gauthier[28,29] first noted the increase of ion pair thermalization lengths with photon energy for non-polar liquids. The $\varphi_{\text{fi}}(0)$ values as well as thermalization lengths obtained by this method were found to be in good agreement with the photoconductivity data (see Fig. 6). The agreement gives strong confirmation of the scheme given in Fig. 7.

In concluding this section, it can be noted that the Onsager model provides an interpretation of all observed features of free ion photogeneration in non-polar liquids. If we rely only on it, however, it is difficult to conclude that the fundamental assumptions of Onsager's theory are valid for the majority of ion pairs. The point is that the possibility of quantitative verification of the theory by experiments dealing only with free ions is principally limited for two reasons: (1) there is a lack of knowledge about the initial thermalized ion pair distribution; (2) if the distribution is really broad, most ion pairs (with small thermalization lengths) do not contribute to the measured signal (see Section IV. E). The latter concerns not only photoconductivity methods but also the field-dependent fluorescence quenching method, since only a change of fluorescence due to a change in escape efficiency is informative in this method.

It should be noted that the free ion yield characterizes the ion pair efficiency of escape which, according to the Onsager model, is independent of the nature of charge particles and their mobility. Photoionization study by means of other methods, described below, requires a knowledge of properties of real states for ion pairs in liquids.

III. THE NATURE OF THERMALIZED ION PAIRS

As mentioned above, Onsager's theory deals with thermalized pairs of ions which undergo mutual diffusion. Usual application of the theory implies the existence of an initial ion pair produced just after charge particle thermalization, which starts the diffusion motion. In other words, it is assumed that thermalization of charge particles is instantaneous and the ion pairs do not geminately recombine and are not affected by any applied field before thermalization. These assumptions simplify interpretation of ionization processes, since in this case initial ion pair distribution is expected to be isotropic in the applied field and the quantum yield of a thermalized ion pair is associated with that of ionization.

The question as to whether this picture is true of photoionization in liquids is closely connected with the nature of charge particles produced by ionization in liquids and with the dynamics of their thermalization. In the most investigated case where photoionization of a solute molecule A occurs, a pair consisting of a cation A^+ and excess electron e^- is produced. Behaviour of the pair is mainly determined by the motion of excess electrons as these are the most mobile particles.

A. Quasi-Free and Localized Electrons

In non-polar liquids as well as in polar liquids and amorphous solids the energetics and time dependence of many electron transfer processes require that the electron–medium interactions play an important role. In a wide range of liquids, strong local interactions lead to electron localization in a group of molecules. In frozen non-polar glasses such localized or trapped electrons, e^-_{tr}, have been observed by the optical absorption and ESR methods.[5] In some hydrocarbon liquids the similar optical absorption band of excess electrons is registered on the nanosecond[74–77] or picosecond[95,96] timescale, indicating that the electron spends some time in the localized state. Another confirmation of the excess electron localized states in the molecular liquids is given by observations of the electron photoexcitation effect on the ion pair escape probability[43–46] and the photoconductivity induced by the photoexcitation of trapped electrons.[97–99]

Prior to localization, the electron states are referred to as quasi-free e^-_{qf} or extended states. The e^-_{qf} state for thermalized excess electrons has been suggested to explain the high-drift mobility of electrons, μ_e, compared to that of ions, observed at first for rare gas liquids. For non-polar molecular liquids the first observation of the high mobility of electrons was made independently by several groups in 1969.[100–104] Since then a large amount of experimental data concerning excess electron transport in non-polar liquids has been gathered.[78,105,106]

One of the surprising features of excess electron mobility in non-polar liquids is its large variation with structure of molecules composing the liquids. This dependence is very apparent in liquid hydrocarbons where mobilities range from $70\,cm^2\,V^{-1}s^{-1}$ for neopentane to $1,3 \times 10^{-2}\,cm^2\,V^{-1}s^{-1}$ for *trans*-decaline at room temperature. Large differences have also been found for the temperature and field dependencies of mobility in liquids which would seem to exhibit very similar physical properties. No uniform theory exists which would describe all the data although some models have been proposed and recently discussed.[3,78]

Without going into detail, three types of liquid can be distinguished, depending on electron mobility; they are referred to as 'high', 'middle' and 'low' mobility liquids:

1. The higher ($\gtrsim 10^2\,cm^2\,V^{-1}s^{-1}$) values of μ_e have been associated with e^-_{qf} transport limited by scattering processes. Hall mobility measurements in tetramethylsilane, carried out recently [107], confirm the model.
2. The lower values of μ_e ($0.1 < \mu_e < 10^2\,cm^2\,V^{-1}s^{-1}$) are thought to be due to repetitive e^-_{qf} trapping by shallow traps. In the simplest case of the model, the experimentally measured average mobility can be presented as $\mu_e = \mu_0 \tau_{loc}(\tau_{loc} + \tau_{tr})^{-1}$, where τ_{loc} and τ_{tr} are the lifetimes of electrons in quasi-free and localized states and μ_0 is the e^-_{qf} mobility, assumed to be far larger than for e^-_{tr} electrons.
3. In low mobility liquids ($< 0.1\,cm^2\,V^{-1}s^{-1}$) the excess electrons appear to spend most of their lifetime in localized states, and the transport can be described either by a hopping process or a trap-controlled e^-_{qf} motion.

Table I demonstrates some excess electron properties and TMPD photoionization parameters in commonly used laboratory solvents which represent these three types of non-polar liquid.

In view of the above, the simple scheme (1) of geminate recombination should be replaced by the following scheme:

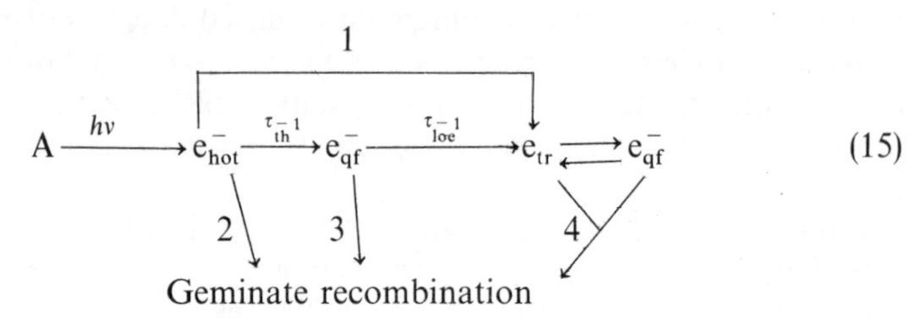

(15)

which includes two states of a thermalized excess electron: e^-_{qf} and e^-_{tr}. Moreover, the absence of any reliable ratios between thermalization time, τ_{th}, localization time, τ_{loc}, and the time of geminate recombination, τ_{gem}, does not allow us a priori to exclude from the scheme the following three processes: (1) localization of a hot electron e^-_{hot} (i.e. a quasi-free electron with the excess kinetic energy above thermal) prior to thermalization, (2) geminate recombination of e^-_{hot} prior to thermalization and (3) geminate recombination of e^-_{qf} before its localization.

Radiation chemists usually describe geminate recombination processes in non-polar liquids using the average values of excess electron mobility (i.e. they take into account only process 4 of scheme 15) and assume that the thermalized ion pair yield equals the gas-phase free ion yield or the yield of ions which may be scavenged by acceptor molecules at a high concentration. It implies the following assumptions:

$$\tau_{th} \ll \tau_{gem} \tag{16}$$

$$\tau_{loc} \ll \tau_{gem} \tag{17}$$

which remove the role of processes 1, 2 and 3.

However, theoretical estimates of τ_{th} reported recently[3] show that the thermalization process can in fact take an appreciable time. The very long times of nanoseconds required, according to the estimates, for electron thermalization in the rare gas liquids have recently been confirmed by experiment.[108] The thermalization time decreases substantially as the complexity of molecules increases, but for liquid neopentane and tetramethylsilane appear to be about 1 ps.[3] The time is comparable or even greater than the time of geminate recombination ($\tau_{gem} \approx 10^{-13}$ s, if $\tau_{gem} = \varepsilon B^3/3e\mu_0$, $B \approx 100$ Å and $\mu_0 \approx 10^2\ \mathrm{cm^2\,V^{-1}s^{-1}}$; see also Section IV.B). This speaks against the cancel process 2 of scheme (15).

Electron localization preceding complete degradation of the excess kinetic energy, i.e. process 1, appears to occur in electron-attaching liquids such as

CCl_4. For electron-non-attaching liquids like hydrocarbons this process was originally suggested by Freeman.[109,110] Recently, the assumption has been supported by measurements of photoconductivity induced by photo-excitation of trapped electrons in low mobility hydrocarbon liquids.[97,99]

Of course, when processes 1, 2 and 3 of scheme (15) are included the kinetics analysis of geminate recombination becomes very complicated. As will be discussed in Section IV.A, one of the important parameters required for a kinetics description is the product of mobility and localization time for quasi-free electrons. As discussed below, measurements of photoconductivity due to photoexcitation of e_{tr}^- electrons in liquids seem to enable the value to be estimated for low and middle mobility liquids.

B. Electron Localization Study

It is not our intent to review data on the electron state evolution from the quasi-free state to the fully equilibrate solvated state but rather only those which are necessary to analyse the kinetics of geminate recombination in the simple scheme (15). In the scheme with one trap level, it is important to know the time of initial localization of e_{qf}^- and e_{hot}^-. In non-polar liquids the dynamics of the localization is thought to be too fast to be investigated by time-resolved spectroscopy. Recent picosecond measurements[95,96] have shown that infrared absorption associated with electron localization by a preexisting trap or configuration fluctuation occurs practically immediately after electron ejection into liquids. The other approach to experimental study of electron localization can be based on transient photoconductivity measurements.

It can be expected that optical absorption of e_{tr}^- electrons at wavelengths $\lesssim 1000$ nm in hydrocarbon liquids, like that supposed in glasses[111–115], results in generation of quasi-free electrons which are then localized again after or before complete thermalization, i.e.

$$h\nu + e_{tr}^- \longrightarrow e_{hot}^- \longrightarrow e_{qf}^- \xrightarrow{\tau_{loc}^{-1}} e_{tr}^- \quad (18)$$

If it is the case, measurements of conductivity caused by photo-detrapping of localized electrons provide direct information on the extent of electron drift in the applied field during the time τ' of the photoinduced transition from one trap to an other.

Such a photoconductivity has been extensively investigated[113–115] in low-temperature hydrocarbon glasses where e_{tr}^- electrons are the long-lived particles and stationary techniques can be used. In hydrocarbon liquids, measurements of the conductivity are hindered by a short lifetime of e_{tr}^-, high average excess electron mobility and photoionization of anions produced by electron scavenging. Only recently, the measurements have been attempted for some liquids at low temperatures.[97–99] Trapped electrons were produced in

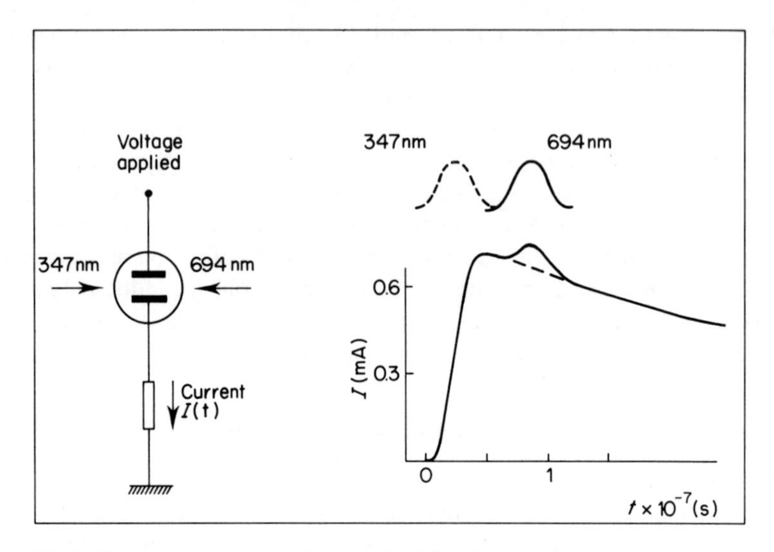

Fig. 8. The measurements of photoconductivity due to photoionization of localized electrons e_{tr}^- in liquid 2,2,4-trimethylpentane.[98] The electrons were produced by 347-nm laser pulse ionization of anthracene solute and excited by the 694-nm laser pulse, the delay time between the pulses being 60 ns. The second pulse action is seen to result in an additional spike of current attributed to ionization of e_{tr}^-.

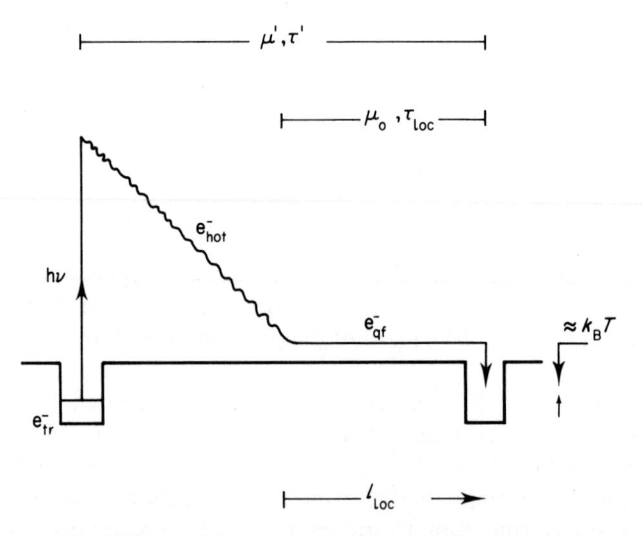

Fig. 9. The mechanism of photoassisted electron transition from one trap to another, discussed in the text where designations are given.

the liquids either by high-energy radiation[97] or by two-photon ionization of solute aromatic molecules.[98,99] For example, Fig. 8 shows the transient current induced by two consecutive laser pulses (347 nm and then 694 nm) in anthracene solution in liquid 2, 2, 4-trimethylpentane. The second, 694-nm light pulse is seen to result in a photocurrent spike with a much shorter decay time than that for excess electrons. The spike was associated[98] with the current caused by photoionization of e_{tr}^-.

Provided that both the photoionization cross-section σ and concentration of trapped electrons are known, the integral of the photocurrent allows an estimation of the product $\mu'\tau'$ to be made, where μ' is the photoliberated electron mobility averaged over the time τ' (see Fig. 9). Assuming that the quantum yield of the photoionization of e_{tr}^- in hydrocarbon liquids, like that in glasses[115], equals unity, σ has been given[97-99] the known value of the absorption cross-section for e_{tr}^-. The concentration of e_{tr}^- was derived from the electron current just before the second light pulse using a known average mobility μ_e of excess electrons. The $\mu'\tau'$ values obtained in this way for some liquids are summarized in Table II. For methylcyclohexane and *n*-hexane the values are seen to be much smaller than those for 2, 2, 3-trimethylpentane. This is consistent with current thinking that branching of the C_n chain in the liquid molecule leads to some decrease of the extent of electron localization.

It should be emphasized that $\mu'\tau'$ values obtained relate to quasi-free electrons with the initial kinetic energy above the thermal. For example, at $\lambda = 700$ nm the energy is $\varepsilon_0 = h\nu - E'_{tr} \approx 1.3$ eV $\gg k_B T$, where the optical depth of the electron trap is taken to be $E'_{tr} \approx 0.5$ eV, like that for hydrocarbon glasses.[115] Therefore, the measured values of $\mu'\tau'$ can be presented as the sum of electron

TABLE II

Quasi-free electron drift displacements before localization measured, $\mu'\tau'$, and estimated, $\mu_{hot}\tau_{th}$, and localization length, l_{loc}, estimated for hydrocarbon liquids.

Liquid	T(K)	$\mu'\tau'$ ($cm^2 V^{-1}$)	$\mu_{hot}\tau_{th}$ [a] ($cm^2 V^{-1}$)	l_{loc} [b] (Å)	μ_e ($cm^2 V^{-1} s^{-1}$)
Isooctane	170	3.8×10^{-11} [c]	2.7×10^{-12}	180	1.1 [d]
Mixture of isooctane and *n*-hexane (hexane mole fraction 0.38)	185	0.6×10^{-11} [e]	—	76	5.4×10^{-2} [e]
Methylcyclohexane	179	4.7×10^{-13} [e]	—	21	5.5×10^{-4} [f]
n-Hexane	180	$(4 \pm 2) \times 10^{-13}$ [g]	2.8×10^{-13}	16	1×10^{-3} [d]

[a] Obtained from Eq. (21) using $G = 90$ Å and $G = 30$ Å for isooctane and *n*-hexane respectively, these G values were taken from Ref. 22 for photon energies 5.7 eV (isooctane) and 5.9 eV (*n*-hexane).
[b] The upper limit of localization length given by Eq. (20) with experimental $\mu'\tau'$ values.
[c] Ref. 98.
[d] Ref. 78.
[e] Ref. 99.
[f] Ref. 52.
[g] Ref. 97.

drift displacements before and after thermalization:

$$\mu'\tau' = \mu_{hot}\tau_{th} + \mu_0\tau_{loc} \tag{19}$$

where μ_{hot} is the hot electron mobility averaged for the time τ_{th}, and μ_0 and τ_{loc} are the mobility and the localization time for thermalized quasi-free electrons. The obtained data on $\mu'\tau'$ allow an estimate to be made of the diffusive length, l_{loc}, that e_{qf}^- travels before localization:

$$l_{loc} = \sqrt{\frac{6k_B T}{e}\mu_0\tau_{loc}} \le \sqrt{\frac{6k_B T}{e}\mu'\tau'} \tag{20}$$

For liquid *n*-hexane it was found that $l_{loc} < 16$ Å (see Table II). Such value of l_{loc} seems to be too small for thermalized quasi-free electrons. It is usually assumed[116] that the mean free path for momentum transfer λ cannot be smaller than κ^{-1}, where κ is the wavenumber of electrons in a conduction band. For thermalized quasi-free electrons with an effective mass equal to the electron mass, $\kappa = (15\text{ Å})^{-1}$ at 180 K. Since a diffusive length of e_{qf}^- before localization is expected to be greater than λ, it is doubtful whether $l_{loc} < 16$ Å for thermalized quasi-free electrons. Thus Balakin and Yakovlev[97] have concluded that localization of hot electrons in low electron mobility liquids occurs prior to the complete thermalization.

This conclusion[97] is confirmed by the estimate of hot electron drift displacements based on the supposition advanced by Mozumder and Magee[117] that the e_{hot}^- electron with the initial energy $\varepsilon_0 \lesssim 0.4$ eV (in a 'subvibration' range) loses its energy by small portions and suffers a large number N of scattering events with momentum transfer before thermalization. In this case, the distribution function of electron thermalization distances is expected to be of Gaussian form, and $\mu_{hot}\tau_{th}$ is given by the sum [97] of the electron drift displacements

$$\mu_{hot}\tau_{th} = \frac{1}{2}\frac{e}{m}\sum_{i=1}^{N}\left(\frac{\lambda}{v_i}\right)^2 = \frac{3}{8}\frac{eG^2}{\bar{\varepsilon}} \tag{21}$$

where v_i and $\varepsilon_i = mv_i^2/2$ are the average velocity of the electron and its energy corresponding to the *i*th scattering event, λ is the effective mean free path for momentum transfer assumed to be independent of ε_i, $G = \lambda\sqrt{\frac{2}{3}N}$ is the thermalization length parameter for Gaussian distribution and $\bar{\varepsilon}^{-1} = N^{-1}\sum_{i=1}^{N}\varepsilon_i^{-1}$ is the average reciprocal energy. If we suppose that energy loss per scattering act is independent of ε_i, we obtain $\bar{\varepsilon}^{-1} = \varepsilon_0^{-1}\ln(\varepsilon_0/k_B T) \approx (0.12\text{ eV})^{-1}$ for $\varepsilon_0 = 0.4$ eV and $T = 180$ K. For this value $\bar{\varepsilon}^{-1}$ and $G \approx 30$ Å derived[22] from TMPD—*n*-hexane photoionization data at $h\nu - I_{th} \approx 1$ eV, Eq. (21) leads to $\mu_{hot}\tau_{th} \approx 3 \times 10^{-13}\text{ cm}^2\text{ V}^{-1}$. The fairly good agreement of the estimate with experimental values listed in Table II for low mobility liquids (*n*-hexane and methylcyclohexane) suggests that hot electrons in the liquids are mainly localized before or just after thermalization. This appears to be not the case for 'middle mobility' liquids like isooctane. The experimental value $\mu'\tau'$ for isooctane much exceeds (~ 10 times) the estimate of $\mu_{hot}\tau_{th}$ and does not lead to difficulties connected with the thermalized quasi-free electron free path, as it does for *n*-hexane. For this reason, the value $\mu'\tau'$ obtained for isooctane has been mainly attributed[98] to thermalized quasi-free electron motion with $\mu_0\tau_{loc} \approx \mu'\tau' \approx 4 \times 10^{-11}\,cm^2\,V^{-1}$. According to Eq. (20) the $\mu_0\tau_{loc}$ value corresponds to $l_{loc} = 180\,Å$. This rather large length is consistent with the trapping model usually considered for 'middle mobility' liquids.

Results obtained for mean drift displacements of hot electrons before localization make very doubtful inequalities (16) and (17), especially for photoionization at photon energies near the ionization threshold. For example, a crude estimate of the probability P_{loc} that hot electrons produced by TMPD ionization in liquid *n*-hexane with photon energy $h\nu = 5.9$ eV will be localized prior to geminate recombination gives $P_{loc} \approx 0.3$ for the obtained values $\mu'\tau' \approx 4 \times 10^{-13}\,cm^2\,V^{-1}$ and the thermalized length parameter $G = 30\,Å$[22] if $P_{loc} \simeq \tau_{gem}/\tau'$, where $\tau_{gem} = \varepsilon G^3/3e\mu'$. In the case of TMPD in liquid isooctane, the estimate leads to $P_{loc} \approx 0.02$ for $\mu'\tau' = 4 \times 10^{-11}\,cm^2\,V^{-1}\,s^{-1}$ and $G = 66\,Å$[22] at $h\nu = 5.4$ eV.

Two conclusions important for the Onsager model can be drawn if the time of quasi-free electron localization is actually larger than that of geminate recombination and if hot electron drift displacement in the applied field is appreciable. Firstly, in the Onsager model the geminate recombination dynamics for the majority of charge pairs produced by ionization can not be considered by using the average excess electron mobility which determines only the rate of process 4 but not process 2 or 3 in scheme 15. This problem will be discussed in Section IV.A. Secondly, a relatively small external electric field can destroy the spherical symmetry of the initial distribution of thermalized ion pairs. This effect discussed at first by Silinsh and coworkers [6] for molecular crystals will now be considered.

C. Whether the Separation Distance Distribution is Isotropic

The drift displacement of hot electrons during thermalization in the applied field can be one possible reason leading to anisotropy of the initial thermalized charge pair in non-polar liquids and appears to be important for the field dependence of escape probability. For low mobility liquids the displacement $\Delta = \mu_{hot}\tau_{th}E$ is expected to be relatively small. Even for the external field strength $E = 10^5\,V\,cm^{-1}$, $\mu_{hot}\tau_{th}E$ equals 4 Å at the assumed value of $\mu_{hot}\tau_{th} = 4 \times 10^{-13}\,cm^2V^{-1}$. Therefore, for rude estimation the field-dependent distribution function $f^*(r, \theta, E)$ for thermalized pairs may be presented as

$$f^*(r, \theta, E) = f(\sqrt{r^2 + \Delta^2 - 2\Delta r \cos\theta}) \approx f(r) - \beta(r)E\cos\theta \qquad (22)$$

where $f(r)$ is the distribution function at $E = 0$, $\beta(r) = \mu_{hot}\tau_{th}[df(r)/dr]$ and θ is the orientation of the pair with respect to the applied field E.

The original Onsager formula[55] for the probability $\phi(r, \theta, E)$ that an isolated pair escapes geminate recombination enables the average escape probability

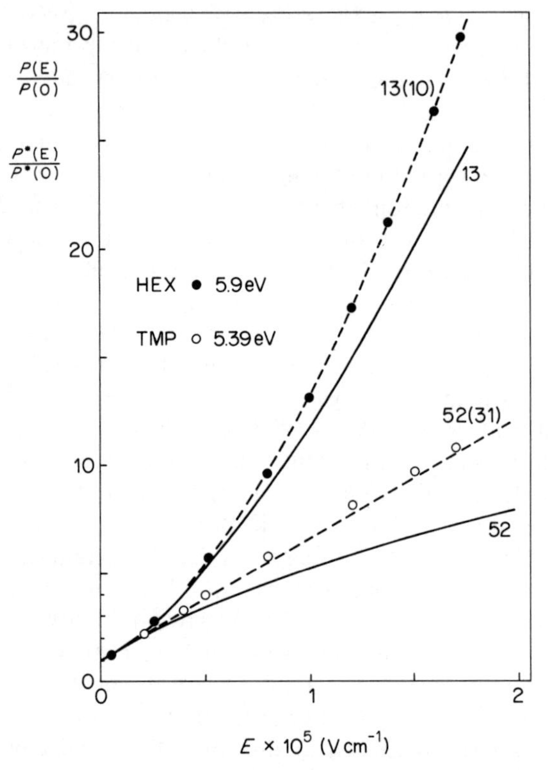

Fig. 10. The effect of field-induced distribution function anisotropy on the best-fit thermalization lenghts derived from photoconductivity data. Electric field dependence of theoretical values of $P(E)/P(0)$ for the isotrop EXP distribution function $f(r)$ (solid lines) and values of $P^*(E)/P^*(0)$ for the field-dependent function (22) (dashed lines) are compared with the photocurrent ratio $I(E)/I(0)$ (points) for TMPD solution in *n*-hexane (HEX) and 2,2,4-trimethylpentane (TMP) at photon energies shown.

Experimental points are from Refs. 20 and 22. For every solvent both $P^*(E)/P^*(0)$ and $P(E)/P(0)$ curves were obtained at the same thermalization length parameter B(Å), shown as the number near the curve; the number in brackets near the dashed lines is the parameter B(Å) for the allowed field-independent EXP function which fits the same experimental data.[22]

The $P^*(E)/P^*(0)$ curves have been calculated[118] using Eq. (23) and $\mu_{hot}\tau_{th}$ values from Table II.

$P^*(E)$ to be obtained for the $f^*(r, \theta, E)$ distribution function:

$$P^*(E) = P(E) - E \int_0^\infty \mathrm{d}r\, r^2 \beta(r) \int_0^\pi 2\pi \cos\theta\, \phi(r, \theta, E) \sin\theta\, \mathrm{d}\theta \qquad (23)$$

where $P(E)$ is the escape probability usually determined for the field independent function $f(r)$. The following results can be obtained[118] for the distribution functional form (22) using Eq. (23).

In the low-field limit, $P^*(E)$ increases linearly with E and has the same slope to intercept ratio as that for field-independent isotropic distribution, i.e.

$$\frac{P^*(E)}{P^*(0)} = 1 + \frac{e r_c E}{2 k_B T}$$

because the correction caused by the distribution anisotropy is quadratically dependent on E.

Another expected result is that the ratio $P^*(E)/P^*(0)$ increases with E more rapidly than $P(E)/P(0)$. Therefore, the comparison of experimental photocurrent ratios $I(E)/I(0)$ or of free ion quantum yield ratios $\varphi_{fi}(E)/\varphi_{fi}(0)$, with theoretical values $P^*(E)/P^*(0)$, has to give a more broad best-fit function $f(r)$ than that for $P(E)/P(0)$ values. This conclusion is demonstrated by Fig. 10 which compares photoconductivity data for TMPD solutions in n-hexane and 2, 2, 4-trimethylpentane, with theoretical values calculated by using Eq. (23) and expected $\mu_{hot}\tau_{th}$ values. The best-fit thermalization parameters B for the function form (22) are seen to be really larger than those obtained[22] for the field-independent function: 13 Å instead of 10 Å for n-hexane at 5.9 eV and 52 Å instead of 31 Å for 2, 2, 4-trimethylpentane at 5.3 eV. In the case of n-hexane, the difference is not too large. Nevertheless, for $B = 13$ Å the escape probability in the zero field is a factor of 3 greater than that for $B = 10$ Å. It is perhaps worth pointing out that the quantum yields of ion pair formation φ_0 obtained[22] by using the field-independent exponential function for the same system exceed unity by a factor of ~ 3 at some wavelengths. Since φ_0 values have been derived as $\varphi_0 = \varphi_{fi}/P$, the field effect appears to provide a way out of the impasse.

Finally, let us come back to the slope of the dependence of the free ion quantum yield ratio $\varphi_{fi}(E)/\varphi_{fi}(0)$ on the field E at low E, which has been shown to conform very well to the Onsager prediction. Concerning the question posed above as to whether diffusive motion governs the fate of a geminate pair in early stages of its life, it is of interest to note that in the low-field limit the right-hand side of Eq. (22) is expected to be a rather general function form for ion pair distance distribution and to be independent of the mechanism of charge particle motion. Let us grant then for the moment that the diffusion description of thermalized ion pair behaviour is good only for times $t > t^*$ after pair generation at $t = 0$, where t^* is less than the ion pair separation time. But is enough large the most of pairs being recombined by the moment t^*. If in the low field the ion pair separation distance distribution at the moment t^* has the form of Eq. (22), the field dependence of escape probability has also to be the same as in the Onsager model. This indicates that, at least in this case, the slope to intercept value for dependence $\varphi_{fi}(E)$ is insensitive to ion pair behaviour in the early stages of geminate recombination. The ion pairs which survive these stages can be only a small part of all the initial pairs and have a separation distribution essentially more broad than the initial pairs.

The case considered above can explain the surprisingly good quantitative agreement of Onsager's theory with experimental data on the electric-field effect on the free ion yield for those non-polar liquids where quasi-free electrons are expected to compose initial thermalized pairs of charges (i.e. in 'middle mobility' and high mobility liquids). In these liquids an electron is unlikely can be described like a point charge, as assumed in Onsager's theory. The assumption is especially important for early stages of geminate recombination where correlated charge particles are expected to be separated mainly by a small distance r. The point is that the Onsager theory requires the charge particle size and free path to be less than the distance r' corresponding to the Coulomb energy change equal to $k_B T$, with r' increasing with r as $r' = r^2/r_c$. For example, $r' = 12$ Å for $r = 60$ Å and $r_c = 290$ Å. This means that the theory is not applicable to thermalized quasi-free electrons at $r < 60$ Å because their size is not less than $\lambda_D/2\pi \sim 10$ Å where λ_D is the de Broglie wavelength. Consequently, the Onsager theory is not applicable for > 90 per cent. of initial ion pairs produced by photoionization of TMPD in liquid isooctane if the ion pair distribution has exponential form with the length[22] parameter of 20 Å at $h\nu = 5.17$ eV.

It is clear that the situation will be less dramatic for later stages of geminate recombination when an ion pair distribution becomes broader.

IV. THE DYNAMICS OF GEMINATE RECOMBINATION

The Onsager calculation has since been and remains extensively applied to photoionization phenomena in non-polar liquids. However, very little is known about the time evolution of the neutralization process of geminate pairs produced by photoionization in these liquids. Data on the geminate recombination kinetics would help to elucidate the time and ion pair separation distance limits for an adequate description of ionization processes using the diffusion equation approach. Moreover, the data would provide some additional information on the initial distribution of ion pairs and their nature. Unfortunately, the geminate recombination is too fast to experimentally study the whole process using direct methods.

In view of current thinking about two states of excess electrons in non-polar liquids (i.e. e_{qf}^- and e_{tr}^-) and the conclusions of the preceding section about the possibility of quasi-free electrons geminately recombining before localization (or even before complete thermalization), the ion pair time evolution can be divided into three stages distinguished by characteristic times:

1. Separation on free ions
2. Geminate recombination after electron localization, i.e. process 4 in scheme (15)
3. Geminate recombination without preceding electron localization, i.e. processes 2 and 3 in scheme (15)

The first is the slowest stage. Although, as discussed by Warman,[3] the actual time taken to become a free ion is a rather difficult parameter to define, it can be approximately estimated as $\tau_{esc} = r_c^2/D$ for the diffusion model, where D is the sum of the diffusion coefficients of ions. For example, $\tau_{esc} \approx 3 \times 10^{-9}$ s in a liquid of dielectric constant 2 at room temperature and $D = 2 \times 10^{-3}\ \mathrm{cm^2\,s^{-1}}$, as in n-hexane liquid.

Even if an electron suffers localization in a liquid, its geminate recombination should be expected to be extremely fast. If $r_0' = 50$ Å is the separation distance of a cation-trapped electron pair and the average excess electron mobility $\mu_e = 0.1\ \mathrm{cm^2\,V^{-1}\,s^{-1}}$ (as for n-hexane at room temperature), the crude estimate of geminate recombination time gives $\tau_{gem} = \varepsilon r_0^3/3e\mu_e = 6 \times 10^{-12}$ s. This oversimplified estimate agrees well with recent measurements of the recombination kinetics of geminate cation-trapped electron pairs generated in liquid n-hexane by picosecond laser.[64] Using similar estimates, one can conclude that trapped electron geminate recombination in liquids with electron mobility of or greater than $1\ \mathrm{cm^2\,V^{-1}\,s^{-1}}$ (as in isooctane) as well as quasi-free electron geminate recombination should be in a subpicosecond range and therefore inaccessible for direct observation. Studies of these processes using some indirect methods are rather ambiguous.

A starting point for recent theoretical analyses of the time evolution of thermalized ion pairs has been the Onsager model. Unfortunately, there is no available theory that would describe the kinetics of geminate recombination before electron thermalization.

A. The Time-Dependent Onsager Problem

The Smoluchowski equation, with a Coulomb potential, external electric field and a term, $K_s[S]n$, representing charge particle scavenging by acceptors with a uniform concentration $[S]$, is used as a mathematical basis for consideration of the geminate ion pair recombination kinetics in the Onsager model:

$$\frac{\partial n}{\partial t} = -\operatorname{div}\mathscr{T} - K_s[S]n \tag{24}$$

$$\mathscr{T} = -\mathscr{D}\operatorname{grad} n + \mu(\mathbf{E}_c + \mathbf{E})n$$

where $n(r, \theta, t)$ is the probability density that a pair of ions has a separation distance r and an orientation θ with respect to the applied field $\mathbf{E}$, D and μ are the summary diffusion coefficient and the mobility of the ions respectively, K_s is the scavenging rate constant and $\mathbf{E}_c$ is the Coulomb field.

Recently, the time-dependent solution of Eq. (24) corresponding to a general boundary condition at the origin has been discussed in detail by Hong and Noolandy in a series of works[119–123] which can be viewed as complementary to the numerical methods developed earlier by Freed and Pederson[124] and by Abell, Mozumder and Magee.[125]

In general, the Smoluchowski equation allows only a numerical solution. An interesting point is that, for long times, the time dependence of the total fraction of ion pairs still presented at time t (or the survival probability $W(t)$) can be given by the same functional form independently of the initial distribution function and the escaped ion fraction $W(\infty)$. The form recently worked out by Van den Ende and coworkers[126] is

$$\frac{W(t)}{W(\infty)} = 1 + \frac{0.6}{\tau^{0.6}} \tag{25}$$

where $\tau = tD/r_c^2$, for $W(t) < 4W(\infty)$ if $W(\infty) < 0.2$. In the limit of infinite time the empirical relationship (25) agrees with a reciprocal square root dependence of $W(t)$ on time, found in early treatments by Mozumder:[127,128]

$$\frac{W(t)}{W(\infty)} = 1 + \left(\frac{r_c^2}{\pi D t}\right)^{1/2} \tag{26}$$

The long time prediction of diffusion model compares well with extensive data for ionization by high-energy radiation in non-polar liquids. The data summarized by Hummel[2] and Warman[3] have been obtained by means of various measurements: optical and microwave absorption of ions and localized electrons, and the scavenging yield of electrons or positive ions.

It should be noted that Eq. (24) describes the time dependence of the separation distance distribution for pairs of charge particles with time-independent mobilities. This is the very case which has since been considered for the kinetics of geminate recombination in non-polar liquids. However, if we suppose e_{qf}^- and e_{tr}^- states for thermalized electrons in liquids, we must modify the equation. In general, this produces large additional mathematical difficulties when describing kinetics as an effective mobility becomes time dependent.

To simplify the problem we can assume that e_{qf}^- localization is very fast with respect to geminate recombination of e_{tr}^- electrons, and after the first localization an electron moves with an average excess electron mobility μ_e. In this case, a distribution of electron localization lengths $f_{tr}(r)$ can be used as an initial distribution for cation-electron pairs described by Eq. (24) with $\mu = \mu_e$, provided that cation mobility is negligibly small. As the starting point for estimates of $f_{tr}(r)$ in the framework of the Onsager model, we can assume the e_{qf}^- motion to be described by Eq. (24) with a mobility, μ_0, of e_{qf}^- electrons and consider electron localization as scavenging these electrons by traps with $K_s[S] = \tau_{loc}^{-1}$.

This approach enables us to connect the initial distribution of thermalized ion pairs, $f(r)$, involved in an interpretation of free ion generation in the Onsager model, with the distribution function, $f_{tr}(r)$, for initially localized electrons, using $\mu_0\tau_{loc}$ values derived from photoconductivity data (see Section III.B). Indeed, $f_{tr}(r)$ is given by

$$f_{tr}(r) = \int_0^\infty K_s[S]n_s(r,t)\mathrm{d}t = \int_0^\infty K_s[S]n(r,t)\mathrm{e}^{-K_s[S]t}\mathrm{d}t \tag{27}$$

where $n_s(r,t)$ is the solution of Eq. (24), which, as pointed out by Mozumder,[127] is directly related to the solution $n(r,t)$ without a scavenger by $n(r,t) = n_s(r,t)\exp(+K_s[S]t)$. According to Eqs. (24) and (27), $f_{tr}(r)$ is determined by the ratio $\mu/K_s[S]$ equaled to $\mu_0\tau_{loc}$ in the model used.

The numerical solution of the problem [61] has shown that the functional form of $f_{tr}(r)$ can be very different from that of $f(r)$ and, consequently, the electron localization should be taken into account if we want to derive information about $f(r)$ from experimental kinetic data on geminate recombination. The result is demonstrated by Fig. 11 which shows the dependence $F_{tr}(r) = 4\pi r^2 f_{tr}(r)$ obtained [61] using the experimental value of $\mu_0\tau_{loc}$ and the exponential form of $f(r)$ suggested from photoconductivity data for TMPD in isooctane. It can be seen that there are 'burning' ion pairs with small distances

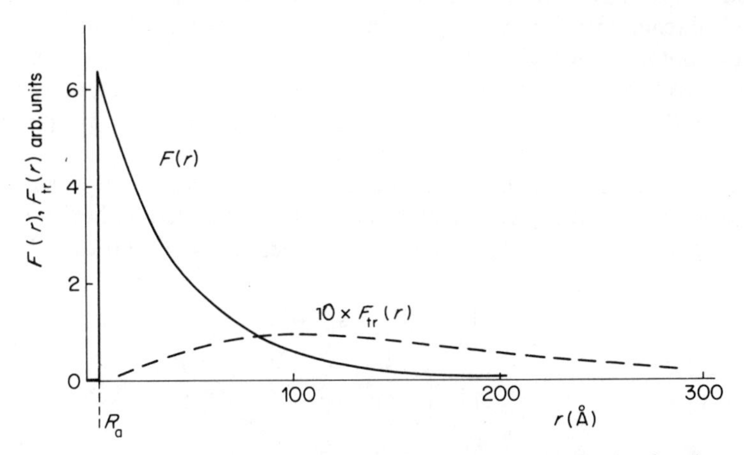

Fig. 11. Demonstration of the difference between forms of distribution functions for initial thermalized ion pairs $F(r) = 4\pi r^2 f(r)$ and trapped electron-cation pairs $F_{tr}(r) = 4\pi r^2 f_{tr}(r)$. The EXP form of $f(r)$ was taken with $B = 40$ Å, like that assumed[22] for TMPD photoionization in isooctane at $h\nu = 5.5$ eV. Function $f_{tr}(r)$ was obtained from Eq. (27) by numerical solution of the Smoluchowski equation (24) with the initial condition $n(r,0) = f(r)$ and parameters: $\mu = \mu_0$, $K_s[S] = \tau_{loc}^{-1}$ and $\mu_0\tau_{loc} = 4 \times 10^{-11}\,\mathrm{cm^2\,V^{-1}}$ given by Table II for isooctane. (Reproduced by permission of the American Institute of Physics.)

during electron localization and broadening of the initial distribution of cation-trapped electrons relative to that for thermalized pairs.

Many research groups have attempted to investigate the fast stages of geminate recombination in non-polar liquids. For trapped electron-cation geminate recombination, the available direct time-resolved data were obtained by means of optical absorption measurements[76,77] only for ionization by high-energy radiation in low mobility liquids, where the first half-life time, $\tau_{1/2}$, of ion pairs can turn out to be in the microsecond region of times at low temperatures. Using pulsed electron beam excitation, evidence has recently been obtained[130] that the $\tau_{1/2}$ value of such pairs is less than 100 ps in liquid *n*-hexane at room temperature.

For photoionization in non-polar liquids, available experimental data on geminate recombination kinetics have been relatively few and obtained only by means of indirect methods, discussed below.

B. Electron Scavenging Effect

Investigation of the electron scavenging effect on ionization processes at high scavenger concentration remains the only method which allows a study of extremely fast geminate recombination stages involving high mobility hot and thermalized quasi-free electrons, i.e. reactions 2 and 3 in scheme (15). True, the process of understanding the scavenging effect is by no means complete.

The basis of the scavenging method is that the probability of ion pair chemical reactions with scavengers depends on the ion pair lifetime. Therefore one can obtain information about a lifetime distribution of ion pairs in liquids without scavengers by determining the scavenged ion yield as a function of the scavenger concentration $[S]$. The mathematical operation proposed by Hummel[129] relies on the condition that the probability that a species undergoes a reaction with solute S during the time t is $1-\exp(-K_s[S]t)$. If $F(t)\,dt$ is the fraction of all reactive species having a lifetime between t and $t+dt$, then the fraction of all species scavenged at a concentration $[S]$ is

$$f(s) = \int_0^\infty F(t)[1-\exp(K_s[S]t)]\,dt$$
$$= 1 - \int_0^\infty F(t)\exp(-K_s[S]t)\,dt$$

Since free ions have a much longer lifetime than geminate-recombining ions and can be scavenged by acceptors at vanishingly small concentration, one can write for photoionization in liquids containing ion scavengers

$$1 - \frac{\varphi_s([S]) - \varphi_{fi}}{\varphi_0} = \int_0^\infty F(t)\exp(-K_s[S]t)\,dt \tag{28}$$

where $\varphi_s([S])$ is the quantum yield of scavenged ions and $F(t)$ represents the lifetime distribution of geminate recombining ion pairs being equal $-\,\mathrm{d}W/\mathrm{d}t$.

If the functional dependence on $[S]$ in the left-hand side of Eq. (28) is known explicitly, $F(t)$ can be defined by the Laplace transformation. Extensive data exist for the scavenging yield, $g_s([S])$, of electrons produced by radiolysis of non-polar liquids. A remarkable feature of the data is that the yield for many different solvents and solutes can be fitted by a single empirical formula of Warman, Asmus and Schuler[131] was:

$$\frac{g_s([S]) - g_{fi}}{g_0} = \frac{\sqrt{\alpha[S]}}{1 + \sqrt{\alpha[S]}} \tag{29}$$

where g_0 and α are empirical parameters and g_{fi} is the free ions radiation yield, being separately measurable. As mentioned in the previous section, the result for a low $[S]$ concentration limit conforms to the long time-limiting reciprocal square root dependence of the survival probability on time, given by the diffusion approach Eq. (26). As pointed out recently by Tachiya[132] and by Crumb and Baird,[133] the exact form of $g_s([S])$ dependence predicted by Eq. (26) allows connection of the scavenging parameters in Eq. (29) with experimental values of the scavenging rate constant K_s and diffusion coefficients of the ions:

$$\frac{\alpha g_0^2}{g_{fi}^2} = \frac{r_c^2 K_s}{D} \tag{30}$$

Comparison of the left-hand side of Eq. (30) with the right, carried out[134,135] for some solutes in cyclohevane, has shown that the experimentally determined value of $\alpha g_0^2/g_{fi}^2$ is in all cases only a factor of 2 to 3 times greater than values calculated as $r_c^2 K_s/D$. In general, the result may probably be considered as satisfactory justification of the diffusion approach for the dynamics of ion pairs in non-polar liquids. Moreover, according to the recent calculation of Warman [3], this discrepancy between theory and experiment can be almost fully due to deviation of the concentration dependence given by Eq. (29) from that predicted by the diffusion theory for concentrations less than practically used. This deviation leads to that the value of α, obtained to give overall agreement between Eq. (29) and the experimental curve, has to be different from α derived from Eq. (30) for the low concentration limit.

It is worth emphasizing that this agreement between experiment and diffusion theory concerns only radiation produced ion pairs and low concentrations of scavengers. In the case there are little doubt that observed scavenging effect is caused by reactions of scavengers with thermalized ions. As concerns photoionization in non-polar liquids, the available data apply to the relatively high scavenger concentration and have a rather questionable interpretation.

1. Electron Scavenger Fluorescence Quenching

Wu and Lipsky[32] were the first to find that, in the presence of a low concentration ($< 0.1\ M$) of an electron scavenger, the fluorescence of a solute (TMPD) in some non-polar liquids is quenched for excitation energies above the photoionization threshold more severely than for below. It has also been found that the probability for the enhanced quenching exhibits a dependence on the solvent very similar to that observed for the rate constant of thermal electron reactions with these same solutes. An analogous effect was also observed[33,38] for pyrene solutions in some liquid hydrocarbons.

These results are consistent with the assumption that the quencher scavengers the geminate electrons. However, the electron scavenging probability $P(S)$ derived from the data[32] turned out to unequivocally be a better fit of the Stern–Volmer expression, i.e.

$$P_{\mathrm{SV}}(S) = \frac{K_{\mathrm{a}}[S]}{1 + K_{\mathrm{a}}[S]} \tag{31}$$

than the WAS equation (29). Since the Stern–Volmer equation is known to relate a quencher reaction with species having an exponential lifetime description, Wee and Lipsky[32] and Lee and Lipsky[34] have advanced an intriguing assumption that the scavenger acts not on the ion pair but rather on some neutral precursor state S_e of the geminate pairs, which was presumed to be a large-orbit, Rydberg-like state:

(Electron scavenging)

$$\mathrm{A} \underset{}{\overset{h\upsilon}{\rightleftarrows}} \mathrm{S_e} \xrightarrow{} \mathrm{e^-_{hot}} \rightarrow \mathrm{e^-_{qf}} \rightarrow \mathrm{e^-_{tr}} \rightleftarrows \mathrm{e^-_{qf}} \tag{32}$$

(Scavenging processes 5, 6, 7, 8 upward from S_e, e^-_{hot}, e^-_{qf}, and $e^-_{tr} \rightleftarrows e^-_{qf}$; geminate recombination processes 2, 3, 4 downward from e^-_{hot}, e^-_{qf}, and $e^-_{tr} \rightleftarrows e^-_{qf}$.)

(Geminate recombination)

With the geminate recombination scheme (15) in mind, this means that process 5 in scheme (32) dominates other possible processes of electron scavenging, i.e. processes 6, 7 and 8 in the scheme. At first, this suggestion was supported by evidence that the quencher (perfluoro-*n*-hexane) not only acted to reduce the fluorescence but also served to reduce the steady state photocurrent.[34] However, more detailed investigations[37] have shown that photocurrent quenching can no longer be considered to support this view and suggested that hot electron reactions with scavengers (i.e. process 6 in scheme 32) is also important.

With such a view, the problem, however, is to discover why scavengers prefer to react with electrons of S_e states but not with electrons of ion pairs. In the opinion of Lipsky and Coworkers, the only reason to ignore the latter process is that the concentration dependence of the scavenging probability derived from experiment differs from the 'square root' equation (29) known for geminate ions produced by radiation. However, should the scavenging probability for electrons produced by low-energy photons be described by the equation?

As mentioned above, the WAS equation (29) is predicted by diffusion theory only in the low $[S]$ concentration limit. The region of scavenger concentrations in which the Smoluchowski equation gives concentration dependence of electron scavenging probability $P(S)$ independent of the form of initial ion pair distribution has recently been discussed by Van den Ende and coworkers [3, 135]. Applying the Laplace transform approach to the empirical equation (25) for the ion pair survival probability at long times they have derived the expected theoretical concentration dependence

$$\varphi_s([S]) = \varphi_{fi}\left[1 + 1.33\left(\frac{r_c^2 K_s[S]}{D}\right)^{0.6}\right] \tag{33}$$

corresponding to the scavenging probability of geminate-recombining ion pairs

$$P_{EWH}(S) = 1.33\frac{\varphi_{fi}}{\varphi_0}\left(\frac{r_c^2 K_s[S]}{D}\right)^{0.6} \tag{34}$$

The upper limit of the concentration region where Eq. (34) is true corresponds approximately to the lower limit of the ion pair life time, at which the empirical Eq. (25) for the longtime survival probability is valid; it can be determined via the relationship

$$\frac{\varphi_s([S])}{\varphi_{fi}} \lesssim 4 \tag{35}$$

From Eqs. (34) and (35) one can readily derive the concentration condition

$$[S] < 4\frac{D}{K_s r_c^2} \tag{36}$$

For typical 'good' electron scavengers in low mobility hydrocarbon liquids at room temperature $D/K_s \approx 3 \times 10^{-14}\,\mathrm{cm}^2\,M$[136] and from Eq. (36) we find that the Smoluchowski equation predicts a certain form of $P(S)$ dependence only for $[S] < 10^{-3}\,M$. (The estimate agrees with the calculation of radiation yield of methyl radicals formed due to electron scavenging by methyl bromide in liquid cyclohexane and isooctane.[3])

Unfortunately, the electron scavenger fluoroscence quenching method used in recent measurements[32,34] is sensitive to relatively large scavenger concentrations, $[S] > 10^{-2}\,M$. Therefore, the discrepancy found out by Wu and

Lipsky between different forms of concentration dependences of scavenging probability for photoionization and radiation conditions can not indicate that the diffusion approach is, in principle, inapplicable for the former case. On the contrary, as Fig. 12 shows, the theoretical expression (34) is satisfactorily conjugated with $P(S)$ dependences obtained[32,34] from fluorescence quenching

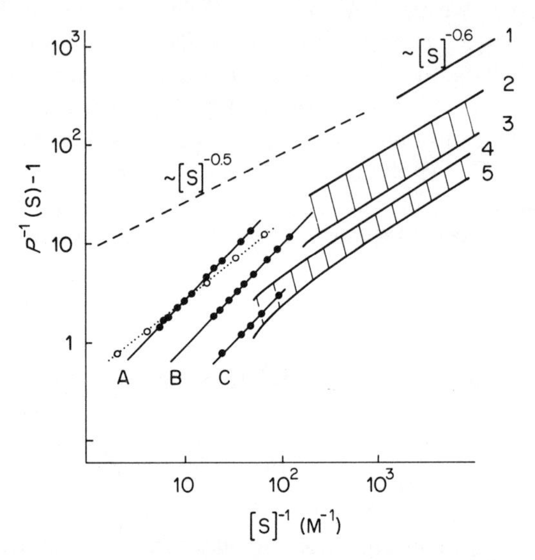

Fig. 12. Comparison of the geminate electron scavenging probability $P(S)$ predicted by the diffusion theory with experiment (From Ref. 61). Solid points present probabilities derived from data[32,34] on TMPD fluorescence quenching by electron scavengers: perfluoro-*n*-hexane in *n*-hexane at $\lambda = 213$ nm (A); CCl_4 in 2,2,4-trimethylpentane at $\lambda = 220$ nm (B); CCl_4 in tetramethylsilane at $\lambda = 220$ nm (C). Curves 1 to 5 show $P_{EWH}(S)$ dependences obtained from Eq. (34) at D values corresponding to mobilities from Table I and $K_s = 1 \times 10^{12}\,M^{-1}\,s^{-1}$ for case A just as for typical scavengers in liquid *n*-hexane (1), $K_s = 1 \times 10^{13}\,M^{-1}\,s^{-1}$ (2 and 3) and $K_s = 7 \times 10^{13}\,M^{-1}\,s^{-1}$ (4 and 5) for cases B and C.[136] The upper and lower curves for B and C were calculated using the highest and smallest ratio φ_{fi}/φ_0 published[22] for TMPD ionization in those same solvents at wavelengths indicated.

Open circles present the result [61] of numerical solution of the Smoluchowski equation for the GAUS initial distribution function with parameter $G = 30$ Å obtained[22] for TMPD ionization in *n*-hexane at 210nm, with $K_s = 1 \times 10^{12} M^{-1} s^{-1}$ and μ_e from Table I for *n*-hexane. The upper dashed line shows the concentration dependence $P_{WAS}(S)$ given by Eq. (29). The parameter α was taken to conjugate the dependence with curve 1 at small concentrations $[S]$. (Reproduced by permission of the American Institute of Physics.)

experiments and, without any adjustable parameters, provides surprisingly good agreement with absolute values $P(S)$ for the systems CCl_4 in isooctane and in tetramethylsilane.

For the case of PFC_6 in n-hexane, the concentration region where Eq. (34) is valid does not quite overlap with that which is experimentally accessible, if a typical rate constant ($10^{12}\ M^{-1}\ s^{-1}$) for electron scavenging in the solvent and inequality (36) are assumed (see Fig. 12). Nevertheless, the numerical solution of the Smoluchowski equation gives [61] very good agreement (within ~ 30 per cent) with all experimental values $P(S)$, if one supposes the Gaussian form of an initial ion pair distribution with the parameter $G = 30$ Å derived[22] from photoconductivity data for the same excitation energy in n-hexane.

On the other hand, as Fig. 12 shows for the system, the empirical WAS formula (29) for scavenging radiation-produced ions leads to far lower $P(S)$ values than those observed. There are two possible ways to rationalize, within the framework of the diffusion approach, the apparent difference in forms of concentration dependences of $P(S)$ at large $[S]$ for electrons produced by photons and for electrons produced by high-energy radiation in liquids. First, in the later case, as mentioned, multiple ionizations often occur in small volumes of liquids and result in a primary population of interacting ion pairs. These are expected to recombine at first more rapidly than isolate pairs with the same separation distance distribution. Furthermore, in the radiation case, ionization involves a large spectrum of excitation energies and thus, quite possibly, another form of the thermalization length distribution as compared to ionization by low-energy photons.

Recently, the model which involves the interaction of a quencher with an electron of an ion pair has been supported by Lee and Lipsky[37] with the additional information obtained via investigations of the effect of scavengers to reduce the photocurrent in TMPD–non-polar systems. It was found that the current quenching parameter K_i can be very different from the recombination fluorescence quenching parameter K_a at the same excitation energy. These results are not consistent with the earlier Wu and Lipsky model in which the solute acts most importantly on the highly excited but metastable state and suggest that fluorescence quenching is mainly due to scavenging of both thermal and hot (or epithermal) electrons. In this case, as pointed out by Lee and Lipsky,[37] the increased sensitivity of the fluorescence intensity to the presence of quencher at a higher excitation energy is naturally explained in terms of the effect of excitation energy to increase the thermalized geminate pair separation distance and thereby enhance the probability of thermal electron scavenging. If epithermal electron scavenging is assumed then photocurrent quenching as a result of reduction of the pair's escape probability can be understood. This process was suggested some time ago by Mozumder and Tachiya[137] to explain the reduction of free ion yield in irradiated neopentane on addition of small electron scavenger concentrations.

Thus, all three time evolution stages mentioned in the beginning of this section (or processes 6, 7 and 8 in scheme 32) seem to be responsible for scavenging processes studied by the fluorescence quenching method. Provided that scavenging of ion pairs but not their precursor is dominant, data on $P(S)$ dependences obtained by this method allow an estimate to be made of the first half-life of electron–cation pairs as

$$\tau_{\text{gem}} = (K_s[S]_{1/2})^{-1} \tag{37}$$

where $[S]_{1/2}$ is the scavenger concentration at which $P(S) = 0.5$.

If we use the average electron scavenging rate constant as K_s in Eq. (37), we derive, from data[32,37] presented in Fig. 12, $\tau_{\text{gem}} \approx 10^{-12}$ s, being of the same order of magnitude as values obtained earlier by Piciulo and Thomas[33] for the lifetime of ion neutralization at photolysis of some non-polar liquids. The values may be associated with the lifetime of geminate pairs only for high mobility liquids where electron localization does not seem to determine the rate of scavenging processes and transport. For middle and especially for low mobility liquids the values appear to be the upper limit of τ_{gem}, because quenchers can act on quasi-free electrons with the rate constant which may be significantly greater than the average constants K_s.

C. Photoassisted Dissociation of Geminate Ion Pairs

As mentioned in Section III, it is generally supposed that an electron ejected in low or middle mobility non-polar liquids can be localized and, in this case, displays optical absorption in the visible and IR regions of the spectrum. The absorption is associated with a transition to a mobile state or the photoionization of e_{tr}^-. This is clearly demonstrated by observation of the conductivity following photoionization of trapped electrons in liquids.[51,97–99] If, as assumed for glasses,[111–115] photoionization of e_{tr}^- electrons in liquids occurs due to a bound-free transition into a mobile state, one should expect that the photoliberated electron has initially an excess kinetic energy more than thermal. In the framework of the Onsager model, this means that photoexcitation of trapped electrons belonging to geminate ion pairs has to result in an increase in the dissociation probability of the pairs. It is clear that studying the dependence of the effect on the delay time between ion pair generation and trapped electron photoexcitation can provide information about ion pair dynamics in liquids.

The effect of photoexcitation of e_{tr}^- electrons (by 400 to 2000 nm light) on the probability of geminate pair separation has been observed at first in 3-methylpentane glass γ-irradiated at 77 K.[138,139] In this case, geminate trapped electrons are rather long-lived ($\sim 10^3$ s) species and a steady state light source could be used for their excitation. Lately, the increase of the yield of free ions generated by 347-nm light pulses in anthracene solution in liquid hexane[43] and methylcyclohexane[44] under additional laser pulse excitation at 694 nm

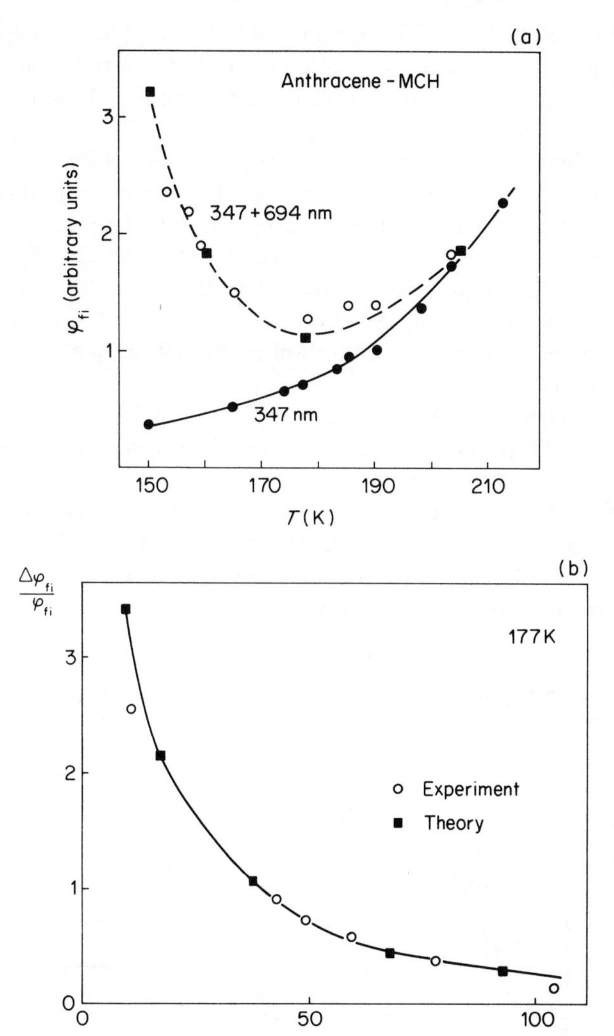

Fig. 13. Effect of additional excitation by the 694-nm laser pulse on the quantum yield φ_{fi} of free ions produced by the 347-nm pulse in anthracene solution in liquid methylcyclohexane. (a) The temperature dependence of φ_{fi} for photoionization by only the 347-nm pulse (●) and cooperative ionization by the 347- and 694-nm consecutive pulses (○).[46] The 694-nm light flux $\tau_p L_{694} = 0.2$ J cm^{-2}; the delay time between pulses $t_d = 50$ ns. (b) Relative quantum yield increase (○)[44] as a function of the delay time t_d at $T = 177$ K, $\tau_p L_{694} = 0.1\,\mathrm{J\,cm^{-2}}$. Points (■) show theoretical results obtained[46] by using Eq. (42) for the EXP initial distribution function with $B = 16$ Å and for thermalization length parameters $a_m = 36$ Å and $a_m = 50$ Å (B).

has been observed by Lukin, Tolmachev and Yakovlev. During preparation of the present paper the similar effect of additional photoexcitation on the free ion yield in an anthracene–hexane system was reported by Braun and Scott,[45] who used a picosecond laser.

Figure 13 demonstrates the effect observed[44,46] in anthracene solution in liquid methylcyclohexane. It can be seen that the combined photoexcitation generates free ions up to ten times that at only 347-nm light excitation. On the basis of strong temperature dependence of the relative increase $\Delta\varphi_{fi}/\varphi_{fi}$ of free ions by 694-nm light and a decay rate of $\Delta\varphi_{fi}/\varphi_{fi}$ with a delay time between 347- and 694-nm light pulses the effect has been attributed to an increased probability of electron–cation separation by excitation of electrons trapped near positive ions.

In view of current thinking about rapid thermalization of low-energy electrons ejected into molecular liquids, one should expect that photoexcitation of an electron trapped near a positive ion A^+ does not result in free ion generation at once but leads at first to formation of a new geminate ion pair $[A^+, e_{tr}^-]'$ if geminate recombination does not occur (see Fig. 14). In this connection, the following scheme of processes was used[43,44] to interpret the action of an additional light qunatum $h\nu_a$:

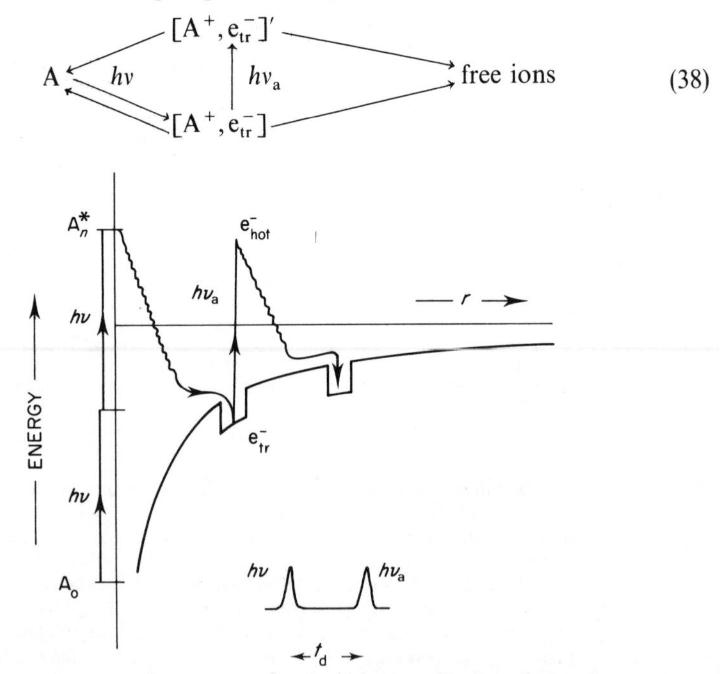

Fig. 14. Illustration of the effect of geminate electron photoexcitation on the free ion quantum yield. Absorption of an additional photon ($h\nu_a$) by an electron trapped near a sibling positive ion leads to formation of a hot electron e_{hot}^- with a higher probability of escape.

Qualitatively the scheme explains the observed increase of $\Delta\varphi_{fi}/\varphi_{fi}$ with delay time or temperature decrease as being due to increasing the fraction of ion pairs surviving at the time of additional light action. However, the quantitative analysis of processes in the scheme is very complicated, involving photoliberated electron thermalization and localization in the Coulomb field of positive ions. In order to analyse a physical picture of processes occurring in scheme (38) using a simple model, Lukin, Tolmachev and Yakovlev[44,46] have assumed that the photoinduced transition $[A^+, e_{tr}^-] \xrightarrow{h\nu_a} [A^+, e_{tr}^-]'$ can be divided into two stages in accordance with processes shown by Fig. 9:

$$[A^+, e_{tr}^-] \rightarrow [A^+, e_{qf}^-] \rightarrow [A^+, e_{tr}^-]' \quad (39)$$

In the first stage, a photoliberated electron becomes thermalized and is thought to be insensitive to the positive ion electric field. In the second stage, after thermalization of the electron, its motion is diffusion-like.

In the model, the separation distribution function $f_1(r, t)$ for pairs of thermalized charge particles $[A^+, e_{qf}^-]$ formed just after single simultaneous photoionization of all trapped electrons at time t can be presented as

$$f_1(r, t) = \int_0^\infty da \int_0^\infty d\theta 2\pi a^2 b(a) \sin\theta f_{tr}(\sqrt{r^2 + a^2 - 2ra\cos\theta}, t) \quad (40)$$

where $b(a)$ is the thermalization length distribution function for photoliberated electrons and $f_{tr}(r, t)$ is the distance distribution function for ion pairs $[A^+, e_{tr}^-]$ just before e_{tr}^- excitation.

Within the diffusion approach, it is clear that the escape probability is enhanced if the distribution $f_{tr}(r, t)$ is replaced by $f_1(r, t)$, because an isotropic form of $b(a)$ means that photoliberated electrons have infinite temperature during the thermalization time. Analysis of this enchanced probability of escape is very simple in the low-a limit. In this case, Eq. (40) is transformed into

$$f_1(r, t) \approx f_{tr}(r, t) + a_m^2 \left[\frac{1}{3r} \frac{\partial}{\partial r} f_{tr}(r, t) + \frac{1}{6} \frac{\partial^2}{\partial r^2} f_{tr}(r, t) \right] \quad (41)$$

where $a_m = [4\pi \int_0^\infty a^4 b(a) da]^{1/2}$ is the mean, squared thermalization length.

Using the Onsager formula for the separation probability and Eq. (41), an increase of the probability caused by a short light pulse is given[46] by

$$\Delta P(t_d, a, L) = L\tau_p \sigma \frac{2\pi a_m^2}{3} \int_0^\infty \left(\frac{r_c}{r}\right)^2 f_{tr}(r, t_d) e - r_c/r \quad (42)$$

where σ is the cross-section of e_{tr}^- photoionization, L is the additional light intensity, τ_p is the light pulse duration and t_d is the delay time between pulses of ion pair generation and e_{tr}^- excitation. Equation (42) predicts a linear dependence of the free ion yield increase $\Delta\varphi_{fi}$ on L. Such dependences $\Delta\varphi_{fi}(L)$ are actually observed in anthracene solution in liquid n-hexane and methylcyclohexane[43,44] up to $\tau_p L \approx 0.2$ J cm^{-2}.

Measurements of $\Delta\varphi_{fi}$ as a function of delay time t_d allow the time evolution of the cation-trapped electron separation distance distribution $f_{tr}(r, t)$ to be studied. For example, Fig. 15 shows dependences of $\Delta\varphi_{fi}/\varphi_{fi}$ on t_d calculated by using Eq. (42) and the Smoluchowski equation for some initial distribution functions. These were all obtained for the same $\sigma L \tau_p a_m^2/6r_r^2$ value taken as unity.

Two important conclusions can be drawn from Fig. 15, as recently pointed out by Tolmachev and coworkers [46]. Firstly, the long-time behaviour of $\Delta\varphi_{fi}/\varphi_{fi}$ as a function of time is in fact independent of the form of the initial separation distribution function for $t \gtrsim 0.1\, r_c^2/D$. The result could be expected in view of the long-time dependence of the geminate pair survival probability discussed above (Section IV. A). The more suprising observation is that the absolute values of $\Delta\varphi_{fi}/\varphi_{fi}$ at long times are the same for different distribution functions. It suggests that not only $W(t)/W(\infty)$, as Eq. (25) predicts, and also the function $f_{tr}(r, t)/W(\infty)$ are not influenced by the average initial distance of ion separation. The validity of this supposition is illustrated by the insert of Fig. 15, which shows $f_{tr}(r, t_d)/W(\infty)$ dependencies as a function of r obtained at $t_d = 2.7 \times 10^{-2}\, r_c^2/D$ by using the Smoluchowski equation for different forms of $f(r) = f_{tr}(r, 0)$.

A corollary of this latter observation is that measurements of the $\Delta\varphi_{fi}/\varphi_{fi}$ relation at long t_d enable us to estimate the thermalization length a_m of photoliberated electrons using Eq. (42) without needing to know the initial ion pair distribution. Using this approach $a_m = 43 \pm 7$ Å has been obtained[46] for hot electrons produced by photoexcitation ($\lambda = 694$ nm) of trapped electrons in liquid methylcyclohexane at $T = 177$ K. This estimate was derived from $\Delta\varphi_{fi}/\varphi_{fi}$ data for the delay time $t_d = 50$ ns ($\approx 2 \times 10^{-2} r_c^2/D$), with the known cross-section of e_{tr}^- optical absorption σ and the electron diffusion coefficient D.

A reasonable value of the thermalization length obtained by utilizing Eq. (42) without any adjustable parameters appears to support the model above for the photoassisted ion pair dissociation and the diffusion approach of ion pair time evolution at long times of $t > 10^{-2}\, r_c^2/D$. Moreover, as Fig. 13 shows, the model allows a satisfactory description of experimental data on temperature and delay-time dependencies of $\Delta\varphi_{fi}/\varphi_{fi}$ for times down to $3 \times 10^{-4}\, r_c^2/D$ at 150 K.

Recent picosecond data obtained by Braun and Scott[45] for the anthracene–hexane system make it possible to judge the ion pair behaviour at times $> 3 \times 10^{-3}\, r_c^2/D$ at room temperature and $> 3 \times 10^{-5}\, r_c^2/D$ at 214 K. On the basis of estimates of the time of $\Delta\varphi_{fi}(t_d)$ initial half-decreases ($\leqslant 9$ and 70 ps for 296 and 214 K respectively), Braun and Scott[45] have also concluded that diffusion of a localized electron and its parent cation determines the lifetime of the geminate pairs.

To date, no direct experimental studies of the ion pair kinetics able to

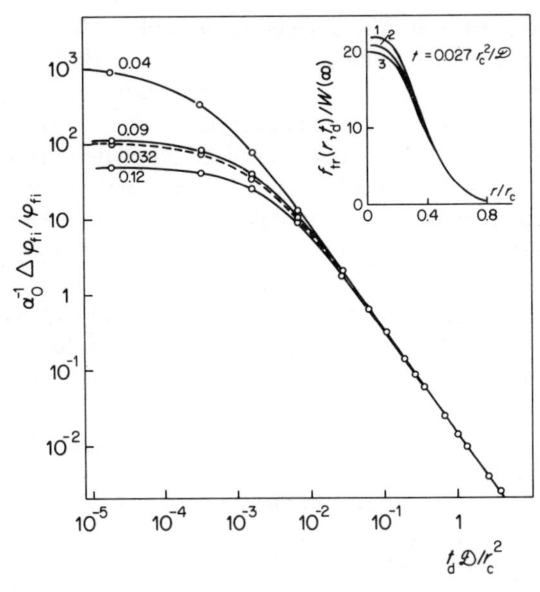

Fig. 15. The relative increase $\Delta\varphi_{fi}/\varphi_{fi}$ of the free ion quantum yield due to additional photoexcitation of geminate electrons as a function of the delay time t_d between the e_{tr}^- generation pulse and the photoexcitation one; $\alpha_o = L\sigma\tau_p a_m^2/(6r_c^2)$. The $\Delta\varphi_{fi}/\varphi_{fi}$ values were calculated[46] from Eq. (42) with the function $f_{tr}(r,t_d)$ taken as the solution of Eq. (24) provided that at $t_d = 0$ the function $f_{tr}(r,0)$ has GAUS (solid lines) or EXP (dashed line) form. The numbers near the curves are the ratios G/r_c or B/r_c of the initial distribution functions $f_{tr}(r,0)$.
Insert: The spatial distribution function of trapped electrons at $t_d = 0.027\ r_c^2/D$ for different initial functions $f_{tr}(r,0)$; GAUS with $G = 0.04r_c(1)$, $G = 0.012r_c(3)$ and EXP with $B = 0.032r_c(2)$.

definitely distinguish between an exponential and Gaussian initial distribution function have been reported for ionization by both radiation and low-energy photons in non-polar liquids. As pointed out recently by Warman,[3] even if the time dependence of the ion pair survival probability, $W(t)$, is known, the possibility of flexibility in the choice of the total ion yield can make decisions about the 'correct' distribution very difficult. Unfortunately, it seems to be the case for the method described above. As Fig. 14 shows for exponential and Gaussian initial distributions curves, $\Delta\varphi_{fi}/\varphi_{fi}$ versus t_d are practically indistinguishable if $\Delta\varphi_{fi}/\varphi_{fi}$ values are the same at $t_d \to 0$.

D. Transient D.C. Photoconductivity Caused by Geminate Ion Pairs

As mentioned in Section II, d.c. conductivity measurement is the widespread method used to study photoionization in non-polar liquids. Until recently the method has been applied only to free ion detection. Moreover, it is usually accepted that in non-polar liquids and solids transient d.c. conductivity signals are due to only charge carriers escaped geminate recombination, and a contribution of geminately recombined carriers is not appreciable. In contrast to this opinion, the dominant contribution of geminate ions to transient d.c. conductivity has been assumed for some experiments in solids[140–142] and liquids[50–54] under conditions when the current time resolution was expected to be less than the geminate recombination time.

Using the diffusion approach to describe the motion of geminate ions in the external electric field allow us to predict in detail the kinetics of d.c. conductivity induced by a short ion pair generation pulse if the initial pair separation distribution is known. Analysis of the problem has revealed[53,54,141,142,160] some interesting features of the kinetics and, as should be expected, its dependence on functional forms of ion pair distribution. There follows a short discussion of possibilities provided by d.c. conductivity measurements for studying ion pair time evolution in non-polar liquids.

1. The Diffusion Model Predictions

Let N ion pairs be generated at an initial moment ($t = 0$) in the volume between two parallel electrodes taking part in the usual d.c. circuit (see the insert in Fig. 16). Even if the sum of all ion pair dipoles $\mathbf{Q} = \sum_{i=1}^{N} e\mathbf{r}_i$ equals zero at $t = 0$, i.e. an initial ion pair separation distribution is spherically symmetric, the motion of ions in the external electric field $\mathbf{E}$ leads to the appearance of $\vec{Q}$ and its change with time. The latter results in a rise of current $I(t)$ (or polarization current) in the circuit:

$$I(t) = \frac{1}{d}\frac{\mathrm{d}Q_{\mathrm{E}}}{\mathrm{d}t} \tag{43}$$

where Q_{E} is the projection of $\mathbf{Q}$ in the $\mathbf{E}$ direction and d is the distance between electrodes. Taking into account the Smoluchowski Eq. (24), one can rewrite Eq. (43) as

$$I(t) = \frac{eN}{d}\int_0^{\infty} \mathrm{d}r \int_0^{\pi} \sin\theta \, \mathrm{d}\theta \, 2\pi r^3 \cos\theta n(r, \theta, t) \tag{44}$$

where $n(r, \theta, t)$ and θ are the same as in Eq. (24).

Analysis of the current kinetics is simplified in the low-temperature limit if one can neglect the diffusion motion of ions (i.e. assume $D = 0$ in Eq. 24) and consider only their drift in the summary Coulomb and applied field, which is described by the equation

$$\frac{\mathrm{d}\mathbf{r}_{\mathrm{i}}(t)}{\mathrm{d}t} = \mu(\mathbf{E}_{\mathrm{c}} + \mathbf{E}) \tag{45}$$

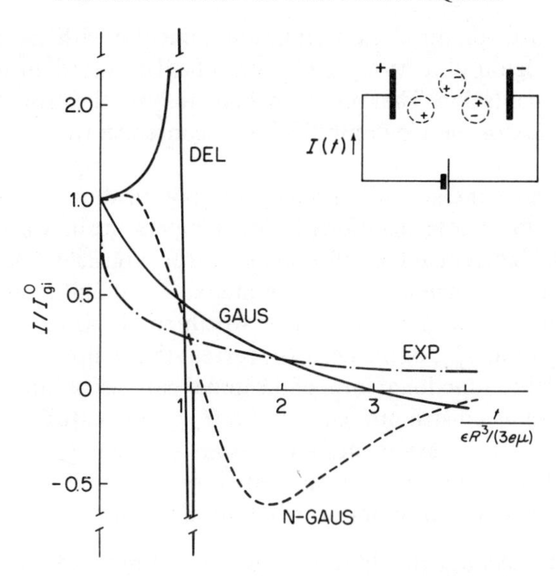

Fig. 16. The transient current $I(t)$ due to geminate pairs generated at moment $t = 0$ in the volume between electrodes in the circuit shown by the insert. The current was calculated[53] by using the solution of Eq. (45) in the low E limit for the different initial distribution functions indicated (see Section II.E). R is the parameter of a distribution, i.e. r_o, G, B or M depending on the function form. $I^o_{gi} = e\mu EN/d$ is the current due to N free ions, N being the number of initially produced ion pairs.

In this case, Eqs. (43) and (45) permit[141,142] an analytical solution for $I(t)$ at $E \rightarrow 0$ to be obtained. Figure 16 shows the calculated dependence $I(t)$ for four trial functional forms of initial ion pair distribution discussed in Section II. These are all normalized to the current $I^0_{gi} = e\mu EN/d$ which would appear if all initial ion pairs were converted into free ions. Some interesting features emerge from Fig. 16:

1. At earlier stage of geminate recombination, the polarization current is approximately equal to I^0_{gi} – as if an ion motion was not influenced by a Coulomb field. It is easily rationalized because Eqs. (43) and (45) give

$$I(t) \sim (E_{ca} + E) \qquad (46)$$

   where E_{ca} is the average projection of the Coulomb field, acting on ions, in the $\mathbf{E}$ direction, and $E_{ca} = 0$ for an isotropic ion pair distribution.
2. At short times, the current kinetics for delta distributions significantly differs from that for 'broad' distributions. In the former case, a current at first increases with time in contrast to the latter case. Physically, the current increase can be understood if one takes into account the fact that ion pairs,

for which a Coulomb field orientation coincides with the external field, reach the region of a stronger Coulomb field earlier than those with the opposite orientation. That results in an increase of the projection averaged over all pairs of the Coulomb field and, consequently, an increase of the current.

3. For all initial pair distributions, a negative polarization current is observed. In the case under consideration (the low-temperature limit), the appearance of a negative current at the constant applied electric field is natural, because all ion pairs will geminately recombine at $E \to 0$ and, consequently, the current integral has to be zero. It is worth emphasizing that the relative maximum value I^-_{max} of the negative current (normalized to I^0_{gi}) is strongly influenced by a functional form of the initial pair distribution, but is almost independent of a distribution parameter (I^-_{max} values are 6×10^{-3}, 0.09 and 0.6 respectively for exponential, Gaussian, and 'narrow Gaussian' distributions). That would be a desirable situation to experimentally distinguish between the assumed initial distribution functions.

Taking into account the diffusion motion of ions greatly complicates the analysis of the current kinetics. Only recently the kinetics has been considered for some initial conditions [53, 54, 160]. By way of illustration, Fig. 17

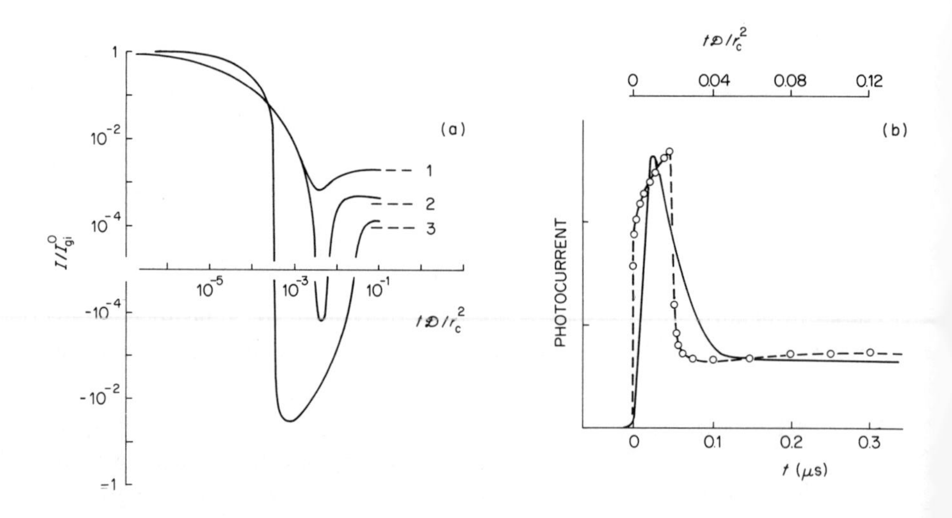

Fig. 17. The polarization current $I(t)$ calculated[54] by using Eq. (44) and numerical solution of the Smoluchowski equation. (a) An infinitely short pulse of generation of ion pairs at $t = 0$ with EXP initial pairs distribution (curves 1 and 2) at parameter $B = 0.037r_c$ and $E = 10k_BT/er_c$ for curve 1 and $E = 1k_BT/er_c$ for curve 2, and with GAUS distribution at $G = 0{\cdot}062r_c$ and $E = 1k_BT/er_c$ (curve 3), the dashed lines show free ion currents I_{fi}. (b) A rectangular (50-ns) pulse of ion pair generation with EXP initial distribution at $B = 20$ A, $E = 1.5 \times 10^4$ V cm^{-1} and $\mu = 6 \times 10^{-4}$ cm^2 V^{-1} s^{-1} as for the electron mobility in liquid methylcyclohexane at 177 K[51,52] (open points). The solid curves represents d.c. transient photoconductivity observed[52] in TMPD solution in methylcyclohexane for 177 K after excitation by a 347-nm laser pulse.

represents three solutions obtained[54] from Eq.(44) by numerical solution of the Smoluchowski equation for exponential and Gaussian distributions at practical temperatures and electric fields. In comparison with Fig. 16 it can be seen that, just as should be expected, the diffusion motion of ions forms a constant component of current I_{fi} at long times, $t \gtrsim r_c^2/D$, which is due to the transformation of ion pairs into free ions. (For the generalized current $I(t)/I_{gi}^0$ this component equals the escape probability P.) In addition, the negative polarization current becomes more broad and small due to diffusion motion. The depth of the minimum on the curve $I(t)$ is affected by the applied field E (as seen from Fig. 17a) and the negative current can vanish at a large enough E.

Another result, expected by analogy with the case when diffusion motion was neglected, is that the initial polarization current I_a approximately equals I_{gi}^0. Since $I_{gi}^0 \gg I_{fi}$ it demonstrates again that, as pointed out earlier,[140-142] the geminate ions can make the main contribution to the transient current in non-polar condensed media.

Figures 16 and 17a show that the function form of the polarization current kinetics depends noticeably on the form of initial pair distribution. Therefore, measurements of the current kinetics can help to distinguish between assumed distributions or, if it is known, to verify the diffusion model predictions of the ion pair time evolution.

However, a problem which has remained to date is how to obtain evidence for the experimental observation of a polarization current. Apart from registration of a negative current which is, unfortunately, not pronounced for the usually preferred exponential form of pair distribution, the evidence can be based on three predictions of the diffusion theory:

1. As indicated by Fig. 17, at times $t < 0.1 r_c^2/D$ one should expect unusual behaviour of the transient current induced by a short pulse of ion pair generation, which consists in appearance of a prompt current component when charge carrier mobilities are independent of time. The component can already be well pronounced for a duration of the generation pulse of $\lesssim 0.1\ r_c^2/D$ (see Fig. 17b).
2. The ratio of the current amplitude I_0 to the usual free ion current, I_{fi}, measured at $t \gtrsim r_c^2/D$, must increase with lowering of the temperature because the value of φ_{fi} decreases and

$$\frac{I_0}{I_{fi}} \approx \frac{I_{gi}^0}{I_{fi}} = \frac{\varphi_0}{\varphi_{fi}} \tag{47}$$

3. Equation (47) predicts that the ratio I_{fi}/I_0 must increase with increasing electric field strength as $\varphi_{fi}(E)$, with I_0 being proportional to E. The prediction appears to be the strongest evidence of the polarization nature of a transient current. It means an unusual dependence of the current pulse form on E, at times being much shorter than the charge carrier flight-time through the distance between electrodes.

2. *Transient D.C. Conductivity Measurements*

As already mentioned, in the diffusion model the only necessary condition for the geminate ion d.c. conductivity to be observed is the time resolution of the current registration $t \lesssim r_c^2/D$ after a short ion pair generation pulse. This condition is easily realized for amorphous non-polar molecular solids, in particular at low temperatures. It is the very case where geminate pairs have been drawn up relatively long ago[140–142] for interpretation of d.c. conductivity.

For low mobility non-polar liquids the conductivity measurement on a timescale $t < r_c^2/D$ is also accessible at present. Taking, for example, liquid *n*-hexane at room temperature, the escape time for primary ions ($\mu_e = 0.09\,\mathrm{cm^2V^{-1}s^{-1}}$) is found to be $r_c^2/D = 3.8\,\mathrm{ns}$. If primary ions convert into molecular ions by scavenging, the escape time for the ions ($\mu \sim 10^{-3}\,\mathrm{cm^2V^{-1}s^{-1}}$) can increase up to $4\,\mu\mathrm{s}$ in various non-polar liquids at room temperature. For the latter case, Sauer and coworkers[50] have recently observed d.c. conductivity signals following pulsed photoionization of solutions of anthracene in some dielectric liquids, and revealed unusual properties.

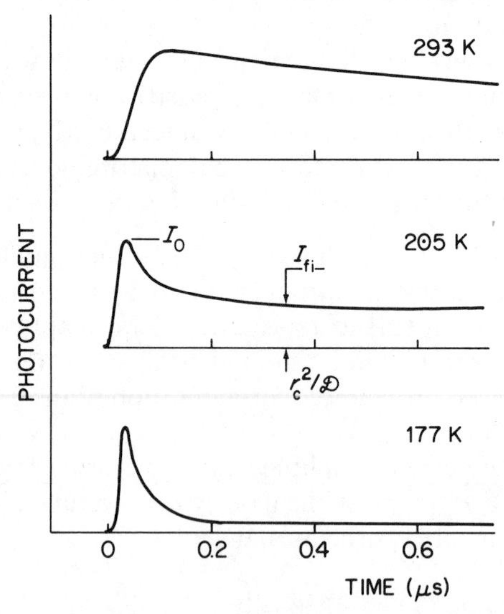

Fig. 18. The d.c. photoconductivity transients due to the 347-nm laser pulse ionization of TMPD in liquid methylcyclohexane for different temperatures indicated. [52] The light pulse finishes at $\tau_p \approx 50$ ns. Lowering temperature results in an increase of the characteristic time r_c^2/D taken to form free ions from ion pairs, from 0.5 ns at 293 K up to $3.5\,\mu\mathrm{s}$ at 177 K. A spike of photocurrent is seen to appear at ≈ 205 K when time r_c^2/D (shown by the arrow on the time scale) becomes more than τ_p.

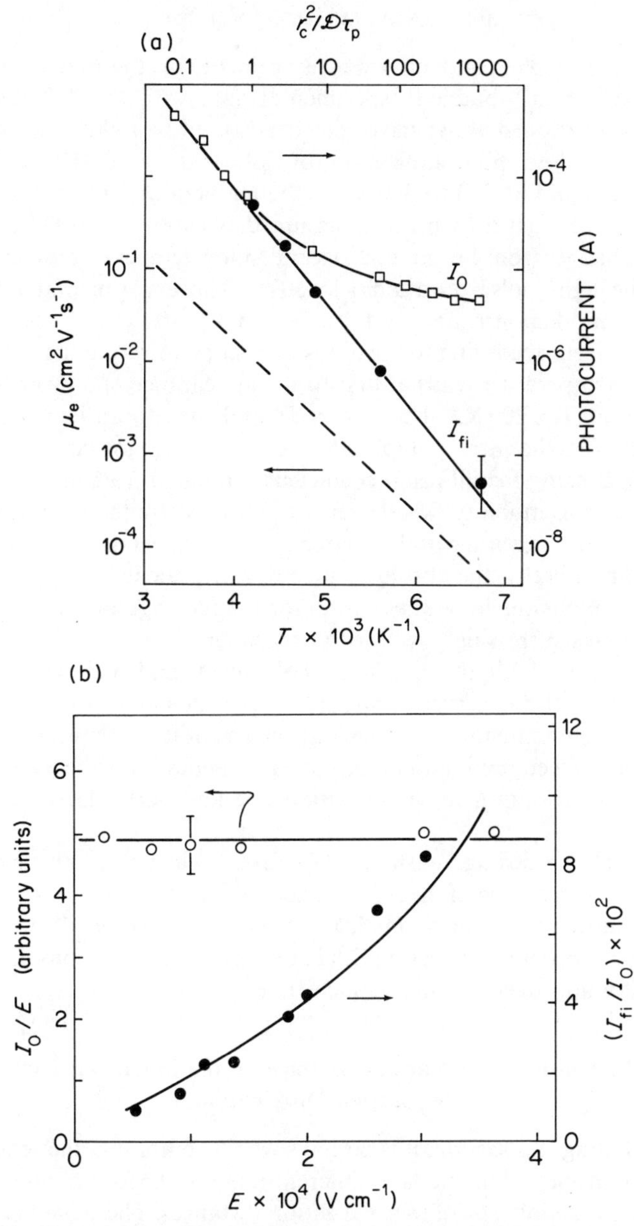

Fig. 19. Parameters of d.c. photoconductivity induced by 347-nm laser pulse ionization of TMPD in liquid methylcyclohexane. [52] (a) Temperature dependences of the current amplitude I_0, the current I_{fi} due to free ions (excess electrons and positive ions) measured at $t \geqslant r_e^2/D$ and the excess electron mobility μ_e. The diffusion coefficient of electron D was estimated as $\mu_e k_B T/e$. The upper scale indicates the $r_c^2/D\,\tau_p$ values. (b) The applied electric field dependence of I_0 and I_{fi} at 148 K (points ○ and ●). The curve is the field dependence of the escape probability $P(E)$ calculated from the Onsager formula (3) for the EXP initial distribution of electron thermalization lengths with parameter $B = 15$ Å.

More recently, d.c. conductivity measurements on the timescale $t < r_c^2/D$ and comparison of obtained experimental data with the diffusion theory predictions mentioned above have been carried out by Lukin and coworkers [51, 52] for pulsed photoionization of solutions of TMPD in methylcyclohexane liquid at 150 to 300 K. In the temperature interval, the escape time r_c^2/D for primary ions, being, as assumed, of cations $TMPD^+$ and excess electrons, changes from 1 ns up to 20 μs and at low temperatures considerably exceeds the light pulse duration $\tau_p \approx 50$ ns. The experimental results obtained[51,52] are demonstrated by Figs. 18 and 19.

As can be seen, according to the diffusion theory, lowering the temperature from room temperature leads at first to the appearance of a prompt current component at $T \approx 200$ K, when $\tau_p \approx r_c^2/D$, and then to an increase in ratio I_0/I_{fi}, where I_0 is the current amplitude and I_{fi} is the free ion current measured at $t \geq r_c^2/D$, D being the diffusion coefficient extimated from an experimental value of electron mobility. Moreover, as predicted by the theory, the ratio I_{fi}/I_0 is found to increase with field strength, with I_0 being proportional to E (see Fig. 19b). Finally, the absolute value of I_0/I_{fi} calculated[54] by numerical solution of the Smoluchowski equation for the rectangular form of the light pulse compares surprisingly well with experiment (see Fig. 17b).

On the basis of all these, Lukin, Tolmachev and Yakovlev[52,54] have concluded that the observed unusual d.c. photoconductivity behaviour is really due to geminate ion pair polarization. In addition, the lack of negative current or a marked minimum in the current kinetics leads them to prefer the exponential distribution to the Gaussian one and particularly to the delta distributions.

Because of the strong sensitivity of current kinetics to the initial pair distribution, it seems that if the d.c. conductivity measurements could be taken during a shorter time period (at least down to $t \approx 10^{-4} r_c^2/D$) it would be possible to more definitely distinguish between trial distributions if, of course, the diffusion approach is valid at such times.

E. Sensitivity of Various Methods to the Initial Ion Pair Separation Distribution

In concluding this section, it is perhaps worth comparing the sensitivity of possible experimental methods to different regions of r for an initial ion pair distribution function $f(r)$ of thermalization distances. These methods are (or can be) based on the following measurements:

1. Field-dependent free ion yield
2. Temperature-dependent free ion yield
3. Electron scavenging probability
4. Photoassisted ion pair dissociation
5. Transient d.c. conductivity.

Let us consider contributions of different spatial regions to a measured signal value M. We take as an example a low mobility liquid like n-hexane ($\mu_e = 0.09\,\mathrm{cm}^{-2}\mathrm{V}^{-1}\mathrm{s}^{-1}$) at room temperature and an exponential distribution for initial thermalized pairs with the parameter $B = 16\,\text{Å}$ which can be assumed from data[22] on ionization TMPD in n-hexane by photon energy $\approx 6.5\,\mathrm{eV}$.

For methods 1 to 4, Fig. 20 shows dependencies $\mathrm{d}M/\mathrm{d}r$ normalized to unity in maxima. For methods 1 and 2 based on free ion yield measurements, the M values were estimated from Eq. (3) at $E = 0$ and $E = 2 \times 10^5\,\mathrm{V\,cm}^{-1}$. (The latter value is near to the highest field strengths used for such measurements in liquids.) In this case, $\mathrm{d}M/\mathrm{d}r \sim r^2 f(r) P(r, E)$ and only ion pairs with a rather large initial r contribute to the signal. To estimate the smallest distance r_m which is still accessible by a method, we arbitrarily adopt that a method is insensitive to ion pairs which contribute less than 3 per cent to the measured signal. The regions $r < r_m$ are indicated in Fig. 20 by dashed lines. As can be

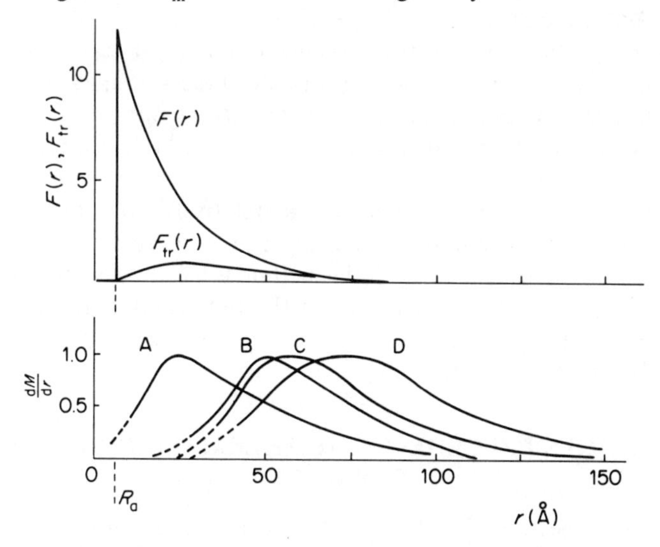

Fig. 20. The comparison of sensitivity of some experimental methods to different separation distances r of initial thermalized ion pairs. Contributions to signals M measured were calculated for n-hexane at room temperature and the exponential initial distribution function $f(r)$ with parameter $B = 16\,\text{Å}$ assumed from data[22] on TMPD photoionization in n-hexane by photon energy of 6.5 eV).

Upper curves show the functions $F(r) = 4\pi r^2 f(r)$ and $F_{tr} = 4\pi r^2 f_{tr}(r)$, where $f_{tr}(r)$ is the initial distribution of trapped electron-cation pairs, which was obtained by numerical solution of the Smoluchowski equation (24) at $\mu = \mu_o$, $K_s[S] = \tau^{-1}{}_{loc}$ and $\mu_o\tau_{loc} = 3 \times 10^{-13}\mathrm{cm}^3\mathrm{V}^{-1}$ given by Table II as the upper limit of $\mu_o\tau_{loc}$ for n-hexane. Dashed lines correspond to regions which contribute 3 per cent. to signal values measured. Lower curves are given for scavenging (A), photoassisted dissociation (B) and free ion measurements at $E = 0(D)$ and $E = 2.10^5$ $\mathrm{V.cm}^{-1}$(C).

seen, for methods 1 and 2 (curves D and C) the signals are due almost entirely to ion pairs with $r > 42$ Å and $r > 36$ Å, which correspond to a small fraction of all initial pairs (12 and 17 per cent respectively).

Method 4 is on principle sensitive only to localized electrons. To obtain the initial distribution function $f_{tr}(r)$ for these electrons, it was accepted, as in Section IV.A, that quasi-free electron motion is described by Eq. (24) with $K_s[S]$ equated to the reciprocal localization time τ_{loc}^{-1} and $\mu_0 \tau_{loc} = 3 \times 10^{-13}\,cm^{-2}\,V^{-1}$ as the upper limit of $\mu_0 \tau_{loc}$. (see Table II). For this method Eq. (42) was used $(dM/dr \sim r^{-2} f_{tr}(r, t_d) \exp(-r_c/r))$. The plot of dM/dr is shown in Fig. 20 for the highest time resolution when only primary trapped electrons are photoexcited, i.e. $t_d = 0$. In this case, r_m equals 27 Å and corresponds to about a half of all the initial trapped electron–cation pairs or ~ 8 per cent of all thermalized pairs.

For the scavenging method, $[S] = 3 \times 10^{-1}\,M$ was used an an upper limit of scavenger concentrations which do not still destroy a thermalized pair distribution in the pure liquid. It was adopted that $K_s = 10^{12}\,M^{-1} s^{-1}$ and dM/dr is proportional to a distribution function $F_{tr}(r)$ of scavenged electrons. Figure 20 shows that in this case almost all distances are responsible for the signal, although the probability of electron scavenging is rather small (0.2).

As concerns the transient d.c. conductivity method, it appears that its usefulness in studying the ion pair time evolution in liquids with $\mu > 10^{-4} cm^2 V^{-1} s^{-1}$ is very limited by a relatively low time resolution $(\sim 10^{-9} s)$ of photo-current measurements. For the pure liquid used above as an example $(\mu_c = 0.09\,cm^2 V^{-1} s^{-1})$ the method is hardly sensitive to geminate ion pairs in general. In this case, when the time resolution does not limit transient current measurements (e.g. at enough low temperatures) the method seems to provide information about all localized electrons.

V. STAGES PRECEDING THE THERMALIZED ION PAIR FORMATION

As noted in Section II, data on field-dependent measurements of photocurrent and fluorescence quenching suggest that separation distances of thermalized ion pairs generated by photoionization in liquid hydrocarbons increase when the photon energy increases at least up to ~ 1.5 eV above the photoionization threshold. This conclusion is in agreement with the following generally accepted model (called below the thermalization model). Photoabsorption of a solute molecule leads to ejection of a hot electron e_{hot}^- with an excess kinetic energy into the medium, the thermalization distance being the larger the higher initial excess energy. In the model, electron thermalization in the vicinity of its parent cation is evidently the stage preceding the thermalized geminate pair formation. It is the stage that principally distinguishes photoionization in the thermalization model from a photoinduced charge transfer between a donor and acceptor in solutions, on the one hand, and from an impurity state–conduction band transition in crystalline solids, on the other hand.

Along with this model other geminate pair formation mechanisms which exclude the stages of e^-_{hot} thermalization has been advanced. Peterson and coworkers[18] have proposed that an excited state of the solute molecule persists for a sufficiently long time to transfer charge to acceptor states of a non-polar liquid, which have been designated as 'tail states' and are associated with random potential wells induced by density fluctuation.[143] According to another model suggested recently by Warman [161], the photoexcitation of aromatic solutes in alkane solvents can result in the solvent ionization due to electron transfer to the solute molecule. However, to date, it is difficult to determine contributions of the mechanisms.

Nevertheless, for the most experimentally investigated region of excitation energies in which $h\nu - I_L \gtrsim 0{,}5$ eV, the thermalization model seems to be more preferable. It permits a simple interpretation of the observed increase of thermalization lengths with photon energy and with decreasing the electron scattering efficiency in different non-polar liquids (see Section II.E). In addition, as has been noted by many authors,[17,22,25] thermalization lengths for photoelectrons at $h\nu - I_L \approx 1$eV do not differ markedly from those for secondary electrons produced by γ-and X-ray absorption, in the latter case the thermalization model being generally accepted.

However, in this model the problems which remain are how the solvent affects the quantum yield of initial ion pairs and from what state the ejection of e^-_{hot} into the liquid proceeds. These questions are now briefly discussed.

A. The Initial Quantum Yield of Geminate Ionpairs

The primary quantum yield φ_0 of geminate pairs could be determined directly in photoconductivity and electric field-dependent fluorescence quenching experiments if values of the external field E such that the escape probability $P(E) = 1$ were accessible. However, even at the highest field, 2×10^5 V cm^{-1}, reached in liquid phase conductivity studies, the average escape probability $P(E)$ does not exceed 0.2 for most of the systems studied (only for TMPD in tetramethylsilane at $h\nu > 5.7$eV $P(E)$ does reach the value of 0.5 at $E = 2 \times 10^5$ V cm^{-1} [122]). Direct determination of φ_0 by measurement of geminate pair optical absorption is doubtful for the following reason. Only those ion pairs may be observed by this method, the electrons of which are localized prior to geminate recombination, since the geminate recombination of quasi-free electrons seems to be extremely fast ($< 10^{-12}$ s). However, the fraction of these electrons may be, as shown previously, essentially less than 1. This would suggest that at present there are no methods permitting direct determination of φ_0.

The main feature of available indirect data on φ_0 for solute molecules in non-polar liquids is that the magnitudes of φ_0, as well as in polar liquids, are of the order of unity at an excitation energy exceeding the ionization threshold by at least 1 eV. At first, this has been clearly demonstrated by the investigations of Holroyd and Russel of temperature dependences of φ_{fi} for TMPD ionization in various non-polar liquids. Later, the conclusion was supported by data on field dependences of φ_{fi} derived from studies of recombination fluorescence[27] and photoconductivity[22] as well as by data on recombination fluorescence quenching by electron scavengers.[32,33,38] It should be emphasized here that absolute values of φ_0 estimated by these methods significantly depend on the assumptions used. In particular, for the method based on $\varphi_{fi}(E)$ measurements, φ_0 values depend upon the type of ion pair distribution chosen and can differ by a factor of 10 for exponential and Gaussian forms.[22]

Figure 21 shows some data on ion pair quantum yields for ionization TMPD in *n*-hexane, isooctane and tetramethylsilane. Two interesting points may be emphasized from these results:

1. The quantum yield φ_0 depends on the excitation energy and the $\varphi_0(h\nu)$ spectrum has two maxima near 5.7 and 7 eV. The first maximum has been

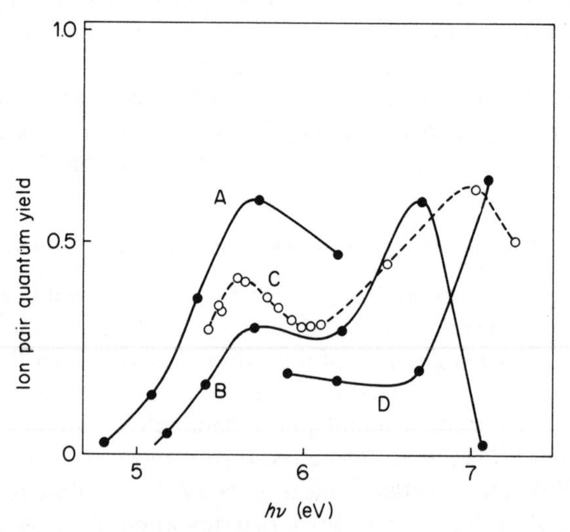

Fig. 21. Initial quantum yield of ion pairs produced by one-photon ionization of TMPD in tetramethylsilane (A), isooctane (B and C) and *n*-hexane (D). For cases A, B and D, data from field-dependent photoconductivity measurements were obtained[22] for the exponential form distribution assumed. For case C, data were derived from scavenger fluorescence quenching measurements[32] provided that scavengers interact with ion pairs but not their precursors, as assumed by Lipsky and coworkers.[32,34]

also observed in the yields of escaped electrons from TMPD for all three solvents.[17,18,22]

2. Values of φ_0 appear to be relatively insensitive to the nature of the solvent, at least for $h\nu \gtrsim 5.4\,\text{eV}$ (i.e. $h\nu - I_L \gtrsim 0.7\,\text{eV}$ for isooctane), and to be markedly less than unity for $h\nu < 5.6\,\text{eV}$.

The latter results suggests that autoionization of A_n^* is not predominant for excitation energies exceeding the ionization threshold by $< 1\,\text{eV}$ and spectral dependences $\varphi_0(h\nu)$ are determined by competition between autoionization and interval conversion of molecular excitation. At least the former process seems to be dependent on the nature of the solvent (through the energy of ionization continuum V_0). Therefore, one should expect that φ_0 values must be sensitive to solvents, especially in a region near the ionization threshold. Unfortunately, to date, there are no reliable data to quantitatively compare φ_0 values in this region for different non-polar liquids.

It should be noted that a conclusion similar to point 2 (the ion pair quantum yield is insensitive to the matrix) has been drawn[144] from photoionization studies of several solutes in low-temperature glasses of a differing chemical nature. This result, together with data on quantum yields of decomposition of solvent molecules sensitized by the solute, has allowed the authors[144] to assume that both reactions (ionization and decomposition) occur from a common intermediate state which belongs to the highly excited charge transfer manifold of states.

B. The Nature of Ion pair Precursors

Absorption spectra of solutes above the photoionization threshold are, as a rule, independent of the nature of solvents and similar to the spectra in gases. In particular, the TMPD absorption spectrum in the vapour and condensed phase shows no evidence of an environmental effect other that the usual dispersive type of red shift attendant in solutions of valence-like states.[12,18] This suggests that the higher valence-like state A_n^* localized on the molecular frame of the solute is initially formed.

The next step in the thermalization model is thought to be analogous to an autoionization. Along with it, as has often been assumed,[40,144–151] the formation of an intermediate state S_e in which an electron is in direct contact with solvent molecules may precede the ejection of e^-_{hot} into the liquid. The role of this 'semiionized' or Rydberg-like state in the mechanism of ionization has been indicated by many authors.[40,144–151] The state S_e is thought to be short lived and either to be internally converted to the ground state or to form a pair in which the electron retains some residual kinetic energy.

Very little is known now about the early stages of photoionization in liquids. Several aspects of these stages have been discussed lately:

1. Braun, Scott and Albrecht[23] have studied the multiphoton ionization of TMPD in 2, 2, 4-trimethylpentane solvent and compared the relative free charge carrier yields φ_{fi} for two-photon and one-photon TMPD excitation. Despite differences in the one-photon and two-photon absorption spectra and differences in the Franck–Condon states A_n^* initially reached via these two preparation modes, the quantum yields φ_{fi} are accurately proportional from 5.1 to 5.6 eV. This result indicates that different A_n^* states must nevertheless lead to some common precursor of the thermalized ionpair without a significant relaxation of A_n^*. According to Braun, Scott and Albrecht,[23] a likely candidate for the precursor states is the set of vibronic states reached by rapid internal conversion of the Franck–Condon states A_n^*.
2. As mentioned in Section IV.B, from data on TMPD fluorescence quenching by electron scavengers it has been suggested by Wu and Lipsky[32] that the scavenger acts on some neutral excited state S_e of TMPD. This state has been presumed to be a large-orbit, Rydberg-like state of TMPD resulting from the directly generated Franck–Condon states A_n^* of TMPD via radiationless transition. Rydberg states of solute molecules in the condensed phase are well indentified by absorption spectroscopy in the case of a lightly doped simple insulator, such as solid rare gases.[152] The Wannier model has been successfully applied to high-lying hydrogenic states observed in these matrices.[152,15] The nature of analogous excitations in molecular non-polar liquids, and especially in organic liquids, is not yet elucidated.

   The indentification by Lipsky and coworkers[32,34] of an intermediate state as a Rydberg-like state of TMPD in liquid hydrocarbons was mainly based on the observation that quenching of S_e by electron scavengers obeys the same Stern–Volmer type of equation so typical for the quenching behaviour of most neutral excited states, instead of a typical ion pair scavenging law observed for high-energy radiation. This divergence, as shown in section IV.B, can not be considered to support this view. Moreover, assuming a Rydberg-like state of quenching by scavengers, it is difficult to rationalize the observed increase of quenching efficiency with an increase of photon energy.[32] On the other hand, for the reaction between a scavenger and ion pair, such behaviour is readily explained in terms of the effect of excitation energy to increase the geminate pair separation distance.
3. The absorption spectra of some solute molecules assigned to direct optical transitions from the ground state to the low-lying Rydberg-like states in non-polar and polar organic solvents have been reported. Such absorption was observed[149,150,154–157] at a photon energy below the photoionization threshold for tetraaminoethylene molecules having a low ionization potential in the gas. The assignment of the spectra to transitions to the Rydberg states was based on sensitivity of the absorption band to temperature and its blue shift in the solution compared with gas absorption.

In addition, the Rydberg nature of the tetraaminoethylene fluorescence state has been deduced from the characteristic dependence of fluorescence spectra on the solvent rigidity and polarity [149,150,154–157].

Recently, using polarized two-photon fluorescence excitation techniques to study the two-photon states of benzene, Scott and Albrecht[158] have found a new resonance band lying from 6 to 7 eV. This two-photon resonance is undetectable in pure liquid benzene but appears in the spectrum of dilute benzene solutions in alkanes. The polarization behaviour of this new feature as well as its sensitivity to solvent and temperature has been interpreted in terms of a two-photon promotion of an electron from the ground state into the 3s Rydberg orbital of benzene in non-polar liquids. Also the role of Rydberg states in photoionization has been postulated by Böttcher and Schmidt [162] and Holroyd and coworkers [163] for interpretation of maxima in the photoconductivity spectra of non-polar liquids.

It should be emphasized that observations described above likely concern the low-lying Rydberg levels possessing compact orbitals. It has been assumed[153,158,159,163] that the Rydberg orbitals do not extend far enough into the solvent to allow dielectric screening, as required by the Wannier model, i.e. the Rydberg transition energies in liquid hydrocarbons appear to be more like those of the gas phase. Such an assumption is supported by observation of the rather small shift of the band corresponding to the Rydberg level (0.09 eV for benzene[158] and 0.06 eV for tetraaminoethylene[157]) as the solvent environment is changed from *n*-hexane to tetramethylsilane. The shift of levels for hydrogenic states lying near the ionization threshold (near the bottom of the conduction band) is expected to be close to a change of the V_0 level, equalling 0.6 ± 0.05 eV for these solvents.[71] As to Rydberg states with large orbitals, there is no experimental data indicating their existence in organic liquids.

VI. SUMMARY

In the review we mainly discussed the photoionization processes following electron ejection into a non-polar liquid. Qualitative features of the processes are clear enough today. The most considerable distinction of their kinetics in non-polar liquids, compared with polar, consists in more intensive geminate recombination in the former case. Considering this recombination leads to the following supposition. Although the photoionization efficiency determined from photoconductivity or charged particle absorption spectra is, as a rule, far lower in non-polar liquids than in polar, nevertheless the initial quantum yield of ion pairs in both systems is about the same.

A good qualitative description of the principal features of geminate recombination of ion pairs photogenerated in non-polar liquids has been given by the Onsager model. The strongest quantitative verification of the model is in good accordance of model predictions with the experimental field dependence of the relative free ion yield in low electric fields. However, this result does not seem to be sufficient to conclude that the Onsager model is fit for a quantitative description of behaviour of the majority of geminate ion pairs. The detailed information on the kinetics of geminate recombination is needed to verify the model. On the other hand, interpretation of the kinetics data in the framework of the Onsager model will not be very successful if one uses average mobilities of charge particles and does not consider the dynamics of thermalization and localization of quasi-free electrons in liquids.

References

1. Mozumder, A., in *Advances in Radiation Chemistry* (Eds. M. Burton and J. L. Magee), Vol. 1, p. 1, Wiley-Interscience, New York, London, Sydney, Toronto, 1969.
2. Hummel, A., in *Advances in Radiation Chemistry* (Eds. M. Burton and J. L. Magee), p. 1, Wiley, New York 1974.
3. Warman, J. M., in *The Study of Fast Processes and Transient Species by Electron Pulse Radiolysis* (Eds. J. H. Baxendale and F. Busi), p. 520, Reidel, Dordrecht, 1982.
4. Schmidt, W. F., in *Photoconductivity and Related Phenomena* (Eds. J. Mort and D. M. Pai), p. 335, Elsevier, New York 1976.
5. Kevan, L., *Adv. Rad. Chem.*, **4**, 181 (1974).
6. Silinsh, E. A., Kolesnilov, V. A., Muzikante, I. J., and Balode, D. R., *Phys. Stat. Solidi (b)*, **113**, 379 (1982).
7. Enek, R. G., and Pfister, G., in *Photoconductivity and Related Phenomena* (Eds. J. Mort and D. M. Pai), p. 215, Elsevier, New York, 1976.
8. Borsenberger, P. M., Contois, L. E., and Hoesterey, D. C., *J. Chem. Phys.*, **68**, 637 (1978).
9. Tjutnev, A. P., Vannikov, A. V., and Saenko, V. S., *High Energy Chem. (USSR)*, **17**, 3 (1983).
10. Freeman, G. R., *Canad. J. Chem.*, **55**, 1797 (1977).
11. Jarnagin, R. C., *Acc. Chem. Res.*, **4**, 420 (1971).
12. Takeda, S. S., Houser, N. E., and Jarnagin, R. C., *J. Chem. Phys.*, **54**, 3195 (1971).
13. Holroyd, R. A., *J. Chem. Phys.*, **57**, 3007 (1972).
14. Alchalai, A. A., Tamir, M., and Ottolengi, M., *J. Phys. Chem.*, **76**, 2229 (1972).
15. Nakato, Y., Ozaki, M., and Tsubomura, H., *J. Phys. Chem.*, **76**, 2105 (1972).
16. Nakato, Y., Chioda, T., and Tsubomura, H., *Bull. Chem. Soc. Jap.*, **47**, 3001 (1974).
17. Holroyd, R. A., and Russel, R. L., *J. Phys. Chem.*, **78**, 2128 (1974).
18. Peterson, S. N., Yaffe, M., Schulc, J. A., and Jarnagin, R. C., *J. Chem. Phys.*, **63**, 2625 (1975).
19. Nakato, Y., and Tsubomura, H., *J. Phys. Chem.*, **79**, 2135 (1975).
20. Sethi, D. S., Choi, H. T., and Braun, C. L., *Chem. Phys. Lett.*, **74**, 223 (1980).
21. Holroyd, R. A., and Jicha, D. L., *Radiat. Phys. Chem.*, **20**, 259 (1982).
22. Choi, H. T., Sethi, D. S., and Braun, C. L., *J. Chem. Phys.*, **77**, 6021 (1982).
23. Braun, C. L., Scott, T. W., and Albrecht, A. C., *Chem. Phys. Lett.*, **84**, 248 (1981).
24. Holroyd, R. A., Preses, J. M., and Zevos, N., *J. Chem. Phys.* **79**, 483 (1983).
25. Bullot, J., Cordier, P., and Gauthier, M., *Chem. Phys. Lett.*, **54**, 77 (1978).
26. Bullot, J., Cordier, P., and Gauthier, M., *J. Chem. Phys.*, **69**, 1374 (1978).
27. Bullot, J., Cordier, P., and Gauthier, M., *J. Chem. Phys.*, **69**, 4903 (1978).

28. Bullot, J., Cordier, P., and Gauthier, M., *J. Phys. Chem.*, **84**, 1253 (1980).
29. Bullot, J., Cordier, P., and Gauthier, M., *J. Phys. Chem.*, **84**, 3516 (1980).
30. Baird, J. K., Bullot, J., Cordier, P., and Gauthier, M., *J. Chem. Phys.*, **74**, 1692 (1981).
31. Braun, C. L., Scott, T. W., and Albrecht, A. C., *Chem. Phys. Lett.*, **90**, 81 (1982).
32. Wu, K. C., and Lipsky, S. J., *J. Chem. Phys.*, **66**, 5614 (1977).
33. Piciulo, P. L., and Thomas, J. K., *J. Chem. Phys.* **63**, 3260 (1978).
34. Lee, K., and Lipsky, S., *Radiat. Phys. Chem.*, **15**, 305 (1980).
35. Siomos, K., Kourouklis, G., Christophorou, L. G., and Carter, J. G., *Radiat. Phys. Chem.*, **15**, 313 (1980).
36. Siomos, K., Kourouklis, G., Christophorou, L. G., and Carter, J. G., *Radiat. Phys. Chem.*, **17**, 75 (1981).
37. Lee, K., and Lipsky, S., *J. Phys. Chem.*, **86**, 1985 (1982).
38. Borovkova, V. A., Kirjuchin, Yu. I., Sinitzina, Z. A., Romashov, L. V., and Bagdasaryan, Ch. S. *High Energy Chem. (USSR)*, **15**, 225 (1981).
39. Richards, J. T., and Thomas, J. K., *Trans. Farad. Soc.*, **66**, 621 (1970).
40. Richards, J. T., West, G., and Thomas, J. K., *J. Phys. Chem.*, **74**, 4137 (1970).
41. Kirjuchin, Yu. I., Borovkova, V. A., Sinitzina, Z. A., and Bagdasaryan, Ch. S., *High Energy Chem. (USSR)*, **13**, 509 (1979).
42. Borovkova, V. A., Kirjuchin, Yu. I., Ramashov, L. V., Sinitzina, Z. A., and Bagdasaryan, Ch. S., *Chem. Phys. (USSR)*, **N1**, 84 (1982).
43. Lukin, L. V., Tolmachev, A. V., and Yakovlev, B. S., *Chem. Phys. Lett.*, **81**, 595 (1981).
44. Lukin, L. V., Tolmachev, A. V., and Yakovlev, B. S., *High Energy Chem., (USSR)*, **16**, 415 (1982).
45. Braun, C. L., and Scott, T. W., *J. Phys. Chem.*, **87**, 4776 (1983).
46. Tolmachev, A. V., Lukin, L. V., and Yakovlev, B. S., *Chem. Phys. USSR* (1985), in press.
47. Simos, K., and Christophorou, L. G., *Chem. Phys. Lett.*, **72**, 43 (1980).
48. Simos, K., Kourouklis, G., and Christophorou, L. G., *Chem. Phys. Lett.*, **80**, 504 (1981).
49. Scott, T. W., Braun, C. L., and Albrecht, A. C., *J. Chem. Phys.*, **76**, 5195 (1982).
50. Sauer, Jr., M. C., Trifunac, Al. D., Cooper, R., and Meisel, D., *Chem. Phys. Lett.*, **92**, 178 (1982).
51. Lukin, L. V., Tolmachev, A. V., and Yakovlev, B. S., *Chem. Phys. Lett.*, **99**, 16 (1983)
52. Lukin, L. V., Tolmachev, A. V., and Yakovlev, B. S., *Chem. Phys. USSR* (1985), in press.
53. Novikov, G. F., and Yakovlev, B. S., *High Energy Chem. USSR*, in press.
54. Tolmachev, A. V., and Yakovlev, B. S., *Chem. Phys. Lett.* (1985), in press. Yakovlev, B. S., Lukin, L. V., Tolmachev, A. V1., in: Conference Record., Eighth Internat. Confer. Conduct. and Breakdown in Dielect. Liquids, Pavia-Italya, Eds.: Molinary, G., and Viviany, A., Sponsored by IEEE Electric. Insulat. Society, p. 332 (1984).
55. Onsager, L., *Phys. Rev.*, **54**, 554 (1938).
56. Terlecki, J., and Fiutak, J., *Internat. J. Radiat. Phys. Chem.*, **4**, 469 (1972).
57. Knights, J. C., and Davis, E. A., *J. Phys. Chem. Solids*, **35**, 543 (1974).
58. Pai, D. M., and Enek, R. C. *Phys. Rev.*, **B11**, 5169 (1975).
59. Hummel, A., and Allen, A. O., *J. Chem. Phys.*, **46**, 1602 (1967).
60. Dodelet, J.-P., Fuochi, P. G., and Freeman, G. R., *Canad. J. Chem.*, **50**, 1617 (1972).
61. Tolmachev, A. V., and Yakovlev, B. S. (in press).

62. Bernas, A., Gauthier, M., and Grand, D., *J. Phys. Chem.*, **76**, 2236 (1972).
63. Bernas, A., Gauthier, M., Grand, D., and Pazlant, G., *Chem. Phys. Lett.*, **17**, 439 (1972).
64. Bernas, A., Blais, J., Gauthier, M., and Grand, D., *Chem. Phys. Lett.*, **30**, 383 (1975).
65. Bullot, J., and Gauthier, M., *Canad. J. Chem.*, **55**, 1821 (1977).
66. Lyons, L. E., *J. Chem. Soc.*, **1957**, 5001.
67. Lyons, L. E., and Mackie, J. C., *Proc. Chem. Soc.*, **71** (1962).
68. Springett, B. E., Jortner, J., and Cohen, M. H., *J. Chem. Phys.*, **48**, 2720 (1968).
69. Davis, H. T., Schmidt, L. D., and Minday, R., *Phys. Rev.*, **A3**, 1077 (1971).
70. Raz, B., and Jortner, J., *Chem. Phys. Lett.*, **4**, 155 (1969).
71. Holroyd, R. A., Tames, S., and Kennedy, A., *J. Phys. Chem.*, **79**, 2857 (1975).
72. Nakato, Y., Ozaki, M., Egava, A., and Tzubomura, T., *Chem. Phys. Lett.*, **9**, 615 (1971).
73. Born, M., *Z. Phys.*, **1**, 45 (1920).
74. Baxendale, J. H., Ball, C., and Wardman, P., *Chem. Phys. Lett.*, **12**, 347 (1971).
75. Richards, J. T., and Thomas, J. K., *Chem. Phys. Lett.*, **10**, 317 (1971).
76. Gillis, H. A., Klassen, N. V., Teather, G. G., and Lokan, K. H., *Chem. Phys. Lett.*, **10**, 481 (1971).
77. Gillis, H. A., Klassen, N. V., and Woods, R. J., *Canad. J. Chem.*, **55**, 2022 (1977).
78. Schmidt, W. F., *Canad. J. Chem.*, **55**, 2197 (1977).
79. Sowada, U., and Holroyd, R. A., *J. Phys. Chem.*, **84**, 1150 (1980).
80. Balakin, A. A., Lukin, L. V., Tolmachev, A. V., and Yakovlev, B. S., *Optics and Spectroscopy (USSR)*, **50**, 161 (1981).
81. Casanovas, J., Grob, R., Delacroix, D., Gueffucci, J. P., and Blanc, D., *J. Chem. Phys.*, **75**, 4661 (1981).
82. Pope, M., and Swenberg, C. E., in *Electronic Processes in Organic Crystals*, p. 206, Oxford University Press, Oxford, 1982.
83. Sano, H., and Mozumder, A., *J. Chem. Phys.*, **66**, 689 (1977).
84. Abell, G. C., and Funabashi, K., *J. Chem. Phys.*, **58**, 1079 (1973).
85. Casanovas, J., Grob, R., Blanc, D., Brunet, G., and Mathien, J. P., *J. Chem. Phys.*, **63**, 3673 (1975).
86. Schmidt, W. F., and Allen, A. O., *J. Chem. Phys.*, **52**, 2345 (1970).
87. Mozumder, A., and Magee, J. L., *J. Chem. Phys.*, **47**, 939 (1967).
88. Funabashi, K., and Kajiwara, T., *J. Phys. Chem.*, **76**, 2726 (1972).
89. Schiller, R., and Vass, Sz., *Intern. J. Radiat. Phys. Chem.*, **7**, 193 (1975).
90. Comizroli, R. B., *Photochem. Photobiol.*, **15**, 399 (1972).
91. Brocklehurst, B., *Nature (London)*, **221**, 921 (1969).
92. Brocklehurst, B., *Chem. Phys. Lett.*, **28**, 357 (1974).
93. Chen, D. H., and Kevan, L., *Molec. Cryst. Liquid Cryst.*, **9**, 183 (1969).
94. Willard, J. E., *Molec. Cryst. Liquid Cryst.*, **9**, 135 (1969).
95. Kenney-Wallace, G. A., in *Picosecond Phenomena* (Eds. S. V. Shank, E. P. Ippen and S. L. Shapiro), p. 208, Springer-Verlag, Berlin, Heidelberg, New York, 1978.
96. Kenney-Wallace, G. A., Hall, G. E., Hunt, L. A., and Sarantidls, K., *J. Phys. Chem.*, **84**, 1145 (1980).
97. Balakin, A. A., and Yakovlev, B. S., *Chem. Phys. Lett.*, **66**, 299 (1979).
98. Balakin, A. A., Lukin, L. V., Tolmachev, A. V., and Yakovlev, B. S. *High Energy Chem. (USSR)*, **15**, 123 (1981).
99. Lukin, L. V., Tolmachev, A. V., and Yakovlev, B. S., *Reports Acad. of Science, USSR. Chemical Series.* (1985), in press.
100. Tewari, P. H., and Freeman, G. R., *J. Chem. Phys.*, **49**, 954, 4394 (1968).
101. Minday, R. M., Schmidt, L. D., and Davis, H. T., *J. Chem. Phys.*, **50**, 1473 (1969).

102. Schmidt, W. F., and Allen, A. O., *J. Chem. Phys.*, **50**, 5037 (1969).
103. Conrad, E. E., and Silverman, J., *J. Chem. Phys.*, **51**, 450 (1969).
104. Minday, R. M., Schmidt, L. D., and Davis, H. T., *J. Chem. Phys.*, **54**, 3112 (1971).
105. Hummel, A., and Schmidt, W. F., *Radiat. Res. Rev.*, **5**, 119 (1974).
106. Davis, H. T., and Brown, R. G., *Adv. Chem. Phys.*, **31**, 329 (1975).
107. Munor, N. C., and Ascarelli, G., *Chem. Phys. Lett.*, **94**, 235 (1983).
108. Sowada, U., Warman, J. M., and De Haas, M. P., *Chem. Phys. Lett.*, **90**, 239 (1982).
109. Freeman, G. R., and Fayadh, J. M., *J. Chem. Phys.*, **43**, 86 (1965).
110. Freeman, G. R., *J. Chem. Phys.*, **46**, 2822 (1967).
111. Burton, M., and Funabashi, K., *Molec. Cryst. Liquid Cryst.*, **9**, 153 (1969).
112. Hamill, W. H., *J. Chem. Phys.*, **53**, 473 (1970).
113. Kevan, L., *J. Phys. Chem.*, **76**, 3830 (1972).
114. Willard, J. E., *J. Phys. Chem.*, **79**, 2966 (1975).
115. Novikov, G. F., and Yakovlev, B. S., *Intern. J. Radiat. Phys. Chem.*, **1**, 479 (1975).
116. Ioffe, A. F., and Regel, A. P., *Progr. Semicond.*, **4**, 237 (1960).
117. Mozumder, A., and Magee, J. L., *J. Chem. Phys.*, **47**, 939 (1967).
118. Lukin, L. V., and Yakovlev, B. S., *High Energy, Chem. USSR* (1985), in press.
119. Noolandi, J., and Hong, K. M., *Chem. Phys. Lett.*, **58**, 575 (1978).
120. Hong, K. M., and Noolandi, J., *J. Chem. Phys.*, **68**, 5163 (1978).
121. Hong, K. M., and Noolandi, J., *J. Chem. Phys.*, **68**, 5172 (1978).
122. Hong, K. M., and Noolandi, J., *J. Chem. Phys.*, **69**, 5026 (1978).
123. Noolandi, J., and Hong, K. M., *J. Chem. Phys.*, **70**, 3230 (1979).
124. Freed, J. H., and Pederson, J. B., *Adv. Magn. Reson.*, **8**, 1 (1976).
125. Abell, G. C., Mozumder, A., and Magee, J. L., *J. Chem. Phys.*, **56**, 5422 (1972).
126. Van den Ende, C. A. M., Luthjens, L. H., Warman, J. M., and Hummel, A., *Radiat. Phys. Chem.*, **19**, 455 (1982).
127. Mozumder, A., *J. Chem. Phys.*, **55**, 3020 (1971).
128. Mozumder, A., *J. Chem. Phys.*, **48**, 1659 (1968).
129. Hummel, A., *J. Chem. Phys.*, **49**, 4840 (1968).
130. Jonah, C. D., *Radiat. Phys. Chem.*, **21**, 53 (1983).
131. Warman, J. M., Asmus, K.-D., and Schuler, R. H., *J. Phys. Chem.*, **73**, 931 (1969).
132. Tachiya, M., *J. Chem. Phys.*, **70**, 4701 (1979).
133. Crumb, J. A., and Baird, J. K., *J. Phys. Chem.*, **83**, 1130 (1979).
134. Mozumder, A., and Magee, J. L., *Radiat. Res.*, **28**, 203 (1966).
135. Van der Ende, C. A. M., Ph.D. Thesis, University of Leiden, The Netherlands, September 1981.
136. Allen, A. O., Gangwer, R. A., and Holroyd, R. A., *J. Phys. Chem.*, **79**, 25 (1975).
137. Mozumder, A., and Tachiya, M., *J. Chem. Phys.*, **62**, 979 (1975).
138. Ametov, K. K., Novikov, G. F., and Yakovlev, B. S., *Solid State Phys. (USSR)*, **18**, 3720 (1976).
139. Yakovlev, B. S., Ametov, K. K., and Novikov, G. F., *Radiat. Phys. Chem.*, **11**, 77 (1978).
140. Frankevich, E. L., and Balabanov, E. I., *Phys. Stat. Sol. (USSR)*, **7**, 1667 (1965).
141. Yakovlev, B. S., and Novikov, G. F., *Intern. J. Radiat. Phys. Chem.*, **7**, 679 (1975).
142. Ametov, K. K., Novikov, G. F., and Yakovlev, B. S., *Radiat. Phys. Chem.*, **10**, 43 (1977).
143. Mott, N. F., and Davis, E. A., *Electronic Processes in Non-Crystalline Materials*, Chap. 2, Clarendon, Oxford, 1971.
144. Bagdasaryan, Kh. S., Kirjukhin, Yu. I., and Sinitzina, Z. A., *Chem. Phys. Lett.*, **57**, 417 (1978).
145. Bagdasaryan, Kh. S., *J. Mendeleev Soc. II (USSR)*, **11**, 216 (1966).

146. Johnson, P. M., and Albrecht, A. C., in *The Chemistry of Ionization and Excitation* (Eds. G. Johnson and G. Seoles), p. 91, University of Newcastle upon Tyne, Taylor-Francis, 1967.
147. Potashnik, R., Ottolenghi, M., and Bensasson, R., *J. Phys. Chem.*, **73**, 1912 (1969).
148. Ottolenghi, M., *Chem. Phys. Lett.*, **12**, 339 (1971).
149. Nakato, Y., *J. Am. Chem. Soc.*, **98**, 7203 (1976).
150. Nakato, Y., Nakane, A., and Tsubomura, H., *Bull. Chem. Soc. Japan*, **49**, 428 (1976).
151. Sinitzina, Z. A., Kirjukhin, Yu. I., and Bagdasaryan, Kh. S., *Rep. USSR Acad. Sci.*, **215**, 1414 (1974).
152. Jortner, J., and Gaathon, A., *Canad. J. Chem.*, **55**, 1801 (1977).
153. Katz, B., Brith, M., Sharf, B., and Jortner, J., *J. Chem. Phys.*, **50**, 5195 (1969).
154. Muto, Y., Nakato, Y., and Tsubomura, H., *Chem. Phys. Lett.*, **9**, 597 (1971).
155. Nakato, Y., Ozaki, M., and Tzubomura, H., *J. Phys. Chem.*, **76**, 2105 (1972).
156. Nakato, Y., Ozaki, M., and Tzubomura, H., *Bull. Chem. Soc. Japan*, **45**, 1299 (1972).
157. Nakato, Y., and Tzubomura, H., *J. Phys. Chem.*, **79**, 2135 (1975).
158. Scott, T. W., and Albrecht, A. C., *J. Chem. Phys.*, **74**, 3807 (1981).
159. Resca, L., and Resta, R., *Phys. Rev.*, **B19**, 1683 (1979).
160. Schmidt, K. H., *Chem. Phys. Lett.*, **103**, 129 (1983).
161. Warman, J. M., *Chem. Phys. Lett.*, **92**, 181 (1982).
162. Böttcher, E. H., and Schmidt, W. F., Proceeding 5th Jihany Symp. Radiat. Chem., Siófok, 19–24 Sept. 1982, vol. 1, Budapest, Akad. Kiáds, p. 427, (1983).
163. Holroyd, R. A., Press, J. M., Böttcher, E. H., and Schmidt, W. F., *J. Phys. Chem.*, **88**, 744 (1984).

Photodissociation and Photoionization
Edited by K. P. Lawley

PHOTOELECTRON SPECTROSCOPY OF EXCITED STATES

KATSUMI KIMURA
Institute for Molecular Science, Okazaki 444, Japan

CONTENTS

I. INTRODUCTION

The electronic structures of numerous chemical species in their ground states have been determined by gas-phase photoelectron spectroscopy. For many years the radiation sources used were limited to frequencies in the VUV or X-ray regions. Recently, a new type of photoelectron spectroscopy has been developed to study excited state molecules in the gas phase by means of stepwise multiphoton ionization, using visible, UV and VUV lasers. This technique will be reviewed in this paper. Use of laser radiation is an efficient method for promoting molecules to electronically excited states, from which they undergo ionization transitions to produce photoelectrons. Therefore, the combination of a photoelectron spectroscopic technique with an excitation-ionization laser system makes it possible to study molecular excited states.

The two-step multiphoton ionization scheme may be represented, in general, by

$$M \xrightarrow{mh\nu} M^* \xrightarrow{nh\nu'} M^+ + e^- \quad (1)$$

where M* is a specific excited state in resonance with the m-photon energy of the excitation laser (ν), M^+ is an ionic state produced from the resonant excited state (M*) by an additional n-photon absorption from the ionization laser (ν')

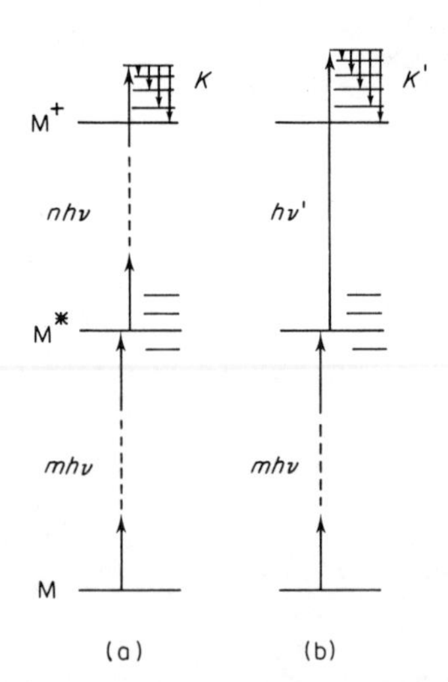

Fig. 1. Schematic energy level diagram relevant to two-step multiphoton ionization. (a) One-colour $(m+n)$ resonant ionization. (b) Two-colour $(m+1)$ resonant ionization. The photoelectron energy is shown by K.

and e^- is the corresponding photoelectron which carries new information about the intermediate excited state. This two-step multiphoton process (1) is schematically shown in Fig. 1a.

In the simplest experimental arrangement, one uses a single laser and tunes its wavelength so as to produce a specific excited state. This process is described by

$$M \xrightarrow{mh\nu} M^* \xrightarrow{nh\nu} M^+ + e^- \tag{2}$$

Then the total photon energy is $(m+n)h\nu$, which exceeds the first ionization potential of M. In the future the process which will probably become the most utilized one in excited state photoelectron spectroscopy is the single-photon ionization of the excited state M* by the second laser (ν'), as shown in Fig. 1b. For one-colour or two-colour experiments, a single laser or two different lasers are needed respectively.

Laser multiphoton ionization of a molecule should be enhanced at a laser frequency at which the photon energy ($mh\nu$) is in exact resonance with one of its specific excited states. Indeed, such enhancement has been widely observed in various ion-current spectra (refer to the review paper by Johnson, 1980). The multiphoton ionization spectrum displayed in a recording of the total ion current as a function of the laser wavelength is often called an 'MPI spectrum'. In this paper we will use the term 'MPI ion-current spectrum' to distinguish it from the 'MPI photoelectron spectrum'.

The MPI ion-current spectrum, shown schematically in Fig. 2, in general consists of many peaks, which correspond to m-photon allowed excited states. From such spectra one can obtain information about resonant intermediate excited states, but no information is thus provided about final states of the ions. Numerous MPI ion-current spectroscopic studies have been reported; these have recently been summarized by Johnson (1980), Johnson and Otis (1981), Antonov and Letokhov (1981), Bekov and Letokhov (1983) and

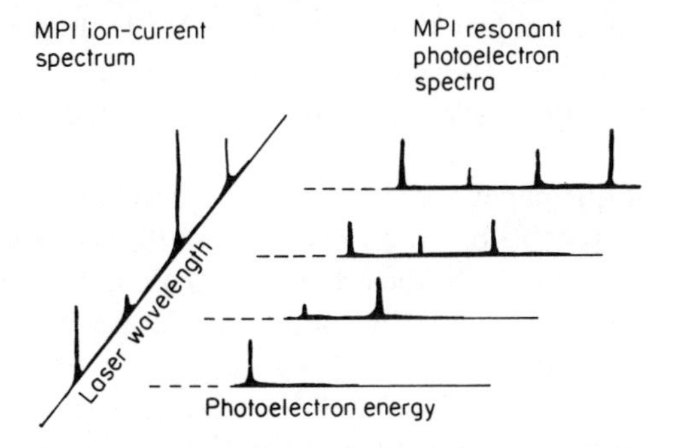

Fig. 2. Schematic drawing showing the relationship between (a) an MPI ion-current spectrum and (b) an MPI photoelectron spectrum.

Letokhov (1983). Mass spectra of ions generated by resonant multiphoton ionization have also been recorded. These studies were summarized by Robin (1980), Lichtin, Zandee and Bernstein (1981) and Antonov and Letokhov (1981). To investigate the mechanisms of ionic fragmentation, identification of the ionic states produced in resonant multiphoton ionization is very important.

To obtain an excited state photoelectron spectrum, one analyses the photoelectrons emitted in process (1) or (2) in terms of their kinetic energy and their angular distribution. Since 1979 the author and his coworkers have been developing in this laboratory a technique based on resonant multiphoton ionization (Achiba *et al.*, 1981a; Achiba, Shobatake and Kimura, 1980) independently of the programmes in the United States by Reilly (Meek, Jones and Reilly, 1980) and by Compton (Compton *et al.*, 1980; Miller and Compton, 1981a, b). Since then, this technique has been applied to obtain photoelectron spectra of many excited states of atoms and of molecules (as will be summarized in Table II).

The principles and applications of conventional photo-electron spectroscopy on molecules have been compiled comprehensively in books by Turner and coworkers (1970), Siegbahn and coworkers (1969), Eland (1974), Rabalais (1977) and Berkowitz (1979). Ionization potential data have been summarized by Turner and coworkers (1970), Siegbahn and coworkers (1974) and Kimura and coworkers (1981). The handbook of Kimura and coworkers (1981) includes HeI photoelectron spectra and ionization potentials of about 200 molecules, with *ab initio* assignments for about 150. The available ionization potential data should prove quite helpful for interpreting excited state photoelectron spectra.

From a technical point of view, in order to practise excited state photoelectron spectroscopy, one must record first the MPI ion-current spectra since the photoelectron spectra should be obtained at those laser wavelengths at which significant ion-current peaks appear. MPI ion-current spectra primarily provide information about the energy levels and population of excited states of a neutral molecule, while MPI photoelectron spectra contain information about the energy levels and populations of ionic states which are produced from a selected excited state, as illustrated schematically in Fig. 3.

The purposes of this paper are (1) to describe the characteristics of excited state photoelectron spectra based on resonant multiphoton ionization, (2) to review coherently the MPI photoelectron spectroscopic studies which have been published thus far and (3) to point out future applications of this technique for investigating molecular excited states. Since we focus our attention on molecular excited state photoelectron spectroscopy associated with resonant multiphoton ionization, we will not include MPI ion-current and mass studies and the 'resonant ionization spectroscopy' (RIS) developed for one-atom detection (refer to the review paper by Hurst and coworkers,

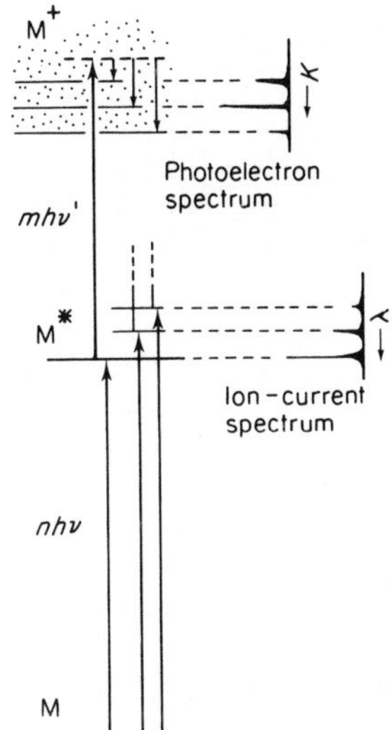

Fig. 3. Schematic energy level diagram relevant to the MPI ion-current and photoelectron spectra. The ion-current peaks correspond to the excited states (M*), while the photoelectron peaks correspond to the ionic states. The photoelectron peak intensity depends on the ionization transition probability between the excited and the ionic state.

1979), and the VUV photoelectron studies of singlet oxygen and vibrationally excited molecules produced by discharge (refer to the review paper by Dyke, Jonathan and Morris, 1979).

II. CHARACTERISTICS OF EXCITED STATE PHOTOELECTRON SPECTROSCOPY

Let us first compare excited state photoelectron spectroscopy of resonant multiphoton ionization with conventional VUV single-photon photoelectron spectroscopy. Specifically, HeI (584 Å) photoelectron spectroscopy, which explores photo-electron bands up to 21.22 eV, is concerned with direct single-electron ionization from ground states. The several possible electronic configurations of the ionic states thus produced are described in terms of single-electron losses with respect to the ground state molecules, as shown schematically in Fig. 4. (When autoionization occurs, two-electron ionization should be taken into account, in which one electron is ejected while the other is excited within the ion.) On the other hand, in excited state photoelectron spectroscopy based on resonant multiphoton ionization, the resonant excited state may be considered as an initial state of ionization. Therefore, as far as direct single-electron ionization is concerned, the various ionic states produced from the ground state are not always obtainable from excited states, also shown schematically in Fig. 4. For example, ionization of a Rydberg electron gives rise to the ground state cation, which is also obtainable by direct single-electron ionization from the ground state. However, ionization of a valence electron from the same Rydberg excited state gives rise to a higher

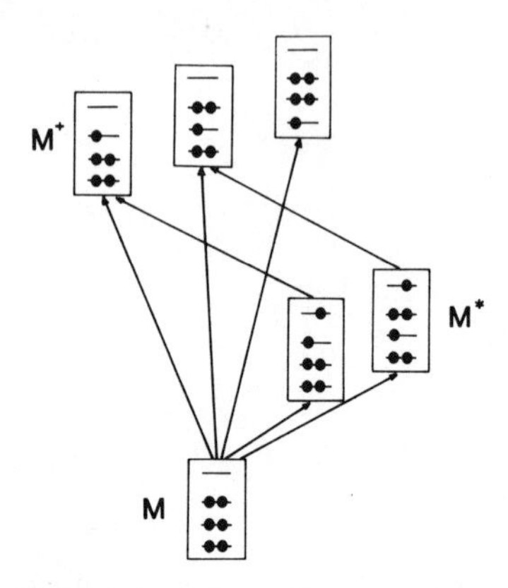

Fig. 4. Schematic electron configurations of M, M* and M^+, relevant to direct single-electron ionization of the ground state and of excited states. The first three ionic states are obtainable from the ground state by single-electron ionization, but are not always allowed from the excited states. The arrow indicates the possible single-electron ionization transition.

ionic state that cannot be produced by direct single-electron ionization from the ground state. Such a higher ionic state may be regarded as having been formed by two-electron excitation with respect to the neutral ground state.

Let us consider a single-photon ionization of the resonant excited state, since this would be the most promising case for resonant MPI photoelectron spectroscopy. The probability of such an ionic transition may be expressed by

$$P \propto \langle \psi_{e''} | \mathrm{M} | \psi_{e'} \rangle | \langle \psi_{v''} | \psi_{v'} \rangle |^2 \tag{3}$$

as in conventional single-photon ionization, where $\psi_{e''}$ and $\psi_{v''}$ are the electronic and vibrational wavefunctions of the initial state and $\psi_{e'}$ and $\psi_{v'}$ are those of the final state. The electronic part $\langle \psi_{e''} | \mathrm{M} | \psi_{e'} \rangle$ largely depends on the resonant excited state subject to the electronic selection rule. On the other hand, the vibrational part $| \langle \psi_{v''} | \psi_{v'} \rangle |^2$ (the Franck–Condon factor) reflects the difference in equilibrium geometry between the excited state (v″) and the ionic state (v′). Therefore, in general, the ionization cross-sections of excited states differ considerably from that of the ground state. Even if the cross-sections are similar, the vibrational structures are different from each other, as shown schematically in Fig. 5. In other words, the photoelectron vibrational structure is characteristic of the resonant excited state. This fact will make it possible for us to track short-lived states.

Another remarkable feature of photoelectron spectroscopy based on multiphoton ionization is that the resonant excited states can be highly

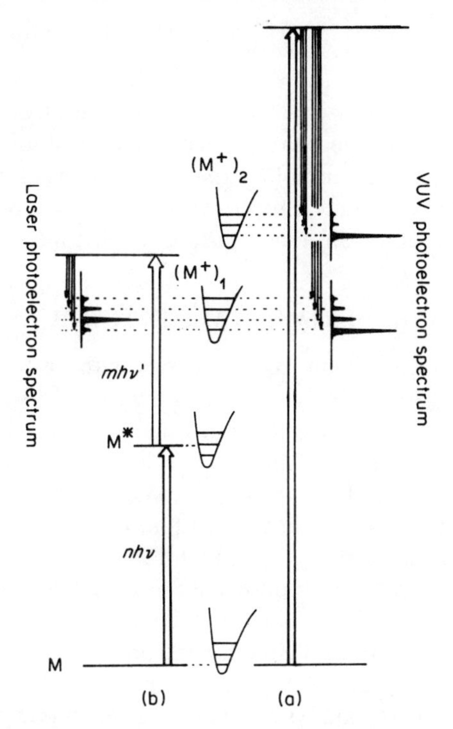

Fig. 5. Schematic energy level diagram relevant to (a) VUV single-photon ionization from the ground state and (b) laser two-step ionization through an excited state. Differences in photoelectron vibrational structure between (a) and (b) are illustrated schematically.

selected, not only with respect to vibrational levels but also for rotational levels by specific laser excitation. It is, therefore, a highly selective analytical diagnostic.

There are four categories of topics which can be investigated by excited state photoelectron spectroscopy with resonant multiphoton ionization:

1. Identification of higher excited states for molecules and characterization of their electronic structure, molecular structure and dynamic behaviour. (One-photon forbidden excited states can also be studied by two-photon excitation techniques.)
2. Formation of specific ionic states and photoelectron spectroscopic studies of their electronic structure, molecular structure and dynamic behaviour. Thus, previously unknown ionic states produced from specific excited states can be studied.
3. Selective excitation of an excited state to autoionizing states and photoelectron spectroscopic studies of autoionization.

4. Investigation of transient species produced by dissociation from an excited state or an ionic state.

It should also be emphasized that MPI photoelectron spectroscopy is applicable to the higher excited states which cannot be studied by fluorescence spectroscopy. Furthermore, use of a supersonic expansion makes it possible not only to study van der Waals clusters with an MPI photoelectron spectrum but also to record high-resolution spectra of cooled molecules.

III. EXPERIMENTAL TECHNIQUES

Excited state photoelectron spectroscopy, with resonant multiphoton ionization, in the gas phase is carried out with an *ns* tunable laser for irradiating the molecules to specific excited states. With a laser source, pulsed photoelectrons of rather low kinetic energies are generated, in marked contrast to conventional HeI and X-ray photoelectron spectroscopic measurements, in which continuously emitted photoelectrons with much higher kinetic energies are produced.

The principal part of the laser photoelectron spectrometer is illustrated in Fig. 6. The apparatus consists of (1) a laser system for excitation and ionization, (2) a nozzle beam source for introducing a gas sample into the ionization region, (3) a spectrometer vacuum chamber with a pumping system, (4) a photoelectron energy analyser, (5) other detection devices for recording the total ion-current and mass spectra, (6) magnetic-field shielding which is very important in photoelectron measurements and (7) a data-acquisition system for photoelectron spectra, as well as for ion-current and mass spectra.

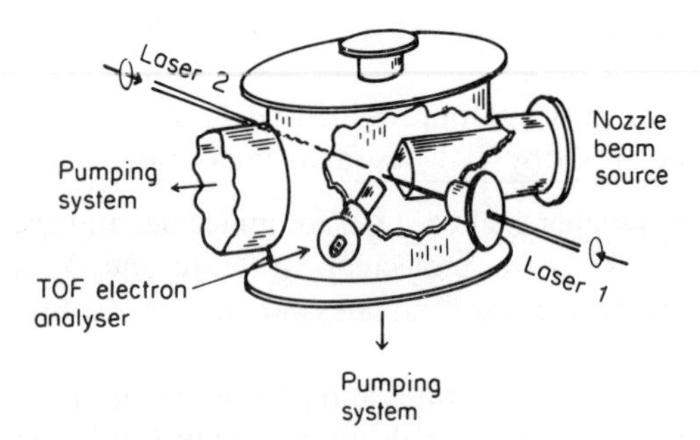

Fig. 6. Schematic drawing of the apparatus for measuring excited state photoelectron spectra with laser sources. A μ-metal shielding (not shown here) is contained inside the main chamber.

A. Lasers for Excitation and Ionization

Tunable dye lasers pumped by Nd–YAG lasers, a nitrogen laser and an XeCl excimer laser have been used for the excitation and ionization sources in the photoelectron spectroscopic studies (listed in Table II). In most of these studies a single laser was used in each experiment. So far only a few studies have been reported in which two lasers were employed. In their two-colour experiments, Miller and Compton (1982) used an excimer-pumped dye laser for the excitation source and a 308-nm XeCl excimer for the ionization source, Achiba *et al.* (1983a) used two kinds of dye lasers. In a typical experiment, a *ns* pulsed, dye laser is focused by an appropriate lens to a small waist coincident with the molecular beam. According to Compton *et al.* (1980), the power density at the focused spot ($\leq 20\,\mu$m) is of the order of 5×10^9 W cm^{-2}. This is typical of the power densities in most photoelectron spectroscopic experiments. In order to generate a photoelectron spectrum with a reasonable S/N ratio it is necessary to accumulate 1000 to 10 000 pulsed photoelectron signals, although it largely depends on the spectral resolution desired and the ionization cross-section.

B. Electron Energy Analyser

Two types of electron energy analysers have been used in the studies reported so far; one is a time-of-flight (TOF) electron velocity analyser and the other is a spherical electrostatic analyser. In the former, 12-, 45- and 50-cm flight tubes have been used by Achiba and coworkers (1980, 1981a), Meek, Jones and Reilly (1980) and Kruit, Kimman and van der Wiel (1981) respectively. For photoelectron measurements with electrostatic analysers, a 160° spherical sector and a hemispherical analyser have been used by Compton and coworkers (1980) and Pratt *et al.* (1983a, b) respectively.

An electron 'parallellizer' was coupled with a 50-cm TOF electron analyser by Kruit, Kimman and van der Wiel (1981) and Kimman, Kruit and van der Wiel (1982). In this method all electrons with a finite component of initial velocity in the direction of the detector, i.e. 50 per cent of the total, are 'parallellized' in a diverging magnetic field, along the axis of the TOF electron analyser tube, and their energy is derived from their time-of-flight analyser (Kruit, Kimman and van der Wiel, 1981).

The photoelectron kinetic energy K may be calculated from the following relationship:

$$\sqrt{K}(\text{eV}) = \{0.0169L(\text{cm})T(\mu\text{s})^{-1}\}^2 \qquad (4)$$

where L is the flight length in centimetres and T is the flight time in microseconds. This relationship is shown for two cases ($L = 12$ and 40 cm) in

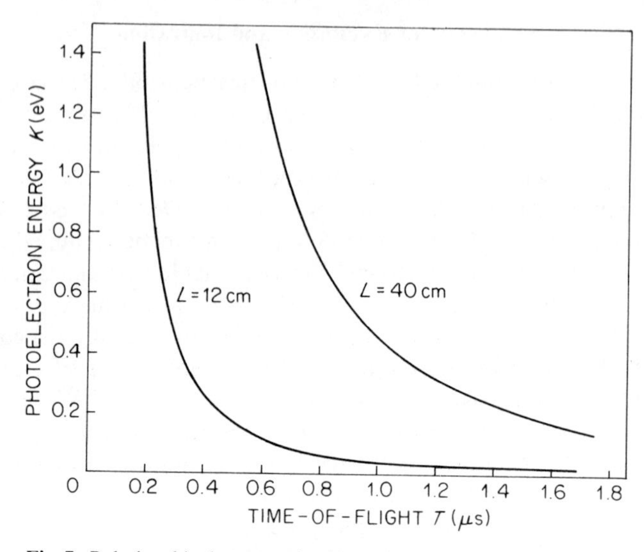

Fig. 7. Relationship between the photoelectron energy (K) and the time of flight (T) for a TOF electron analyser. The curves show the relationship for flight lengths, $L = 12$ and 40 cm.

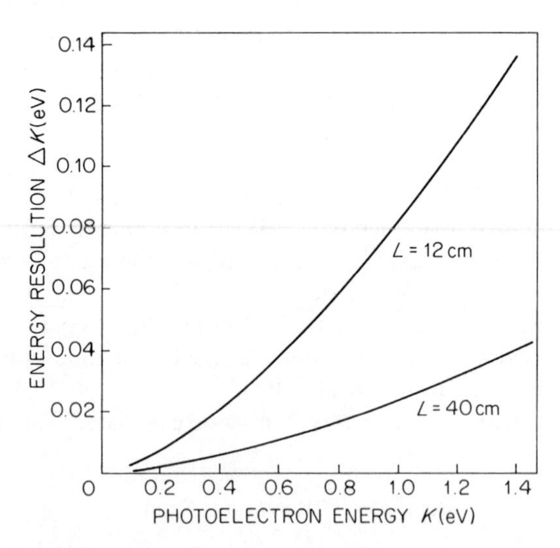

Fig. 8. Relationship between the photoelectron energy resolution (ΔK) and the photoelectron energy (K) for a TOF electron analyser. The two curves are obtained at flight lengths of $L = 12$ and 40 cm.

Fig. 7. The energy resolution is then given by

$$\Delta K(\text{meV}) \simeq 100\Delta T(\text{ns})\{K(\text{eV})\}^{3/2}L(\text{cm})^{-1} \tag{5}$$

where ΔT(ns) is the time resolution of the detection system. This relation is shown for $L = 12$ and 40 cm in Fig. 8. It is clear that energy resolution is high, especially for low-energy electrons. For instance, for a tube length of 12 cm, with $\Delta T \simeq 10$ ns the energy resolutions of the photoelectrons of $K = 0.2$ and 0.4 eV are about 9 and 25 meV respectively.

C. Nozzle Beam Source

An effusive nozzle beam can be used for single molecules under normal conditions. However, a supersonic nozzle beam technique is required to

TABLE I
Laser wavelengths (λ) and photoelectron kinetic energies useful for photoelectron energy calibration.[a]

Atom	λ(nm)	K(eV)	Resonant state	Ionic state	($m+n$) MPI process
Xe	363.9	1.50	$5d[2\frac{1}{2}]^{\circ}_{J=3}$	$Xe^{+}(^{2}P_{3/2})$	3 + 1
Xe	370.5	1.17	$5d[3\frac{1}{2}]^{\circ}_{J=3}$	$Xe^{+}(^{2}P_{3/2})$	3 + 1
Kr	370.7	2.72	$5s[1\frac{1}{2}]^{\circ}_{J=1}$	$Kr^{+}(^{2}P_{3/2})$	3 + 2
Xe	440.9	1.93	$6s[1\frac{1}{2}]_{J=1}$	$Xe^{+}(^{2}P_{3/2})$	3 + 2
		0.63		$Xe^{+}(^{2}P_{1/2})$	3 + 2
Fe	447.7	0.44	$e^{5}D_{J}\, J = 4$	$Fe^{+}(a^{6}D_{J})\, J = \frac{9}{2}$	2 + 1
		0.21	4	$Fe^{+}(a^{6}F_{J})\, J = \frac{9}{2}$	2 + 1
		0.14	4	$\frac{7}{2}$	2 + 1
		0.09	4	$\frac{5}{2}$	2 + 1
		0.05	4	$\frac{3}{2}$	2 + 1
Fe	448.0	0.44	$e^{5}D_{J}\, J = 3$	$Fe^{+}(a^{6}D_{J})\, J = \frac{7}{2}$	2 + 1
		0.25	3	$Fe^{+}(a^{6}F_{J})\, J = \frac{9}{2}$	2 + 1
		0.18	3	$\frac{7}{2}$	2 + 1
		0.13	3	$\frac{5}{2}$	2 + 1
		0.10	3	$\frac{3}{2}$	2 + 1
Fe	463.4	0.16	$e^{7}D_{J}\, J = 4$	$Fe^{+}(a^{6}D_{J})\, J = \frac{9}{2}$	2 + 1
		0.11	4	$\frac{7}{2}$	2 + 1
	464.9	0.18	3	$\frac{9}{2}$	2 + 1
		0.13	3	$\frac{7}{2}$	2 + 1
		0.10	3	$\frac{5}{2}$	2 + 1
	465.9	0.21	2	$\frac{9}{2}$	2 + 1
		0.15	2	$\frac{7}{2}$	2 + 1
		0.12	2	$\frac{5}{2}$	2 + 1
		0.09	2	$\frac{3}{2}$	2 + 1
		0.08	2	$\frac{1}{2}$	2 + 1

[a]The values shown for Xe and Kr are from Sato, Achiba and Kimura (1984a) and those for Fe from Nagano and coworkers (1982).

investigate cooled molecules and van der Waals complexes. Recently, a pulsed, supersonic nozzle beam was used in studies of some van der Waals complexes (see also Table II; Sato, Achiba and Kimura, 1984b; Fuke *et al.*, 1983).

D. Calibration for the Photoelectron Energy

Calibration of the photoelectron energy scale can be made with reference species, the most common one being xenon, which has ionization potentials of 12.130 eV ($^2P_{3/2}$) and 13.436 eV ($^2P_{1/2}$) (Turner *et al.*, 1970). The kinetic energies expected for photoelectrons emitted at laser wavelengths of 363.9, 370.5 and 440.9 nm are summarized in Table I. With 363.9- and 370.5-nm radiation, Xe is ionized by (3 + 1) resonant ionization through the $5d[2\frac{1}{2}]^{\circ}_{J=3}$ and $5d[3\frac{1}{2}]^{\circ}_{J=3}$ states respectively. With 440.9-nm radiation, Xe is ionized by (3 + 2) resonant ionization through the $6s[1\frac{1}{2}]^{\circ}_{J=1}$ state.

Iron atoms are also recommended for reference species, because the atomic Fe is easily produced from $Fe(CO)_5$ in the gas phase at room temperature. Many sharp photoelectron peaks appear at specific laser wavelengths (Nagano *et al.*, 1982). Therefore, iron pentacarbonyl is a good reference sample for photoelectron energy calibration in the low-energy region, as shown in Table I. The photoelectron kinetic energies expected at several laser wavelengths for atomic iron are also summarized in this table.

E. Example of Apparatus

The experimental apparatus constructed in this laboratory (Achiba *et al.*, 1981a; Achiba, Shobatake and Kimura, 1980) for excited state photoelectron spectroscopy is briefly described below. The entire experimental system is shown schematically in Fig. 9. It consists of a vacuum chamber, a pulsed

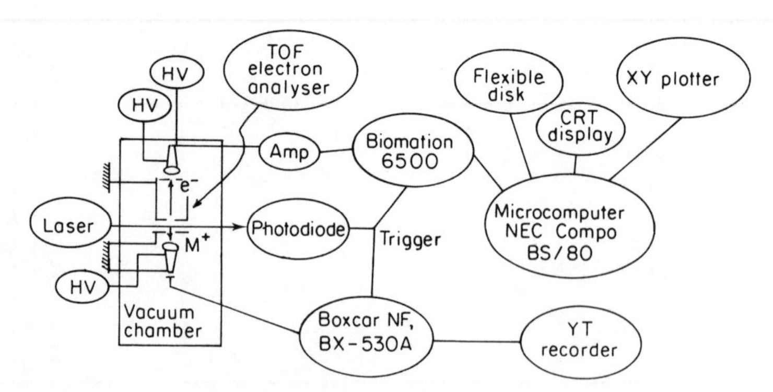

Fig. 9. A block diagram for photoelectron measurements with a pulsed laser (Achiba, Sato and Kinura, 1983a).

nozzle beam sample inlet source, a TOF electron analyser, a total ion-current detector, a pulsed tunable dye laser and a data-acquisition system.

Radiation from an Nd–YAG pumped dye laser is focused by a 10 to 25-cm focal length lens onto the ionization region, located about 10 mm downstream from the nozzle tip. To generate the desired laser wavelengths, e.g. around 380 nm, radiation from the dye laser and the fundamental line of Nd–YAG laser are mixed by using a wavelength extension system (Sato, Achiba and Kimura, 1983). In two-colour experiments, two dye lasers irradiate the beam from opposite directions (Achiba, Sato and Kimura, 1983a).

Two kinds of nozzles are used: one is of stainless steel with an Mo aperture ($d = 0.1$ mm) for producing a continuous jet, while the other is a pulsed injector with an orifice of 0.3 mm in diameter. The open time of the pulsed nozzle is 1 ms; it has a repetition rate of 10 Hz. The vacuum apparatus is shown schematically in Fig. 10; it consists of three chambers. The ionization region is

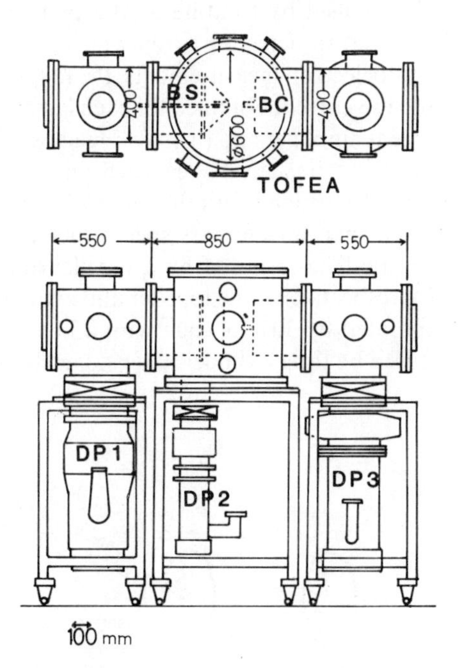

Fig. 10. Schematic of the apparatus for excited state photoelectron spectroscopy, developed in this laboratory (Achiba, Shobatake and Kimura, 1980). BS: beam source, BC: beam catcher, TOFEA: time-of-flight electron analyser, DP1: oil diffusion pump (5000 l^{-1}s), DP2: oil diffusion pump (1500 l^{-1}s), DP3: oil diffusion pump (3500 l^{-1}s). A μ-metal shielding is contained in the main vacuum chamber, surrounding the ionization region and the TOF electron analyser.

evacuated by two oil diffusion pumps with liquid nitrogen buffles to maintain the pressure in the electron analyser at about 1×10^{-5} torr.

Photoelectron kinetic energy measurements are carried out with a TOF electron analyser, in which photoelectrons travel through a magnetically shielded 12-cm long (field-free) drift tube to a channeltron located at the terminus of the tube. A laser pulse detected by a fast pindiode triggers the sweep of a transient recorder (500 MHz per channel, 2 ns per channel) into which the photoelectron signals generated by the channeltron electron multiplier are fed. Signal averaging and data storage are performed by a microcomputer-controlled data-acquisition system.

In order to investigate the angular dependence of the photoelectrons emitted by resonant multiphoton ionization, intensity measurements are carried out at various angles (θ). Relative intensities at a range of angles are recorded relative to the intensity at $\theta = 0°$. The direction of the laser polarization vector is changed by rotating a ($\frac{1}{2}$) λ plate with respect to the direction of the photoelectron detector, as shown schematically in Fig. 11.

Prior to making photoelectron measurements, the MPI ion-current spectra are recorded. The total ion current obtained for each laser shot is converted to a voltage by a fast current amplifier and then averaged by a boxcar integrator. The output signals from the boxcar are recorded on a strip chart recorded. Mass spectra of ions are determined with the same TOF analyser used for the photoelectron measurements. In the mass spectrometry mode, the ions are accelerated by an electric field imposed by a repeller and an extractor, set typically at 10 and 5 V respectively. The repeller and extractor are the disks of gold-plated brass separated by 30 mm, each with a hole 10 mm in diameter. The ions are then focused by three cylindrical lenses introduced into the TOF analyser tube.

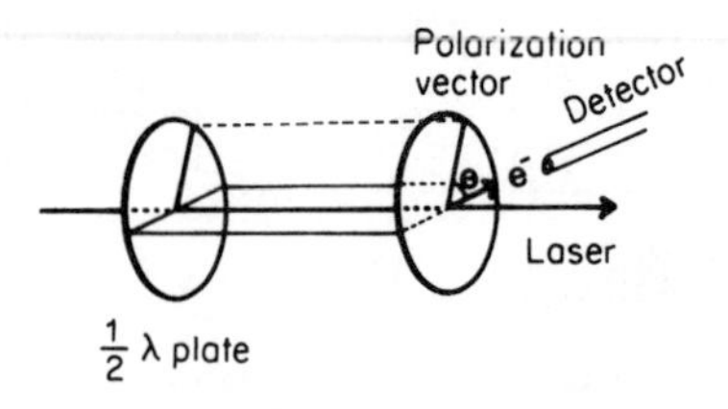

Fig. 11. Schematic drawing of the measurement of photoelectron angular distributions with laser radiation. The direction of the polarization vector is changed by rotating a $\lambda/2$ plate with respect to the direction of the photoelectron detection.

IV. BASIC INFORMATION OBTAINED FROM EXCITED STATE PHOTOELECTRON SPECTROSCOPY

A. Photoelectron Energy

The photoelectron spectrum of a molecule provides information on the energy levels of its ionic states—in particular, those which are electronically and vibrationally excited. The kinetic energy of photoelectrons, determined from a signal maximum, is related to energy of the ionic state by

$$K = E + nh\nu' - K \tag{6}$$

where K is the photoelectron kinetic energy, $nh\nu'$ the n-photon energy, E is the energy of the excited state from which ionization takes place and I is the energy of the ionic state. The conservation relation (6) is shown in Fig. 12. The energies E and I may be expressed in terms of

$$E = E_0 + \Delta E \tag{7}$$

$$I = I_0 + \Delta I \tag{8}$$

where E_0 and I_0 are the 0–0 energies of the excited and the ionic state respectively, and ΔE and ΔI are the internal energies of the excited and the ionic state respectively. The information obtained from Eq. (6) is essentially the same as that obtained from conventional VUV photoelectron spectra, if the same ionic states are involved.

The intensity of the signals recorded in excited state photoelectron spectroscopy depend sensitively on the excited state which is selected as the initial state for ionization. In other words, the ionization cross-sections for producing ionic states depend largely on the electronic and vibrational wavefunctions, not only of the final, ionic states but also of the initial, excited states. Since electronic and vibrational selection rules for ionization transition apply, molecular excited states are ionized selectively. Thus one can obtain

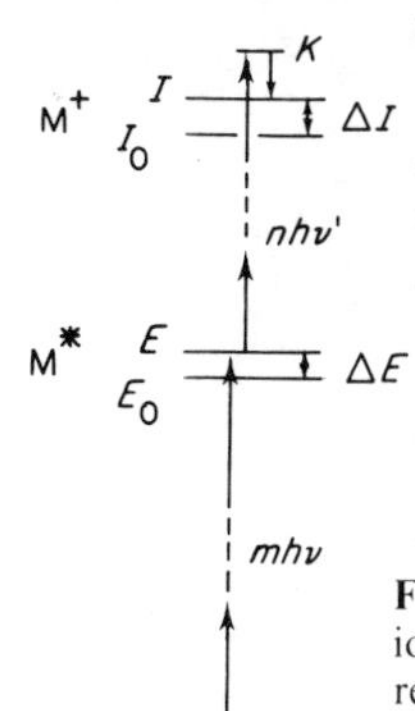

Fig. 12. Schematic energy level diagram relevant to two-step multiphoton ionization. The internal energies of M^* and M^+ are shown by ΔE and ΔI respectively, the 0—0 energies of M^* and M^+ are shown by E_0 and I_0 respectively and the photoelectron energy is given by K.

new information about the intermediate excited states when the ionic states are well identified or, if the excited states are well characterized, one can obtain new information about the ionic states.

B. Ionization Selectivity

Ionization selectivity has recently been found in the spectrum of xenon (Sato, Achiba and Kumura, 1984). Its $5d[2\frac{1}{2}]^{\circ}_{J=3}$ Rydberg state with the $^2P_{3/2}$ core is in exact resonance with the three-photon energy of a laser wavelength of 363.9 nm; ionization occurs by absorption of one more photon. In this case, formation of the two ionic states, $^2P_{3/2}$ and $^2P_{1/2}$, is energetically possible, but the photoelectron spectrum obtained as a result of the (3 + 1) absorption at this wavelength consists of a single peak only due to $Xe^+(^2P_{3/2})$, with a kinetic energy of 1.50 eV, as shown in Fig. 13. No peak due to Xe ($^2P_{1/2}$) was found. The authors concluded that removal of the Rydberg election does not induce relaxation of the core elections. This result is evidence that single-electron ionization takes place mainly from this Rydberg state when exposed to intense 363.9-nm radiation. Ionization selectivity is an important characteristic of excited state ionization, in sharp contrast to conventional VUV photoionization. In a HeI photoelectron spectrum of Xe, it is well known that two peaks due to both ionic states of Xe^+ ($^2P_{3/2}$ and $^2P_{1/2}$) were observed (Turner and coworkers, 1970).

Another example of ionization selectivity was found in NO (Achiba, Sato and Kimura, 1983b). In the resonant multiphoton ionization of this molecule through the excited B state, the photoelectron spectra indicated that the ground state of the cation NO^+ was not produced by direct ionization from

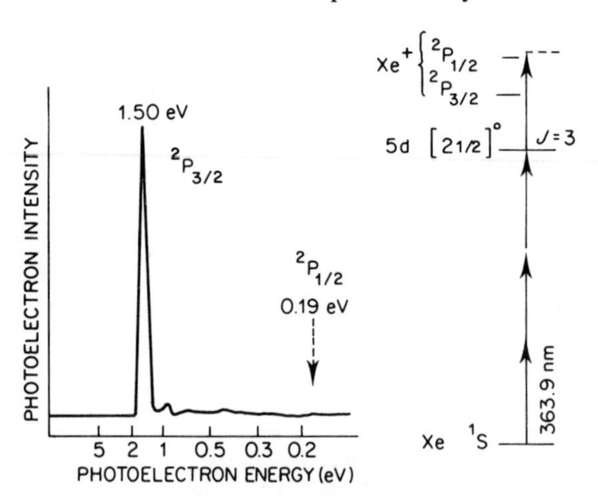

Fig. 13. The energy diagram relevant to (3 + 1) resonant multiphoton ionization of Xe through the $5d[2\frac{1}{2}]^{\circ}_{J=3}$ Rydberg state and the resulting photoelectron spectrum.

the B state, but was produced only by autoionization through higher resonant excited states. The ground state $X(\ldots 3\sigma^2 1\pi^4)$ of NO^+ cannot be generated by direct single-electron ionization from the excited B state $(\ldots 3\sigma^2 1\pi^3 2\pi)$. This ionization selectivity provides an insight into the mechanism of autoionization. In general, ionization selectivity will become more and more important in studies of excited state photoelectron spectroscopy.

Let us now consider other cases in which single-electron ionization selectivity presents a clue to the identification of either neutral excited states of ionic states. A planar carbonyl compound, such as formaldehyde ($H_2C{=}O$), has n and π orbitals which are perpendicular to each other. Its ground state is designated by $\psi(\pi^2 n^2)$ and the $n\pi^*$ and $\pi\pi^*$ excited states are designated by $\psi(\pi^2 n\pi^*)$ and $\psi(\pi n^2 \pi^*)$ respectively. Removal of an electron from the n* or π^* orbitals gives rise to the ionic state $\psi^+(\pi^2 n)$ or $\psi^+(\pi n^2)$ respectively. For single-electron ionization the $\psi^+(\pi^2 n)$ ionic state can be produced from the excited state of $\psi(\pi^2 n\pi^*)$, but not from $\psi(\pi n^2 \pi^*)$. Similarly, the $\psi^+(\pi n^2)$ ionic state can be produced from $\psi(\pi n^2 \pi^*)$ but not from $\psi(\pi^2 n\pi^*)$. Namely,

$$\psi(\pi^2 n^2) \begin{cases} \nearrow \psi(\pi^2 n\pi^*) \rightarrow \psi^+(\pi^2 n) \\ \searrow \psi(\pi n^2 \pi^*) \rightarrow \psi^+(\pi n^2) \end{cases} \tag{9}$$

Such correlations between the neutral excited states and the ionic states permit unique assignments to be made of the excited or the ionic states. Another example is provided by the van der Waals complex, A—B. Ionization of this complex can occur through two kinds of resonant excited state, designated by $\psi(A^*{-}B)$ and $\psi(A{-}B^*)$. These give rise to the ionic states $\psi(A^+{-}B)$ and $\psi(A{-}B^+)$ respectively. Namely,

$$\psi(A{-}B) \begin{cases} \nearrow \psi(A^*{-}B) \rightarrow \psi(A^+{-}B) \\ \searrow \psi(A{-}B^*) \rightarrow \psi(A{-}B^+) \end{cases} \tag{10}$$

Again, for direct single-electron ionization, the ionic states $\psi(A^+{-}B)$ and $\psi(A{-}B^+)$ are produced from $\psi(A^*{-}B)$ and $\psi(A{-}B^*)$ respectively, but are not produced from $\psi(A{-}B^*)$ and $\psi(A^*{-}B)$ respectively.

To date no experimental studies have been reported which demonstrate the two different ionization processes (9) and (10). These are interesting subjects for investigation by resonant MPI photoelectron spectroscopy. A recent laser photoelectron study of the Ar—NO van der Waals complex has indicated that the $(2+1)$ resonant multiphoton ionization through the excited state Ar—NO* (C Rydberg state) gives rise to the ground state ion Ar—NO^+ (Sato, Achiba and Kumura, 1983).

C. Spin-Orbit Splitting

Ejection of an electron from a degenerate orbital of a closed-shell atom or molecule produces an ion in an electronic state characterized by a non-zero

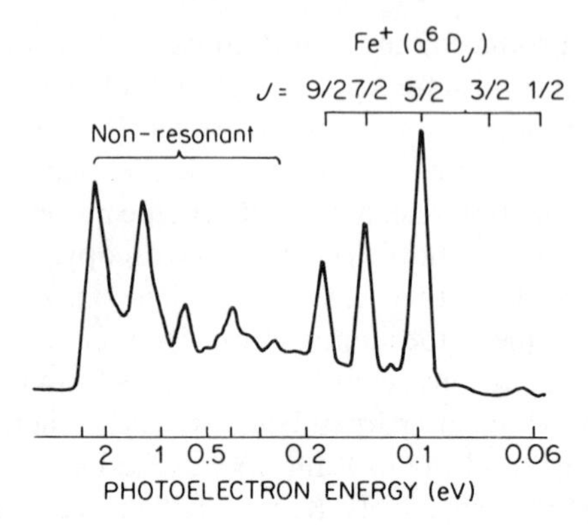

Fig. 14. The photoelectron spectrum obtained for (2 + 1) resonant ionization of atomic Fe through the $J = 3$ level of the e^7D_J excited state, at a laser wavelength of 464.9 nm (Nagano *et al.*, 1982). The Fe atoms were produced by photodecomposition of $Fe(CO)_5$ and then ionized at the same laser.

orbital angular momentum and electron spin angular momentum. The degeneracy of this ionic state is split by interaction (i.e. spin-orbit coupling) between spin and orbital angular momenta. Examples for spin-orbit splitting in HeI photoelectron spectroscopy have been reviewed in the book of Rabalais (1977). Similar phenomena should be observed in excited state photoelectron spectroscopy.

Spin-orbit splitting in the photoelectron spectra of Fe^+ generated from excited states of Fe has been demonstrated (Nagano *et al.*, 1982). In that study, Fe atoms were produced by photodecomposition of $Fe(CO)_5$ with a visible laser, and then ionized through resonant excited states by the same laser. Removal of a 5s electron from the e^7D_J excited state, whose electron configuration is $5d^6[^5D]4s[^6D]5s$, gives rise to the a^6D_J ionic state. A photoelectron spectrum obtained at the $J = 3$ level of the e^7D_J state of Fe is shown in Fig. 14. The three designated peaks are due to the $J = 5/2$, $7/2$ and $9/2$ levels of $Fe^+(e^6D_J)$. This photoelectron spectrum shows that the strongest peak appears at $J = J' - \frac{1}{2}$, where J' is the J number of the e^7D_J excited state. This relationship is a consequence of the angular momentum conservation rule.

D. Vibrational Structure

As in conventional VUV photoionization, the most intense ionization transition occurs from the initial excited state to a point on the ionic potential surface where the nuclear configuration is identical to that of the initial state; the process is a vertical ionization. When the equilibrium geometry of the ionic

TABLE II

Summary of excited state photoelectron spectroscopic studies associated with resonant multiphoton ionization.

Substance	Resonant excited state	MPI process	Reference
		Atoms	
Kr	$5s[1\frac{1}{2}]^{\circ}_{J=1}$	$3h\nu + 2h\nu$	Sato, Achiba and Kimura (1984a)
Xe	$6s[1\frac{1}{2}]^{\circ}_{J=1}$	$3h\nu + 2h\nu$	Compton and coworkers (1980)
		$3h\nu + h\nu'$	Miller and Compton (1982)
		$3h\nu + nh\nu (n \geq 2)$	Kruit and coworkers (1983)
		$3h\nu + 2h\nu$	Sato, Achiba and Kimura (1984a)
	$5d[2\frac{1}{2}]^{\circ}_{J=3}$, $5d[3\frac{1}{2}]^{\circ}_{J=3}$	$3h\nu + h\nu$	Sato, Achiba and Kimura (1984a)
Fe	$e^7D_{J=2,3,4}$, $e^5D_{J=3,4}$	$2h\nu + h\nu$	Nagano and coworkers (1982)
		Molecules	
H_2	B, $v=7$, $J=2, 4$	$3h\nu + h\nu$	Pratt, Dehmer and Dehmer (1983)
NO	A, $v=0 \sim 3$	$2h\nu + 2h\nu$	Miller and Compton (1981a)
	$v=0$	$2h\nu + 2h\nu'$	Miller and Compton (1982)
	$v=0$, $J=6\frac{1}{2}$	$2h\nu + 2h\nu$	Kimman, Kruit and van der Wiel (1982)
	B, $v=9$, Q_{22}, Q_{11}	$2h\nu + h\nu$	Achiba, Sato and Kimura (1983b)
	C, $v=0, 1$	$2h\nu + h\nu$	Miller and Compton (1981a)
	$v=0$, $N=2, 4, 8$	$3h\nu + 2h\nu$	White and coworkers (1982)
	F, $v=0, 1$	$3h\nu + h\nu$	Achiba, Sato and Kimura (1983a)
	$H(H'^2)$, $v=0, 1$	$3h\nu + h\nu$	Achiba, Sato and Kimura (1983a)
CO	A, $v=1 \sim 3$	$2h\nu + 2h\nu$ (or $3h\nu$)	Pratt and coworkers (1983a)

TABLE II (Contd.)

Substance	Resonant excited state	MPI process	Reference
H_2S	4A − 3	$3h\nu + h\nu$	
	4A − 1	$3h\nu + h\nu$	
	3C	$3h\nu + h\nu$	Achiba and coworkers (1982)
	3D	$3h\nu + h\nu$	
	4A − 1	$3h\nu + h\nu$	Miller and coworkers (1982)
NH_3	A, $v = 1, 2$	$2h\nu + 2h\nu$	
	B, $v = 7 \sim 11$	$3h\nu + h\nu$	Glownia and coworkers (1982)
	C′, $v = 2 \sim 5$	$3h\nu + h\nu$	
	D, $v = 0, 2, 8, 9$	$3h\nu + h\nu$	
	C′, $v = 0 \sim 5$	$3h\nu + h\nu$	Achiba and coworkers (1982)
CH_3I	A_1, $v = 0, 1$	$2h\nu + h\nu$	Kimura, Sato and Achiba (1983)
	E, $v = 0, 1$	$2h\nu + h\nu$	
Benzene	$S_1(B_{2u})$	$2h\nu + 2h\nu$	Achiba and coworkers (1981b)
			Achiba and coworkers (1983c)
	$S_1(B_{2u})$	$h\nu + h\nu$	Meek, Jones and Reilly (1980)
Toluene	S_1	$h\nu + h\nu$	Meek, Long and Reilly (1982)
Chlorobenzene	S_1	$h\nu + h\nu$	Anderson, Redev and Zare (1982)
Phenol	S_1	$h\nu + h\nu$	Yoshiuchi and coworkers (1983)
Naphthalene	S_1, S_2	$h\nu + h\nu'$	Hiraya and coworkers (1983)
		van der Waals complexes	
Ar—NO	C, $v = 0 \sim 3$	$2h\nu + h\nu$	Sato, Achiba and Kimura (1983)
H_2O-phenol	S_1	$h\nu + h\nu$	Fuke and coworkers (1983)
$(H_2O)_2$-phenol	S_1	$h\nu + h\nu$	

state is similar to that of the neutral excited state, the $\Delta v = 0$ ionization transition is predominant. This occurs especially for Rydberg excited states since thier equilibrium geometries resemble those of the corresponding ionic states, and mainly single photoelectron peaks are expected to appear. Indeed, only single photoelectron peaks have been observed for the Rydberg C state of NO (Miller and Compton, 1981a; White *et al.*, 1982), the Rydberg F and H(H′) states of NO (Achiba, Sato and Kimura, 1983a) and the Rydberg C′ state of NH_3 (Achiba, Sato and Kimura 1983a; Achiba *et al.*, 1981a; Glownia *et al.*, 1982).

When the equilibrium geometry of the ionic state is considerably different from that of the excited state from which ionization takes place, the photoelectron band consists of a prominent vibrational progression. In the case of direct ionization, the vibrational structure is governed by Franck–Condon factors between the excited and the ionic states. Vibrationally resolved photoelectron spectra have been reported for some excited states of toluene (Meek, Long and Reilly, 1982), chlorobenzene (Anderson, Redev and Zare, 1982), carbon monoxide (Pratt, Poliakoff and Dehmer, 1983a) and methyl iodide (Kimura, Sato and Achiba, 1983c).

Photoelectron spectra obtained from a specific vibronic level of an excited state thus provides new information on vibrational modes of ionic states. Furthermore, photoelectron vibrational structure obtainable from a series of vibronic levels of an excited state gives new information about equilibrium geometry of the excited state.

The studies of excited state photoelectron spectra reported so far are summarized in Table II, which have been obtained with the resonant MPI technique. Several of these photoelectron studies will be described in more detail in the next chapter.

V. EXCITED STATE PHOTOELECTRON SPECTRA

A. Atoms

1. Xenon

The reasonantly enhanced MPI photoelectron spectrum of Xe has been reported first by Compton and coworkers (1980) for the (3 + 2) resonant multiphoton ionization through the $J = 1$ level of the $6s[1\frac{1}{2}]^{\circ}_{J}$ state at 40.9 nm. The two peaks were attributed to the $^2P_{3/2}$ and $^2P_{1/2}$ ions. Recently, higher resolution spectra were measured at the same laser wavelength by Miller and Compton (1982), Kruit and coworkers (1983) and Sato, Achiba and Kimura (1984a). The spectrum obtained by Sato, Achiba and Kimura (1984a) is shown in Fig. 15.

The formation of the two ionic states (Fig. 15) is rather surprising for a

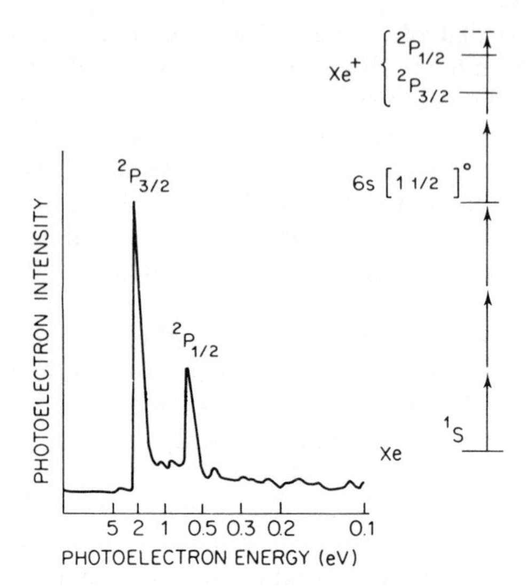

Fig. 15. The photoelectron spectrum obtained for (3 + 2) resonant multiphoton ionization of Xe through the $6s[1\frac{1}{2}]^{\circ}_{J=1}$ Rydberg state.

single-electron ionization, as mentioned in Section IV.B, since only the $^2P_{3/2}$ ionic state is allowed from the $6s[1\frac{1}{2}]^{\circ}_{J}$ Rydberg state with a core of $^2P_{3/2}$. Compton and coworkers (1980) have suggested that the formation of the $^2P_{1/2}$ ionic state may be due either to the configuration interaction of the 6s state with the 6s′ state or to the resonance (or near-resonance) of the fourth photon with some Rydberg state which has a $^2P_{1/2}$ core. Recently, Sato, Achiba and Kimura (1984a) suggested on the basis of a theoretical calculation that the $^2P_{1/2}$ ion formation is not due to configuration mixing between the 6s and the 6s′ state and is probably due to a near-resonance (in the four-photon state) with a $^2P_{1/2}$ character.

Two-colour experiments for the resonant (3 + 1) ionization of Xe through the $J = 1$ level of the $6s[1\frac{1}{2}]^{\circ}_{J}$ Rydberg state have been reported by Miller and Compton (1982). They observed an additional peak which is attributed to ionization from the 6s state by one photon of the second colour (308 nm). Using higher laser power densities (10^{11} W cm^{-2}), Kruit and coworkers (1983) have observed two above-threshold ionization photoelectron peaks through this 6s Rydberg state.

The $5d[2\frac{1}{2}]^{\circ}_{J=3}$ and $5d[3\frac{1}{2}]^{\circ}_{J=3}$ Rydberg states of Xe are in resonance with three-photon energies at 363.9 and 370.5 nm respectively, according to an ion-current study by Aron and Johnson (1977, 1980). Sato, Achiba and Kimura (1984a) have obtained photoelectron spectra of multiphoton ionization through these 5d Rydberg states. The photoelectron spectrum obtained at 363.9 nm is shown in Fig. 15. As mentioned in Section IV.B, single-electron transition ionization should give rise to only the $^2P_{3/2}$ ionic state.

2. *Krypton*

The MPI ion-current spectrum of Kr shows a sharp peak at a laser wavelength of 370.0 nm, which is attributed to the three-photon resonance with the $J = 1$ level of the $5s[1\frac{1}{2}]_J^{\circ}$ state. At this laser wavelength, Kr is ionized by at least five photons. Sato, Achiba and Kimura (1984a) have obtained a photoelectron peak due to the $^2P_{3/2}$ ionic state. No signals corresponding to the $^2P_{1/2}$ ionic state have been detected, even though the transition is energetically possible. Therefore, a single-electron ionization transition is the main process in this resonant ionization of Kr. This result is different from that for the Xe 6s state.

3. *Iron*

Atomic iron is readily produced by photodecomposition of gaseous iron pentacarbonyl and ferrocene by visible laser radiation. An energy level

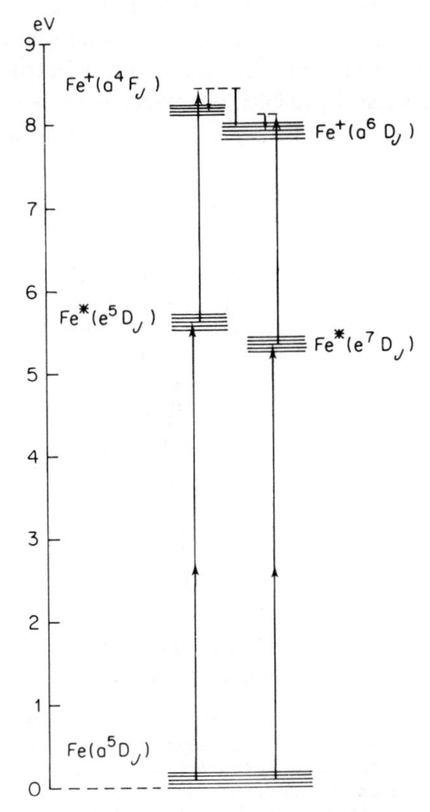

Fig. 16. Energy level diagram for Fe and Fe^+, relevant to (2 + 1) resonant multiphoton ionization through the e^5D_J and the e^7D_J excited state.

diagram for Fe and Fe^+ is shown in Fig. 16. Nagano and coworkers (1982) have obtained an MPI ion-current spectrum for Fe atoms produced by photodecomposition of $Fe(CO)_5$. The ion-current spectrum consists of sharp peaks at 463.4, 464.9 and 465.9 nm, which may be interpreted in terms of $\Delta J = 0$ transitions of the two-photon excitation $Fe^*(e^7D_J) \leftarrow\leftarrow Fe(a^5D_J)$, where $J = 2$, 3 and 4. Another sequence consisting of five sharp peaks was found at 447.7, 448.0, 448.1, 448.2 and 448.3 nm by Leutwyler, Evan and Jortner (1981), interpreted in terms of $\Delta J = 0$ transitions of the two-photon excitation $Fe^*(e^5D_J) \leftarrow\leftarrow Fe(a^5D_J)$, where $J = 0, 1, 2, 3$ and 4 respectively. At these wavelength, the (2 + 1) two-step ionization takes place.

The photoelectron spectrum obtained at the $J = 3$ level of the e^7D_J excited state (Fig. 14) shows two types of photoelectron bands which are denoted as 'a^6D_J' and 'non-resonant'. These can be distinguished experimentally; the photoelectron bands due to $Fe^+(a^6D_J)$ are sensitive to the laser wavelengths, whereas the non-resonant peaks are insensitive to the wavelength but are very sensitive to the laser power. The strongest photoelectron peaks have been found at $J = J' - \frac{1}{2}$, where J' is the J number of the resonant e^7D_J state. Upon ionization through the resonant a^5D_J state, photoelectron bands due to the a^6D_J and a^4F_J ionic states have been observed, as shown in Fig. 17 (Nagano *et al.*, 1982). Transition to the $Fe^+(a^4F_J)$ ionic state, which has the electron

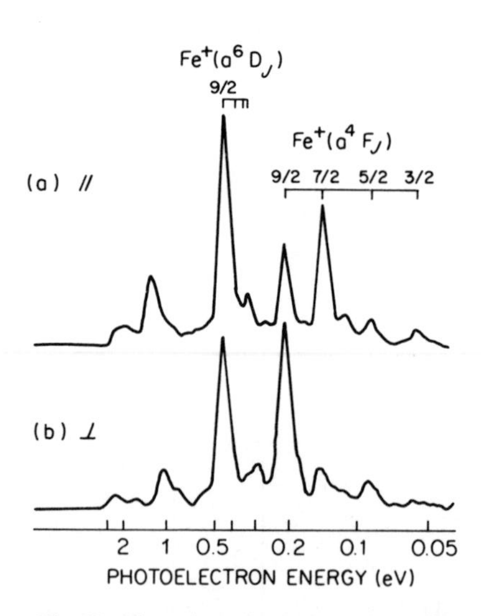

Fig. 17. The photoelectron spectra obtained for (2 + 1) resonant multiphoton ionization of Fe through the $J = 3$ level of the e^5D_J excited state, at a laser wavelength of 447.7 nm, with (a) the parallel polarization and (b) the perpendicular polarization.

configuration (3d) , is not allowed by single-electron ionization from the resonant e^5D_J state. Observation of such a satellite band seems to be due to resonant autoionization. The strongest peak in this case appears at $J = J' + \frac{1}{2}$, where J' is the J number of the resonant e^5D_J state. The relationships, $J = J' \pm \frac{1}{2}$, are a consequence of angular momentum conservation. In Fig. 17, the photoelectron spectra obtained by the parallel and the perpendicular polarization are compared, showing a large angular distribution difference. This angular dependence has been interpreted in terms of cosine-square distribution (Nagano *et al.*, 1982).

B. Small Molecules

1. H_2

Pratt, Dehmar and Dehmar (1983b) have reported photoelectron spectra obtained by (3 + 1) resonant multiphoton ionization of H_2 through specific rotational levels of the B, $v = 7$ state, indicating resolved rotational structure, which provides information on the partial waves of the ejected photoelectron. The allowed rotational levels of the H_2^+ ion are determined from various selection rules of ionization transition: a$\not\leftrightarrow$s, $= \leftrightarrow -$, $J = 0, \pm 1$, and $N(H_2, X) - J(H_2^*, B) = \pm 1$ and ± 3. The photoelectron spectra indicate that the $N = 3$ and 5 rotational levels of H_2^+ are observed by pumping the R(3) transition, while only the $N = 1$ and 3 levels are observed by pumping the P(3) transition. No even rotational levels have been detected. An additional peak appearing in the spectrum has been tentatively interpreted by these authors as arising from photoionization of $H^*(n = 2)$ formed by photodissociation of H_2 in the B state.

2. NO

a. The A state

Photoelectron spectra for (2 + 2) resonant ionization of NO through the Rydberg A state, obtained at different vibrational levels ($v = 0 - 3$), have been reported by Miller and Compton (1981a), Ebata *et al.* (1982). This MIP process is described by

$$NO(X^2\Pi) \xrightarrow{2h\nu} NO^*(A^2\Sigma^+) \xrightarrow{2h\nu} NO^+(X^1\Sigma^+) + e^-$$

The laser wavelengths are 452.6, 436.0, 409.3 and 391.2 nm for the $v = 0$, 1, 2 and 3 of the A state respectively. Two types of bands appear in each photoelectron spectrum: one is the higher energy peak due to $\Delta v = 0$ transition and the other is the lower energy peak near zero energy. The origin of the zero-energy band has not yet been identified. In the conventional HeI photoelectron spectrum of NO, an extended progression, which is governed by Franck–Condon factors between the ground state of NO and that of NO^+,. was observed in the first ionization (Turner *et al.*, 1970). However, in the above (2 + 2) resonant ionization process, $\Delta v = 0$ transitions are dominant, since the N—O distance of the Rydberg A state of NO is almost the same as that of the ground state NO^+ ion. (In this case, the Franck–Condon factors are less than 10^{-3} for $\Delta v \neq 0$ transitions.) Kimman, Kruit and van der Wiel (1982) have repeated these experiments, using their electron 'parallelizer' analyser, by selecting a specific rotational level ($J = 6\frac{1}{2}$) of the A, $v = 0$ state of NO. As a result, they have observed photoelectron peaks due to all the energetically possible vibrational levels ($v = 0$ to 5) of the ground state NO^+ ion, and assigned the $v = 0$ and the $v > 0$ peaks to direct ionization and electronic (or vibrational) preionization respectively.

b. The B state

This is a valence-type excited state with the electronic configuration expressed by $\ldots 1\pi^3 2\pi^2$. Since the ground state ion has the configuration $\ldots 1\pi^4$, single-electron transition of the B state does not produce a ground state NO^+ ion. Therefore, photoionization of the B state is a model case for investigating mechanisms of autoionization. From an MPI ion-current spectrum associated with the $v = 9$ level of the B state in the 375 to 382-nm region, Achiba, Sato and Kimura (1983b) have found several rotational lines with anomalous intensities in addition to many normal rotational ones which follow a Boltzmann distribution. They also measured photoelectron spectra and photoelectron angular distributions for both the normal and the anomalous intensity lines. A photoelectron spectrum recorded at the normal rotational line (e.g. Q_{11}, $J = 1\frac{1}{2}$) shows that the two main peaks due to $v = 1$ and 2 levels of the ground state NO^+ ion are observed with an intensity ratio of 3:7. The formation of the ground state ion is presumed to be due to autoionization taking place through higher valence-type excited states, probably the $I^2\Sigma^+$ or $B'^2\Delta$ state that is allowed from the B state by one-photon absorption. The branching ratio reflects Franck–Condon factors between the autoionized state and the ground state NO^+ state. A photoelectron spectrum recorded at the anomalous intensity line (e.g. Q_{22}, $J = 6\frac{1}{2}$) showed a newly observed strong peak due to the $v = 0$ ion as well as the $v = 1$ and 2 ion peaks whose relative intensities are similar to those obtained at a normal-intensity rotational line. Since the $v = 0$ photoelectron peak shows no angular dependence, and a new peak due to the $v = 6$ ion appears at a higher laser power, Achiba, Sato and Kimura (1983b) have proposed a relatively long-lived discrete level (possibly $4\,d\delta$, $v = 6$ state) as a possible candidate for the third-photon, autoionizing state. The appearance of the $v = 6$ photoelectron peak is explained by an additional one-photon absorption of the autoionizing state.

c. The C state

This state is $3p\pi$ Rydberg in character. Miller and Compton (1981a) have reported photoelectron spectra for (3 + 2) resonant multiphoton ionization through the C state with the $v = 0$ and 1 vibrational levels, indicating that only $\Delta v = 0$ ionization transition occurs. This is expected because the C state of NO has almost the same equilibrium distance as the ground state of NO^+. From an analysis of high-resolution absorption spectra, Lagerqvist and Miescher (1958) indicated that the lower rotational levels of the C, $v = 0$ state are strongly perturbed by rotational levels of the nearly degenerate B, $v = 7$ valence state. White and coworkers (1982) have obtained photoelectron spectra for (3 + 2) resonant ionization of several rotational levels of the C, $v = 0$ state, which is a good example for study of Rydberg–valence interactions in small molecules. The photoelectron spectra obtained by these investigators show a vibrational progression but a single Franck–Condon pattern. Each spectrum is divided into two types of transitions, one of which is characterized by a strong $v = 0$ ion peak expected from the $\Delta v = 0$ transition of the C state and the other is a progression leading to the $v = 3$, 4 and 5 ions. The intensity ratio of the $v > 3$ to $v = 0$ vibrational states is strongly dependent on the pumped rotational level. Since the Franck–Condon factors between the C, $v = 0$ level and the $v > 3$ levels of the ground state ion are vanishingly small, the observed $v > 3$ photoelectron peaks cannot be due to direct ionization. However, the B, $v = 7$ state has considerable Franck–Condon overlap with the $v > 3$ levels of the NO^+ ion. The $\Delta v = 0$ ionization reflects the Rydberg character of the mixed C–B state, whereas the progression of the $v = 3$, 4 and 5 of the NO^+ ion reflects the non-Rydberg character of the mixed C–B state.

d. The F and H(H′) states

Achiba, Sato and Kimura (1983a) have obtained photoelectron spectra for (3 + 1) resonant ionization of NO through the Rydberg F and H(H′) states with the $v = 0$ and 1 vibrational levels. Each spectrum shows mainly a single peak assigned to $\Delta v = 0$ ionization transition. Such results are reasonable, since an electron is ejected from the Rydberg orbital. They have also obtained the photoelectron angular distributions which may be expressed approximately by the cosine-square formula. For instance, the photoelectron angular distribution of the (3 + 1) resonant ionization of the F, $v = 1$ level yielded a value of $\beta = 1.0$ if the formula given by Eq. (12) in Section VI.

3. CO

Pratt and coworkers (1983a) have reported photoelectron spectra of CO for (2 + 2) or (2 + 3) resonant ionization through the $v = 1$ to 3 levels of its excited

A state. The ionization process is described by

$$CO(X^1\Sigma^+, v=0) \overset{2h\nu}{\longleftrightarrow} CO^*(A^1\Pi, v=1,2,3) \overset{2h\nu \text{ or } 3h\nu}{\longleftrightarrow} CO^+(X^2\Sigma^+, v=0 \text{ to } 10)$$

$$\text{and} \quad CO^+(A^2\Pi, v=0,1) + e^-$$

The photoelectron spectrum obtained at the $v = 1$ bandhead of the A state of CO shows a single vibrational progression due to the ground state ion, consisting of a strong peak due to the $v = 0$ ion and many weak peaks due to its higher vibrational levels. The observed branching ratios do not match the calculated Franck–Condon factors. This suggests that the branching ratios are influenced by accidental resonances at the four-, five- and six-photon levels. In the photoelectron spectrum obtained at the $v = 2$ bandhead of the A state of CO, the $v = 0$ and 1 peaks of the excited A state of CO^+ appear in addition to the vibrational peaks of the ground state ion, although the transition leaving the ion in the A state involves a two-electron transition.

4. H_2S

An ion-current spectrum of H_2S obtained by Achiba and coworkers (1982) in the 420 to 455-nm laser wavelength region is shown in Fig. 18. They have obtained photoelectron spectra for (3 + 1) resonant multiphoton ionization of H_2S through its Rydberg states, denoted by 4A-3(0, 0, 0), 4A-1(1, 0, 0) 3C(0, 0, 0) and 3D(0, 0, 0). The 4A Rydberg states are produced by exciting an electron from the $2b_1$ non-bonding orbital to the $4pa_1$ Rydberg orbital, while the 3C and 3D Rydberg states are obtained by exciting an electron from the $2b_1$ non-bonding orbital to the $3da_1$ and the $3da_2$ Rydberg orbital respectively. The photoelectron spectra obtained for the above four Rydberg states

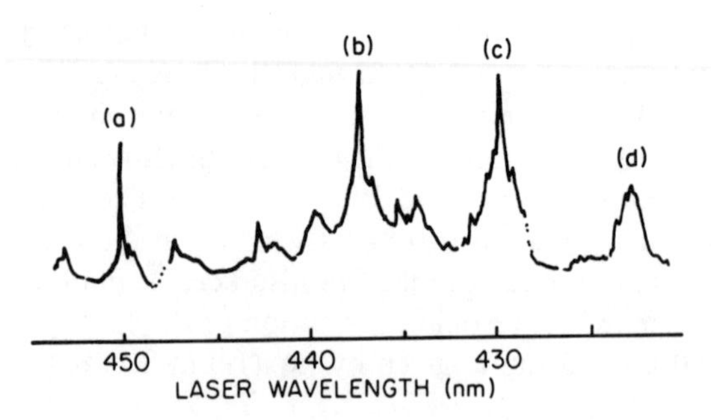

Fig. 18. An MPI ion-current spectrum of H_2S in the laser wavelength 420 to 455 nm. The main peaks are due to 3 + 1) resonant ionization through the Rydberg states: (a) 4A-3(0,0,0), (b) 4A-1(1,0,0), (c) 3C(0,0,0) and (d) 3D(0,0,0) (Achiba *et al.*, 1982).

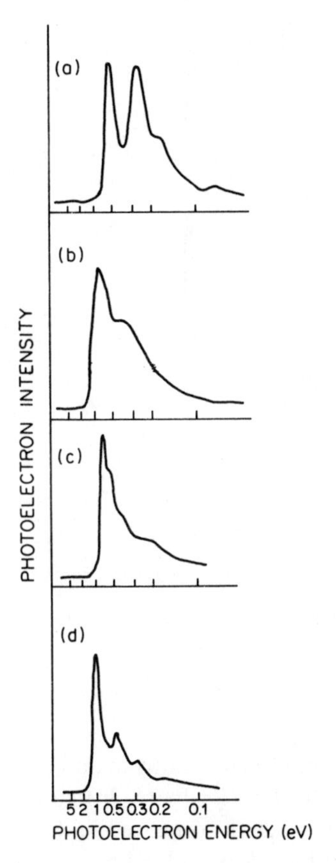

Fig. 19. Photoelectron spectra of the four Rydberg states of H_2S, obtained at the MPI ion-current peaks shown in Fig. 18 (Achiba *et al.*, 1982).

are shown in Fig. 19. Miller and coworkers (1982) have also reported a photoelectron spectrum for the 4A-1 state. On the basis of the photoelectron spectroscopic study, it has been suggested that formation of HS^+ and S^+ ions results mainly from additional photon absorption of the ground state H_2S^+ ion at the $v = 0$ and 1 levels respectively (Achiba *et al.*, 1982).

5. *NH_3*

Achiba, Sato and Kimura (1983a) have obtained photoelectron spectra for (3 + 1) resonant multiphoton ionization of NH_3 through the $v = 0$ to 5 levels of the Rydberg C′ state. They showed that in each case a sharp prominent peak is observed with some weaker bands, as shown in Fig. 20. According to an ion-

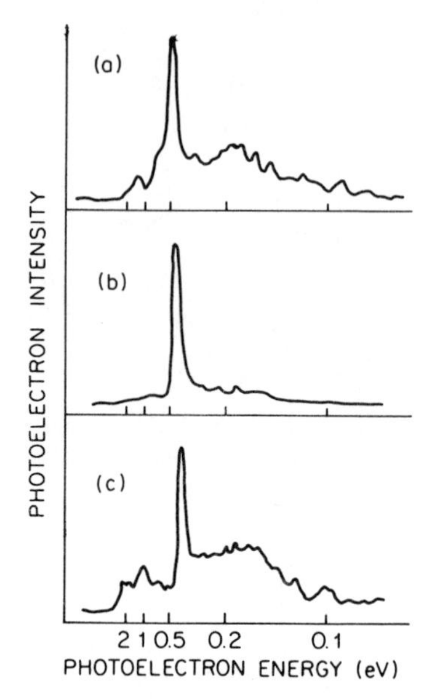

Fig. 20. Photoelectron spectra of (3 + 1) resonant ionization of NH_3 through its Rydberg C′ states, obtained by tuning the laser wavelength to the $v = 0$, 2 and 4 vibrational levels of the C′ state (Achiba, Sato and Kimura, 1983a).

current spectroscopic analysis by Glownia and coworkers (1980), the C′ state of NH_3 is regarded as one of the Rydberg (3p) states which converge to the first ionization limit. The sharp peaks observed in the photoelectron spectra have been assigned by Achiba, Sato and Kimura (1983a) to $\Delta v = 0$ ionization transitions. Using a hemispherical electrostatic analyser, Glownia and coworkers (1982) have also obtained photoelectron peaks with zero kinetic energy, in addition to the above-mentioned $\Delta v = 0$ peaks. However, any zero kinetic energy photoelectrons have been detected in the experiments of Achiba, Sato and Kimura (1983a), even though photoelectrons were accelerated in the TOF electron analyser.

C. Large Molecules

1. Benzene

Achiba and coworkers (1983c) measured photoelectron spectra for benzene under collision-free conditions in such a way that the benzene molecule is ionized by four photons through various two-photon allowed vibronic levels of the $^1B_{2u}$ excited state. The energy diagram relevant to the resonant multiphoton ionization is shown in Fig. 21. Two examples of the photo-

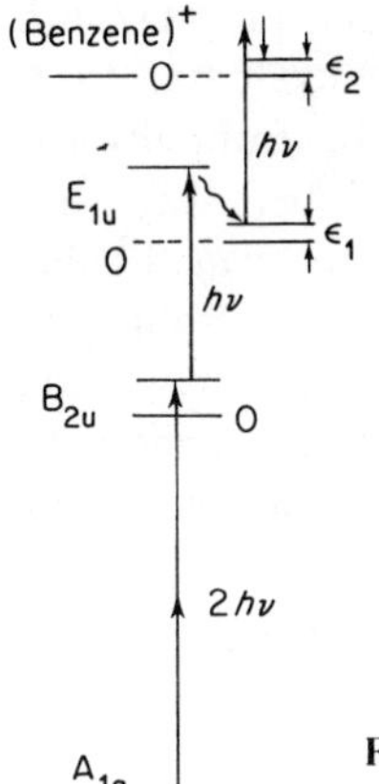

Fig. 21. Energy level diagram relevant to (2 + 1 + 1) resonant ionization of benzene through the $^1B_{2u}$ and $^1E_{1u}$ states.

electron spectra are shown in Fig. 22a and b, indicating single, prominent, somewhat broad bands which may be attributed to $\Delta v = 0$ ionization transitions. The prominent peak is shifted with increasing one-photon energy by $K = h\nu - C$, as shown in Fig. 22c, where K is the photoelectron energy, $h\nu$ is the one-photon energy and C is a constant. This one-photon relationship indicates that the third-photon level is in resonance with another real excited state. Therefore, the overall ionization process is described by (2 + 1 + 1). It has also been indicated that fast intramolecular vibrational relaxation occurs at the third-photon state within the $^1E_{1u}(\pi\pi^*)$ state, and subsequent

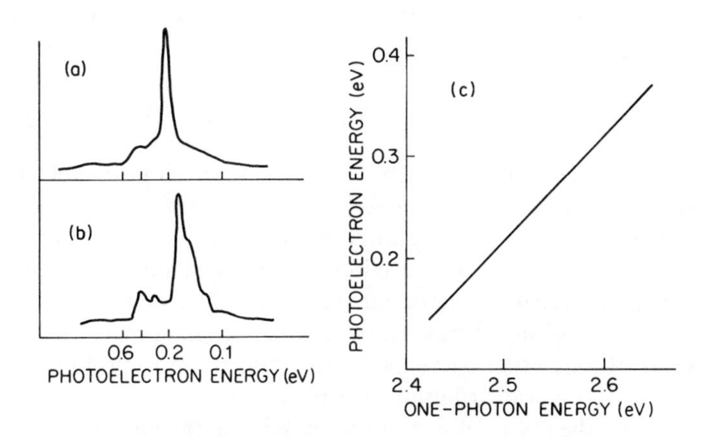

Fig. 22. Photoelectron spectra obtained through (a) the 14^1_0 and (b) the $14^1_0 1^1_0$ level of the $^1B_{2u}$ state of benzene. (c) The linear relationship with a slope of 1.0 between the photoelectron energy and the one-photon energy (Achiba *et al.*, 1983c).

ionization takes place from vibrationally relaxed levels. The relaxation within the $^1E_{1u}$ state is confirmed by the fact that its 0–0 energy evaluated from the photoelectron energies is almost identical to that available from absorption spectroscopy. The lifetimes of the vibrationally relaxed $^1E_{1u}$ states are suggested to be in the order of 10^{-11} to 10^{-12} s or longer (Achiba *et al.*, 1983c).

Photoelectrons ejected as a result of (1 + 1) resonant ionization through ten different vibronic levels of the S_1 state of C_6H_6 and three of C_6D_6 have been obtained by Long and coworkers (1983). They have assigned the ν_{16} mode to 319 and 278 cm^{-1} in C_6H_6 and C_6D_6 respectively, and the ν_4 mode tentatively to 415 and 351 cm^{-1} respectively.

2. *Toluene*

Meek, Long and Reilly (1982) have obtained vibrationally resolved photoelectron spectra for (1 + 1) resonant ionization of toluene. In their work the laser was tuned to three particular vibronic levels of the S_1 state of toluene: the 0–0 origin, the 6^1_0 level, and the 12^1_0 level. When the laser is tuned to the S_1–S_0 vibronic origin, photoelectrons appear preferentially in the first peak, indicating that production of vibrationally unexcited ions is dominant. When the laser is in resonance with the $6b^1_0$ transition, the strongest photoelectron peak has its threshold approximately 480 cm^{-1} to lower energy than the much weaker origin peak. Furthermore, when the laser is tuned to the 12^1_0 level (932 cm^{-1} above the S_1 origin), the most intense photoelectron peak lies about 960 cm^{-1} lower than the first peak and is ascribed to ions produced with one quantum in the vibrational mode 12 of the ion. This behaviour represents evidence for a preference of $\Delta v = 0$ ionization transitions from excited vibronic states of toluene, as suggested in previous work by Duncan Dietz and Smalley (1981).

3. *Chlorobenzene*

Anderson, Redev and Zare (1982) have reported vibrationally resolved photoelectron spectra obtained by (1 + 1) resonant ionization of chlorobenzene through four different vibrational levels of the $^1B_2$ state. In ionization from the vibrationless level, ions are formed almost exclusively in the ground vibrational state. When the resonant state has a quantum of the 18b mode excited, the largest group of ions is formed with one quantum of the 18b mode. In ionization through the 6b level of the $^1B_2$ state, the most prominent feature of the ion vibrational distribution is a progression of the 6a mode. In the ionization from the $6b^1_0$ level, a short progression in the 6b mode of the ion is observed as well as excitation of the 12 mode and the 9a/9b mode. The vibrational frequencies in the chlorobenzene cation have been found to be 18b(b_2) 266 cm^{-1}, 6a(a_1) 419 cm^{-1}, 6b(b_2) 451 cm^{-1}, 12(a_1) 661 cm^{-1}, 9a(a_1)

or 9b(b_2) 1145 cm^{-1}, slightly smaller than those in the neutral ground and the neutral excited states.

4. *Benzaldehyde*

According to Long and coworkers (1983), the photoelectron spectra obtained by 258.9-nm irradiation of benzene and benzaldehyde are apparently identical. These results suggest that the benzene cation is generated by (1 + 1) resonant ionization of benzaldehyde. They have also indicated that the benzene formation by decomposition of benzaldehyde in its excited S_2 state occurs within the 2-ns laser pulse. This process appears to be faster than the rate of photoionization of the S_2 excited state benzene. From an analysis of kinetic data on benzaldehyde photolysis, Berger, Goldblatt and Steel (1973) have reported that intersystem crossing populates two triplet states which then decompose with lifetimes of ~ 20 ns and 2.5 μs.

5. *Naphthalene*

Hiraya and coworkers (1983) have studied vibrationally resolved photoelectron spectra for (1 + 1) resonant ionization through several vibronic levels of the S_1 and S_2 states of naphthalene cooled in a supersonic free jet. Two-colour experiments were carried out in that work. They have obtained information about vibrational modes excited in the ionic state. It has also been indicated that electronic relaxation from the S_1 to the S_2 state is a very fast process compared with the ionization process; no photoelectrons from the S_2 state are observed.

D. Van der Waals Complex

1. *Ar—NO*

Sato, Achiba and Kimura (1984b) have obtained photoelectron spectra for (2 + 1) resonant ionization of Ar—NO van der Waals molecules through its Rydberg C state. The Ar—NO complex was produced in a supersonic free jet. The ion-current spectrum in the 383 to 384-nm laser wavelength region shows a vibrational progression attributed to the Rydberg state which may be represented by Ar—NO*(C). This spectrum is slightly shifted to the longer wavelength compared with the spectrum of the Rydberg C state of free NO. From the photoelectron spectra obtained for the Ar—NO*(C) excited state, the adiabatic ionization potential of Ar—NO has been determined to be 9.148 $\pm$ 0.005 eV. It has also been indicated from the MPI ion-current and photoelectron spectra and the C state has the dissociation energy of $D_e = 0.058 \pm 0.001$ eV, while the ground state Ar—NO^+ ion has $D_e = 0.132 \pm 0.005$ eV.

2. H_2O–Phenol

Photoelectron spectra of the S_1 state of the 1:1 van der Waals complex formed between H_2O and phenol in a supersonic free jet have been studied by Fuke and coworkers (1983). The 0_0^0 peak of the S_1 state of this complex in its MPI ion-current spectrum is shifted to the longer wavelength by $356\,cm^{-1}$ than that of free phenol. From the photoelectron spectrum obtained for the 0_0^0 level of the complex S_1 state, it has been indicated that its adiabatic and vertical ionization potentials are 8.10 and 8.29 eV respectively. It has also been found that the H_2O–phenol complex is stabilized by 0.47 eV compared with the free phenol molecule.

VI. PHOTOELECTRON ANGULAR DISTRIBUTION

The angular distribution of photoelectrons emitted by $(m+n)$ resonant multiphoton ionization of an isotropic ensemble irradiated with linearly polarized laser light is expressed by the following formula (refer to the review paper by Berry, 1976):

$$I(\theta) = \sum_{i=0}^{n+m} a_i \cos^{2i}\theta \tag{11}$$

where θ is the angle between the polarization vector of the incident photon and the direction of the photoelectron. The a_i's are parameters that depend on properties of the initial, the intermediate and the final states as well as on the partial waves of the photoelectron. The coefficients (a_i) are also functions of the intensity and linewidth of the laser, according to Dixit and Lambropoulos (1983). In the $(n+m)$ resonant MPI process, the n-photon absorption produces a partially aligned excited species at the resonant intermediate state.

The photoelectron angular distributions in two-photon resonant ionization of atomic Ti and Na have been studied experimentally and theoretically by Berry and his coworkers (Duncanson *et al.*, 1976; Edelstein *et al.*, 1974; Hanson *et al.*, 1980; Strand *et al.*, 1978). Kaminski, Kessler and Kollath (1980); measured the angular distribution and spin polarization of photoelectrons produced by $(1+1)$ resonant ionization of Cs through the 7^2P excited state. Recently, Feldmann and Welge (1982) measured the photoelectron angular distributions of two- and three-photon ionization of Sr.

A complete kinematic analysis of photoelectrons from polarized targets with total angular momentum $J = \frac{1}{2}$ has been reported by Huang (1982). For an atomic photoionization process, a complete measurement of photoelectrons from one subshell can, in principle, yield as many as seventeen independent dynamical quantities for the electronic dipole transition (Huang, 1981, 1982).

The photoelectron angular distributions obtained for the $(3+1)$ and $(3+2)$

resonant ionizations of Xe and Kr show obvious deviations from the cosine-square distribution, reflecting the spatial alignment of the total electronic angular momentum in the resonant Rydberg state (Sato, Achiba and Kimura, 1984a). The photoelectron angular distributions in the (3 + 1) resonant ionization of Xe through the $5d[2\frac{1}{2}]^{\circ}_{J=3}$ and $5d[3\frac{1}{2}]^{\circ}_{J=3}$ may be represented by $(m+n)=4$ in Eq. (11). However, in the (3 + 2) resonant multiphoton ionization of Xe and Kr through s-type Rydberg states, the photoelectron angular distributions are approximated by $(m+n)=2$ in Eq. (11) instead of $(m+n)=5$ (Sato, Achiba and Kimura, 1984a). This may be due to the isotropic distribution of the s-type Rydberg electrons.

In single-photon ionization of randomly oriented excited states, as in conventional VUV photoelectron spectroscopy, the photoelectron angular distribution is given by

$$I(\theta) \propto 1 + \frac{\beta}{2}(3\cos^2\theta - 1) \tag{12}$$

for polarized light (Cooper and Zare, 1968), where β is the asymmetric parameter ranging between -1 and 2. In the (3 + 1) resonant ionization of NO and NH_3 through Rydberg excited states, Achiba, Sato and Kimura (1983a) have indicated that the angular distributions approximately follow the cosine-square function (12). Their results suggest that these Rydberg excited states are not spatially aligned during successive ionization processes.

Various aspects of the photoelectron angular distribution in a strongly driven resonant multiphoton ionization process have been discussed by Dixit and Lambropoulos (1981). They indicated that the angular distribution changes drastically with increasing intensity as a result of a.c. Stark shifting of the levels. The angular distribution of photoelectrons from polarized targets exposed to polarized radiation has been analysed in the dipole approximation by Klar and Kleinpoppen (1982). The photoelectron angular distribution in three-photon ionization of metastable 2S helium has been calculated by Olsen and coworkers (1978) in terms of generalized cross-sections. For non-resonant multiphoton ionization, Fabre and coworkers (1981), Leuchs and Smith (1982) and Kruit, Kimman and van der Wiel (1981) have reported photoelectron angular distributions, measured for Xe or Na.

VII. CONCLUSIONS

As seen from the previous chapters, the excited state photoelectron spectroscopy of laser multiphoton ionization and the conventional VUV photoelectron spectroscopy of single-photon ionization provide different information in many aspects, but both kinds of information are complementary to each other. The ionic state information available from conventional

VUV photoelectron spectra as well as from *ab initio* theoretical calculations should be important for interpreting the photoelectron spectra of excited states. The well-identified ionic states available from conventional photoelectron spectroscopy can be used for testing the assignment of excited state photoelectron spectra, and therefore these are useful for the diagnosis of excited states. When the well-identified excited states are selected as the initial states of ionization, one may study ionic states that cannot be produced from conventional VUV ionization. Even if the final ionic states are the same, the ground state ionization and the excited state ionization yield different cross-sections and thus different populations.

In photoelectron spectroscopy with synchrotron radiation one can obtain photoelectron spectra as a function of photon energy. The photon-energy dependence of the photoelectron spectrum is important for investigation of mechanisms of photoionization such as autoionization. The excited state photoelectron spectroscopy is also a powerful method for studying the mechanism of autoionization, since one may observe a photoelectron spectrum at the specific autoionizing state that is selectively produced by further photoabsorption from a specific excited state. In this sense, the photoelectron spectroscopy with synchrotron radiation is also complementary to laser MPI photoelectron spectroscopy.

Spectral assignments in MPI ion-current spectroscopy have been made, usually without considering which ionic states should be formed as a result of multiphoton ionization. The identification of the resonant excited state is not unambiguous in MPI ion-current spectroscopy. More direct information on the resonant excited state can be obtained by a photoelectron energy analysis, although tentative assignments of MPI ion-current spectra are always necessary prior to photoelectron measurements.

The ideal case in resonant multiphoton ionization is the process represented by $(mh\nu + h\nu')$ which can be performed with two-colour experiments. It is necessary to scan the wave-length of the second laser in order to distinguish 'autoionization' from 'direct ionization'. If the absorption of two or more photons is required for the transition from a resonant excited state to an ionization continuum, the situation becomes more complicated because of the possibility of resonance with a higher excited state.

Fujimura and Lin (1981) have indicated from a theoretical study that there are two types of $(m + n)$ ionization processes: one is the simultaneous process in which all the photons $(n + m)$ are simultaneously absorbed and the other is the sequential process in which ionization takes place from the resonant state by the n-photon absorption. At the normal laser power density (5×10^9 W cm^{-2}), as in the most experiments reported so far, the main process is considered to be the sequential one. However, at a higher laser power density, the simultaneous process may also be important in MPI photoelectron spectroscopy.

The TOF electron analyser has a good resolution for the low-energy photoelectrons which are mostly produced in laser multiphoton ionization. The energy resolution in this case is considerably higher compared with HeI photoelectron spectroscopy. Therefore, the resonant MPI photoelectron spectroscopy has a high ability for molecular spectroscopy of excited states. A nanosecond laser has so far been used for excitation and ionization. However, use of a picosecond laser will make it possible to follow fast phenomena involving the excited states of picosecond lifetimes. The technique of excited state photoelectron spectroscopy is a highly selective analytical diagnostic for excited state molecules, and is applicable to the higher excited states which cannot be studied by fluorescence spectroscopy. The use of a supersonic expansion in the photoelectron spectroscopy provides large possibilities of studying molecular clusters as well as cooled molecules. The technique described in this paper has a bright prospect of developing the new field of application.

Acknowledgements

The author thanks colleagues Dr Y. Achiba and Mr K. Sato for their remarkable contribution to the development of excited state photoelectron spectroscopy in this laboratory.

The author also wishes to thank Professor S. H. Bauer of Cornell University for his helpful suggestions in the preparation of the manuscript during his stay at the Institute for Molecular Science.

References

Achiba, Y. Hiraya, A. and Kimura, K. (1984). *J. Chem. Phys.,* **80**, 6047

Achiba, Y., Sato, K., Shobatake, K., and Kimura, K. (1983a). *J. Chem. Phys.*, **78**, 5474.

Achiba, Y., Sato, K., and Kimura, K. (1983b). *Abstracts of the Symposium of Molecular Structure and Molecular Spectroscopy*, Chem. Soc. of Japan, Sendai, p. 572; *J. Chem. Phys.* (to be published).

Achiba, Y., Sato, K., Shindo, Y., and Kimura, K. (1981a). *Annual Review*, p. 105, Institute for Molecular Science, Okazaki.

Achiba, Y., Sato, K., Shobatake, K., and Kimura, K. (1981b). *J. Photochem.*, **17**, 53.

Achiba, Y., Sato, K., Shobatake, K., and Kimura, K. (1982). *J. Chem. Phys.*, **77**, 2709.

Achiba, Y., Sato, K., Shobatake, K., and Kimura, K. (1983). *J. Chem. Phys.* **79**, 5213.

Achiba, Y., Shobatake, K., and Kimura, K. (1980). *Annual Review*, p. 100, Institute for Molecular Science, Okazaki.

Anderson, S. L., Redev, D. M., and Zare, R. N. (1982). *Chem. Phys. Lett.*, **93**, 11.

Antonov, V. S., and Letokhov, V. S. (1981). *Appl. Phys.*, **24**, 89.

Aron, K., and Johnson, P. M. (1977). *J. Chem. Phys.*, **67**, 5099.

Bekov, G. I., and Letokhov, V.S. (1983). *Appl. Phys.*, **B30**, 161.

Berger, M., Goldblatt, L. L., and Steel, C. (1973). *J. Am. Chem. Soc.*, **95**, 1717.

Berkowitz, J. (1979). *Photoabsorption, Photoionization and Photoelectron Spectroscopy*, Academic Press, New York.

Berry, R. S. (1976). 'Two-photon processes', in *Electron and Photon Interactions with Atoms* (Eds. H. Kleinpoppen and M. R. C. McDowell), p. 559, Plenum.
Compton, R. N., Miller, J. C., Carter, A. E., and Kruit, P. (1980). *Chem. Phys. Lett.*, **71**, 87.
Cooper, J., and Zare, R. N. (1968). *J. Chem. Phys.*, **48**, 942.
Dixit, S. N., and Lambropoulos, P. (1981). *Phys. Rev. Lett.*, **46**, 1278.
Duncan, M. A., Dietz, T. G., and Smalley, R. E. (1979). *Chem. Phys.*, **44**, 415.
Duncan, M. A., Dietz, T. G., and Smalley, R. E. (1981). *J. Chem. Phys.*, **75**, 2118.
Duncanson, Jr., J. A., Strand, M. P., Lindgard, A., and Berry, R. S. (1976). *Phys. Rev. Lett.*, **37**, 987.
Dyke, J. M., Jonathan, N., and Morris, A. (1979). 'Vacuum ultraviolet photoelectron spectroscopy of transient species', in *Electron Spectroscopy* (Eds. C. R. Brundle and A. D. Baker), Academic Press, New York.
Ebata, T., Abe, H., Mikami, N., and Ito, M. (1982). *Chem. Phys. Lett.*, **86**, 445.
Edelstein, S., Lambropoulos, M., Duncanson, J., and Berry, R. S. (1974). *Phys. Rev.*, **A9**, 2459.
Eland, J. H. D. (1974). *Photoelectron Spectroscopy*, Wiley-Halsted, New York.
Fabre, F., Agostini, P., Petite, G., and Clement, M. (1981). *J. Phys. B.*, **14**, L677.
Feldmann, D., Krautwald, J., Chin, S. L., von Hellfeld, A., and Welge, K. H. (1982). *J. Phys. B.*, **15**, 1663.
Feldmann, D., and Welge, K. H. (1982). *J. Phys. B.*, **15**, 1651.
Fujimura, Y., and Lin, S. H. (1981). *J. Chem. Phys.*, **78**, 6468.
Fuke, K., Yoshiuchi, H., Kaya, K., Achiba, Y., Sato, K., and Kimura, K. (1984). *Chem. Phys. Lett.*, **108**, 179.
Glownia, J. H., Riley, S. J., Colson, S. D., Miller, J. C., and Compton, R. N. (1982). *J. Chem. Phys.*, **77**, 68.
Glownia, J. H., Riley, S. J., Colson, S. D., and Nieman, G. C. (1980). *J. Chem. Phys.*, **73**, 4296.
Hansen, J. C., Duncanson, Jr., J. A., Chien, R. L., and Berry, R. S. (1980). *Phys. Rev.*, **A21**, 222.
Hiraya, A., Achiba, Y., Sato, K., Mikami, N., and Kimura, K. (1984). To be published in *J. Chem. Phys.*
Huang, K.-N. (1981). *Bull. Am. Phys. Soc.*, **26**, 1301.
Huang, K.-N. (1982). *Bull. Am. Phys. Soc.*, **27**, 40.
Hung, K.-N. (1982). *Phys. Rev. Lett.*, **48**, 1811.
Hurst, G. S., Payne, M. G., Kramer, S. D., and Young, J. P. (1979). *Rev. Modern Phys.*, **51**, 767.
Johnson, P. M. (1980). *Acc. Chem. Rev.*, **13**, 20.
Johnson, P. M., and Otis, C. E. (1981). *Ann. Rev. Phys. Chem.*, **32**, 139.
Kaminski, H., Kessler, J., and Kollath, K. J. (1980). *Phys. Rev. Lett.*, **45**, 1161.
Kimman, J., Kruit, P., and van der Wiel, M. J. (1982). *Chem. Phys. Lett.*, **88**, 576.
Kimura, K., Katsumata, S., Achiba, Y., Yamazaki, T., and Iwata, S. (1981). *Handbook of HeI Photoelectron Spectra of Fundamental Organic Molecules–Ionization Energies, Ab Initio Assignments, and Valence Electronic Structure for 200 Molecules*, Japan Scientific Societies Press, Tokoyo and Halsted Press, New York.
Kimura, K., Sato, K., and Achiba, Y. (1983). *Abstracts of the Symposium of Molecular Structure and Molecular Spectroscopy*, p. 538, Chem. Soc. of Japan, Sendai.
Klar, H., and Kleinpoppen, H. (1982). *J. Phys. B*, **15**, 933.
Kruit, P., Kimman, J., Miller, H. C., and van der Wiel, M. J. (1983). *J. Phys. B: At. Mol. Phys.*, **16**, 937.
Kruit, P., Kimman, J., and van der Wiel, M. J. (1981). *J. Phys. B*, **14**, L597.

Lagerqvist, A., and Miescher, E. (1958). *Helv. Phys. Acta*, **31**, 221.
Lambropoulos, P. (1980). *Appl. Opt.*, **19**, 3926.
Letokhov, V. S. (1983). *Nonlinear Laser Chemistry*, Springer-Verlag, Berlin.
Leuchs, G., and Smith, S. J. (1982). *J. Phys. B*, **15**, 1051.
Leutwyler, S., Even, U., and Jortner, J. (1981). *J. Phys. Chem.*, **85**, 3026.
Lichtin, D. A., Zandee, L., and Bernstein, R. B. (1981). 'Potential analytical aspects of laser multiphoton ionization mass spectrometry', in *Lasers in Chemical Analysis* (Eds. G. M. Heiftje, J. C. Travis and F. Elytle), The Humana Press, New Jersey.
Long, S. R., Meek, J. T., Harrington, P. J., and Reilly, J. P. (1983). *J. Chem. Phys.*, **78**, 3341.
Martin, Jr., E. A., and Mandel, L. (1976). *Appl. Opt.*, **15**, 2378.
Meek, J. T., Jones, R. K., and Reilly, J. P. (1980). *J. Chem. Phys.*, **73**, 3503.
Meek, J. T., Long, S. R., and Reilly, J. P. (1982). *J. Phys. Chem.*, **86**, 2809.
Miller, J. C., and Compton, R. N. (1981a). *J. Chem. Phys.*, **75**, 22.
Miller, J. C., and Compton, R. N. (1981b). *J. Chem. Phys.*, **75**, 2020.
Miller, J. C., and Compton, R. N. (1982). *Chem. Phys. Lett.*, **93**, 453.
Miller, J. C., Compton, R. N., Carney, T. E., and Baer, T. (1982). *J. Chem. Phys.*, **76**, 5648.
Nagano, Y., Achiba, Y., Sato, K., and Kimura, K. (1982). *Chem. Phys. Lett.*, **93**, 510.
Olsen, T., Lambropoulos, P., Wheatly, S. E., and Rountree, S. P. (1978). *J. Phys. B: At. Mol. Phys.*, **11**, 4167.
Pratt, S. T., Dehmer, P. M., and Dehmer, J. L. (1983b). *J. Chem. Phys.*, **78**, 4315.
Pratt, S. T., Poliakoff, E. D., Dehmer, P. M., and Dehmer, J. L. (1983a). *J. Chem. Phys.*, **78**, 65.
Rabalais, J. W. (1977). *Principles of Ultraviolet Photoelectron Spectroscopy*, Wiley-Interscience, New York.
Robin, M. B. (1980). *Appl. Opt.*, **19**, 3941.
Sato, K., Achiba, Y., and Kimura, K. (1983). *Abstracts of the Symposium of Molecular Structure and Molecular Spectroscopy*, p. 716, Chem. Soc. of Japan, Sendai.
Sato, K., Achiba, Y., and Kimura, K. (1984). *J. Chem. Phys.* (in press).
Siegbahn, K., Allison, D. A., and Allison, J. H. (1974). 'ESCA-photoelectron spectroscopy', in *Handbook of Spectroscopy* (Ed. J. W. Robinson), Vol. I, p. 257, CRC Press, Florida.
Siegbahn, K., Nordling, C., Johansson, G., Hedman, J., Heden, P. F., Hamrin, K., Gelius, U., Bergmark, T., Werme, L. O., Manne, R., and Baer, Y. (1969). *ESCA Applied to Free Molecules*, North-Holland, Amsterdam, London.
Turner, D. W., Baker, A. D., Baker, C., and Brundle, C. R. (1970). *Molecular Photoelectron Spectroscopy, A Handbook of He 854 Å* Spectra, Interscience, London, New York.
White, M. G., Seaver, M., Chupka, W. A., and Colson, S. D. (1982). *Phys. Rev. Lett.*, **49**, 28.

Photodissociation and Photoionization
Edited by K. P. Lawley

PREDISSOCIATION OF POLYATOMIC VAN DER WAALS MOLECULES*

KENNETH C. JANDA

Division of Chemistry and Chemical Engineering,
California Institute of Technology,
Pasadena, California 91125, USA

CONTENTS

I. INTRODUCTION

This chapter is a progress report on how much is understood about van der Waals molecule predissociation dynamics. The field is receiving increasing attention as it becomes widely realized that van der Waals molecules provide tractable model systems for state-to-state studies of intramolecular energy redistribution. The Δj predissociation of triatomics like ArH_2 can be described

*Contribution No. 6917.

with arbitrary accuracy by close-coupling calculations on empirical potential energy surfaces.[1] Distorted wave calculations reproduce the qualitative trends for more complicated triatomics like HeI_2.[2] By gradually increasing the complexity of molecules being studied, it should be possible to isolate the effects of structure and exit channels on the predissociation rate. This is what makes the van der Waals systems so special. In the near future, it is conceivable that intramolecular energy transfer rates for a series of molecules like ArHCl, $(HCl)_2$, HCl:HF, HCl:ClF, HCl:C_2H_4 and HCl:C_2F_4 will be known as a function of vibrational and rotational quantum numbers.

Predissociation studies have played a key role in understanding chemical dynamics since the first observation of line broadening in S_2 was made by Henri and Teves.[3] Henri and Teves gave a semiclassical explanation that the S_2 excited state lifetime is so short that the rotational period is not well defined or quantized. In 1929 Rice published a perturbation theory for describing predissociation.[4] To illustrate just how long ago these developments occurred, a letter following that of Henri and Teves berates the editor of *Nature* for not allowing the word 'scientist' to be used to describe people who do science![5]

In 1933, Rosen published a study of HO_2 vibrational predissociation which presaged much of the current activity on triatomic molecules.[6] His thinking was developed by noting the analogy between predissociation and the Auger effect. Except for his choice of a coordinate system, his analysis of HO_2 predissociation is the same distorted wave approximation so successfully applied to HeI_2 by Beswick and Jortner.[2] Rosen's analysis of the deuterium isotope effect was the precursor for the present momentum gap law.[7]

Progress to 1981 was extensively reviewed by Beswick and Jortner[2] and by Levy[8] in Volume 47 of this series. At that time the vast majority of reliable data on weakly bound molecule predissociation consisted of studies of the rare gas–I_2 systems by Levy[8]. Beswick and Jortner's analysis was able to mimic Levy's data for HeI_2 on such aspects as (1) the $\Delta v = -1$ channel being dominant, (2) the rate increasing quadratically with v and (3) correct rate constants being obtained for a reasonable potential energy surface approximation. Also, by 1981, the theoretical evidence from several groups predicted that the $V \rightarrow T$ predissociation mechanism would predict very long lifetimes for molecules like ArHCl ($v = 1$).[2,7,9]

Experimental progress since the 1981 reviews has been dramatic. McKeller[10] has made detailed studies of linewidth versus energy level for the ArH_2 system. High energy levels of HeI_2 have been studied.[11,12] $NeCl_2$[13] and $NeBv_2$[14] have been observed and lifetime studied as a function of v. $NeCl_2$ in $v = 1$ lives for at least 10^{-5} s![15] Pine and Lafferty[16] have recorded a high-resolution HF dimer infrared spectrum which will dramatically increase our understanding of that molecule. Velocity distributions and line widths as a function of bonding partner have been measured for ethylene clusters.[17–23] Infrared linewidths have been measured for a variety of polyatomic molecules.[24–36] Propensity rules for $V \rightarrow V'$, R dissociation pathways on electronic excited surfaces of tetrazene[37–39] and gloyoxal[40–42] have been measured. This new wealth of data provides the basis for this review.

Section II will give an update on triatomic van der Waals molecule predissociation. Agreement between theory and experiment[10] for these systems is so complete that the problem is essentially solved except for adequate determination of the potential energy surfaces for molecules other than ArH_2. As expected from momentum gap agruments, Δv predissociation of ArH_2 proceeds exceedingly slowly. For ArH_2 ($J = 0$, $j = 0$, $v = 1$), the process occurs with maximum probability to $Ar + H_2$ ($j = 4$, $v = 0$) even though only second-order anisotropy terms are included in the Hamiltonian.[43] The $\Delta j = 6$ channel is also populated. It is evident that a low-order perturbation theory would be unable to reproduce this result.

Section III reviews progress on simple dimers. As previously mentioned, Pine and Lafferty have recorded a high-resolution spectrum of HF dimer in the HF stretch fundamental region.[16] These data should provide the basis for theory of the HF dimer comparable in accuracy to that on ArH_2. For now it can simply be said that vibrationally excited HF dimer is much longer lived than predicted by considering quenching of HF vibrations by collision.[44] When the internal HF stretch is excited, it decays on a nanosecond timescale.[16] Decay of the external HF stretch takes longer than 10^{-7} s![45]

Understanding the mechanism of predissociation of polyatomic molecules is still largely a wide-open field. Data for molecules on the ground electronic surface are discussed in Section IV and excited electronic states in Section V. Ewing's propensity rules do provide a qualitative picture to predict which mechanism will be active for simple systems.[46,47] Data available still cover so few of the myriad possibilities that it is hard to guess if a general picture will evolve. For decay on the ground electronic surface it is difficult to probe product states or even to say just when they are produced. Most data simply consist of line-broadening measurements, although product velocity distributions have been measured for several systems.

For ethylene dimer, the most extensively studied example, linewidths indicate that the initially excited state decays on a picosecond timescale for excitation of either the in-plane[16–22] or out-of-plane[20] CH bending or stretching modes.[23] There is no evidence, however, that free ethylene is produced in less than 10^{-7} s. Are there long-lived species with more energy in the van der Waals modes than necessary to dissociate the molecule? Do the initial wavefunctions simply dephase? Is dissociation prompt? When the ν_7 out-of-plane bending mode of ethylene is excited in a cluster, the energy is almost certainly transferred to the van der Waals modes on a timescale consistent with the linewidths. Overall, dissociation probably occurs on a timescale consistent with a statistical theory such as RRKM.

For predissociation on excited electronic surfaces, more information is gained both about the overall rate and the product state distribution. Initial results indicate that symmetry arguments can reasonably predict (or at least justify) which vibrations will participate in the decay mechanism. In these cases, complete dissociation to products can be shown to occur on a nanosecond timescale. It is encouraging for the infrared experiments that there is not yet any serious disagreement between homogeneous lineshapes and measured cluster disappearance rates.

II. TRIATOMIC MOLECULES

The recent surge of interest in van der Waals molecule dynamics is largely due to the experimental studies on ArH_2 and HeI_2. The work of McKellar and Welsh[48-50] provided data for the most completely determined triatomic potential surface of any van der Waals molecule.[51] McKellar's recent linewidth data[10] provide a good test of dynamical theories. The Levy group has provided a wealth of detail on HeI_2 dynamics.[8] In addition to decay rates, they also measure product vibrational state distributions. Although the HeI_2 potential surface is not well determined, this has not discouraged theoretical analysis.[2] It would appear that calculations for HeI_2 are far less sensitive to details of the surface than for ArH_2. This can be crudely explained by the fact that predissociation of ArH_2 is an improbable event which results from high-order coupling of small terms in the Hamiltonian.

The insight gleaned from HeI_2 allowed predictions of rate versus structure for analogous systems. These predictions have recently been confirmed for the molecules $NeCl_2$[13,15] and $NeBr_2$[14]. For Ne bound to Cl_2 ($X\,^1\Sigma$, $v = 1$) the metastable complex lives longer than 10 μs!

One triatomic molecule has long been a subject of lively discussion but, as of now, no hard data exist for its dynamics. When Child[9] reported that the Δv predissociation rate for ArHCl would be slower than $1\ s^{-1}$, there was widespread disbelief among experimentalists.[44] Based on recent studies of $(HF)_2$[16] it now appears that Δv dissociation of ArHCl may be too slow to observe. Recent calculations, however, do predict observable Δj predissociation rates.[52]

A. ArH_2

The recent experimental work by McKellar[10] and theoretical calculations by LeRoy and coworkers[43,53] make the ArH_2 problem the most completely understood example of van der Waals molecule predissociation. Experiment and theory agree that only one state of ArH_2 (actually KrH_2, see below) is broadened enough to observe. This is the $J = 0$ (J is overall angular momentum) state of $v = 1$, $j = 2$ (j is the H_2 internal angular momentum) which decays to Ar and H_2 ($v = 1$, $j = 0$). In this decay, total angular momentum is conserved without the aid of orbital angular momentum ArHD is subject to considerably more line broadening than ArH_2. In the transformation of coordinates to the new centre of mass frame, both the isotropic and the anisotropic parts of the ArH_2 potential surface yield odd Legendre terms in the potential expansion. This opens up the $\Delta j = -1$ decay channel with a smaller energy gap than the $\Delta j = -2$ channel available to ArH_2. As a result, broadening is observed for a variety of ArHD states. Theory and experiment are in remarkable agreement as to the line width as a function of vibrational level.

ArH_2 is the primary model system because the potential energy surface expressed in a Legendre expansion converges rapidly, because the H_2 subunit yields widely spaced energy levels and because the spectrum of the constituents is both sparse and weak, allowing that of the complex to be observed in (relatively!) simple absorption experiments. The well depth, ε, of the first anisotropic term of the Legendre expansion (P_2) is 7.59 cm^{-1}. Since this is less than the H_2 rotational constant, 61.8 cm^{-1}, the molecule is best described with a free internal rotor basis set where:

J = total angular momentum quantum number
l = rotational quantum number of the complex
j = rotational quantum number of the H_2 subunit
v = H_2 vibrational quantum number

Of these quantum numbers, only J is a rigorous quantum number, but the other three provide useful insight and a rather accurate description of the energy levels. In describing ArH_2 vibrational transitions, one labels the bands according to Δj. For instance, $Q(1)$ means $j = 1 \leftarrow 1$, $S(0)$ means $j = 2 \leftarrow 0$. The branch labels refer to Δl which can be ± 1 or ± 3 (see Fig. 2).

The well depth of ArH_2 is about 60 cm^{-1} while the infrared transition energies involved with $\Delta v = \pm 1$ are over 4000 cm^{-1}. As predicted by the momentum gap law and confirmed by close-coupling calculations, the Δv predissociation rate is immeasurably slow. The Δj transitions produce much smaller energy gaps and observable predissociation widths are predicted for ArH_2, ArD_2, KrH_2 and ArHD.[53] The widths are observed by McKellar for KrH_2 and ArHD.[10] For ArH_2, only one state is predicted to be broader than McKellar's instrumental resolution and it is obscured by an H_2 transition.

Note that if the anisotropy of the ArH_2 potential were substantially larger, Δj predissociation would correlate with vibrational predissociation from the van der Waals vibrational modes. This puts ArH_2 predissociation in a fundamentally different class from polyatomic predissociation where line widths are determined (in favourable cases) by either direct $V \rightarrow T$ decay from the covalent modes or by energy transfer from the covalent to the van der Waals modes.

McKellar's recent experiments on ArH_2 are a *tour de force* of conventional spectroscopy. He uses a 5.5-m multipass cell to obtain a total absorption path length of 220 m. The cell is cooled to liquid N_2 temperatures to favour the production of van der Waals molecules. By maintaining a total pressure of ~ 80 torr, pressure broadening is reduced to less than 0.1 cm^{-1}. To the lowest pressures measured, the ArH_2 linewidths were still linearly dependent on gas density. Thus no effects of intrinsic lifetime broadening are observable. Such a spectrum, for the *S*(0) transition of ArH_2, is illustrated in Fig. 1. In contrast, the ArHD transition widths were found to be independent of pressure below ~ 300 torr. As seen in Fig. 2, the ArHD lines for the *S*(0) branch are between three and ten times broader than the 0.1 cm^{-1} instrumental limit for the various members of the N- and T-branch. A plot of the linewidth as a function of the end-over-end angular momentum along with Hutson and LeRoy's calculated values[53] are shown in Fig. 3.

The ArD_2 spectra, like ArH_2, show no signs of predissociation broadening.

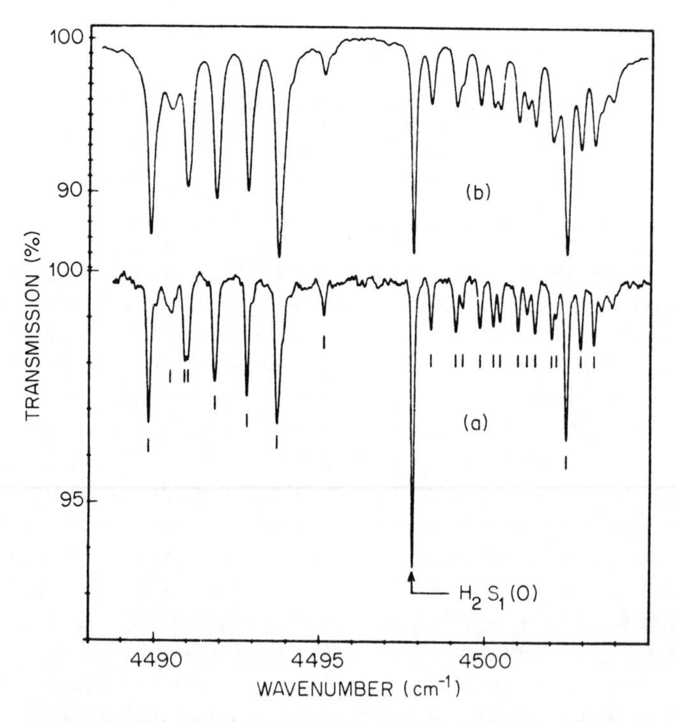

Fig. 1. The central portion of the $H_2{\cdot}Ar$ spectrum which accompanies the *S*(0) transition of H_2 in a *para*-H_2, Ar mixture at 77 K and (a) 60 torr or (b) 153 torr. Spectrum (a) is limited by instrumental resolving power while spectrum (b) shows the effects of pressure broadening. The short vertical lines denote $H_2{\cdot}Ar$ transitions. The intense transition at 4598 cm^{-1} is an H_2 quadrupole allowed transition. (From Ref. 10, Fig. 5. Reproduced by permission of The Royal Society of Chemistry.)

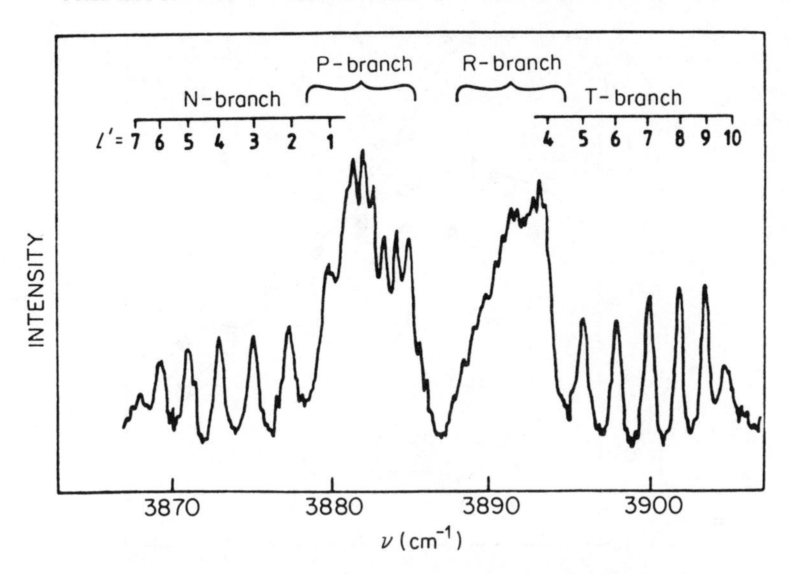

Fig. 2. The $S_1(0)$ transition for HD·Ar at 77 K and 302 torr. (Note that the spectrum is upside-down compared to Fig. 1.) The transition widths do not change if the pressure is lowered. Quantum labels are discussed in the text. (From Ref. 53, Fig. 1, adapted from Ref. 10, Fig. 8. Reproduced by permission of the American Institute of Physics.)

Because of the smaller rotational constants the ArD_2 spectra are more congested. The KrH_2 spectrum is noticeably broadened on one transition which terminates at the $j = 2$, $l = 2$, $J = 0$ level. This level has no angular momentum: it decays to products which have no angular momentum. Unfortunately, the comparable transition was hidden by an H_2 transition in the ArH_2 spectrum and was not well resolved in ArD_2.

LeRoy and coworkers have performed converged close-coupling calculations for ArH_2 and ArHD on a detailed potential energy surface.[53] The ability of these calculations to reproduce the experimental widths, as shown in Fig. 3, is remarkable. Datta and Chu achieved similar accuracy using the complex coordinate coupled channel method.[54] The calculations are accurate enough to consider using transition width data as input to refine the potential surface. In essence, then, the problem is solved. The real joy of theory, of course, is the ability to test for the effects of various perturbations of the model on the result. Also, this rigorously solved model system can be used to test approximation schemes.

In one interesting test of the predissociation physics, Hutson and LeRoy[53] calculated the ArHD widths using only the isotropic terms of the ArH_2 surface. These produce anisotropy in ArHD because of the centre of mass change. They found that this 'loaded sphere' model produced only half of the observed width. The transformed anisotropic part of the ArH_2 potential is equally important. The fact that the ArHD widths are ten to twenty times as

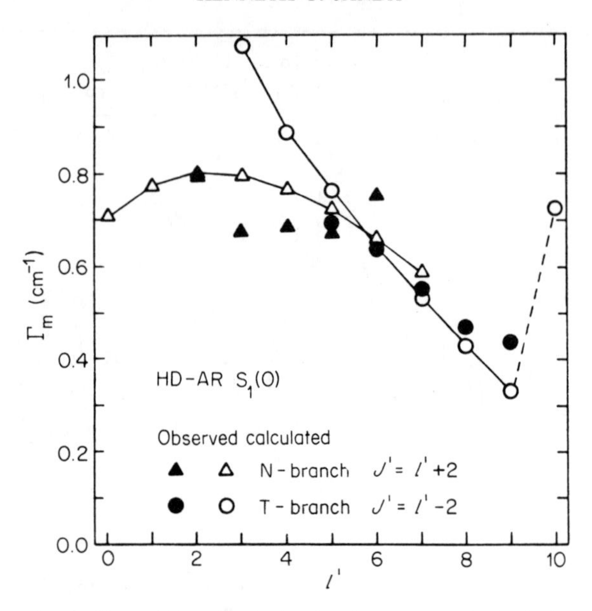

Fig. 3. Comparison of calculated and experimental widths for predissociation states of HD·Ar. The calculated widths (open symbols) are from close-coupling calculations by Hutson and LeRoy. Experimental widths (solid symbols) are from McKellar. Calculated points for $l' = 2$ and 6 do not show because they are under the experimental points. (From Ref. 53, Fig. 2. Reproduced by permission of the American Institute of Physics.)

large as those of comparable ArH_2 states is due to the fact that the first order anisotropy for ArHD is especially strong at small R. The $\Delta j = -2$ partial widths are also considerably larger for ArHD than for ArH_2. These can again be explained as due to the increased anisotropy of the potential. For both $\Delta j = -1$ and $\Delta j = -2$ transition energy gap effects would predict faster rates for ArHD than for ArH_2. It would be very interesting to perform numerical studies to judge the relative importance of energy gap and potential effects on these trends.

The trend shown in Fig. 3 of decreasing width as l' increases is explained by Hutson and LeRoy as a centrifugal effect. Since $\langle R_l \rangle$ increases with l, the molecules experience less of the anisotropy in the potential which has a shorter range that the isotropic part of the potential. For $l' = 8$ and 9 the molecule is above the threshold for pure rotational predissociation (change in l only) but constrained by the centrifugal barrier. Apparently, tunnelling through the barrier is slow. For $l' = 10$, rotational predissociation dominates the spectrum.

The Carley–LeRoy potential[51] used for the Δj predissociation calculations contains terms which couple the H_2 bond length to the van der Waals interaction. For instance, the different average H_2 bond length due to contrifugal stretching has a 10 per cent. effect on Δj predissociation widths of ArH_2. The vibrational anharmonicity between H_2 ($v = 0$ and 1) makes a factor of three difference in the Δj predissociation rate. This coupling of the weak and strong bonds makes the ArH_2 potential the best candidate for calculating Δv predissociation rates. These rates have recently been calculated by Hutson, Ashton and LeRoy.[43] They are far too slow to be observable by present experimental methods, e.g. $\tau(ArH_2\ v = 1,\ J = l = j = 0) = 0.26$ m s.

The Δv calculations confirm several of the qualitative predictions made by Ewing on the basis of a much simpler model. They also show that quantitative calculations on the basis of perturbation theory will prove quite difficult. Especially interesting is a look at the dissociation product channels. For the ArH_2 ($v = 1$, $J = l = j = 0$), for instance, the main H_2 product state is $j = 4$, even though the highest order anisotropy in the potential is P_2; $j = 6$ is the second most probable product! Third-order perturbation theory is necessary to even attempt to reproduce this result.

The partial widths for ArHD Δv predissociation are reproduced in Table I. A bimodal distribution of products is seen. Hutson, Ashton and LeRoy[44] attribute the low j products to high-order coupling of the low-order anisotropy terms, as in the ArH_2 case. The main two product channels, $j = 7$ and 8, are attributed to direct coupling of high-order anisotropy terms (again, these are produced in the coordinate transformation from ArH_2 to ArHD). It was necessary to include many open and closed channels in the calculation due to the high-order coupling.

The close-coupling calculations support the validity of Ewing's momentum gap picture which is illustrated in Fig. 4. In this picture, the slow rate of Δv predissociation is attributed to poor overlap of the bound Ar—H_2 radial

TABLE I
Results of close-coupling calculations for HD·Ar.[a]

E_m(cm^{-1})	3606.322	3863.786
Γ_m(cm^{-1})	15.6×10^{-9}	39.6×10^{-8}
Γ_{mvj}(cm^{-1})		
$(v, j) = (0, 0)$	0.4×10^{-9}	1.2×10^{-8}
$(v, j) = (0, 1)$	1.0×10^{-9}	3.3×10^{-8}
$(v, j) = (0, 2)$	1.2×10^{-9}	4.7×10^{-8}
$(v, j) = (0, 3)$	1.2×10^{-9}	4.8×10^{-8}
$(v, j) = (0, 4)$	1.8×10^{-9}	3.3×10^{-8}
$(v, j) = (0, 5)$	2.0×10^{-9}	1.0×10^{-8}
$(v, j) = (0, 6)$	0.3×10^{-9}	0.3×10^{-8}
$(v, j) = (0, 7)$	3.7×10^{-9}	4.9×10^{-8}
$(v, j) = (0, 8)$	4.0×10^{-9}	14.7×10^{-8}
$(v, j) = (0, 9)$		1.2×10^{-8}

[a]Both resonances are $v = 1$, $J = 0$, $n = 0$. (This table is reprinted with permission from Ref. 43 Copyright © 1982 American Chemical Society.)

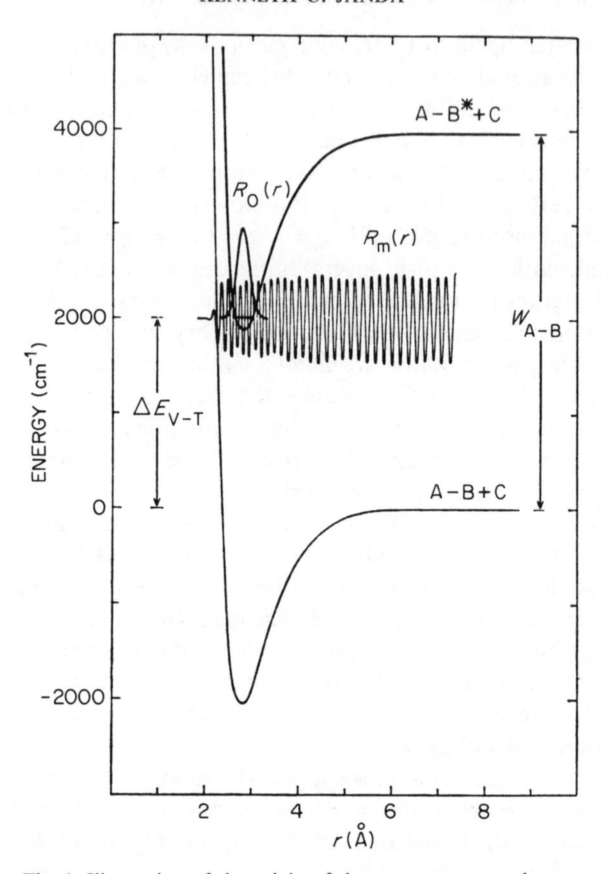

Fig. 4. Illustration of the origin of the momentum gap law as applied to the HF dimer. r is the weak bond coordinate. The lower and upper surfaces are for the HF ground and $v = 1$ excited states respectively. The two surfaces are nearly parallel because the weak and strong modes couple only slightly. The vertical excitation of a covalent stretch produces the weak mode wavefunction $R_0(r)$. For a pure V→T predissociation the final weak mode wavefunction is $R_m(r)$, whose frequency is proportional to the final relative velocity. Since the perturbation between $R_0(r)$ and $R_m(r)$ is expected to be a slow function of r the perturbation matrix element goes rapidly towards zero as the final momentum increases. (From Ref. 46, Fig. 2. Reproduced by permission of The Royal Society of Chemistry.)

wavefunction with the final plane wave. More specifically, the matrix element which puts energy into the reaction coordinate is approximated by $\langle R_0(r)|H'|R_m(r)\rangle$, where $R_0(r)$ is the Gaussian weak stretching wavefunction produced by a vertical transition from the ArH_2 ground state, $R_m(r)$ is the final translational wavefunction and H' is a perturbation which is assumed to be a relatively slow function of r. Since the oscillation frequency of $R_m(r)$ increases linearly with momentum the matrix element goes rapidly towards zero as momentum increases. Indeed, Hutson, Ashton and LeRoy[43] find that the appropriate matrix element integral has a value of 10^{-4} times the value of a partial integral over half a period in $R_m(r)$. They conclude that vibrational predissociation of ArH_2 and ArHD is such an improbable event that very small approximations can introduce huge relative errors in a calculation. This problem should not be quite so bad in other molecules where more channels are nearly in resonance, producing faster rates. For these cases, however, it will be more difficult to perform exact calculations for comparison with simple models because of the huge number of channels involved.

B. ArHCl

Although there is no experimental data on the predissociation of ArHCl, the molecule still receives extensive theoretical consideration. It was Child's prediction[9] of a very long-lived metastable $v = 1$ state which convinced many people that van der Waals molecule vibrational predissociation should be studied. Certainly many experimentalists were skeptical of this prediction. Also, the ArHCl potential energy surface is well characterized[55] by combining microwave, scattering and virial coefficient data as a test of the potential. The ArHCl potential surface is less well determined than that of ArH_2 for two reasons. First, no bound state information is available to help describe the ArClH local minimum. Second, no data are available to correlate the weak bonding attraction with the strong bond length. Any infrared spectra which could be obtained would help alleviate these weaknesses.

Ashton, Child and Hutson[52] have recently performed close-coupling calculations on the Δj predissociation of ArHCl. In its lowest energy levels, the molecule is best thought of as a linear wide amplitude bender, but addition of internal energy quickly overcomes the barrier to internal rotation (estimated to be $40\,cm^{-1}$ above the zero-point energy), making j a useful label for the bending quantum number. The calculations use a model potential with no secondary minimum so the results are not expected to match any future experiment in detail.

At relatively high internal energies ($\geq 200\,cm^{-1}$), Ewing's[46,47] propensity rules qualitatively describe the results. The predominate channel is $\Delta j = -1$ and the decay rate falls in accordance with a momentum gap law if excess energy is added to the weak bond stretch. Also, the rate drops as j increases in qualitative accord with an energy gap law.

For low internal energy levels the product state distribution peaks at $j'' = 0$. Not only is this in disagreement with the propensity rule prediction but also with a perturbation treatment which yields nearly correct total linewidths. Ashton, Child and Hutson[52] attribute this effect to open-channel coupling, in which the HCl fragment loses angular momentum as it leaves the complex.

Presumably the effect does not continue to high j because the level spacings become too large to couple.

Total linewidths predicted are as large as 8 cm^{-1} but are generally less than 1 cm^{-1} for states with small amounts of energy in the weak stretching motion. Can these effects be observed? Long path-length absorption experiments similar to McKeller's on ArH_2 are made difficult by the intense HCl rotational spectrum. Boom and van der Elsken[56] have observed 6 ArHCl transitions between 30 and 45 cm^{-1}, but rotational structure was not resolved and problems of assigning the transitions are immense. Perhaps the most useful next step would be to calculate an approximate spectrum for the molecule to give experimentalists a clue as to what sort of band structure to expect. Molecular beam experiments will probably be necessary to separate the ArHCl spectrum from that of HCl in either the near- or far-infrared.

C. HeI_2

The series of rare gas–iodine clusters has been extensively studied by Levy and coworkers.[8] The work has already received a thorough review in this series.[8] The experiments are more extensive than those on any other van der Waals predissociation. In addition to measuring line broadening in $B \leftarrow X$ excitation spectra, product states were directly monitored by dispersing fluorescence. For HeI_2 and NeI_2, Levy and others measured that $\Delta v = -1$ dominates the predissociation mechanism. Although final rotational states could not be resolved, the dispersed bandshapes indicated only modest deposition of energy into the rotational modes. For ArI_2 the stronger bond strength requires that at least $\Delta v = -3$ be involved in predissociation. For low v this is the major channel. Due to anharmonicity, the $\Delta v = -1$ channels for HeI_2 and NeI_2 and the $\Delta v = -3$ channel for ArI_2 close for high v. By monitoring this effect the D_0 for each complex could be estimated: HeI_2, 14 cm^{-1}; NeI_2, 66 cm^{-1}; and ArI_2, 225 cm^{-1}.

Levy and others were able to measure HeI_2 bands starting from $v = 0$ in the ground electronic state to $3 < v < 35$ in the excited ($B^3\Pi_0$) electronic state (v is the I_2 vibrational quantum number). Line-broadening measurements yield predissociation lifetimes ranging from 220 ps for $v = 12$ to 38 ps for $v = 26$. More recently, Sharfin, Kroger and Wallace extended the measurements to $v = 68$.[11] Up to $v = 45$ the lifetimes are characterized by the simple formula:

$$\Gamma = 0.555 \times 10^{-4} v^2 + 0.174 \times 10^{-5} v^3 (\mathrm{cm}^{-1}) \quad (1)$$

The v^2 dependence is interpreted as due to the anharmonicity effects on the $\Delta v = -1$ channel while the v^3 dependence accounts for the increasing importance of $\Delta v = -2$ for high v. For $v > 45$ the simple relationship breaks down as the $v = -1$ channel closes. The resulting dependence of lifetime on v is shown in Fig. 5.

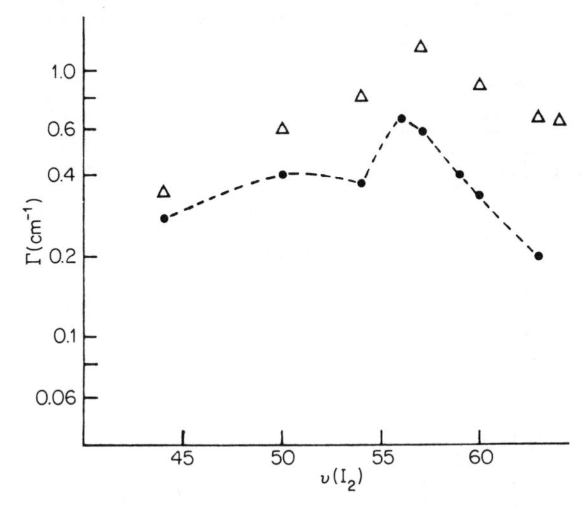

Fig. 5. Dependence of linewidth on vibrational level for HeI_2 predissociation. Below $v = 45$ the $\Delta v = -1$ channel closes and the dependence of Γ on v is no longer monotonic. The experimental points (open triangles) are from Ref. 11. The calculated points are obtained from quasi-classical simulation. (From Ref. 12, Fig. 4. Reproduced by permission of The Royal Society of Chemistry.)

The experimental results can be qualitatively duplicated by distorted wave,[2] close-coupling[2] and quasi-classical[12] calculations on reasonable potential energy surfaces. The $\Delta v = -1$ channel can be seen to be a result of the energy or momentum gap laws. The qualitative behaviour represented in Eq. (1) is due largely to the kinematics of the predissociation while the absolute value of the rate depends very sensitively on the potential.

HeI_2 has a limited value as a prototype for vibrational predissociation in spite of the wealth of experimental data and theoretical interpretation already in existence. The predissociation widths of laser excitation spectra are generally greater than rotational spacings. This severely limits the determination of empirical potential energy surfaces. In fact, calculations on the empirical surface determined by Blazy and coworkers[57] from their spectra lead to predissociation rate estimates which are in error by six orders of magnitude![58]

Segev and Shapiro[58] postulate that this discrepancy results from a misassignment of a vibrational band to the van der Waals stretching motion rather than a librational mode. They propose a B state librational potential and state ordering as illustrated in Fig. 6. The state labelled 1 is primarily T-shaped and that marked 2 is primarily colinear. Blazy and coworkers[57] do indeed observe two closely spaced levels but rule out librational assignments on the basis of selection rules. Perhaps this difficulty can be overcome by mixing the wide-amplitude bending and stretching motions.

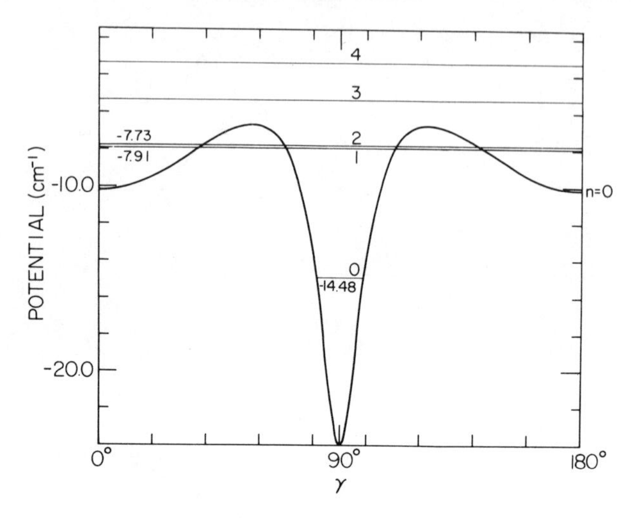

Fig. 6. This effective librational potential for HeI_2 would reproduce the spectrum obtained by Levy and others if the wavefunctions mix the bending and stretching modes to make librational transitions allowed. The state marked '0' carries most of the intensity. Transitions to '1' result mainly in a T-shaped structure while those to '2' result in a linear structure. (From Ref. 58, Fig. 7. Reproduced by permission of the American Institute of Physics.)

Recently, Cross and Valentini[59] have observed I_2 fluorescence after exciting ArI_2 to energies well above the I_2 B state dissociation limit. The observed spectra are consistent with a colinear ArI_2 excited state which undergoes impulsive energy transfer from I_2 to Ar. This suggests that they are exciting a librational mode similar to state 2 of Fig. 6.

D. $NeCl_2$ and $NeBr_2$

To increase the ability to compare experiment and theory for a system analogous to HeI_2 it will be necessary to choose a system for which a potential energy surface can be more completely determined. This can be accomplished with larger rotational constants, less line broadening or both. Complexes of the lighter halogens should meet these criteria. Unfortunately, the $B \leftarrow X$ transition strength is much smaller for light halogens since it is only allowed due to spin-orbit coupling. This limitation is offset by the sensitivity of the free jet laser excited fluorescence technique.

Spectra for both $NeCl_2$[13] and $NeBr_2$[14] have been recorded with 0.15-cm^{-1} resolution. Excitation is observed from the $NeCl_2$ electronic ground state with $v = 0$, 1 and 2. The Cl_2 vibrational quantum is 550 cm^{-1} while the Ne—Cl_2 bond strength is probably less than 70 cm^{-1}. The $v = 1$ and $v = 2$ states still survive the 10^{-5} s travel time from the nozzle to the excitation region. No similar bands have been observed for rare gas–I_2 spectra. This result confirms the significance of momentum gap arguments for these systems.

The rotational contour for the B←X, 11←1 excitation spectrum is shown in Fig. 7. Although rotational structure is observed it is not resolved. The limitation is not due to predissociation of the excited state, however, but to the laser bandwidth. So far there are no experimental data sensitive to the rate of predissociation. The structure of Fig. 7 is consistent with a T-shaped $NeCl_2$ molecule. Two computer-generated bandshapes are shown to illustrate the effect of changing the assumed structure on the bandcontour. Presently, the structure is known to about 10 per cent accuracy. Higher resolution experiments will yield both a rotational determination of the potential energy surface and lifetime as a function of the quantum state.

In $NeBr_2$ spectra homogeneous broadening of the transitions as a function of v is observable for $v > 18$ with the present laser bandwidth. Figure 8 shows the estimates of lifetime versus v. The dependence is seen to be similar to that of

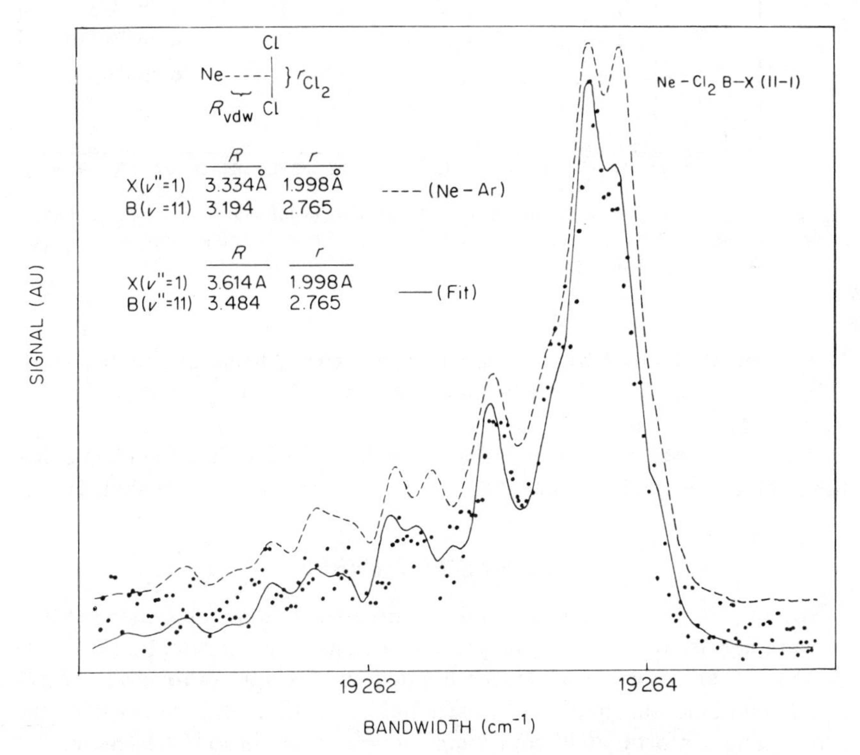

Fig. 7. The rotational contour of the $NeCl_2$ X←B, 11←1 excitation. The dots are data points. The solid line is generated from a T-shaped geometry adjusted to give the best fit to the data. The dashed line shows how the fit changes if the Ne—Cl_2 bond length is shortened by 10 per cent.

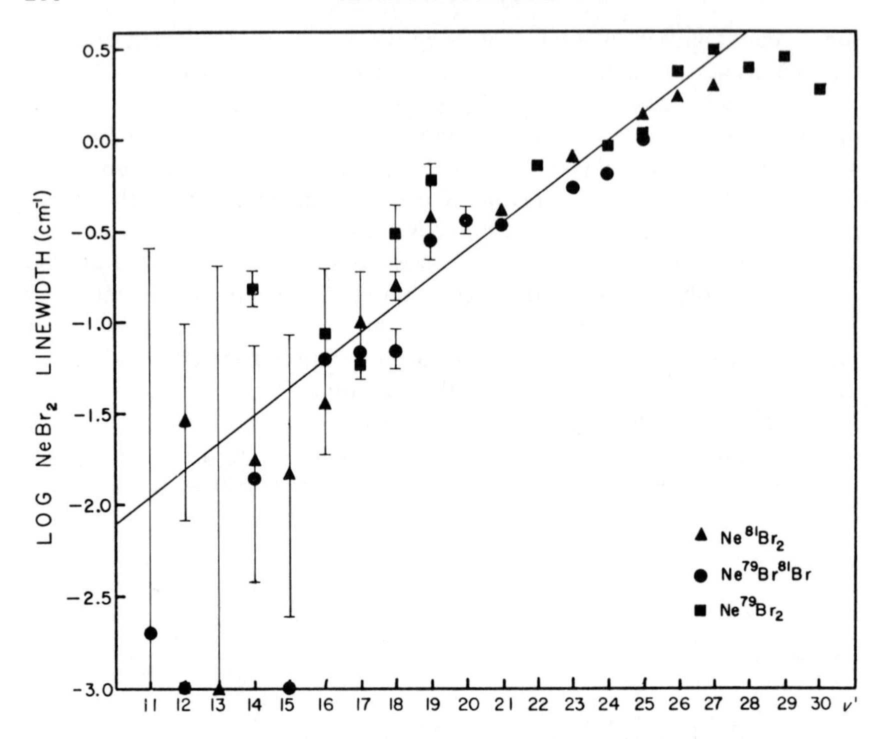

Fig. 8. Linewidth verses v for $NeBr_2$. Below $v = 18$ the linewidths are instrument limited so attempts at deconvolution produce large error bars. Above $v = 27$ the linewidths level off. This probably indicates that the $\Delta v = -1$ channel is closed.

HeI_2 in that the linewidth increases smoothly with v up to $v = 26$ and then levels off. At $v = 26$ the Br_2 B state vibrational spacing is 55 cm^{-1} so it is reasonable that the $\Delta v = 1$ channel closes.

The data for $NeCl_2$ and $NeBr_2$ are still preliminary but they do indicate that these molecules may be amenable to a more complete analysis than HeI_2.

III. SIMPLE DIMERS

Recent progress in the study of dimer predissociation would indicate that these molecules will provide the greatest increase in understanding over the next few years. Various hydrogen-bonded species appear to have spectra which combine the merits of rotational resolution and observable line broadening. Unfortunately, rotational congestion sets in so fast, as molecular complexity increases, that the spectra of a relatively simple dimer like $(N_2O)_2$ may never yield a resolvable structure.

A. HF Dimer

Pine and Lafferty[16] have recently studied the HF dimer spectrum at a Doppler limited resolution of 200 MHz ($7 \times 10^{-3}\,cm^{-1}$) in the region of the HF stretching motions. Two samples of the spectrum are shown in Fig. 9. The experimental technique is long-path absorption (64m) of an HF sample at 2.07 torr and 219 K. A laser spectrometer in which the output of a ring dye laser and an Argon ion laser were mixed provided the infrared radiation. The part of the spectrum illustrated in Fig. 9 represents a small fraction of the transitions observed and assigned by Pine and Lafferty. Complete bands were observed for both the internal and external HF stretching motions. Analysis of these bands yields ground state rotational constants in accord with the microwave spectra of Dyke, Howard and Klemperer.[60] With the analysis of these data, and that of mixed isotopes, the HF dimer potential energy surface should be determined with accuracy comparable to that of ArH_2.

The transitions shown in Fig. 9a involve excitation of the HF stretch in which the proton is not involved in hydrogen bonding. No evidence of broadening beyond the Doppler widths are observed. In contrast the lines of Figure 9b, which involve excitation of the bound proton stretching motion, are clearly broadened. Assignment of the transitions is based on frequency shifts and band polarizations. The fact that the internal stretching state decays faster than the external one is intuitively reasonable in that, in the internal stretch proton, motion couples directly to the weak bond.

Pine and Lafferty estimate that a homogeneous broadening of 30 MHz would have been observed in Fig. 9a. This results in a state lifetime of at least 10^{-8} s. Muenter[45] has measured the ${}^{R}P_0^{+}(2)(J = 2, K = 0, v = 0 \rightarrow J = 1, K = 1, v = 1)$ transition by infrared–microwave double resonance in an electric resonance state selected beam. They observed no homogeneous contribution to a 30-MHz Doppler width. This suggests that the state lifetime is longer than 100 ns. This is certainly the sharpest clearly resolved infrared predissociation transition which has been measured for a weakly bound molecule. For the internal HF stretch, linewidths range from 100 to 600 MHz as J increases. Accurate linewidth determination awaits convolution fitting of those spectra which are not completely resolved (see Fig. 9b). State lifetimes are between 0.5 and 3 ns.

Although the $(HF)_2$ predissociation lifetimes are longer than those of HeI_2 or the empirical estimates of Klemperer and others, they are much shorter (for internal HF excitation) than would be predicted for decay dominated by a $V \rightarrow T$ mechanism. The products must be rotationally excited.[61] This will complicate the theoretical simulation of the data. Since the HF rotational levels are widely spaced, however, close-coupling calculations may be feasible. This molecule will clearly stimulate much activity.

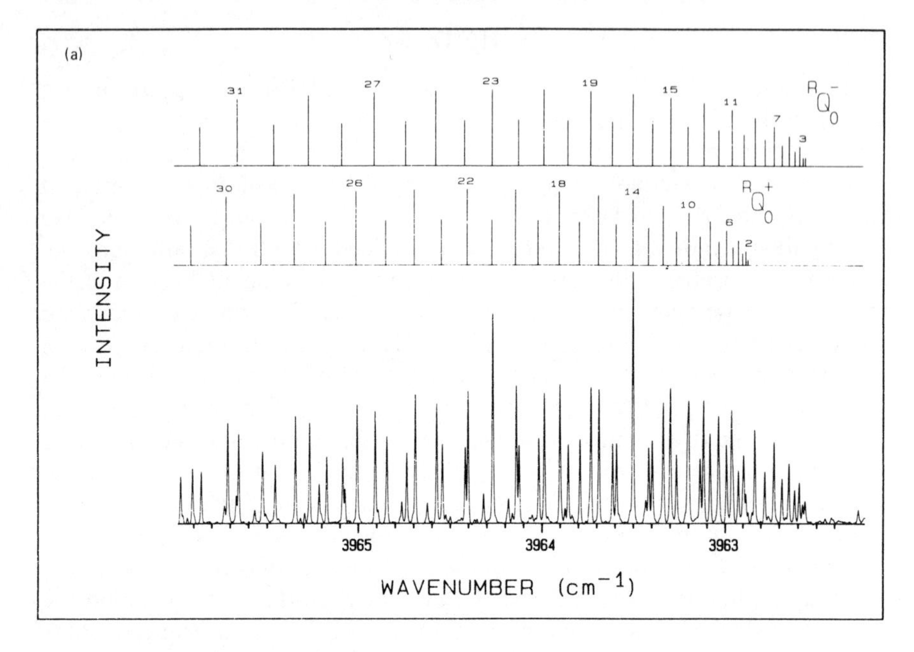

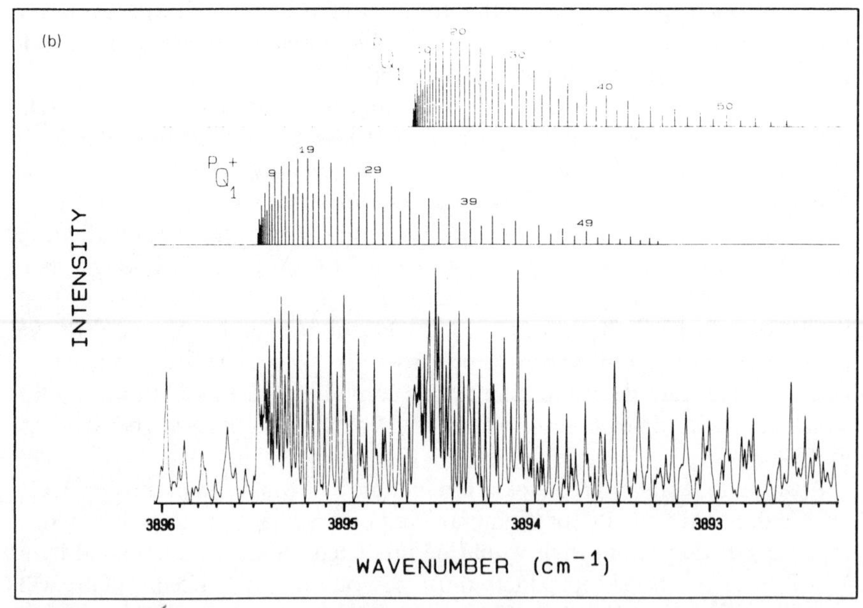

Fig. 9. The (a) $^{R}Q_0$ and (b) $^{P}Q_1$ subbranches of the ν_1 free hydrogen stretching fundamental with calculated stick spectra. These ν_1 lines show no evidence of lifetime broadening. The broad experimental transitions which are not labelled are from the tail of the $^{Q}R_0$ subbranch of the ν_2 hydrogen-bonded proton stretching. From Ref. 16, (Fig. 4. Reproduced by permission of the American Institute of Physics.)

HF dimer spectra have also been studied at low resolution by direct monitoring of products of molecular beam dissociation.[36] In addition to the two bands observed by Pine and Lafferty[16] an intense transition is observed at 3720 cm^{-1}. How to explain the origin of this peak is currently a mystery. If it arises from a directly dissociative transition then its intensity at any particular frequency could be too weak for Pine and Lafferty's high-resolution spectrometer to observe. On the other hand, it is not clear what the nature of such a direct dissociation transition might be since it is to the red of the main spectrum.

Although the HF dimer may represent the most favourable example of its type for complete analysis, similar spectra have recently been recorded for OC:HF[25] and HCN:HF.[26] Spectra for both of these molecules show enough structure to inspire hope of an eventual complete analysis on an empirical potential energy surface.

B. N_2O Dimer and CO_2 Dimer

Gough, Miller and Scoles[62] made the first direct measurements of molecule attenuation by infrared predissociation in their study of the N_2O dimer. The ν_3 mode was excited by a diode laser, while beam attenuation was monitored with a cryogenic bolometer. The spectrum shows evidence of both homogeneous and heterogeneous structure and has yet to be completely characterized. Recently, Ding, Miller and Watts[63] have recorded spectra for the dimer upon exciting the $\nu_1 + \nu_3$ modes with an f-centre laser. From spectral intensities they estimate that homogeneous broadenings must be of the order of 1 cm^{-1}. Morales and Ewing[64] postulate that the decay can only be this fast if the mechanism is of the type $V \rightarrow V'$, T so that the momentum gap is small. Similar data exists for the CO_2 dimer.[65] The prospect for studying these molecules is discouraging because individual excited states will not be resolved since the rotational structure is less than the linewidth. Unless exceedingly cold beams which contain only dimers (as opposed to larger clusters) can be obtained, it is unlikely that these systems will be useful for quantitative theoretical treatment.

IV POLYATOMIC MOLECULES: GROUND ELECTRONIC SURFACE

Given the complexity of the HF dimer spectrum and the lack of useful resolution for the N_2O dimer, it might seem that polyatomic molecule predissociation spectroscopy would be hopeless. In one respect, however, polyatomic molecule spectra can be simpler than those of smaller molecules. As molecular weight and dimensions increase the rotational bands collapse to less than one wavenumber for supersonically cooled samples. For several species lifetime broadening has completely dominated the rotational contour. Structure determinations for the molecules can be performed by microwave spectroscopy. Thus it is still possible to correlate dynamics with structure even though the broad IR transitions yield no rotational structure.

For molecules discussed in this section neither dissociation lifetimes nor product internal states have been directly monitored. One hopes that progress will be made along these lines in the near future. For now we are left with the perplexing question of how to interpret line-broadening information. Until recently all measured infrared lineshapes were greater than one wavenumber wide, indicating maximum state lifetimes of about 5 ps. Since it seems unlikely that all molecules predissociate so quickly it was tempting to postulate that the lineshape does not really reflect intramolecular energy flow from a high-frequency mode to the van der Waals modes. Recently we have found that beams of NeC_2H_4 produced in very dilute expansions yield spectra with observable structure. For excitation of the ethylene wag motion the lifetime of NeC_2H_4 is two orders of magnitude longer than that of the C_2H_4 dimer. This supports the view that lineshapes are determined by the intramolecular relaxation of energy to the van der Waals modes.

Since the spectra discussed in this section are of low resolution ($\Gamma \geq 1\ cm^{-1}$), separation of homogeneous from inhomogeneous broadening is not always straightforward. For very cold beams of molecules with intrinsically broad lineshapes, like $(C_2H_4)_2$, a two-level model for homogeneous broadening is adequate for laser fluencies below $10\ mJ\ cm^{-2}$.[66] When heterogeneous widths are comparable to homogeneous widths, however, Geraedts and coworkers[67] have pointed out that orientational hole burning can make the homogeneous widths appear broader than they actually are. Orientational hole burning results from the fact that different states of a molecule have their populations depleted at different rates due to orientations of the transition moments with respect to the laser polarization.

A. Ethylene Clusters

In an ideal world there would be a laser available to excite every vibrational mode in a molecule so that energy transfer as a function of excitation symmetry could be studied. In the real world most work is done with CO_2 lasers which operate only between 900 and $1100\ cm^{-1}$. Using optical parametric oscillators and f-centre lasers the frequency range above $2500\ cm^{-1}$ is also accessible. Rather than study relaxation as a function of mode for a single molecule we have taken the approach of studying how the relaxation of the ν_7 out-of-plane wag at $950\ cm^{-1}$ is affected by a variety of weak bonding partners.[17,19] Hoffbauer and coworkers[20] and Bomse, Cross and Valentini[21] have measured transitional energy distributions of the products for the dimer dissociation while Fischer, Miller and Watts[23] have measured lineshapes for excitation of C—H stretching modes. Hoffbauer and coworkers[20] have also studied excitation of the C_2D_4 dimer in an in-plane scissor mode and excitation of the mixed dimer $C_2H_4 \cdot C_2D_4$ in each of the two IR active modes accessible to the CO_2 laser.[20] This body of data makes the ethylene system the most completely studied type of polyatomic van der Waals molecule predissociation.

Gentry[68] opened up the area of polyatomic molecule predissociation experiments when he showed that the ethylene dimer could be photodissociated with any line from a pulsed CO_2 laser. We showed[17] that the intrinsic homogeneous width of the transition is no more than 12 cm^{-1} and that a simple two-level plus decay model correctly predicts observation of measurable photodissociation over a very broad frequency range (200 cm^{-1}) at fluences typical of a pulsed laser.[66] Subsequently, Geraedts and coworkers[22] have measured the intrinsic width to be less than 11 cm^{-1} while Hoffbauer and coworkers[20] obtain a value of 17.5. Paradoxically, Gentry[20] performed hole-burning experiments with two lasers which show no residual inhomogeneous contributions to their widths while similar experiments by Reuss[22] indicate some residual heterogeneous contribution even though their intrinsic widths are the narrowest yet measured for this transition. Clearly, linewidth measurements on clusters are not trivial and are best interpreted in a qualitative manner.

In spite of the lack of perfect agreement about the intrinsic linewidths for ethylene dimer excitation at 950 cm^{-1} all measurements do indicate that the initially excited state decays on a subpicosecond timescale. This is quite a different result than what would have been predicted by a simple comparison with ArH_2, ArHCl or HeI_2 and a momentum gap law. The energy gap for this process is probably between 600 and 700 cm^{-1}. This would predict an immeasurably slow rate for a pure $V \rightarrow T$ dissociation. Indeed, three groups have measured that the dissociation products separate at a relative kinetic energy which is only 10 per cent. of the energy gap.[20,21,69] Since there are no low-frequency vibrational modes to account for the energy deficit the product rotational modes must be highly excited. Ewing[70] has calculated that a $V \rightarrow R$, T process could account for decay in the picosecond range. The intrinsic width of each channel would only be about 0.1 cm^{-2}, but many possible channels would add up to broader total widths.

The ν_7 mode of ethylene excited at 950 cm^{-1} is a wide-amplitude hydrogen wag. It is easy to imagine how this motion would efficiently couple to the rotational degrees of freedom. It is not clear, however, how soon after the decay of the excited vibration the two constituents separate. No correlation between the exiting momentum and the excitation laser polarization has been observed.[20,21] This would indicate a long-lived intermediate. However, the very low final-product relative kinetic energy, the breadth of the dimer initial velocity distribution, the possibility of orientational saturation and the floppy nature of the dimer all conspire to make such correlation difficult to observe.

TABLE II
Data for CO_2 laser photodissociation of ethylene clusters.

C_2H_4—	$\omega_0(cm^{-1})$	$\Gamma(cm^{-1})$	$\tau(ps)$	$\langle\mu\rangle^2(10^{-3}D^2)^a$	Ref. No.
Ne	948	<0.5	>0.10	—	See text
Ar	948	<3.0	>1.8	—	See text
Xe	948	9.7	0.51	36	22
HF	975	1.6	3.2	77	18, 19
HCl	964	1.6	3.3	50	18
NO	952	4.5	1.1	51	18
SF_6	944	10	5.1	—	22
C_2F_4	955	5.8	0.9	46	17
C_2D_4	954	14	0.4	44.1	20
C_2H_4	952	11–18	0.5	53–87	17, 20, 22

[a]The values for $\langle\mu\rangle^2$ reported in Ref. 22 are multiplied by a factor of three. Reference 22 neglected the effect of spatial averaging.

The several groups studying dissociation of the ethylene dimer have also measured the lineshape for ethylene bound to Ne, Xe, HF, HCl, NO, SF_6, C_2F_4 and C_2D_4. A summary of lineshifts and widths is given in Table II. It is apparent that the transition linewidth generally increases as molecular weight and complexity increases. The data suggest that the relaxation of the initially excited state is limited by conservation of angular momentum during the intramolecular energy transfer. The data are also consistent with a PDP (pure dephasing) argument insofar as PDP depends on the density of states. However, PDP also depends on occupation numbers in such a way that it is precluded in these molecules at low temperatures. For all predissociation processes for which velocity distributions have been measured very little energy goes into relative kinetic energy of the products. For most of the molecules in Table II only the rotational degrees of freedom can take up the excess energy. If the dissociation is direct, Ewing's propensity rules[47] should apply: i.e. the total change in rotational quantum numbers should be as small as possible. If the dissociation involves an intermediate state with highly excited van der Waals modes, the bending modes which correlate with product rotation must be most active—if the energy went into stretches the dissociation would be prompt. In either case the initial state will decay to states with high local angular momenta. The local angular momenta must cancel in order to conserve the total angular momentum.

To see how the angular momentum constraints apply consider three molecules separately: $(C_2H_4)_2$, $C_2H_4 \cdot HCl$ and $C_2H_4 \cdot Ne$. Spectra for this set of molecules are shown in Fig. 10. For the ethylene dimer, satisfaction of the angular momentum constraints is naturally met by the two ethylene constituents being excited to rotate against each other. For a parallel structure of the dimer it is easy to visualize how an impulsive mechanism of energy

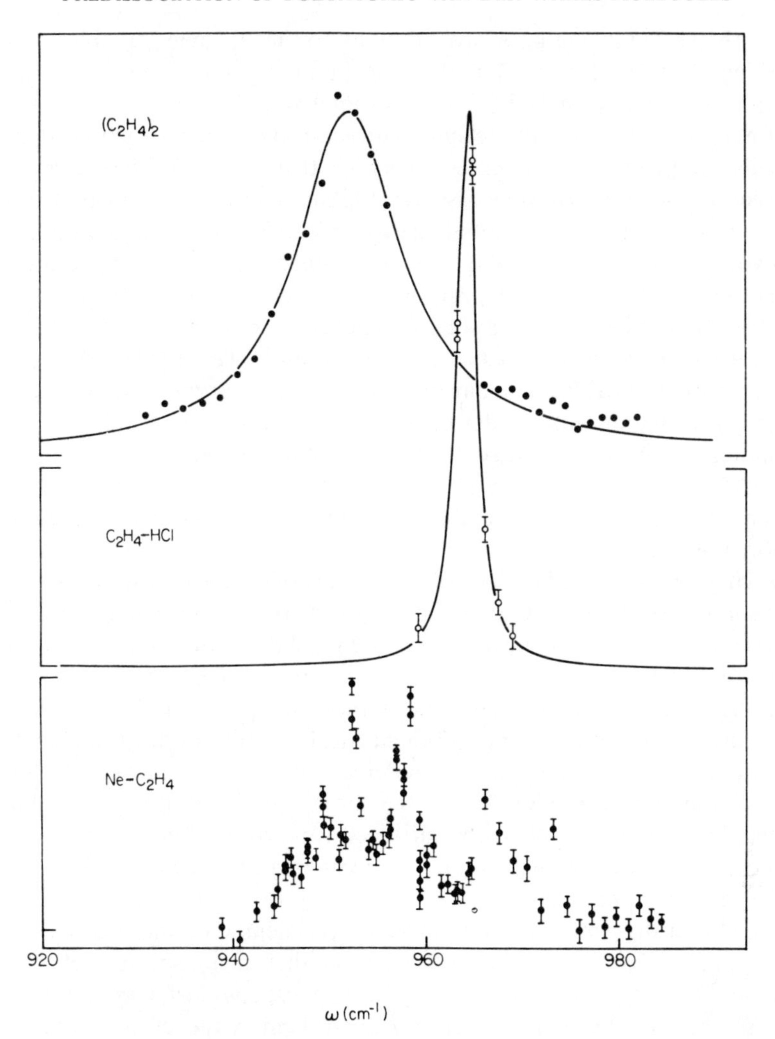

Fig. 10. CO_2 laser dissociation spectra for the C_2H_4 dimer, $C_2H_4{\cdot}HCl$ and $C_2H_4{\cdot}Ne$. This series shows the strongest dependence of linewidth on bonding partner measured for any molecule. The top two spectra are mostly homogeneous while that of NeC_2H_4 clearly is not. The NeC_2H_4 band retains its inhomogeneous breadth even at ~ 1 K because the C_2H_4 constituent is a perturbed rotor whose hyperfine levels do not come into equilibrium.

transfer would produce such an excitation. The final rotational quantum numbers need not be greater than ten so Ewing's angular momentum gap criterion is satisfied.

$C_2H_4{\cdot}HCl$ is less able to satisfy the above criteria for two reasons. In the structure of $C_2H_4{\cdot}HCl$ the HCl proton points between the ethylene carbon atoms thereby reducing impulsive coupling to the ethylene wag. Instead the coupling may be electrostatic since the bond dipoles involved are large. Second, E_{v_7}—E_{Bond} for $C_2H_4 \cdot HCl$ is estimated to be $400\,cm^{-1}$. To divide the final angular momentum evenly between C_2H_4 and HCl would require HCl to accept most of the excess energy with $n = 4$ or 5. The energy gap between $n = 4$ and 5, however, is over $100\,cm^{-1}$ with the result that translation must provide a substantial energy sink unless a fortuitous resonance is involved. Otherwise, orbital angular momentum must compensate for excess angular momentum deposition into the ethylene constituent.

$C_2H_4 \cdot Ne$ will have the largest excess energy since the bond should be very weak. Also, the Ne product can only compensate C_2H_4 angular momentum via generation of orbital angular momentum. These arguments are in good accord with the linewidth ordering $(C_2H_4)_2 > C_2H_4 \cdot HCl > C_2H_4 \cdot Ne$.

Similar qualitative arguments can be extended to the other examples of Table II but the uncertainties in the data and the qualitative nature of the arguments make justification of differences of less than an order to magnitude a risky business.

Having jumped ahead to discuss the conclusions drawn from Table II, it is appropriate to discuss the data. Photodissociation experiments are deceptively simple. A molecular beam is irradiated and the change in beam intensity is measured with either a mass spectrometer or bolometer. Major problems arise, however, in characterizing both the laser power and beam cluster distribution. For pulsed laser experiments the laser power is a strong function of time so that it is difficult to model the power dependence of cluster dissociation. For c.w. CO_2 lasers the mode quality is often a problem, especially if the laser goes through several apertures before reaching the dissociation region. We have found it necessary to monitor the laser power inside the beam chamber.

The cluster distribution is a more severe problem. One wants to make far more dimers than higher clusters since neither mass spectrometers nor bolometers can clearly separate the signal from just one cluster size. Also, to be able to measure linewidths narrower than $1\,cm^{-1}$ the cluster rotational distribution will have to be characterized by a temperature of less than 1 K. Figure 11 shows how difficult it is to meet these constraints for observing the $C_2H_4 \cdot HF$ spectrum. The figure consists of beam depletion spectra for a variety of mixtures of HF, C_2H_4, Ar and He as detected at the $C_2H_4 \cdot HF$ parent ion peak. In the top trace the HF and C_2H_4 make up only 0.1 and 0.2 per cent. of the total gas mixture. The observed spectrum is a broad, mostly homogeneous lineshape only slightly shifted from the $(C_2H_4)_2$ spectrum. Since the molecules being studied must contain at least one HF constituent the spectrum is due to $(C_2H_4)_n \cdot (HF)_m$. As the gas mixture is made more dilute the broad peak disappears and a much narrower one appears shifted by $25\,cm^{-1}$ from the C_2H_4 dimer frequency. This is attributed to $C_2H_4 \cdot HF$.

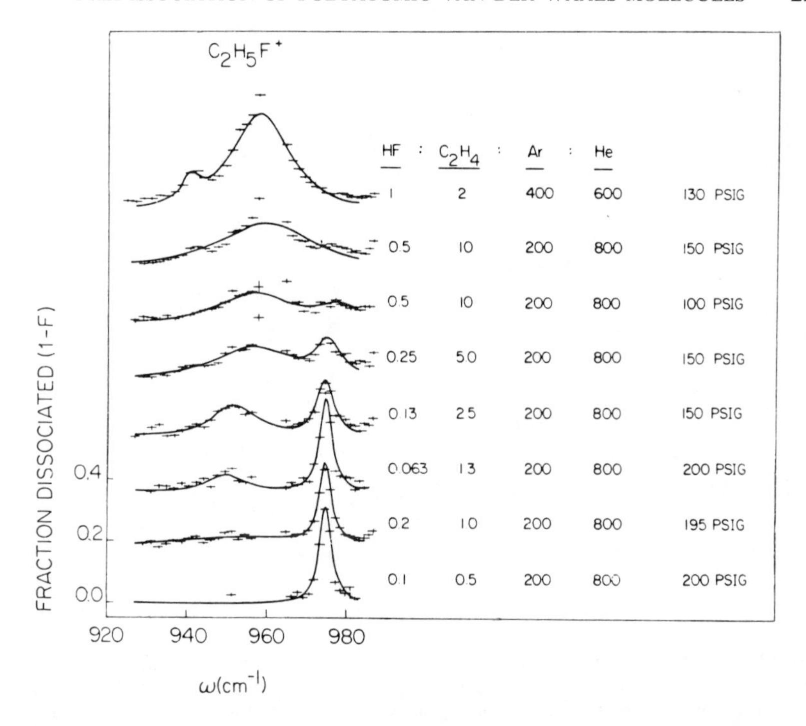

Fig. 11. CO_2 laser dissociation spectra detected at $C_2H_4 \cdot HF^+$ for a variety of gas expansion mixtures. The HF:C_2H_4Ar:He ratios and the total presures are given in the figure. It is apparent that only for exceedingly dilute expansions are spectra recorded which reflect absorption by the simple dimer HF·C_2H_4 as opposed to a higher cluster.

The $25\,\text{cm}^{-1}$ shift is consistent with opposition of the HF and CH bond dipoles. Note that the limiting spectrum is not achieved until the HF and C_2H_4 contribute only 0.0002 and 0.001 partial pressure of the mixture! Most data presented in this review come from much more concentrated mixtures.

Often spectroscopists are protected from assigning transitions to the wrong molecule because they must fit a rather specific Hamiltonian. Predissociation lineshapes not always provide this protection. Most of the ethylene complexes appear qualitatively the same. $C_2H_4 \cdot HCl$ and $C_2H_4 \cdot HF$ at least have a substantial frequency shift. Recently, the $C_2H_4 \cdot Ne$ spectrum has proved to have resolvable structure, as shown in Fig. 12a. To obtain a more complete spectrum both N_2O and CO_2 laser lines were used. This spectrum is clearly inhomogeneous. For Table II the maximum estimated linewidth was approximated as $0.5\,\text{cm}^{-1}$, based on the discontinuous peaks which are apparent. Analysis of the $C_2H_4 \cdot Ne$ spectrum is not complete but a one-dimensional hindered internal rotor Hamiltonian reproduces several of the major features.

Figure 12b gives the hindered rotor scheme and a correlation diagram which shows how the ethylene rotor states (approximated as a symmetric top)

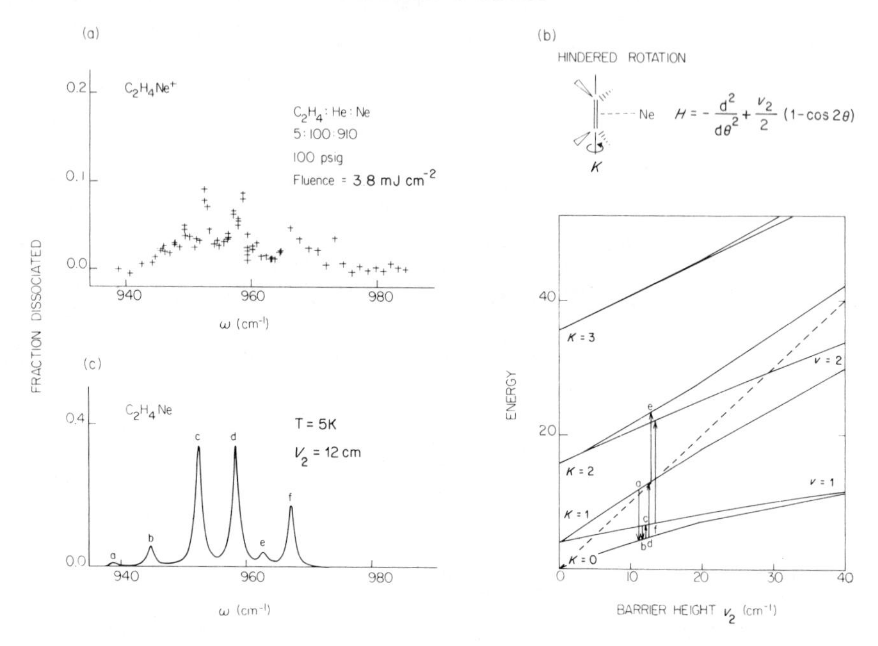

Fig. 12. (a) CO_2 and N_2O laser dissociation spectrum of NeC_2H_4. (b) Hindered rotor model Hamiltonian. (c) Perturbation correlation diagram for the rotor energy levels, and spectral simulation for $v_2 = 12\ cm^{-1}$. Transitions producing peaks a to f are represented in part (c) to illustrate the changes in quantum state.

are perturbed by v_2, the hindering potential. The solid line in Fig. 12a is a simulated spectrum for $v_2 = 12\ cm^{-1}$ and a state population characteristic of a rotational temperature of 5 K, except that nuclear hyperfine is not cooled. The arrows in Fig. 12c show the quantum number changes for the various peaks in the spectrum. At the very least, this NeC_2H_4 spectrum shows that it is possible to do state-resolved infrared photodissociation experiments. Recent experiments on $C_2H_4 \cdot Ar$ have revealed a similar structure. The data given in Table II for the rare gas complexes should be regarded as preliminary.

Also reported in Table II are transition moments for the various bimers. For most of the molecules $\langle \mu \rangle^2$ is somewhat greater than the value for free ethylene, $35.3 \times 10^{-3} D^2$. Cluster formation has two effects on transition moment. Firstly, changes in the force fields will cause the monomer basis states to mix. Secondly, the bonding partner provides a polarizable medium to enhance bond dipoles. For ν_7 the transition is strongly allowed. Any mixing with ν_8, a nearly resonant transition which is IR forbidden, would decrease the ν_7 intensity and shift the fundamental to the blue. Although slight blue shifts are observed for the bimers the transition intensity actually increases. This suggests that any mixing of vibrational levels is masked by the polarizability

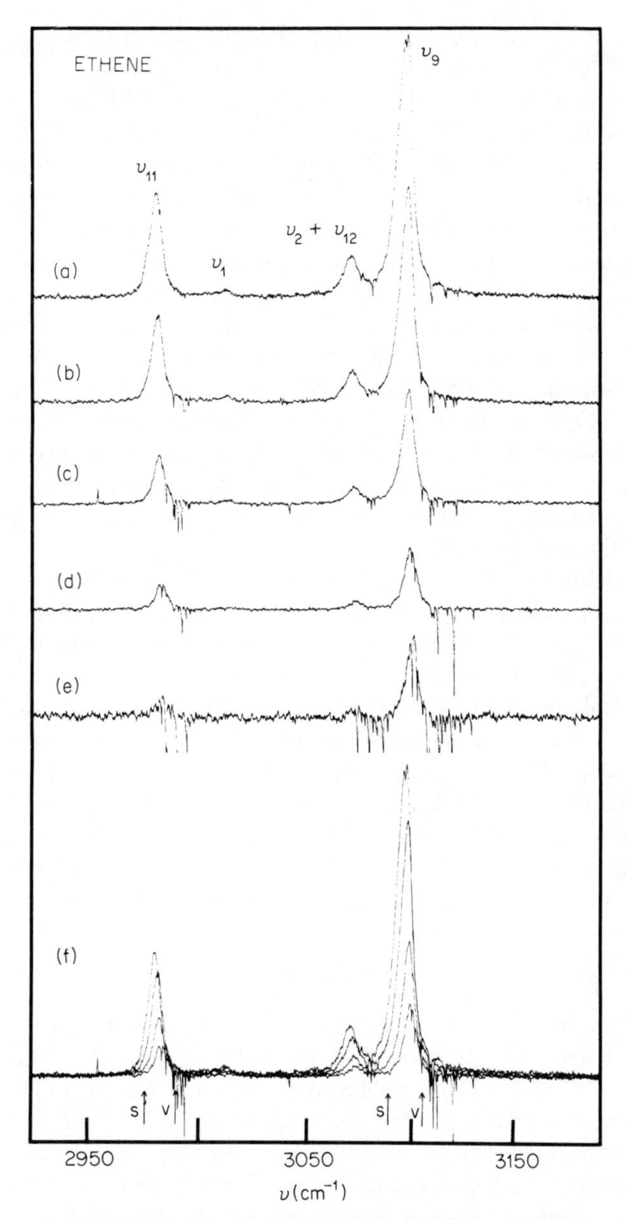

Fig. 13. The ethylene cluster infrared predissociation spectra in the 3000-cm^{-1} region for source pressures (kPa): (a) 2000, (b) 1235, (c) 960, (d) 770 and (e) 505. The five spectra are drawn together in (f) to emphasize frequency shifts. S and V mark the solid- and vapour-phase monomer frequencies. The sharp negative spikes mark monomer rotational structure. (Figure provided by Roger Miller.)

enhancement for the ν_7 mode. Confirmation of state mixing will require the observation of some borrowed intensity in the ν_8 mode.

Also note that both Hoffbauer and coworkers[20] and Casassa, Bomse and Janda[17] measure the $(C_2H_4)_2$ transition moment to be somewhat more than twice that of free C_2H_4. This again suggests that the ν_7 mode is not strongly mixed with IR forbidden transitions, even in the case of dimer formation. For $(C_2H_4)_2$ one might expect frequency shifts due to resonant dipole coupling of the two constituents as observed for $(SF_6)_2$ by Geraedts, Tolte and Reuss.[30] The lack of such a shift suggests that the ethylene planes are parallel but offset so that the angle between the transition moments and the centre of mass vector is near 55°. At this angle resonant dipole interactions go to zero by symmetry.

The measurements of Hoffbauer and coworkers[20] on the ν_{12} mode excitation of $(C_2D_4)_2$ and $C_2H_4 \cdot C_2D_4$ show that simple additivity of transition moments is not always the case. In fact, the transition moment is larger for the mixed bimer than for the dimer. Since ν_{12} is a much weaker transition than ν_7 it is reasonable that it would be more strongly affected by state mixing with strongly allowed transitions.

Fischer, Miller and Watts[23] have recorded ethylene dimer spectra in the CH stretch region. Their spectrum is shown in Fig. 13. Firstly, note that the symmetric ν_1 mode becomes slightly allowed in the dimer. These data may be useful when calculating state mixing upon dimer formation. Secondly, note that the lineshapes of Fig. 13 are well simulated by a 5-cm^{-1} Lorentzian. This suggests that the CH stretching modes also relax on the picosecond timescale. This result would not be predicted by either a direct $V \rightarrow T$ or $V \rightarrow T, R$ predissociation model due to the large momentum gap. It appears that the CH stretch bands may be broadened by redistribution of energy within the vibrational manifold without involving predissociation. Such broadening has been postulated for polyatomic molecules without any weak bonds.[71]

B. SF_6 Clusters

Several groups have examined the spectra of SF_6 bound to various partners. This section will concentrate on the dimer and $Ar \cdot SF_6$ spectra of Geraedts and coworkers.[29,30,72] The larger cluster work of Gough, Knight and Scoles[73] and Melinon and coworkers[74] will be discussed in Section IV.E. SF_6 dimer spectra for excitation of the triply degenerate ν_3 (in the monomer) mode show a 20-cm^{-1} splitting.[29,30] Geraedts, Stolte and Reuss[30] showed that a resonant dipole–dipole interaction satisfactory accounts for the splitting in the dimer and also for higher clusters. Also they were able to monitor the spectrum of isotopically substituted dimers. These data should remove any ambiguity about the assignment of the transitions to the dimer.

To understand the resonant dipole–dipole splitting, consider a simplified picture of the dimer with the two sulphur atoms on the z axis and each ν_3

vibration polarized with components on the x, y and z axes, as illustrated in Fig. 14. The six final wavefunctions are:

$$\phi_1 = \phi_{z1} + \phi_{z2}$$
$$\phi_2 = \phi_{z1} - \phi_{z2}$$
$$\phi_3 = \phi_{x1} + \phi_{x2}$$
$$\phi_4 = \phi_{x1} - \phi_{x2}$$
$$\phi_5 = \phi_{y1} + \phi_{y2}$$
$$\phi_6 = \phi_{y1} - \phi_{y2}$$

By symmetry $\phi_3 = \phi_5$ and $\phi_4 = \phi_6$; ϕ_1, ϕ_3 and ϕ_5 are IR allowed while ϕ_2, ϕ_4 and ϕ_6 are IR forbidden. In ϕ_1 the transition dipoles lie end to end leading to a lower transition energy, while in ϕ_4 and ϕ_5 the dipoles are side by side repelling each other and increasing the transition energy. A more complete analysis shows that the splitting goes as $P_2(\theta)$ and $\langle R^{-3} \rangle$, so that one of these parameters may be estimated if the other is known. Also note that for a dimer transition made up of non-degenerate monomer modes only a bandshift would be expected since the antisymmetric component will always be symmetry forbidden.

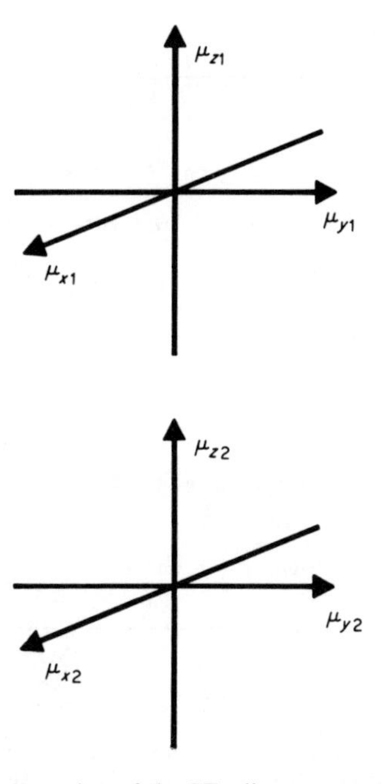

Fig. 14. Representation of the SF_6 dimer ν_3 transition dipoles. Six wavefunctions for SF_6 dimer vibrations will result from appropriate symmetry combinations as described in the text.

In their studies of $(SF_6)_2$ Geraedts and coworkers[67,72] have also explicated the role of orientational hole burning in artifically broadened, apparently homogeneous transitions. The SF_6 dimer provides an excellent case for studying orientational hole burning because one of the transitions is parallel while the other is perpendicular. Differently polarized transitions saturate at different rates. Basically the effect is due to the fact that different M levels of a given state saturate at different laser powers because their transition moments have different orientations with respect to the laser field. One sublevel can be saturated and broadened while another is still not saturated. The experimentalist may be fooled into thinking the line is intrinsically broad since broadening is observed before the transition is completely saturated. In practical terms two protective measures should be taken. Firstly, cross-sections or log $(\Delta I/I)$ should be plotted against frequency to avoid fluence broadening of the spectra.[66] The cross-sections should be measured at two different laser fluences (preferably an order of magnitude different) to avoid orientational broadening.

The orientational theory outlined by Geraedts and coworkers[67] quantitatively applies when $\langle J \rangle \gg \langle K \rangle$. This condition is well met for mild expansions of heavy molecules like $(SF_6)_2$. We have found that a more complete theory is necessary for a molecule like $(C_2H_4)_2$ at 4K. For cold

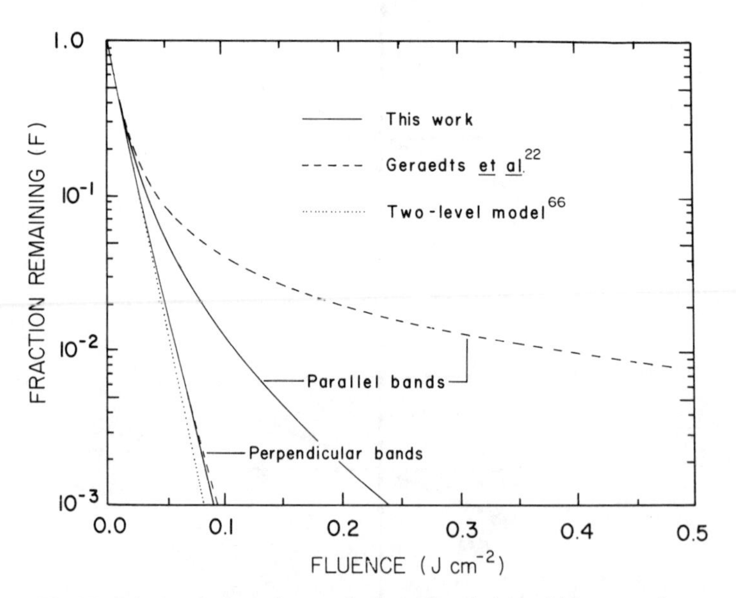

Fig. 15. Calculated dependence of photodissociation yield versus laser fluence for ν_7 excitation of the C_2H_4 dimer. The dotted line results from a simple two-level model. The dashed line for a $J \gg K$ approximation of transition moment orientations and 4 K rotational distribution. The solid lines are the full quantum results for 4 K.

beams it is easiest to apply the fluence-broadening formula to each pair of states and add the intensities, by taking into account the appropriate Hönl–London factors. The difference between the $J \gg K$ and full quantum models is illustrated in Fig. 15.

Geraedts and coworkers report an $(SF_6)_2$ intrinsic linewidths of $1.5\,cm^{-1}$. This implies a lower limit lifetime of about 3.5 ps for the vibrationally excited molecule. All of the caveats applied to the interpretation of ethylene dimer lineshapes as measures of photodissociation dynamics apply even more strongly to the SF_6 dimer because of the much larger bath of vibrational levels in near-resonance with the excited mode. Product state detection would be necessary before a specific mechanism could be proposed.

Geraedts and coworkers also report an $ArSF_6$ spectrum, as do Gough, Knight and Scoles.[73] In the spectra published so far $ArSF_6$ is probably not separated from clusters with several Ar atoms. It is clear, however, that the $ArSF_6$ transition is not split. The spectrum occurs between the two SF_6 dimer lines.

C. OCS Clusters

Hoffbauer and coworkers[27,28] have studied CO_2 laser photodissociation of ArOCS, $(OCS)_2$, $(OCS)_3$ and OCS bound to CD_4, C_2H_6, C_4H_{10} and C_6H_{12}. The OCS ν_2 bend overtone at $1045\,cm^{-1}$ was excited. For ArOCS a narrow ($\sim 1.0\,cm^{-1}$ FWHM) transition is observed at a shift of only $1\,cm^{-1}$ to the red of the free molecule spectrum. In spite of the fact that the width is on the order of laser spacing between lines, two-laser spectra indicate that the transition is nearly homogeneous. The fitted $\langle \mu \rangle^2$ for ArOCS is $2.0 \times 10^{-4} D^2$, only 15 per cent. of the free monomer value. The monomer obtains its intensity from a Fermi resonance. It is possible that the Fermi resonance is broken in ArOCS even though the $2\nu_2$ frequency shift is only $1.0\,cm^{-1}$. Alternatively, most of the ArOCS transition intensity occurs between the lines of the CO_2 laser. For the OCS dimer a $3.7\,cm^{-1}$ wide line is observed at a $3\,cm^{-1}$ red shift from the monomer. The $\langle \mu \rangle^2$ for the dimer is again lower than for the free monomer. For the trimer a two-peak spectrum is observed with each peak about $6\,cm^{-1}$ wide. They are split by $8\,cm^{-1}$. The blue component is near the dimer peak. On average only $8\,cm^{-1}$ of the excitation energy goes to relative translation of dimer dissociation products!

For OCS bound to the series of alkanes the transition frequency is always $1045 \pm 1\,cm^{-1}$ and the widths fall within the range $4.9 \pm 0.9\,cm^{-1}$. The transition moment varies over a factor of two depending on the bonding partner. There is no obvious trend in the variation. From the small range of lifetimes one concludes that the bonding partner modes are only important in determining linewidths up to a certain point after which the extra modes are not involved. Vernon and coworkers[75] came to a similar conclusion in their study of $(C_6H_6)_3$.

D. Hydrocarbon and Halocarbon Clusters

Pendley and Ewing[76] have studied the infrared spectrum of the acetylene dimer by long-path FT–IR absorption of acetylene gas at 215 K and 0.03 to 07 atmospheres. Five bands could be assigned to dimer absorption (three for $(HCCH)_2$; two for $(DCCD)_2$) on the basis of quadradic density dependence. The bands for the D—C stretch of $(DCCH)_2$ and the C≡C stretch of both species showed no obvious affects of line-broadening dynamics under these conditions. The profiles could be theoretically simulated with a staggered parallel dimer structure recommended by Sakai, Koida and Kihara.[77] By contrast, the experimental band at 620 cm^{-1} for $(HCCH)_2$ is significantly narrower than predicted! Pendley and Ewing conclude that the $K=0$ transitions lead to a long-lived vibrationally excited state and that this set of transitions gives the observed narrow band. The $\Delta K = \pm 1$ transitions which broaden the predicted profile are presumed to be homogeneously broadened, so that they appear as a broad, but weak, background in the observed spectrum.

The 620 cm^{-1} mode is described as an in-plane, out-of-phase combination of the acetylene symmetric bend. Unlike the other observed bands, this transition is blue shifted by 8 cm^{-1} from the free molecule. The band shift is in agreement with predictions based on the Kihara potential. Pendley and Ewing also discuss the symmetry of the vibrational decay processes. The initially excited vibration correlates directly with products rotating in the molecular plane, but with cancelling total angular momentum. No explanation was given for the strong dependence of lifetime on the K quantum number. Molecular beam studies to remove ambiguities in the above interpretation will prove quite difficult due to the lack of tunable lasers at the appropriate frequencies.

Vernon and coworkers[75] have recorded spectra and product translational distributions for excitation of $(C_6H_6)_2$ and $(C_6H_6)_3$ at 3000 cm^{-1} with a laser bandwidth of about 3 cm^{-1}. The spectra looked much like liquid benzene spectra. The dimer beams were rather warm (translational temperature = 50 K), so it would be dangerous to rule out heterogeneous structure as an explanation for bandshapes broader than the laser width. Very little translational energy was imparted to the products. As with ethylene dimer dissociation at 950 cm^{-1} no correlation of product momenta with laser polarization is observed. Again, however, the dimers were rather warm and the rotational period would be on the picosecond timescale: lack of polarization may not rule out prompt dissociation. For $(C_6H_6)_3$ it was shown that the photon energy was not shared equally by the products. The dissociation channel was dominated by

$$(C_6H_6)_3 \xrightarrow{h\nu} (C_6H_6)_2 + C_6H_6$$

Since the bond energy is expected to be ~ 750 cm^{-1}, 2300 cm^{-1} of excess energy is deposited in the internal degrees of freedom of the products. If this were shared equally the dimer product would itself rapidly dissociate. This result supports the conclusion of Hoffbauer and coworkers[27,28] that beyond a certain size of molecule the extra modes in a bonding partner do not become involved in the relaxation of the excited molecule (see Section IV.C).

Geraedts[72] reports a dissociation spectrum for $(CF_3Br)_2$ at 1085 cm^{-1} which is clearly inhomogeneous. He interprets the inhomogeneity as resulting from unequal sites of the two molecules in the dimer. Figure 16 shows a $CF_3Br \cdot Ar$ spectrum recorded at Caltech by Francis Celii. The 8-cm^{-1} wide spectrum is clearly not completely homogeneous, but the high peak intensity at low laser power indicates substantial homogeneous broadening. Interpretation of lineshapes for such heavy molecules is probably not useful without product state information.

Geraedts[72] has also reported a spectrum for the SF_4 dimer. Like the SF_6 dimer the SF_4 dimer spectrum shows resonant dipole–dipole splitting, although the red peak is partly out of range of the CO_2 laser. The linewidth of the blue peak is 4.7 cm^{-1}.

Audibert and Palange[34] have recently performed pump-probe experiments on the $(CH_3)_2O \cdot HF$ complex in which the complex is excited in the HF stretch and then the HF population is probed. The experiment is performed on a gas cell filled with 20 torr of HF and 74.4 torr of $(CH_3)_2O$ at room temperature.

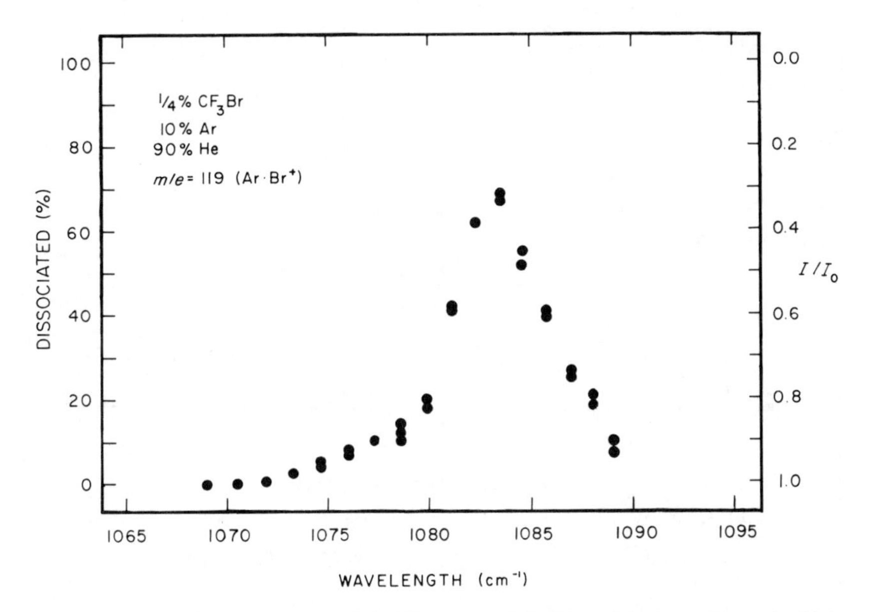

Fig. 16. Predissociation spectrum of $ArCF_3Br$ recorded with a CO_2 laser. From the high fraction of dissociation at any frequency one concludes that the lineshape is largely homogeneous.

From pressure-drop measurements it is estimated that nearly half of the HF is complexed. It is deduced that 70 per cent of the excited complexes decay within 10 ns. It would be very interesting to increase the time resolution of this experiment. It would also be exciting to be able to measure the dissociation product HF rotational distribution.

E. Large Clusters

Several groups have recently attempted to elucidate the analogy between an infrared active molecule in a large cluster and the same molecule isolated in a matrix. In these studies linewidths probably do not relate to dissociation dynamics even if they do reflect decay of the initially excited state. Also, the background state density may be high enough to ensure that pure dephasing processes are considered. So far, no one has worked out how to separate a specific sized cluster for study. Often lineshapes are simply due to a variety of cluster sizes producing a superposition of slightly offset transition frequencies. This effect is well illustrated by the Ar_nSF_6 system studied by Geraedts,[72] Gough, Knight and Scoles[73] and Melinon and coworkers.[74]

Figure 17 shows an example from van den Bergh's laboratory. The beam-

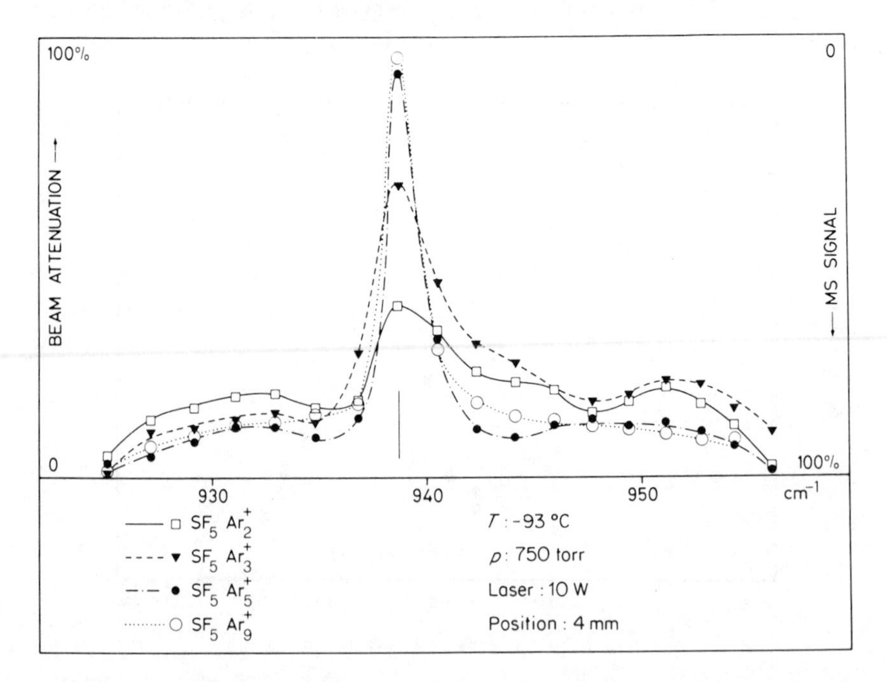

Fig. 17. CO_2 laser photodissociation spectra for an Ar_nSF_6 expansion recorded on a variety of mass peaks. The spectrum tends to sharpen as cluster size increases. The solid vertical line is the SF_6 frequency in an Ar matrix. (Figure provided by Hugh van den Bergh.)

depletion spectrum is measured on mass peaks of $SF_6Ar_n^+$ for $n = 2, 3, 5$ and 9 for a 180 K, 1 bar, 100 μm beam source. The spectrum recorded on low n mass peaks are very broad. Presumably many different neutral clusters fragment to this ion upon electron impact. The spectrum recorded on $n = 9$ contains a prominent peak at a frequency corresponding to SF_6 in an Ar matrix. Is it possible that the broad wings on this spectrum correspond to non-equilibrium clusters with the SF_6 not completely surrounded by Ar atoms?

Similar correspondence between cluster spectra and matrix spectra have been demonstrated for CH_3F in Ar by Gough, Knight and Scoles.[73] Figure 18 shows an attempt by Francis Celii at Caltech to measure the onset of microcrystalline structure. In these spectra a narrow, intense line is observed at 1037 cm^{-1} for detection of $CH_3F \cdot Ar_n^+$ with $5 < n < 13$. For $n = 1$ the transition appears to be homogeneously broadened to 6 cm^{-1} FWHM. For the range $5 < n < 13$ there is apparently an excited state which decays more slowly than the $n = 1$ cluster, even though a much larger background of states exist. For now we do not know if the narrow transition disappears for $n > 13$ or if it simply shifts between the CO_2 laser lines used to obtain dissociation.

Large clusters of HF,[36] H_2O[35] and N_2O[33] have been examined to explore the analogy between clusters and liquids. For $(HF)_3$ one remarkable feature is the disappearance of non-hydrogen-bonded proton stretches. It is clear that a force field based on dimer potentials will not be appropriate for even small clusters, much less the liquid. Vernon and coworkers[35] found that a variety of potentials proposed for H_2O interactions are insufficient to reproduce the small cluster spectra. It is still not clear why hydrogen-bonded spectra are so broad. It is clear that the assumption by Stepanov[78] that predissociation broadening produces the widths is wrong. More likely hydrogen-bond potentials have strong coupling between weak and strong bond vibrations, producing spectra with a dense set of combination bands.

Ding Miller and Watts[63] have performed an analysis of N_2O cluster spectra using a Monte Carlo simulation plus local mode analysis which had previously been applied to liquid water and amorphous ice.[79] Watts found that for the N_2O dimer at 25 K the molecule samples T shaped, parallel and crossed configurations to generate the observed spectrum. This clearly indicates the need to obtain ~ 1 K rotational temperatures if spectroscopy is to be successful. In the fifty-five molecule simulation there is a clear spectral shift of molecules within the cluster from those on the surface. Watts concluded that a given intermediate-size cluster will absorb over a broad bandwidth. It is therefore unnecessary to have a distribution of clusters to have broad bands. Watts also concluded that it might be possible to extract subtle details about the potential from such simulations. In the case of N_2O he concluded that the dipole moment decreases as the bond stretches.

Studies of large clusters probably will not proceed by individual quantitative improvements in our understanding. Instead, careful, but qualitative,

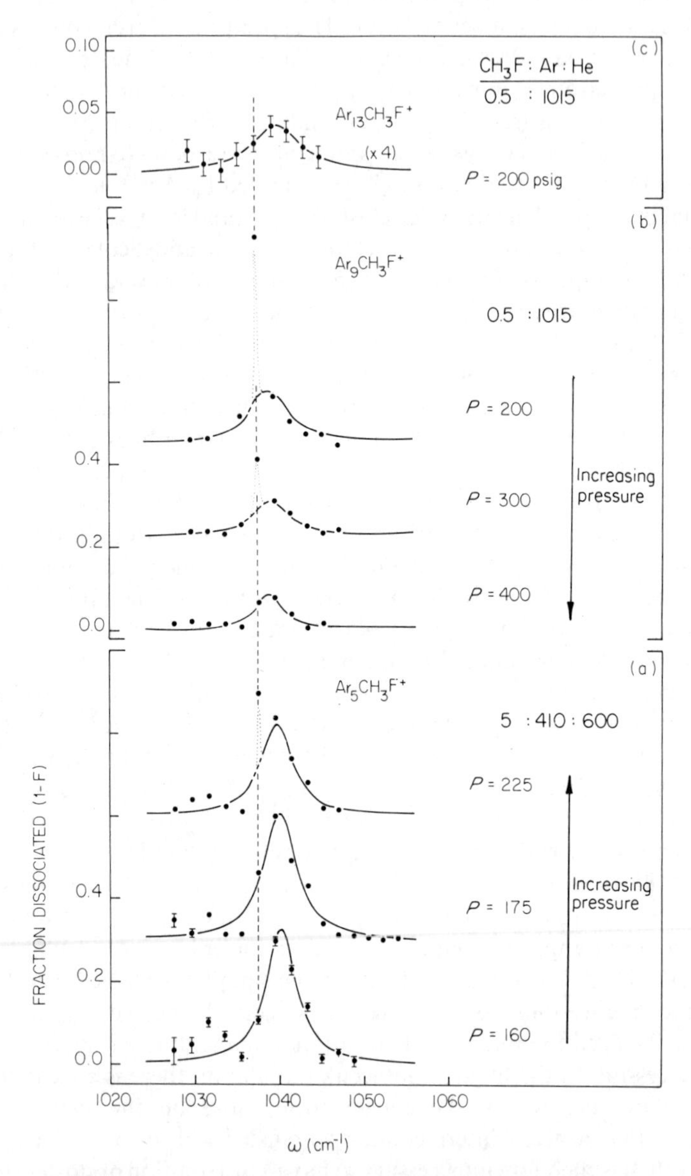

Fig. 18. CO_2 laser photodissociation spectra of clusters detected at mass peaks corresponding to Ar_nCH_3F. For several expansion conditions which produce intermediate-sized clusters an intense spike is observed at 1038 cm^{-1}. It is not observed when the mass peak of $Ar_{13}CH_3F$ or higher is detected at any source pressure.

experiments will have to be simulated by complex computer models such as the one developed by Watts. In this respect there certainly is an analogy between clusters and liquids.

V. POLYATOMIC MOLECULES: EXCITED ELECTRONIC SURFACES

Studying intramolecular relaxation of molecules in the electronic excited state has the huge advantage of optical fluorescence detection. Thus for the same reason that HeI_2 is the best experimentally characterized example of triatomic van der Waals molecule predissociation, glyoxal and tetrazine species are the best characterized examples of polyatomic molecule predissociation. Fluorescence detection not only gives information on final product states of dissociation reactions but provides a real-time clock (as opposed to linewidths) for measuring decay times. For H_2–glyoxal excited to vibronic levels above threshold it is clear that the constituents have actually separated within 1 ns.[80]

The only disadvantage to studying vibrational predissociation on electronic excited states is that one is not sure how the huge background of bath states will affect the dynamics. There are several examples that this effect is non-negligible. Two of them are no observed[8] fluorescence when ArI_2 is excited to $v(I_2) < 10$ and observed phosphorescence when Ar–gloyoxal is excited to the ^1A_u vibrationless state.[81] Recently Brinza[13] has observed that the A and B state curve crossing in Cl_2 perturbs the $NeCl_2$ (B state) bond energy.

A. Glyoxal Complexes

Soep and coworkers[40,41,81] have examined the laser excited fluorescence spectra of Ar, He, H_2 and D_2 bound to glyoxal. The excited transition corresponds to $^1A_u \leftarrow {}^1A_g$ in the free molecule. Six vibrational levels of the excited states are observed: 0, 7^2, 5^1, 8^1, 8^17^2, 8^15^1. ν_7 is a C—H torsional mode at 233 cm^{-1}, ν_5 is an in-plane CCO bend at 509 cm^{-1} and ν_8 is a C—H wag at 735 cm^{-1}.[82] No rotationally resolved spectra of a complex have been recorded so their structures are not known. The predissociation evidence suggests that the bonding partner is above the glyoxal plane.

The evidence that glyoxal complexes predissociate on the nanosecond timescale when excited above the vibrational threshold is simply that all observed fluorescence is from free glyoxal. When the complex 8^1_0 transition is excited the fluorescence lifetime corresponds to the free glyoxal vibrational ground state. When the ^1A_u vibrational ground state of the complex is excited no fluorescence is observed. The bonding partner stimulates internal conversion and intersystem crossing on a timescale fast with respect to the fluorescence lifetime (2.4 μs). Since fluorescence is observed when the complex is excited over the vibrational threshold, predissociation is not only faster than fluorescence ($\tau = 0.8\ \mu s$) but also faster than non-radiative relaxation.

For the 8_0^1 vibrational excitation of H_2–glyoxal a van der Waals stretching vibrational progression is observed with $\omega = 17\ cm^{-1}$, $\omega x = 3\ cm^{-1}$. Substitution with D_2 supports this assignment (no arguments were presented to rule out other assignments). Using a Morse potential formula and the 10-cm^{-1} bandshift the H_2-glyoxal bond strengths are estimated to be $30\ cm^{-1}$ in the ^1A_g state and $40\ cm^{-1}$ in the ^1A_u state. Also observed was a van der Waals hot band from which it was deduced that $T_{vib} \approx T_{rot} \sim$ 'a few K' for the expansion. Unlike Ewing's[87] C_2H_2 dimer IR spectrum where the bandshift (between the free molecule transition and that of the complex) was found to depend on the vibrational symmetry, a constant bandshift is recorded for all observed levels of H_2– glyoxal.

By dispersion of the fluorescence the selection rules for vibrational predissociation can be tabulated as in Table III. Although there are exceptions, most of the data are roughly in accord with a $\Delta v = -1$ selection rule and the momentum gap law. When the $8^1 7^2$ combination is excited the lowest energy single-quantum process dominates the relaxation: $8^1 7^2 \rightarrow 8^1 7^1 + 233\ cm^{-1}$. Similarly, for 7^2 excitation, $\Delta v = -1$ dominates, no $\Delta v = -2$ being observed. For 5^1 excitation, however, the $5^1 \rightarrow 7^1$ channel is competitive with the $5^1 \rightarrow 0$ channel. Also, for $8^1 5^1$ excitation two seperate $\Delta v = 2$ processes are observed: $8^1 5^1 \rightarrow 8^1 7^1$ and $8^1 5^1 \rightarrow 0$.

Examination of the $8^1 5^1$ relaxation leads to several interesting conclusions. For single-quantum relaxation the rate for ν_8 relaxation is a factor of five faster than for ν_5 relaxation, in spite of the unfavourable momentum gap. This can be explained if H_2-glyoxal is a non-planar molecule by the fact that ν_8 is an out-of-plane mode while ν_5 is an in-plane one. This situation is reminiscent of the C_2H_4 dimer ν_7 and ν_{12} comparison. The fact that the $8^1 5^1 \rightarrow 0$ channel is also observed suggests a model wherein ν_8 relaxes to an excited H_2 internal rotor mode. This metastable level must then live long enough for ν_5 to relax. If this

TABLE III
Branching ratios for H_2–glyoxal ^1A_u predissociation.

Excited vibration state	Fractional branching ratios to final vibrational states				
	0	7^1	8^1	5^1	$8^1 7^1$
7^2	—	1.0	—	—	—
8^1	1.0	—	—	—	—
$8^1 7^2$	—	—	—	—	1.0
5^1	0.4	0.6	—	—	—
$8^1 5^1$	0.2	—	0.1	0.5	0.2

All data are from Ref. 81.

model is correct, it is concluded that the actual dissociation time is only a factor of two or three longer than the time for the initially excited vibration to relax to the van der Waals modes. Otherwise the $8^1 5^1 \rightarrow 0$ channel would dominate.

Soep[41] has also measured an interesting isotope effect for relaxation of the 8^1 level. For H_{2n} glyxol the ν_8 vibration relaxes predominantly by the single-quantum $\Delta\nu_8 = -1$ path. For D_2-glyoxal, however, the two-quantum $\Delta\nu_8 = -1$, $\Delta\nu_7 = +1$ path is roughly competitive with the single-quantum relaxation. Beswick[41] interprets this isotope effect as an illustration of the momentum gap law. For a given ΔE the D_2 momentum is higher than the H_2 momentum. D_2 will thus be more selective towards low ΔE pathways than H_2. I would venture that the isotope effect will ultimately prove to illustrate an angular momentum gap law, with the argument being analogous to the linear momentum gap law.

B. S-Tetrazine–Ar

Studies of tetrazine clusters differ in two important ways from those of glyoxal clusters. Firstly, the much richer set of vibrational levels make bookkeeping a real problem. Secondly, the fluorescence lifetime of tetrazine is three orders of magnitude faster than that of glyoxal. Levy and coworkers[37,38] have conquered the complexity of the spectrum with careful spectroscopy and have used the fast clock to quantitatively compare rates for different branching processes in the decay of Ar–tetrazine. The multitude of branching ratios catalogued in this study will not be reproduced here. Instead, the results will be qualitatively contrasted with the previously discussed study of glyoxal complexes.

From the allowed decay channels Haynam and Levy[37] were able to bracket the Ar–tetrazine bond strength $254\ \text{cm}^{-1} < D_0 < 381\ \text{cm}^{-1}$ ($D_0'' - D_0' = -23\ \text{cm}^{-1}$). The same complexation bandshift, $-23\ \text{cm}^{-1}$, was observed for all vibrations except the low-frequency out-of-plane ν_{16} modes: $16a_0^2 = 7.2\ \text{cm}^{-1}$, $16b_0^2 = 15.9\ \text{cm}^{-1}$. As compared to glyoxal the relaxation pathways for most excited vibrations are quite complex, with multiquantum transitions being common. As an example, for the $6a_0^1$ excitation, fluorescence is observed from 6a and $16a^2$ of the complex and 16a of the free molecule. No simple $\Delta v(6a) = -1$ predissociation to the vibrationless free tetrazine is observed. Levy and others have analysed the data as if each final state were reached by an independent mechanism. Ramaekers and coworkers[39] have performed time-resolved studies that show that the data can be correctly explained by a much simpler sequential mechanism:

$$\overline{6a'} \rightarrow \overline{16a^2}$$

$$\overline{16a^2} \rightarrow 16a' + Ar$$

A bar over the vibrational mode indicates the Ar–tetrazine complex. The ring twist mode, ν_{16}, figures prominantly in each of the bands produced upon vibrational predissociation of Ar–tetrazine. Again it is apparent that the out-of-plane modes couple most strongly to van der Waals modes when the structure has the bonding partner out of plane.

Levy and Ramaekers disagree on the rate constants for vibrational predissociation. Levy finds them to be of the order of $10^8\,s^{-1}$ for all modes studied above threshold. Ramaekers finds a value of $2 \times 10^9\,s^{-1}$ for the $16a^2$ mode. While Levy noticed no change in linewidths going from the free molecule to the complex, Ramaekers observed a doubling of the real-time decay rate of the $16a^2$ mode upon complexation. Certainly this discrepancy needs to be reexamined. Ramaeker's arguments have two strong advantages: firstly, they are based on real-time measurements; secondly, the sequential mechanism is much easier to justify than the parallel mechanism.

C. Other Examples

Zevier, Cavasquillo and Levy[83] have studied the complex of *trans*-stilbene bound to one and two helium atoms. They find a He–stilbene bond strength of less than $82\,cm^{-1}$ in the excited state of *trans*-stilbene and $76\,cm^{-1}$ in the ground state. For excitation of an in-plane bend of the complex at $197\,cm^{-1}$ both the He atoms are ejected in a time much shorter than the 2.7-s fluorescent lifetime. For excitation of the two quanta of the in-plane bend only the $\Delta v_b = -1$ and $\Delta v_b = -2$ channels are observed in predissociation; there is very little mode scrambling which is surprising for such a large molecule. Preliminary evidence for Ar cluster predissociation indicates somewhat more mode scrambling.

Hopkins, Powers and Smalley[84] studied Ar and N_2 bound to alkylbenzenes as a time probe to compare vibrational predissociation rates to intramolecular energy flow from the ring to the alkyl chain. They found that Ar did predissociate much faster than the fluorescent lifetime while the N_2 complexes lived longer than the intramolecular energy redistribution time.

Zewail and coworkers[85,86] have been studying the isoquinoline molecule bound to H_2O and methanol. They find bond energies of the order of $1000\,cm^{-1}$. For these very large, more strongly bound molecules it appears that RRKM theory produces good estimates of predissociation times.[86]

VI. EPILOGUE

This review demonstrates that the experimental characterization of van der Waals molecule predissociation is an active, rapidly expanding activity in the chemical physics community. Although the growth in theoretical work has also been rapid, only a taste of that is presented here.

It is fun to speculate what new information will be available for another review three years hence. The work of Pine and Lafferty on the HF dimer would seem to make that molecule a prime candidate for intense study. A detailed potential should be obtained for extensive theoretical investigation. Further experiments should provide the lifetime as a function of the rotational and a variety of vibrational mode excitations. Now that it has been proved that vibrationally excited HF dimer can live for 10^{-9} s it will be fascinating to see just how long it does live. How about ArHF*? Is its lifetime too short for it ever to be measured?

Spectroscopic measurements which give potential energy coupling terms between weak and strong bonds are in very short supply. The magnitude of these terms, of course, will dominate calculations of intramolecular relaxation. Will these numbers be obtained first from high-resolution spectroscopy or from modelling predissociation dynamics? For many molecules such as ArI_2 or the C_2H_4 dimer high-resolution spectroscopy will be vitiated by homogeneous line broadening. Still, even some large molecules such as Ar-S-tetrazine have rotationally resolved spectra, including excited van der Waals vibrational modes, which will yield detailed potential energy surfaces.

Finally, I expect time-resolved pump-probe experiments to obtain real-time dynamics, and rotational product state distributions will soon by available to completely characterize predissociation dynamics for a variety of molecules. All the evidence points towards very fast dissociation of polyatomic molecules excited in bending vibrations. Will the same be true of C—H stretches or are the ethylene dimer lineshapes measured by Miller simply due to dephasing within the C—H manifold?

We still have much to learn.

Acknowledgements

I would like to thank my coworkers at Caltech who have helped keep pace with this rapidly advancing field. Michael Casassa, Colin Western, Dave Brinza, Francis Celii and Barry Swartz have contributed to work in this article. Dwight Evard, Sally Hair, Fritz Thommen and Paul Whitmore have started to gather data for the next review. Their work is supported by the Department of Energy (infrared predissociation) and the National Science Foundation and ACS Petroleum Research Fund (visible laser excited fluorescence). I would also like to thank the Alfred P. Sloan Foundation for the support of my research. I would like to thank the many people who sent preprints and figures and discussed their work with me. In the latter category, Roger Miller, Jörg Ruess and George Ewing have been especially helpful.

References

1. LeRoy, R. J., Corey, G. C., and Hutson, J. M. *Faraday Discuss. Chem. Soc.*, **73**, 339 (1982).
2. Beswick, J. A., and Jortner, J., *Adv. Chem. Phys.*, **47**, 363 (1981).
3. Henri, V., and Teves, M. C., *Nature*, **114**, 894 (1924).
4. Rice, O. K., *Phys. Rev.*, **33**, 748 (1929).
5. Dingle, H., *Nature*, **114**, 897 (1924).
6. Rosen, N., *J. Chem. Phys.*, **1**, 319 (1933).
7. Ewing, G. E., *J. Chem. Phys.*, **71**, 3143 (1979).
8. Levy, D. H., *Adv. Chem. Phys.*, **47**, 323 (1981).
9. Child, M. S., *Faraday Discuss. Chem. Soc.*, **62**, 307 (1977).
10. McKellar, A. R. W., *Faraday Discuss. Chem. Soc.*, **73**, 89 (1982).
11. Sharfin, W., Kroger, P., and Wallace, S. C., *Chem. Phys. Lett.*, **85**, 81 (1982).
12. Beswick, J. A., Delgado-Barrio, G., Villareal, P., and Mareca, P., *Faraday Discuss. Chem. Soc.*, **73**, 406 (1982).
13. Brinza, D. E., Ph.D. Thesis, Caltech, Pasadena, Calif., 1983.
14. Swartz, B. A., Ph.D. Thesis, Caltech, Pasadena, Calif., 1983.
15. Brinza, D. E., Swartz, B. A., Western, C. M., and Janda, K. C., *J. Chem. Phys.*, **79**, 1541 (1983).
16. Pine, A. S., and Lafferty, W. J., *J. Chem. Phys.*, **78**, 2154 (1983).
17. Casassa, M. P., Bomse, D. S., and Janda, K. C., *J. Chem. Phys.*, **74**, 5044 (1981).
18. Casassa, M. P., Western, C. M., Celii. F. G., Brinza, D. E., and Janda, K. C., *J. Chem. Phys.*, **79**, 3227 (1983).
19. Casassa, M. P., Western, C. M., and Janda, K. C., *Chem. Phys.*, **81**, 0000 (1984).
20. Hoffbauer, M. A., Liu, K., Giese, C. F., and Gentry, W. R., *J. Chem. Phys.*, **78**, 5567 (1983).
21. Bomse, D. S., Cross, J. B., and Valentini, J. J., *J. Chem. Phys.*, **78**, 7175 (1983).
22. Geraedts, J., Snels, M., Stolte, S., and Reuss, J., *Chem. Phys.*, submitted for publication 1983.
23. Fischer, G., Miller, R. E., and Watts, R. O., *Chem. Phys.*, submitted for publication 1983.
24. Casassa, M. P., Bomse, D. S., and Janda, K. C., *J. Phys. Chem.*, **85**, 2623 (1981).
25. Kyro, E. K., Shoja-Chaghervaud, P., McMillan, K., Eliades, M., Danzeiser, D., and Bevan, J. W., *J. Chem. Phys.*, **79**, 78 (1983).
26. Kyro, E., Warren, R., McMillan, K., Eliades, M., Danzeiser, D., Shoja-Chaghervaud, P., Lieb, S. G., and Bevan, J. W., *J. Chem. Phys.*, **78**, 5881 (1983).
27. Hoffbauer, M. A., Liu, K., Giese, C. F., and Gentry, W. R., *J. Phys. Chem.*, **87**, 2096 (1983).
28. Hoffbauer, M. A., Giese, G. F., and Gentry, W. R., *J. Chem. Phys.* **79**, 192 (1983).
29. Geraedts, J., Setiadi, S., Stolte, S., and Reuss, J., *Chem. Phys. Lett.*, **78**, 277 (1981).
30. Geraedts, J., Stolte, S., and Reuss, J., *Z. Phys. A*, **304**, 167 (1982).
31. Gough, T. E., Knight, D. G., and Scoles, G., *Chem. Phys. Lett.*, **97**, 155 (1983).
32. Pendley, R. D., and Ewing, G., *J. Chem. Phys.*, **78**, 3531 (1983).
33. Ding, A., Miller, R. E., and Watts, R. O., *Proc. Ninth Int. Conf. Mol. Beams, Freiburg, Germany*, p. 27, 1983.
34. Audibert, M. M., and Palange, E., *Proc. 1983 Conf. on the Dynamics of Molecular Collisions, Gull Lake, Minnesota*, p. D13, *Chem. Phys. Lett.*, **101**, 407 (1983).
35. Vernon, M. F., Krajnovich, D. J., Kwok, H. S., Lisy, J. M., Shen, Y. R., and Lee, Y. T., *J. Chem. Phys.*, **77**, 47 (1982).

36. Vernon, M. F., Lisy, J. M., Kwok, H. S., Krajnovich, D. J., Tramer, A., Shen, Y. R., and Lee, Y. T., *J. Chem. Phys.*, **75**, 4733 (1981).
37. Haynam, C. A., and Levy, D. H., *J. Chem. Phys.*, **78**, 2091 (1983).
38. Braumbaugh, D. V., Kenny, J. E., and Levy, D. H., *J. Chem. Phys.* **78**, 3415 (1983).
39. Ramaeckers, J. J. F., Krijen, L. B., Lips, H. J., Langelaar, J., and Rettschick, K. P. H., *Laser Chem.*, **2**, 125 (1983).
40. Halberstadt, N., and Soep, B., *Chem. Phys. Lett.*, **87**, 109 (1982).
41. Beswick, J. A., Halberstadt, N., Jouvet, C., and Soep, B., *Laser Chem.*, **1**, 77 (1983).
42. Jouvet, C., Sulkes, M., and Rice, S. A., *J. Chem. Phys.*, **78**, 3935 (1983).
43. Hutson, J. M., Ashton, C. J., and LeRoy, R. J., *J. Phys. Chem.*, **87**, 2713 (1983).
44. Klemperer, W., *Ber. Bunsenges, Phys. Chem.*, **78**, 128 (1974).
45. Deleon, R. L., Muenter, J. S., *J. Chem. Phys.*, **80**, 6420 (1984).
46. Ewing, G. E., *Faraday Discuss. Chem. Soc.*, **73**, 325 (1982).
47. Ewing, G. E., *Faraday Discuss. Chem. Soc.*, **73**, 402 (1982).
48. McKellar, A. R. W., and Welsh, H. L., *J. Chem. Phys.*, **55**, 595.
49. McKellar, A. R. W., and Welsh, H. L., *Can. J. Phys.*, **50**, 1458 (1972).
50. McKellar, A. R. W., *J. Chem. Phys.*, **61**, 4636 (1974).
51. LeRoy, R. J., and Carley, J. S., *Adv. Chem. Phys.*, **42**, 353 (1980).
52. Ashton, C. J., Child, M. S., and Hutson, J. M., *J. Chem. Phys.*, **78**, 4025 (1983).
53. Hutson, J. M., and LeRoy, R. J., *J. Chem. Phys.*, **78**, 4040 (1983).
54. Datta, K. K., and Chu, S. I., *Chem. Phys. Lett.*, **95**, 38 (1983).
55. Hutson, J. M., and Howard, B. J., *Mol. Phys.*, **45**, 769 (1982).
56. Boom, E. W., and van der Elsken, J., *J. Chem. Phys.*, **73**, 15 (1980).
57. Blazy, J. A., DeKoven, B. M., Russell, T. D., and Levy, D. H., *J. Chem. Phys.*, **72**, 2439 (1980).
58. Segev, E., and Shapiro, M., *J. Chem. Phys.*, **78**, 4969 (1983).
59. Valentini, J. J., and Cross, J. B., *J. Chem. Phys.*, **77**, 572 (1982).
60. Dyke, T. R., Howard, B. J., and Klemperer, W., *J. Chem. Phys.*, **56**, 2442 (1972).
61. Ewing, G. E., *J. Chem. Phys.*, **72**, 2096 (1981).
62. Gough, T. E., Miller, R. E., and Scoles, G., *J. Chem. Phys.*, **69**, 1588 (1978).
63. Ding, A., Miller, R. E., and Watts, R. O., *Proc. Ninth Int. Conf. on Mol. Beams,* Freiburg, Germany, p. 27, 1983.
64. Morales, D. A., and Ewing, G., *Chem. Phys.*, **53**, 141 (1980).
65. Gough, T. E., Miller, R. E., and Scoles, G., *J. Phys. Chem.*, **85**, 4041 (1981).
66. Casassa, M. P., Celii, F. G., and Janda, K. C., *J. Chem. Phys.*, **76**, 5295 (1981).
67. Geraedts, J., Waayer, M., Stolte, S., and Reuss, J., *Faraday Discuss. Chem. Soc.*, **73**, 375 (1981).
68. Gentry, W. R., *Proc. Eleventh Int. Conf. on the Physics of Electronic and Atomic Collisions, Amsterdam,* 1979.
69. Kraznovich, D., Ph.D. Thesis, University of California, Berkeley, California, 1983.
70. Ewing, G. E., *Chem. Phys.*, **63**, 411 (1981).
71. Bray, R. G., and Berry, M. J., *J. Chem. Phys.*, **71**, 4909 (1979).
72. Geraedts, J., Thesis, Katholieke Univeristeit te Nijmegen, Nijmegen, The Netherlands, 1982.
73. Gough, T. E., Knight, D. G., and Scoles, G., *Chem. Phys. Lett.*, **97**, 155 (1983).
74. Melinon, P., Zellweger, J. M., Philippoz, J. M., Monot, R., and van der Bergh, H., *Proc. Ninth Int. Conf. Mol. Beams,* Freiberg, Germany, 1983.
75. Vernon, M. F., Lisy, J. M., Kwok, H. S., Krajnovich, D. J., Tramer, A., Shen, Y. R., and Lee, Y. T., *J. Phys. Chem.*, **85**, 3327 (1981).
76. Pendley, R. D., and Ewing, G. E., *J. Chem. Phys.*, **78**, 3531 (1983).
77. Sakai, N., Koide, A., and Kihara, T., *Chem. Phys. Lett.*, **47**, 415 (1977).

78. Stepanov, B. I., *Nature*, **157**, 808 (1946).
79. Reimers, J. R., and Watts, R. O., *Chem. Phys. Lett.*, **94**, 222 (1983).
80. Young, L., Hayman, C. A., Morter, C., and Levy, D. H., *Abstracts of the 186th American Chemical Society National Meeting*, Washington, D.C., 1983.
81. Jouvet, C., and Soep, B., *J. Chem. Phys.*, **75**, 1661 (1981).
82. Brand, J. C. D., *Trans. Faraday Soc.*, **50**, 431 (1954).
83. Zwier, T. S., Carrasquillo, E., and Levy, D., *J. Chem. Phys.*, **78**, 5493 (1983).
84. Hopkins, J. B., Powers, D. E., and Smalley, R. E., *J. Chem. Phys.*, **74**, 745 (1981).
85. Felker, P. M., and Zewail, A., *J. Chem. Phys.*, **78**, 5266 (1983).
86. Khundkar, L. R., Marcus, R., and Zewail, A., *J. Phys. Chem.*, **87**, 2473 (1983).
87. Pendley, R. D., and Ewing, G. E., *J. Chem. Phys.*, **78**, 3531 (1983).

Photodissociation and Photoionization
Edited by K. P. Lawley

ION PHOTOFRAGMENT SPECTROSCOPY

JOHN T. MOSELEY

Department of Physics
and
Chemical Physics Institute,
University of Oregon,
Eugene, Oregon 97403, USA

CONTENTS

I. INTRODUCTION

The term 'ion photofragment spectroscopy can be used to refer to any study of ion photodissociation where a photofragment is observed directly as a carrier of the information sought. Virtually all experimental studies take advantage of the charge on the ionized fragment, which allows it to be detected with high efficiency. The ionized fragment may simply be detected in order to determine the fragmentation channels of the photodissociation, may be used to measure the total photodissociation cross-section or may be investigated in more detail, e.g. by a determination of its kinetic energy or angular distribution, or its internal energy. In this review we shall be primarily interested in these latter, more detailed, studies, which attempt to illuminate the dynamics of the dissociation process.

There are many motivations for the study of the photofragmentation of molecular ions. A general observation is the fact that spectroscopic information on molecular ions has proven much more difficult to obtain than the corresponding information on neutral molecules, due primarily to the relatively low densities of ions that can be obtained. The particular case of photodissociation is especially favourable for ion studies, since photofragment ions can generally be detected with nearly unit efficiency. Thus some kinds of photodissociation studies, particularly those involving some detailed measurement on the photofragment, can be carried out more easily on ions than on neutral molecules.

Ion photofragment spectroscopy is particularly adept at investigating transitions from bound to repulsive states—socalled direct dissociations. Such studies typically yield information on the vibrational levels of the bound state, as well as on the shape and symmetry of the dissociative state. In the case of predissociations, photofragment spectroscopy can yield information at a resolution competitive with saturation spectroscopy, allowing the determination of the lifetimes of the dissociated levels as well as information on the predissociation mechanisms. Similarly high resolution can be obtained in two-photon dissociations. In all of these studies, much can be learned about the dynamics of the dissociation process, often related to low-energy collisions between the photofragments. Indeed, photodissociation can be viewed as a 'half-collision', where the system is prepared in a specific state or superposition of states and then the second half of a 'collision' is investigated. Finally, a study of these dissociative processes yields information about molecular interactions in the poorly understood intermediate region where neither the states of the parent molecule nor those of the fragments provide a good description of the system.

Several fairly recent reviews have covered various of the aspects of ion photofragmentation mentioned above. Dunbar (1979) reviewed ion photodissociation without specific emphasis on fragmentation dynamics. Moseley and Durup (1981a) reviewed ion photofragment spectroscopy as performed using fast ion beams. The determination of ion molecular potential curves using photodissociative processes (Moseley, 1983) and the half-collision aspects of photofragment spectroscopy (Moseley, 1982) have also been recently reviewed. The present review will attempt to present a more comprehensive treatment of the photofragmentation of small molecular ions than those previous reviews. Experimental studies using gas-phase, ion trap and ion beam techniques will be considered. Emphasis will be on 'small' molecules and on experiments which have yielded detailed data. Theoretical work which applies to the understanding of specific experimental results will also be covered, but no attempt is made to review the very substantial body of theory related to this subject.

A recent review in this series, 'Photofragment Dynamics' (Leone, 1983), has covered similar ground for neutral molecules. A comparison of that review with this one will reveal that while the objectives of the studies on ions and neutrals are basically the same, the experimental techniques are often quite different, and the information obtained is generally complimentary. Taken together, these two reviews should provide a reasonable understanding of the current state of research into the dynamics of molecular photofragmentation.

II. EXPERIMENTAL TECHNIQUES

Three primary experimental techniques have been used to study ion photofragmentation processes. These are gas-phase techniques, such as the drift tube mass spectrometer, trapped ion techniques, such as ICR and quadrupole traps, and fast ion beams. All of these techniques have been used with lasers as photon sources, and in particularly favourable cases the light from incoherent sources has been used. We will review these techniques only briefly. The reader is referred to the cited literature for experimental details. An interesting point to note in this discussion is that all of these techniques, which have existed for many years, began to be applied to studies of ion photodissociation at about the same time, approximately 10 years ago.

A. Gas-Phase Techniques

All published work on ion photodissociation in the gas phase has been performed using drift tube mass spectrometers (Beyer and Vanderhoff, 1976;

Moseley *et al.*, 1975). However, recently a selected ion flow tube (Adams and Smith, 1975) has been constructed to allow laser interactions with the ions (Hansen, Kuo and Moseley, 1983a), and it is reasonable to expect increasing application of the flow tube to such studies in the future, due to the much higher chemical versatility of the flow tube.

With the drift tube, ions are formed in the gas phase at a fraction of a torr by initial electron processes and subsequent ion–molecule reactions and drift under the influence of a weak electric field through the background gas towards on extraction aperture. The electric field and the variable drift distance can be chosen so that the ions experience sufficient thermal energy collisions to approach thermal equilibrium with the gas molecules. Just before passing through the extraction aperture, the ions intersect the intracavity photons of a chopped laser. The ions that pass through the aperture into the high-vacuum analysis region are mass selected by a quadrupole mass spectrometer and individually detected by an electron multiplier. The quadrupole can be set to observe either the disappearance of the parent molecule or the appearance of suspected photofragments. This technique has been used primarily for the determination of absolute photodissociation and photodetachment cross-sections from the disappearance of the parent ions (see Smith, Lee and Moseley, 1979, and references therein), but has also been useful in some photofragment studies.

B. Trapped Ion Techniques

In many ways, ion-trap techniques are similar to gas-phase techniques. Most photodissociation studies have been performed using ion cyclotron resonance (ICR) traps (Dunbar, 1974; Eyler, 1974; Freiser and Beauchamp, 1974; Richardson, Stephenson and Brauman 1974). The ions are irradiated and photodissociation cross-sections determined from the loss of parent ions and/or the appearance of photofragment ions. The long trapping times allow the use of weak lasers of even monochromatized arc lamps, but the uncertainty in this time, and in the ion–photon overlap, complicate the determination of absolute cross-sections. When used in the double-resonance mode, in which an ion of a chosen mass is rapidly removed from the cell, it is often possible to identify the photofragments of a particular parent ion even when several species are in the cell, and in the presence of ion–molecule reactions. Ion traps have been applied primarily to photodissociation studies of larger ions than will be emphasized in this review. However, there is no fundamental problem in studying small ions using trapping techniques. The review by Dunbar (1979) emphasizes measurements on larger ions using this technique.

A particularly important and ingenious application of the ion-trap and photofragment spectroscopy was used to obtain the radiofrequency spectrum of H_2^+ (Dehmelt and Jefferts, 1962; Jefferts, 1968; Richardson, Jefferts and

Dehmelt, 1968). Although this work preceded all other experiments discussed in this review, it did not lead directly to ion photofragment spectroscopy as a field of study.

C. Fast Ion Beams

Although the two techniques discussed above have been useful in the study of ion photofragmentation, the major developments in this field have come about primarily through the application of fast ion beams to the problem. The pioneering study in this field was that of von Busch and Dunn (1972), who crossed a beam of H_2^+ ions with photons from a powerful arc lamp and measured the total cross-section for photoproduction of H^+ ions, with a resolution of 200 Å.

Application of the laser to this technique opened up many new possibilities, since the low ion density could be at least partially compensated by the high photon density of the laser. First to apply leasers to the spectroscopic study of ions in fast beams were Ozenne, Pham and Durup (1972) and van Asselt, Maas and Los (1974a, 1974b). Again H_2^+ was the ion of choice, but, in contrast to the study of von Busch and Dunn, individual laser lines were used and sufficient photofragments were produced to allow dispersal by an energy analyzer, thus resulting in kinetic energy distributions of the ion fragments. These experiments marked the beginning of ion photofragment spectroscopy.

Due to the versatility of this technique, and to the fact that it is now in use by over ten different research groups in the United States, Europe and Australia, a somewhat more detailed discussion of the experimental techniques will be given. Although the details of the various facilities differ greatly, and the 'fast ion beams' vary in energy from 10 to 10^5 eV, all of them are basically double-mass spectrometers with a provision for laser interaction with the beam between the spectrometers. A typical, and relatively recent, example is the one described by Carrington and Buttenshaw (1981a) and depicted in Fig. 1. The instrument there is actually a commercial double-mass spectrometer, modified to allow lasers to pass either coaxially, as shown, or crossed with the ion beam.

In this as well as the other applications of this technique, ions are extracted from an ion source, accelerated to the desired energy and the species to be studied selected by the first mass spectrometer. In the following laser interaction region, photofragments are produced through either one or two photon processes, as will be discussed later. The laser may be tunable or it may be held at a fixed frequency. In this latter case, a limited tuning range can be obtained in the coaxial beams mode by varying the ion velocity, thus changing the Doppler shift. The photofragments produced can now be selected by a second mass or energy analyzer and detected individually. At fixed laser frequency, the energy analyzer can be scanned to measure the energy distribution of the photofragments; in the crossed beams mode, the laser

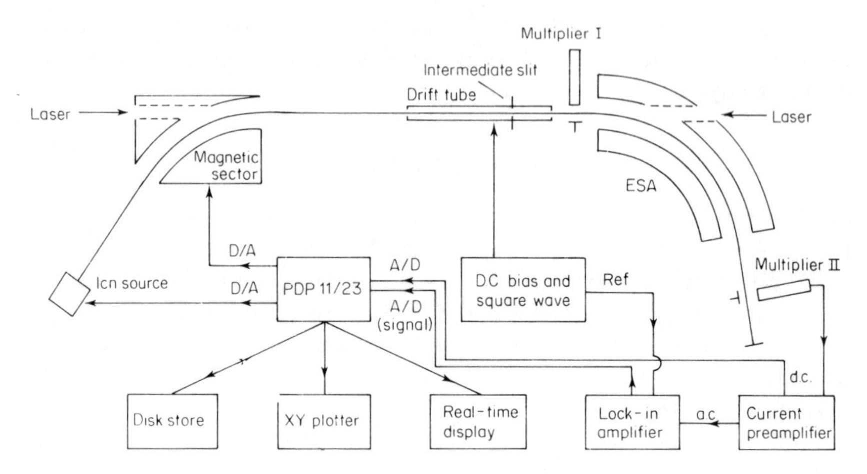

Fig. 1. Typical ion photofragment spectrometer. (from Carrington and Buttenshaw, 1981a. Reproduced by permission of Taylor and Francis, Ltd.)

polarization may be rotated to determine the angular distribution of the photofragments.

If it is desired to have high resolution in the energy and angular distribution measurements, it is necessary to have high angular collimation on both and parent and the photofragment ion beams. These consideration, as well as many others, have been discussed by Huber and coworkers (1977) in an article describing the apparatus at SRI International. This apparatus uses a unique quadrupole bender (Zeman, 1978) to direct the parent beam into the interaction region without destroying the 2 mrad collimation, and another to direct the photofragments into the energy analyzer. It is thus more complex than the apparatus shown in Fig. 1, but has succeeded in obtaining the best energy resolution of any of the existing experiments.

Rather than listing or describing the other facilities here, any unique features will be described in conjunction with discussions of research utilizing the other facilities.

D. Photon Sources

The primary photon sources for the experiments to be described have been lasers. As we shall see, most well-known lasers with powers greater than a few milliwatts have proven useful. Experiments have been performed with fixed and tunable lasers, both pulsed and c.w. Very high resolution experiments have been performed using a single-mode lasers. Except for the initial work on H_2^+, incoherent sources have been useful only in the ICR experiments, where a large number of them have been performed using arc lamp sources.

III. HYDROGEN DIATOMIC IONS

The simplest molecule is the H_2^+ ion; indeed, it is the only molecule which can be studied without concern for electron correlation effects. For this reason it has occupied a unique position in the development of molecular quantum mechanics. A very detailed review of the structure and spectroscopy of this molecule, including an up-to-date discussion of the current state of the theoretical description, has been given by Carrington and Kennedy (1983).

The two lowest potential curves are shown in Fig. 2; while the ground electronic state is bound ($D_0 = 2.6507$ eV), the first excited state is purely repulsive. The lowest bound excited states are predicted to be 11 eV above the ground state and cannot be observed using standard spectroscopic techniques. The $2p\sigma_u$ is easily accessible, however, and thus photodissociation has played an important role in the spectroscopy of H_2^+ and its isotopes.

Dehmelt, Richardson and Jefferts (Dehmelt and Jefferts, 1962; Jefferts, 1968, 1969; Richardson, Jefferts and Dehmelt 1968) were apparently the first to use the observation of ion photofragments in a spectroscopic measurement. Their application was both ingenious and unique; although it could, in principle, be applied to other ions, it has only been used with H_2^+. The experiment is based on the observation that, since the $1s\sigma_g \rightarrow 2p\sigma_u$ transition

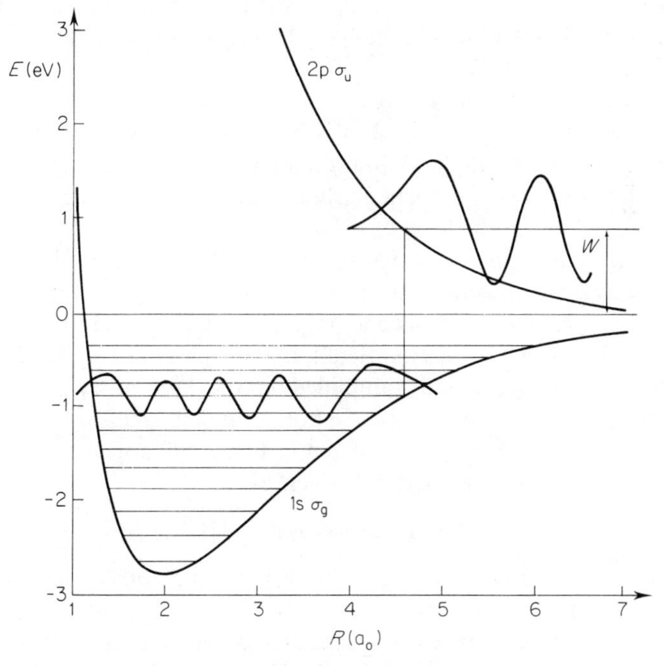

Fig. 2. The two lowest potential curves of the H_2^+ ion. (From Ozenne *et al.*, 1976. Reproduced by permission of North-Holland Publishing Co.)

obeys normal dipole selection rules, the photodissociation rate is proportional to the square of the z component of the total angular momentum. Thus H_2^+ ions held in the collision-free environment of their radiofrequency quadrupole trap and irradiated with polarized light develop a net spatial orientation, as the higher M_F levels are more rapidly photodissociated. Therefore, radio-frequency transitions between hyperfine levels in the molecule affect the net observed photodissociation. The ion trap was set up to hold both H_2^+ and photofragment H^+ ions, which were periodically sampled as the radio-frequency was slowly swept. Hyperfine transitions were detected as changes in the H^+ to H_2^+ ratio.

Determinations of the Fermi contact, nuclear hyperfine and spin-rotation parameters were made for H_2^+ vibrational levels 4 to 8. The measurements provided a very stringent test of the ground state potential curve and of the theoretical understanding of the interactions listed above, and were finally definitively interpreted by Ray and Certain (1977).

Unfortunately, this very elegant series of experiments did not lead to further exploitation of ion photofragment observations as a spectroscopic technique. As previously mentioned, the experiments which precipitated the current interest in this field were the total cross-section measurements of von Busch and Dunn (1972), followed quickly by measurements of the energy distributions of the photofragments using lasers by Ozenne, Pham and Durrup (1972) and van Asselt, Maas and Los (1974a, 1974b). All of these measurements were on H_2^+, HD^+ or D_2^+.

The principle of the photofragment kinetic energy measurements can be discussed with reference to Fig. 2. It is first important to note that typically in fast ion beams virtually all vibrational levels are sufficiently populated to allow observation. H_2^+ is a particularly favourable case; all vibrational levels up through $v = 18$ (of the total of twenty-one levels) have a population of at least 0.1 per cent of the total. Thus we can reasonably consider attempting to photodissociate the molecule from $v = 8$ using roughly 2 eV photons, as illustrated in the figure. A second important fact, as has been discussed elsewhere (Huber *et al.*, 1977) in detail, is that the relatively small centre of mass energies W are greatly amplified in the laboratory system. For a homonuclear diatomic molecule, a photofragment resulting from a dissociative transition such as that indicated in Fig. 2 will appear at $\theta = 0°$ in the laboratory frame with an energy T given by

$$T = T_0/2 \pm (WT_0)^{1/2} + W/2 \qquad (1)$$

where T_0 is the parent ion energy and W is the total kinetic energy of separation in the centre of mass frame. Thus for $T_0 = 3000$ eV, a photofragment corresponding to $W = 0.1$ eV appears in the laboratory frame with an energy of about 1517 eV, compared with 1500 eV for $W = 0$. It is also clear that high angular collimation of the parent ion beam and angular resolution of the photofragments is required so that one does not observe photofragments at 0° in the laboratory that actually came from a transition of larger W and $\theta \neq 0°$. This imposes angular definition requirements in the range of milliradians in the laboratory system if accurate measurement of photofragment energies is desired.

A typical kinetic energy spectrum obtained by Ozenne and coworkers (1976) is shown by the solid line in Fig. 3, for the situation illustrated in Fig. 2. The peaks in the kinetic energy spectrum arise from the various vibrational levels which can be photodissociated at this photon energy. The intensity of these peaks is determined by the population of the vibrational levels, the Franck–Condon factors and the photodissociation cross-sections, as well as an apparatus function describing the effect of the measurement on the actual initial kinetic energy distribution. The dashed line is from a Monte Carlo

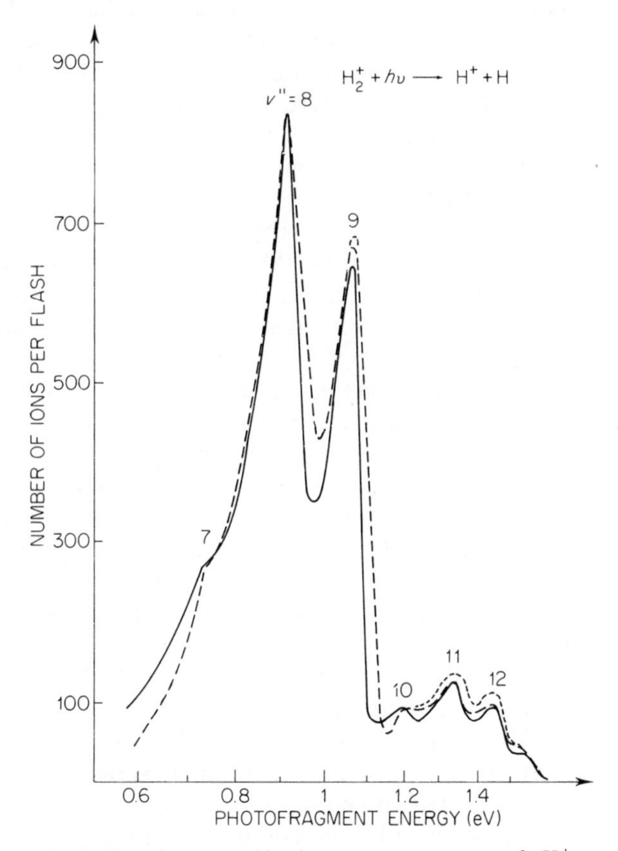

Fig. 3. Photofragment kinetic energy spectrum of $H_2^+ + h\nu \rightarrow H^+ + H$, at the photon energy illustrated in Fig. 2 (From Ozenne *et al.*, 1976. Reproduced by permission of North-Hollant Publishing Co.)

computer simulation of the apparatus function, using Franck–Condon populations and potential curves and cross-sections as discussed in the reference. This work, along with similar studies by van Asselt, Maas and Los (1975) and Thomas, Dale and Paulson (1977a, 1977b), did not particularly contribute to our understanding of the H_2^+ molecule. Rather, they demonstrated using this relatively well-characterized species that kinetic energy measurements on ion photofragments could be used to determined vibrational level spacings and populations, as well as information about the dissociative state. Application to other molecular ions which were less well characterized will be discussed in subsequent sections.

Recently, Carrington and coworkers (Carrington and Buttenshaw, 1981a; Carrington, Buttenshaw and Kennedy, 1983; Carrington, Buttenshaw and Roberts, 1979) have investigated vibration–rotation bands high in the vibrational manifold of HD^+, using a two-photon technique. This is the simplest molecule to have an allowed vibration–rotation spectrum, and precision measurements of such spectra near the dissociation limit can test even the most comprehensive theoretical calculations.

The experiments can be described with reference to Figs. 1 and 4. The two lasers are both CO_2 lasers: laser 1 is operated at low power to induce the rovibrational transition while laser 2 is operated at high power to provide the photofragment signal. Turning is obtained by sweeping the voltage on the drift tube over a total range from 1 to 10 kV, thus effectively covering most of the frequency range from 880 to 1080 cm^{-1}. It is possible to obtain very high resolution in this coaxial laser and ion beam arrangement due to the narrowing of the velocity distribution in an ion beam when it is accelerated. This arises very simply from the fact that a population of ions with an energy spread ΔE that is extracted from a ion source and accelerated to an energy E maintains this energy spread ΔE; however, since $E = mv^2/2$, the velocity spread at an energy $E \gg \Delta E$ must be substantially reduced from its value before acceleration. In fact, an ion beam with a 1-eV energy spread has at a laboratory energy of 3 keV a velocity spread similar to a temperature of less than 1 K. Figure 5 shows a sample of the data obtained, indicating that a resolution of better than 10 MHz was observed. Ten transitions in the $v = 18$ to 16 band were studied, most of them using more than one of the CO_2 laser lines. Similarly, using a CO laser, seven transitions were studied in the $v = 17$ to 14 band. The experimental uncertainty in the first set of measurements was $\pm 0.0005\ cm^{-1}$; in the second it was $\pm 0.001\ cm^{-1}$.

The results described above were compared with the predictions of the best available theoretical calculations (Wolniewicz and Poll, 1980) and found to differ from them typically by a factor of 10 or more greater than the stated experimental uncertainty, and in one case by a factor of 100. Improvements in resolution and sensitivity are planned which should allow a direct determination of the values of the hyperfine constants, and a search for the highest

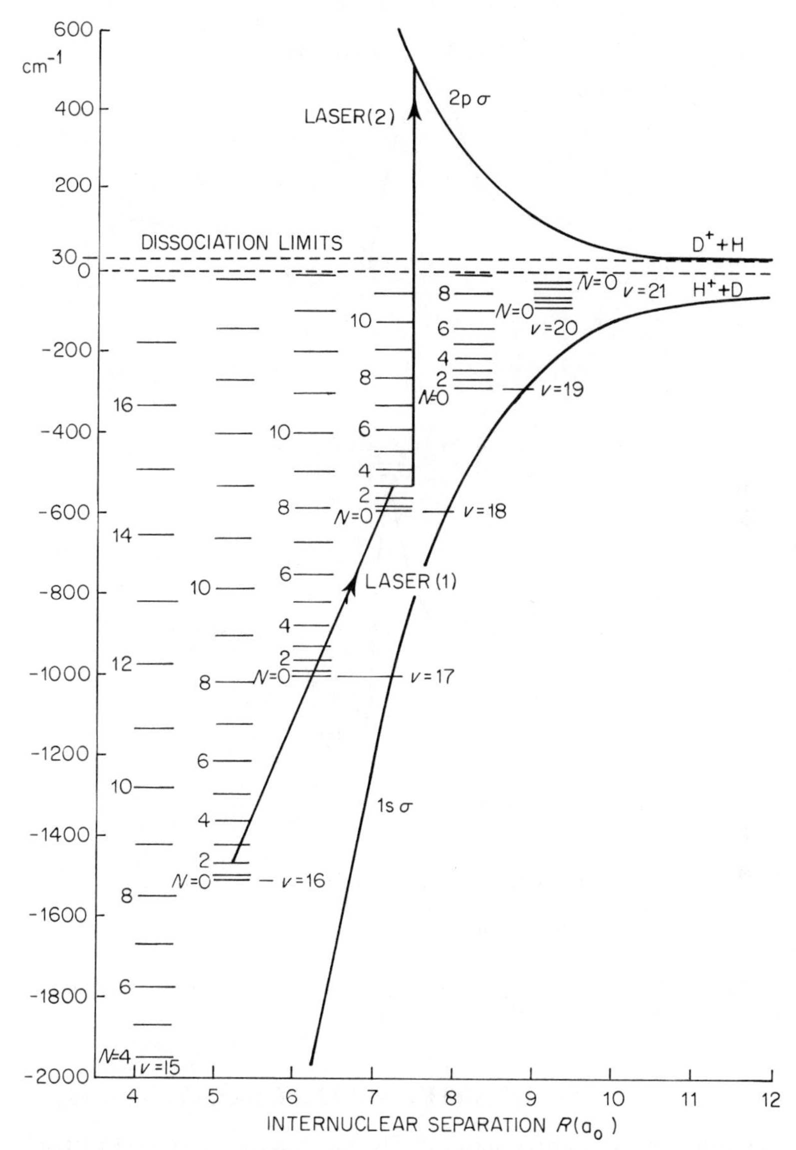

Fig. 4. Energy levels of HD^+ near the dissociation limit. (From Carrington and Buttenshaw, 1981a. Reproduced by permission of Taylor and Francis, Ltd.)

bound level, predicted to be $v = 21$, $N = 1$. This level could exhibit interesting hyperfine effects, since its separation from the lower dissociation limit will be small with respect to the splitting between the two limits, and therefore the charge and electron spin distributions might no longer be symmetric with respect to the centre of the nuclei. In any case, these and future experiments will

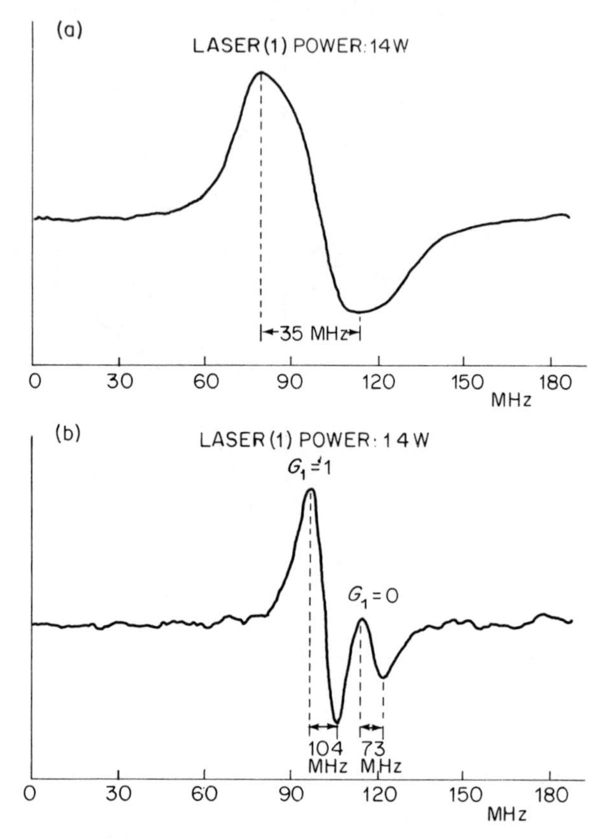

Fig. 5. Example of data obtained on the vibration–rotation spectroscopy of HD^+, using the two-laser scheme illustrated in Fig. 4. (From Carrington and Buttenshaw, 1981a. Reproduced by permission of Taylor and Francis, Ltd.)

provide a severe test of the current theories of non-adiabatic effects in molecules.

IV. QUARTET STATES OF O_2^+

The four lowest quartet states of O_2^+ have been the subject of the most extensive investigations utilizing ion photofragment spectroscopy of any molecule. One good reason for this can be seen in Fig. 6—there are three states which have allowed transitions in the visible from the metastable $a^4\Pi_u$ state of this ion. As a result, this ion undergoes a wide variety of photodissociative processes which can be easily studied, and many of the experimental and theoretical techniques now being applied to other ions were first developed on O_2^+. These states are also of special interest because of their high multiplicity.

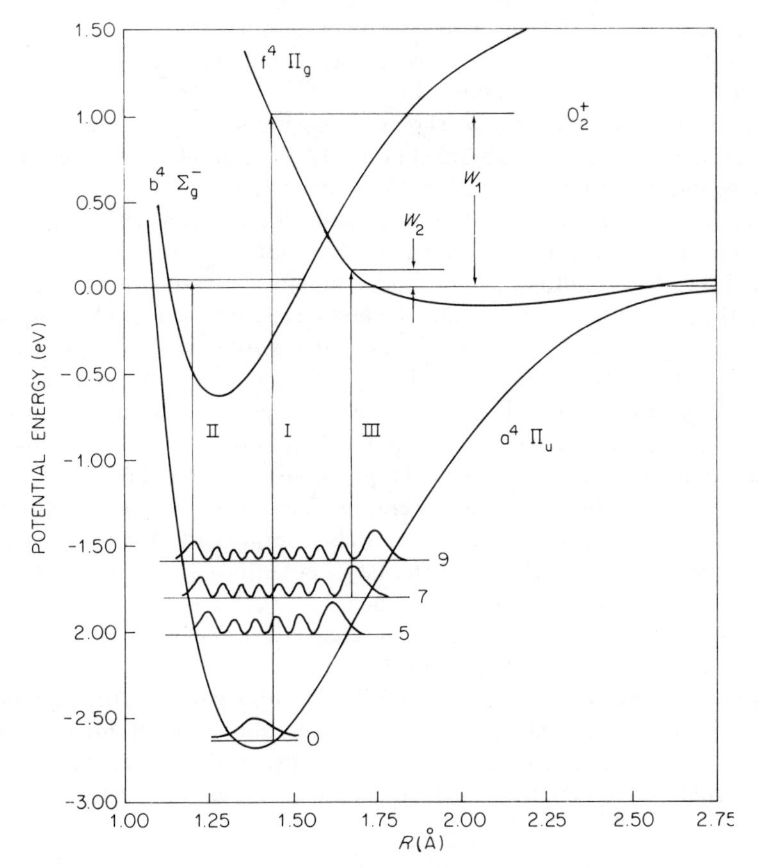

Fig. 6. Potential curves of the three lowest quartet states of O_2^+. In addition, the $d^4\Sigma_g^+$ state is in this region and crosses the b state just below its crossing with the f state (Moseley and Durup, 1981).

The studies described here represent substantially the most detailed investigation of any quarter system, as well as the most extensive application of ion photofragment spectroscopy to any molecule. It should be pointed out that the shape and location of the $f^4\Pi_g$ state shown on Fig. 6, and of the $d^4\Sigma_g^+$ state which is also in this region (Hansen *et al.*, 1982), were essentially unknown prior to the work described below.

A. Determination of Dissociative Potential Curves

In the work on the direct photodissociation of H_2^+, discussed in the previous section, known bound and repulsive potential curves were used to calculate expected photofragment energy distributions, which were compared with the experimental ones, for example in Fig. 3. Clearly in principle this can be turned around—if the ground state potential curve and the photofragment kinetic energy distribution are known, at least a limited range of the repulsive potential curve can be determined.

The solid curve in Fig. 7 shows a kinetic energy spectrum of O^+ photofragments from O_2^+ obtained at 5600 Å. The number above each peak corresponds to the initial vibrational level in the a state; in fact, the separations between these peaks were used to show that this transition originated in the a state (Tabché-Fouhaillé *et al.*, 1976). The vibrational spacing of the a state was well known from photoionization studies and from investigations of emission from the $b^4\Sigma_g^+$ state to the a state, the first negative system of O_2^+. The identity of the repulsive state was established in this work to be the $f^4\Pi_g$ state, from the angular distribution of the photofragments. (A discussion of photofragment angular distributions will be given later in this section.) However, insufficient data were obtained in this early work to allow a determination of the shape of the $f^4\Pi_g$ potential curve. More recently, spectra such as the one shown in Fig. 7 were obtained at twelve wavelengths from 5208 to 7525 Å and were used to accurately determine the dissociative part of this curve (Grieman *et al.*, 1980). The f state potential curve shown in Fig. 6 is drawn to represent this determination.

The dashed curve in Fig. 7 represents a calculated kinetic energy distribution which assumed a 400 K rotational temperature and is not corrected for the apparatus function. Without this correction, the peak of the calculated distribution should lie approximately above the half-intensity point on the experimental distribution. The intensities of the calculated peaks are not normalized to the experimental ones, but are as determined from a best fit

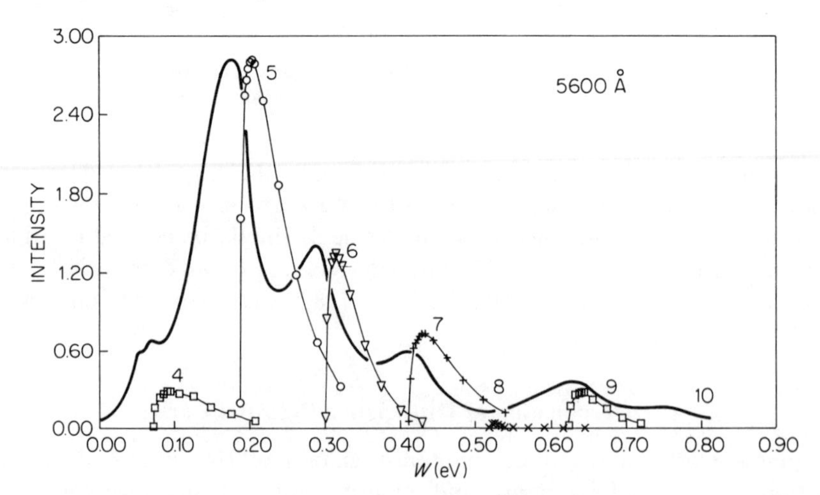

Fig. 7. Kinetic energy spectrum of O^+ photofragments from O_2^+ obtained at 5600 Å (solid curve). The dashed curve is from a calculation described in the text (Grieman *et al.*, 1980).

to all of the spectra, which also resulted in a determination of the relative vibrational populations of the a state levels.

B. Predissociation Spectroscopy

Transitions between the a state and b state levels below the dissociation limit have been intensively studied in emission (Albritton *et al.*, 1977, and references therein). However, b state levels above $v = 3$ had not been definitively observed. In order to study predissociations, the important observable is not the photofragment kinetic energy but the appearance of photofragments (perhaps in a narrow fixed energy range) as a function of wavelength.

The transition labelled II in Fig. 6,

$$O_2^+(a^4\Pi_u) + h\nu \rightarrow O_2^+(b^4\Sigma_g^+) \rightarrow O^+(^4S^O) + O(^3P) \tag{2}$$

was first observed by Moseley and coworkers (1976), identified by Carrington, Roberts and Sarre (1977) and studied in detail by Tadjeddine and coworkers (1978). This study, made at 0.2 Å resolution, covered the (4, 4), (4, 3), (4, 5) and (5, 5) bands, with all rotational structure resolved. The structure of this quartet system is quite complex (Albritton *et al.*, 1977). When fine-structure levels are considered, there are forty-eight allowed transitions that terminate in a single upper rotational level. The low-resolution study referenced above could resolve only twenty-four branches; the highest-resolution normal optical spectroscopy can usually resolve forty branches.

However, it is possible to take advantage of the narrowing of the velocity

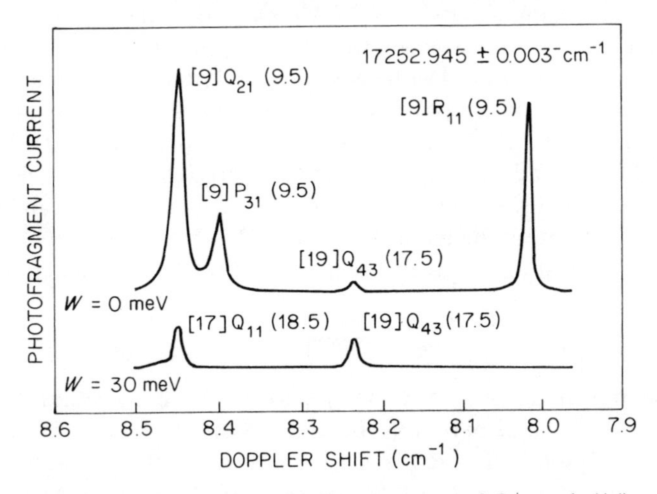

Fig. 8. Velocity-tuned photofragment spectrum of O_2^+ a→b (4,4) transitions near 17261 cm^{-1}. The two scans are for different settings of the energy analyser (Cosby, Moller and Helm, 1983).

distribution in a fast ion beam to resolve all fine-structure levels. Figure 8 shows an example of the data obtained by Cosby and coworkers (1980) in the (4, 4) band using a single-mode dye laser at fixed frequency with modest velocity tuning. The spectroscopic notation, which is explained in detail in the reference, is briefly as follows: P, Q and R give the ΔJ of the transition, the preceding number gives the rotational quantum number N' of the lower state and the following subscript gives the fine structure quantum number F' of the upper state, followed by F'' of the lower state. Note that the two scans are for different settings of the energy analyser, and hence sample different ranges of rotational levels above the dissociation limit. The three transitions at the left of the figure would be blended in normal optical spectroscopy.

Such high-resolution spectra have been obtained using the coaxial beams technique for the (4, 1) and (5, 2) bands (Carre *et al.*, 1980), the (3, 3), (4, 4), (4, 5) and (5, 5) bands (Carrington *et al.*, 1978; Cosby *et al.*, 1980) and the (5, 6), (5, 7), (6, 7), (7, 8) and (8, 9) bands (Hansen *et al.*, 1983). A sample of the data obtained by Hansen and coworkers is shown in Fig. 9. It is clear from these data and from Fig. 8 that the predissociation linewidth is changing rapidly with vibrational level, and that this degrades the resolution for the higher vibrational levels. However, this provides other useful information which will be discussed in the following section.

Hansen and coworkers (1983) have reanalysed all available data on the First Negative bands, including the emission data. Molecular constants were determined for vibrational levels 0 to 9 of the a state and 0 to 8 of the b state using the multiplet Hamiltonians defined by Zare and coworkers (1973) and including the third-order spin-orbit interaction (Brown *et al.*, 1981). This high-order perturbation was found to be very important in the fits to the (4, 4) and (4, 5) bands, which were studied with the highest precision and accuracy, 0.003 $\mathrm{cm^{-1}}$, and determined with good statistical significance for all bands up through the (6, 7) band. Thus the importance of this interaction to an accurate description of the First Negative bands seems to be well established. In addition, Hansen and coworkers determined merged molecular constants and Dunham coefficients, and in general provide the most accurate and comprehensive description of the a and b states of O_2^+ that is currently available.

Another interesting predissociation in the O_2^+ quartet system, indicated in Fig. 7 as III, is

$$O_2^+(a^4\Pi_u) + h\nu \rightarrow O_2^+(f^4\Pi_g) \rightarrow O^+(^4S^0) + O(^3P) \qquad (3)$$

where the transition is now to bound or quasi-bound levels of the f state rather than to the dissociative part of the potential, as in transition I. Transitions of this type that result in the predissociation of two successive vibrational levels ($v' = 2, 3$) of the $f^4\Pi_g$, $\Omega' = \frac{5}{2}, \frac{3}{2}$ states have been observed by Helm, Cosby and Huestis (1980). The existence of a potential barrier (even in the absence of rotation) in this state was suggested theoretically in *ab initio* calculations by

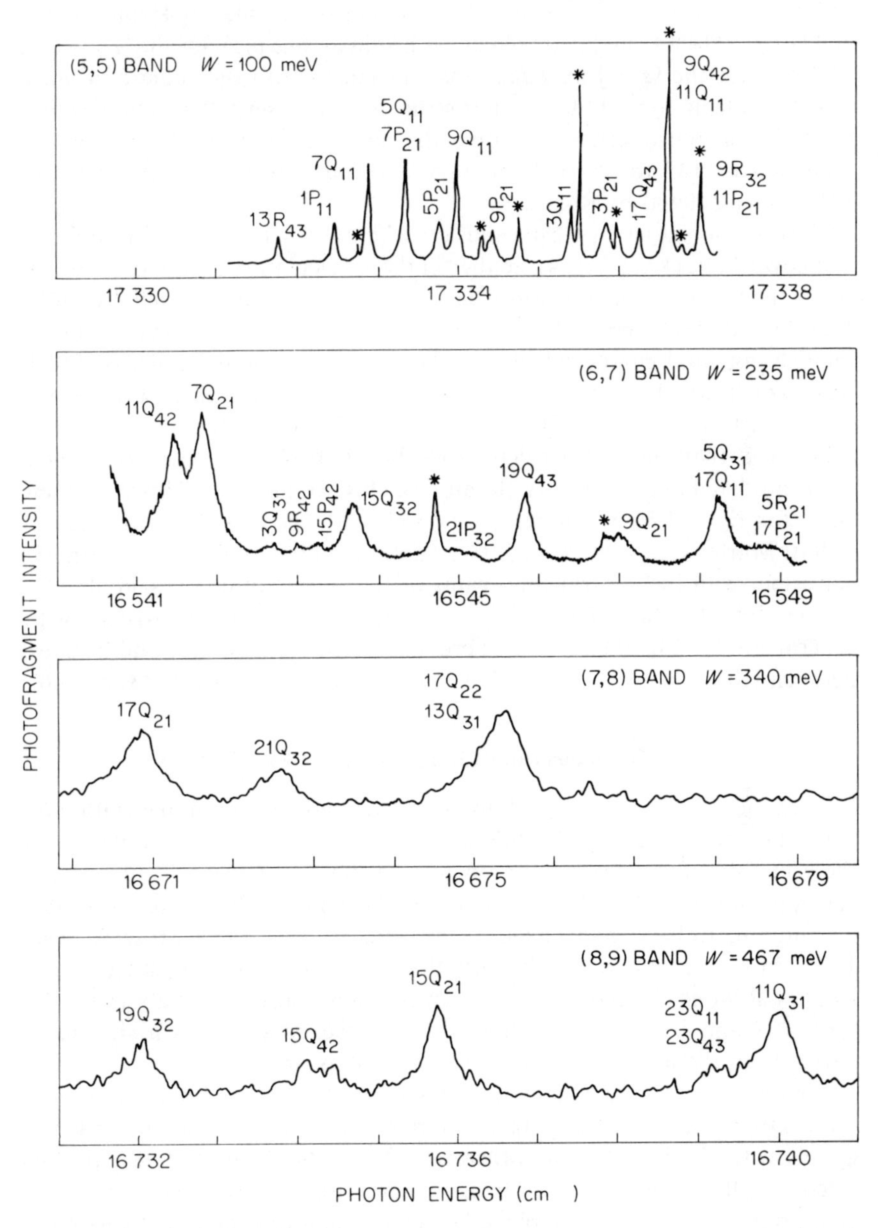

Fig. 9. Predissociation photofragment spectra from higher vibrational bands of O_2^+ a→b transitions (Hansen *et al.*, 1983).

Beebe, Thurlstrup and Andersen (1976) and by Roche (Carre *et al.*, 1980). From the locations of the highest observed quasi-bound levels, Helm, Crosby and Huestis (1980) estimated the barrier height as 48.8 meV for the $\Omega' = \frac{3}{2}$ and 44.7 meV for the $\Omega' = \frac{5}{2}$ in remarkable agreement with the value of 47 meV from the calculations of Beebe, Thurlstrup and Andersen (1976). On the other hand, the difference between the top of the barrier and the bottom of the well is estimated as 142 to 155 meV, in very good agreement with the value of 151 meV calculated by Roche.

However, as shown by Cosby and coworkers (1978, Fig. 9) and by Greiman and coworkers, (1980, Fig. 4), neither of these calculations actually represent either the bound or the dissociative part of this potential very well. Recently, Marian and coworkers (1982) have made a very sophisticated calculation of the O_2^+ potential curves (which may be described as 'large-scale MRD-CI, AO, DZD'), which are in remarkably good agreement with experimentally determined curves. Their $f^4\Pi_g$ curve is clearly in much better agreement with experiment than the earlier calculations, but unfortunately they do not give sufficient detail about the curve to allow a close comparison. However, their value for R_e of 1.86 Å compares very well with the experimental value of 1.79 Å. Marian and coworkers (1982, Table 6) imply a well depth of 80 meV, consistent with the experimental value of 100 meV. Nothing is said about a barrier, but the plot of the curve indicates a small one, consistent with experiment. Finally, the calculated crossing point between the f and b states, between vibrational levels 6 and 7, is in excellent agreement with experiment.

C. Predissociation Lifetimes and Mechanisms

Inspection of Figs. 8 and 9 shows that the predissociation linewidth, and hence lifetime, varies significantly with vibrational level. Measurements of these lifetimes for a wide selection of rotational and fine-structure levels, encompassing vibrational levels 3 to 8, have been published (Carre *et al.*, 1980; Carrington, Roberts and Sarre, 1978; Hansen *et al.*, 1982; Moseley *et al.*, 1979). In general terms, the lifetimes decrease from about 2 ns for the high rotational levels (31 and 33) of $v' = 3$ which were observed to about 0.01 ns for $v' = 7$ and 8. A careful examination of the data reveals a significant variation of the lifetimes on N' and F' as well as on v'.

Four states cross the b state in the region of the observed predissociation. Two of these, the α and the E states (Marian *et al.*, 1982) are spin-forbidden to couple with the b state. The other two states, the d and f states, are not electronically coupled to the b state, but can lead to predissociation through spin-orbit and rotational coupling. We shall therefore attempt to explain the predissociation through coupling with these two states. Figure 10 shows the calculated variation of the relative linewidth (Hansen *et al.*, 1982) for the predissociation of a $^4\Sigma_g^-$ state by a $^4\Sigma_g^+$ state (upper half of the figure) and by a

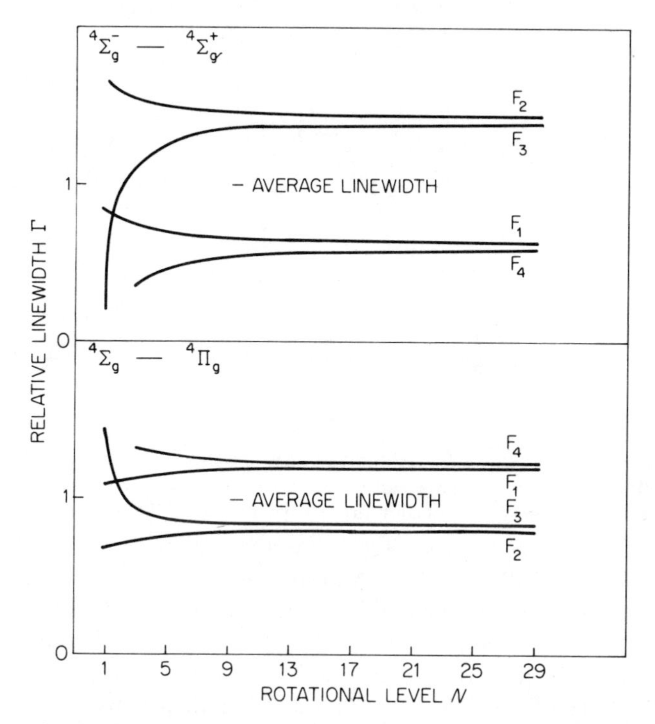

Fig. 10. Calculated variation of linewidths for the predissociation of a $^4\Sigma_g^-$ state by a $^4\Pi_g$ state and by a $^4\Sigma_g^+$ state (Hansen *et al.*, 1982).

$^4\Pi_g$ state (lower half), as functions of N' and F', assuming that only spin-orbit coupling is important. Note that the variation with N' is significant only for small N' and that the dependences on F' are approximately opposite for these two cases. Linewidths obtained from data such as those discussed above can be fit to an appropriate combination of these curves for a given vibrational level in order to determine the relative significance of the d and f states on the predissociation of the b state. Such an analysis has led to the determination of average linewidths and branching ratios to the d and f states for vibrational levels $v' = 4$ to 8 (Hansen *et al.*, 1982), with the results shown in Fig. 11 by the closed circles with error bars. It is of interest to compare these conclusions with theoretical calculations.

Since the $^4\Pi_g$ potential curve has been experimentally determined, relative linewidths calculated for the dissociation of the b state by this state should give reasonably good agreement with the experimental results of Fig. 11a, if the spin-orbit coupling matrix element does not vary significantly with the vibrational level. The crosses in Fig. 11a show the calculated linewidths, assuming a coupling matrix element of 47.6 cm^{-1}. This differs considerably from the calculated value of this matrix element of 18 cm^{-1} by Roche (Carre

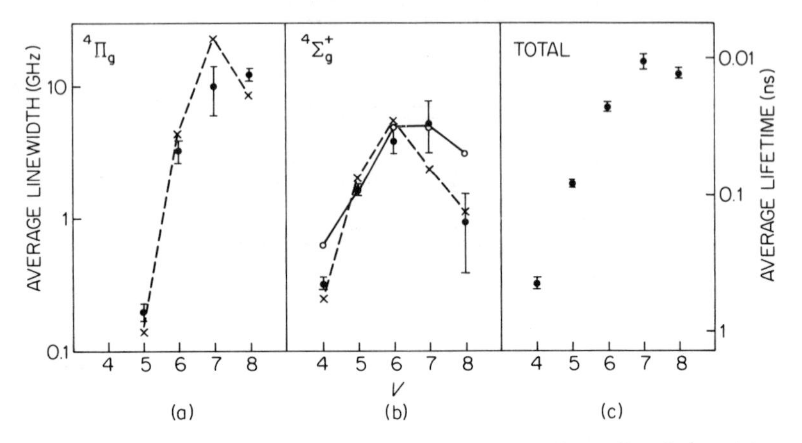

Fig. 11. Summary of experimental results (closed circles) on the predissociation of the b state of O_2^+ by the d and f states, and comparison with calculations (crosses and open circles) described in the text (Hansen *et al.*, 1982).

et al, 1980) and of 29.3 cm^{-1} by Marian and coworkers (1982), who also showed that this matrix element did not vary rapidly with R. This discrepancy is not yet understood, but does not indicate that rotational coupling should have been considered. If rotational coupling was sufficiently large to account for this discrepancy it would have led to a dramatic increase in the linewidth with N', which was not observed experimentally.

A detailed comparison for the d state contribution to the dissociation is more questionable, since there has been no experimental determination of this potential curve. Carre and coworkers (1980) have calculated a potential curve for this state, as well as linewidths for its predissociation of the b state. This calculation, normalized to the experimental linewidth at $v' = 5$, and extended to include $v' = 8$, is represented by the open circles in Fig. 11b. While the comparison between levels 5, 6 and 7 is very satisfactory, it is not so for levels 4 and 8. Better agreement can be obtained by shifting the calculated curve 0.05 Å to larger R, with the result shown by the crosses in Fig. 11b. For the original curve, the matrix element implied by this fit is 16.9 cm^{-1}; for the shifted curve it is 33 cm^{-1}. The calculated matrix element is 11 cm^{-1}; again, the discrepancy is not yet understood.

It is also interesting to compare the experimental results with the linewidths calculated by Marian and coworkers (1982) for the predissociation of the (F_1, $N' = 9$) levels of the b state in vibrational levels 4 to 8. These calculations used both their calculated potential curves and spin-orbit coupling matrix elements, including their variation with R. The comparison is given in Table I. Agreement for levels 4 and 5 is remarkably good, and qualitatively good agreement continues for all of the f state contributions. However, the disagreement for the b state contribution for levels 6, 7 and 8 is severe. The

TABLE I
Comparison of calculated[a] and experimental[b] linewidths for predissociation of the O_2^+ b state levels $v' = 4$ to 8, F_1 and $N = 9$. All linewidths are given in wavenumbers.

$v'(b^4\Sigma_g^-)$	4	5	6	7	8
$d^4\Sigma_g^+$ contribution					
Calculated	0.013	0.043	0.025	0.009	0
Experiment	0.011	0.044	0.183	0.180	0
$f^4\Pi_g$ contribution					
Calculated	0	0.004	0.069	0.179	0.003
Experiment	0	0.006	0.150	0.320	0.370
Total					
Calculated	0.013	0.047	0.094	0.188	0.003
Experiment	0.011	0.050	0.333	0.500	0.400

[a]Marian and coworkers (1982).
[b]Hansen and coworkers (1983).

calculations show it falling rapidly after $v' = 5$, becoming negligible by $v' = 8$. The experimental data seem very clear on the point that the d and f state contributions are comparable and are both quite large for levels 6 and 7, and indicate that the d state contribution is still significant at $v' = 8$.

In spite of some discrepancies, the general agreement of the experimental data and potential curves with the calculations of Marian and coworkers (1982) is remarkably good, considering the complications of these quartet states. Obviously, the experimental data provide a stringent test of calculations of these predissociation processes, and therefore should lead to a better understanding of these phenomena in the future.

D. Branching Ratios and 'Half-Collisions'

It is of interest to view these photodissociations as half-collisions, since in order to understand in detail collisional processes involving atoms of non-zero spin it is necessary to understand all of the couplings between the various potential curves which cross these potential curves at energies less than the collision energy. In a photodissociation, one is able to prepare the molecule in a specific quantum state (or a limited superposition of quantum states), at a specific internuclear separation and with a specific separation energy, and then follow in some detail the fragments as they separate—thus, a 'half-collision'. Direct studies of collisional processes, even those which are differential in both energy and angle, necessarily average over many of these parameters.

In this O_2^+ quartet system, the final state of the half-collision consists of an $O^+(^4S^0_{3/2})$ ion and an oxygen atom in one of the $^3P_{2,1,0}$ fine-structure states.

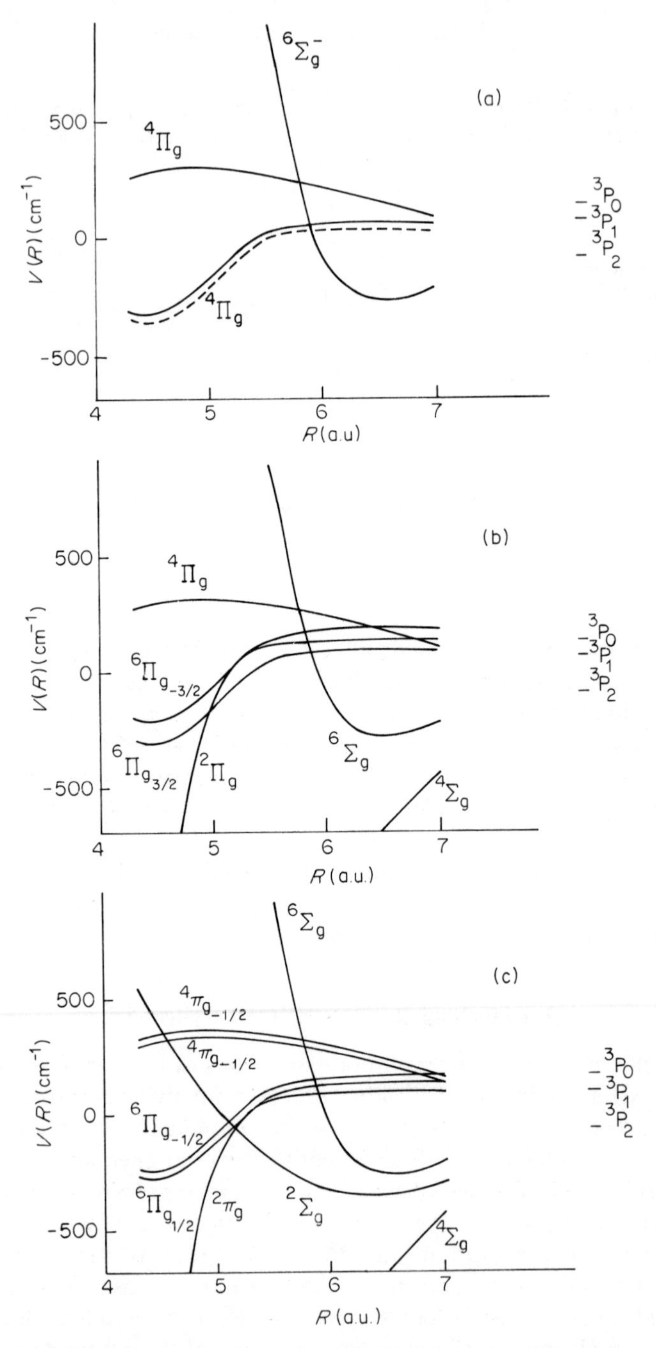

Fig. 12. Potential curves of the lowest gerade states of O_2^+ in the intermediate region between 4.3 and 7.0 a.u. (approx. 2.3 to 3.7 Å) (Durup, 1981).

Experimentally, the branching ratio into these fine-structure states has been determined from the photofragment kinetic energy distributions for many of the photodissociations discussed above (Cosby *et al.*, 1980; Helm *et al.*, 1980; Pernot *et al.*, 1977, 1979; Tadjeddine, 1979). The theoretical challenge is to calculate these branching ratios.

It is useful to distinguish three regions of internuclear separation: (1) the region near the equilibrium internuclear separation where the electronic Hamiltonian dominates and the 'standard' description of the molecule applies, (2) the region at large internuclear separation where the atomic description is valid and (3) the intermediate region where effects such as spin-orbit coupling are of the same order of magnitude as the electronic couplings. Figure 12 shows the gerade states in this intermediate region. An ion is first 'prepared' in one of the dissociating states discussed in the previous sections—quasi-bound levels of the b or f states, or the repulsive part of the f state—and the dissociation followed theoretically through the region shown in Fig. 12, with all relevant couplings calculated, to predict the final result in the atomic region, i.e. the branching into the fine-structure states of the O atom. Durup (1981) has made such a calculation using the concept of molecular diabatic states in the intermediate region. Excellent qualitative, and often quantitative, agreement is found with the experimental observations. A simple adiabatic calculation fails badly for the dissociation of the f state. The success of these calculations indicates that significant progress is being made in the understanding of the molecular couplings which occur in the intermediate region, which is an important step in the detailed understanding of low-energy collisions.

E. Angular Distributions

Angular distributions of photofragments were first treated theoretically by Zare (1964, 1972) and later by a number of others, both quantum mechanically and classically. The work of Dehmelt and Jefferts (1962) on the optical orientation of H_2^+ is also important in this regard. Angular distributions were reviewed in an earlier volume in this series by Bersohn and Lin (1969). As is well known, the angular distribution for photofragments from initially randomly oriented molecules can always be written in the form

$$I(\theta) = \frac{1 + \beta P_2(\cos\theta)}{4\pi} \tag{4}$$

where θ is the angle between the laser polarization and the direction of ejection of fragments in the centre-of-mass frame, $P_2(\cos\theta)$ is the second Legendre polynomial and β, the anisotropy parameter, can take on values between -1 and 2. As β varies from -1 to 2, the angular distribution varies from $\sin^2\theta$ through isotropic ($\beta = 0$) to $\cos^2\theta$. Classically, the anisotropy parameter is given by

$$\beta = 2P_2(\cos\chi)P_2(\cos\varepsilon) \tag{5}$$

where χ is the angle between the dipole transition moment and the internuclear axis and ε is the angle through which the molecular axis rotates from the time the molecule absorbs the photon until the fragments are at infinite separation. Thus, in the simplest case of a fast, direct dissociation of a diatomic molecule, β is equal to 2 if $\Delta\Lambda = 0$ (parallel transition) and to -1 if $\Delta\Lambda = \pm 1$ (perpendicular transition).

The relevant parameters in the quantum mechanical description are vibronic transition moments and upper-state phase shifts for the two or three allowed transitions from a given lower level ($J' = J'' + 1, J''$). Malegat (1980) has developed a semiclassical technique for calculating the anisotropy parameter which yields results in excellent agreement with the quantum mechanical values. The interest of these results is that they provide a practical method to determine the identity of the dissociative potential curve from the angular distributions.

In the case of a predissociation, however, the situation changes dramatically. Now the ΔJ of the transition is specified and the angular distribution turns out to depend on ΔJ rather than on $\Delta\Lambda$ (Zare, 1972). In the limit of large J', one obtains

$$I(\theta) = \frac{3}{16\pi}(1 + \cos^2\theta) \tag{6}$$

for the R- and P-branches, corresponding to $\beta = \frac{1}{2}$, and

$$I(\theta) = \frac{3}{8\pi}\sin^2\theta \tag{7}$$

for the Q-branches, corresponding to $\beta = -1$. A detailed treatment of this has been given by Pernot and coworkers (1979), where the experimental confirmation discussed below is also presented.

Using a photofragment spectrometer with high angular resolution, it is possible to determine the angular distribution from the kinetic energy distribution. The idea is qualitatively clear with reference to Fig. 13. These two spectra were taken for transitions that differed only in upper fine-structure level, and hence in J'. Thus the P_{21} transition corresponds to $\Delta J = -1$ and the Q_{11} to $\Delta J = 0$. Clearly the Q_{11} transition is more oriented along the ion beam direction (i.e. perpendicular to the laser polarization since the laser was necessarily parallel to the ion beam in the coaxial beams arrangement with which these data were obtained) than is the P_{21} transition. However, both transitions necessarily result in essentially the same separation energy W.

Spectra such as those of Fig. 13 could be fitted to spectra computed from a Monte Carlo calculation of ion trajectories in the apparatus such as the one

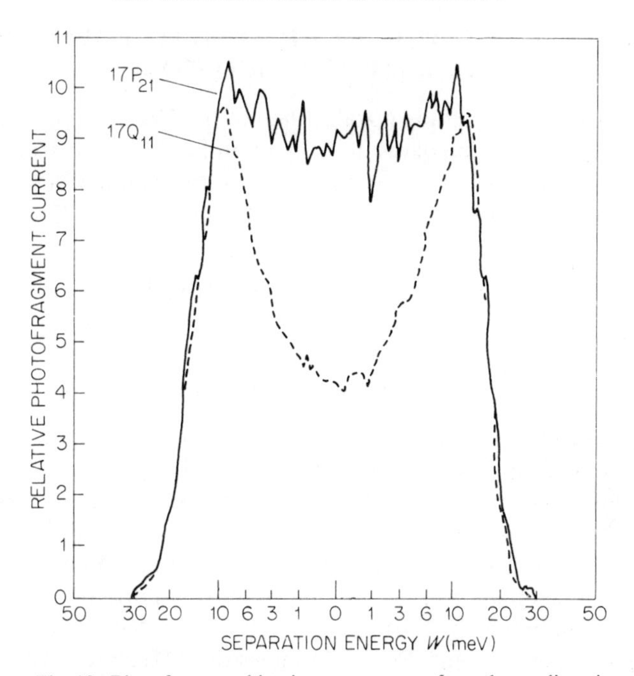

Fig. 13. Photofragment kinetic energy spectra from the predissociation of O_2^+ (b, $v = 4$) for two transitions which differ only in upper level J. The differences in the kinetic energy spectra imply very different angular distributions (Pernot *et al.*, 1979).

used for the calculation on H_2^+ shown in Fig. 3. As input this programme only requires W, β and the size and location of the apertures along the beam path. As expected, W determines primarily the width of the spectrum while β determines the shape around $W = 0$. Thus W and β are only weakly interdependent, and it was possible to determine β to an accuracy of ± 0.05. Over the range of vibrational levels sampled (N' values of 9 to 20, including normal and isotopic ions), it was found that $\beta(\Delta J = 1) = 0.55$, $\beta(\Delta J = 0) = -0.97$ and $\beta(\Delta J = -1) = 0.45$.

Recently, Broyer and coworkers (1981) have treated the problem of the angular distribution of photofragments in the density matrix formalism, and applied this to the evolution of the excited state under the influence of an external magnetic field. Beswick (1979) had previously predicted rather complex angular distributions from the individual M_J sublevels in a magnetic field, using the Zare formalism. The matrix density formalism introduces only slight modifications in Beswick's theory. Attempts by Broyer and coworkers (1981) to observe the predicted highly oscillatory behaviour were thwarted by an inability to resolve individual M_J levels. However, differences in the kinetic energy spectra for 'high' and 'low' M_J showed that this theoretical treatment is at least qualitatively correct.

V. PHOTODISSOCIATION OF CH^+ AND ITS FORMATION BY RADIATIVE ASSOCIATION

A recently completed study (Graff, 1983) of the CH^+ ion provides a good example of the application of results from photofragment spectroscopy to the determination of collision properties and of the half-collision aspects of photofragment spectroscopy. The collisional interaction of interest here is the radiative association reaction

$$C^+(^2P_{1/2}) + H(^2S) \rightarrow CH^+(A^1\Pi) \tag{8}$$

followed by

$$CH^+(A^1\Pi) \rightarrow CH^+(X^1\Sigma^+) + h\nu \tag{9}$$

The suspected importance of this reaction in the interstellar medium has prompted numerous theoretical attempts to calculate the rate coefficient (Abgrall, Guisti-Suzor and Roueff, 1976, and references therein). It is generally agreed that this coefficient is so small that direct experimental measurement cannot be used to check the soundness of the various theoretical approaches.

The mechanism that has been considered most important in enhancing the radiative association is the existence of quasi-bound shape resonances due to the rotation of the molecule. Figure 14 shows the $A^1\Pi$ state of CH^+ for $J = 1$ and $J = 32$. The actual determination of these potential curves is a part of the

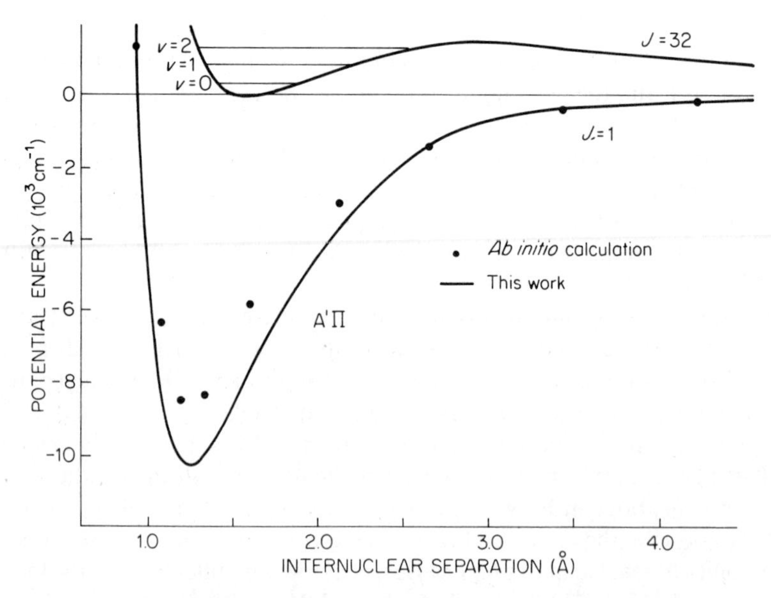

Fig. 14. The $A^1\pi$ potential curve of CH^+, drawn for $J = 1$ and $J = 32$, to illustrate the formation of quasi-bound states behind a rotational barrier (Helm *et al.*, 1982a).

work to be described below. The point to be made here is that for this $J = 32$ level there exist three vibrational levels above the dissociation limit behind the rotational barrier. Atoms approaching along this potential curve with the appropriate energy and angular momentum will 'see' these shape resonances, and thus have a substantially increased collision complex lifetime, increasing the probability of radiative stabilization. However, the number, energy and width of these resonances depends crucially on the depth of the $A^1\Pi$ well and on the long-range shape of the potential curve. None of these had been determined by the previous optical spectroscopic experiments on CH^+; nor had the shape resonances been observed.

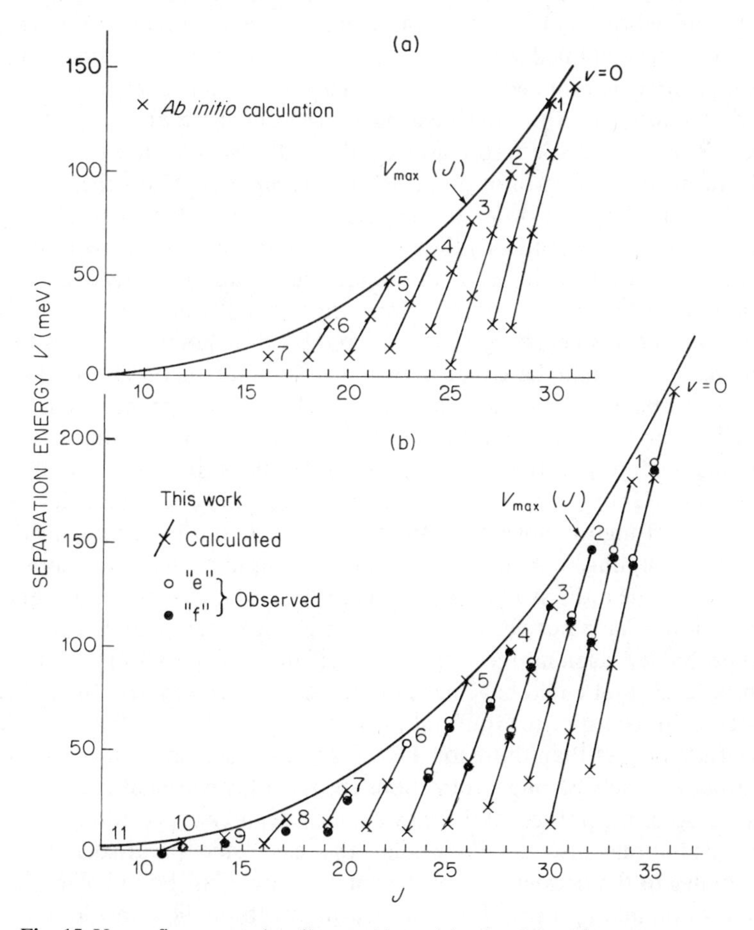

Fig. 15. Upper figure: rotationally quasi-bound levels of the $A^1\Pi$ state determined using a theoretical potential curve. Lower figure: quasi-bound levels determined using the experimental potential curve and comparison with the levels observed in photodissociation (Helm *et al.*, 1982a).

Ion photofragment spectroscopy provides an ideal way to observe these rotationally quasi-bound levels. It is possible to excite them from the ground $X^1\Sigma^+$ state and to observe the photofragment C^+ ions which result when the molecule dissociates by tunnelling through the indicated barrier. Near threshold photodissociation in CH^+, which may have been attributable to this mechanism, was first observed by Carrington and Sarre (1979) and was studied more extensively by Cosby, Helm and Moseley (1980). However, in neither case was it possible to actually assign the observed transitions. Subsequently, a much more extensive data set has been obtained and analysed by Helm and coworkers (1982a). They were able to identify thirty-two quasi-bound levels, grouping into Λ-doublet components for rotational quantum numbers $12 \leq J \leq 35$ and vibrational quantum numbers $0 \leq v \leq 10$. Figure 15 shows the location of the identified levels in energy, as well as levels calculated using the potential curve determined in this work. The dissociation energy of CH^+ was also determined. It is thus now possible to perform an accurate calculation of the effect of these shape resonances on the radiative association rate.

However, in the course of this work it became clear that other processes must be considered in order to accurately determine the association rate. Figure 16 shows schematically the potential curves that can be involved in this association, with the spin-orbit splitting in the C^+ atom greatly exaggerated. In the interstellar medium, the C^+ ions are essentially all in the lower $^2P_{1/2}$ state, and thus it is necessary to understand how the potential curves arising from this limit might be coupled to the $A^1\Pi$ state, in order to allow the radiative stabilization to take place. The $A^1\Pi$ curve determined as discussed above leads to the conclusion that there are twelve rovibrational levels which are bound with respect to the $^2P_{3/2}$ limit, but lie above the $^2P_{1/2}$ limit. A recent calculation (Graff *et al.*, 1983b) concludes that most of these levels are sufficiently strongly coupled to states dissociating to the lower limit, primarily by radial coupling, but also by rotational coupling, that they should be important in the radiative association. Unfortunately, transitions terminating in these levels have not yet been identified in the experimental spectra. The most probable reason for this is that the oscillator strengths for the transitions to these levels that could have been observed over the wavelength studied so far are, again according to Graff and coworkers, typically more than an order of magnitude weaker than those for the transitions observed to the rotationally quasi-bound levels. Stronger transitions exist in other wavelenth regions and a detailed experimental test of these calculations can be expected in the future. The calculations also conclude that the electronic predissociation also contributes to the dissociation of the rotationally quasi-bound levels, and is often the dominant process. Lifetime measurements on the transitions already observed could test this prediction.

Band, Freed and Kouri (Band and Freed, 1981; Band, Freed and Kouri, 1981) have recently presented a general theory describing the differential cross-

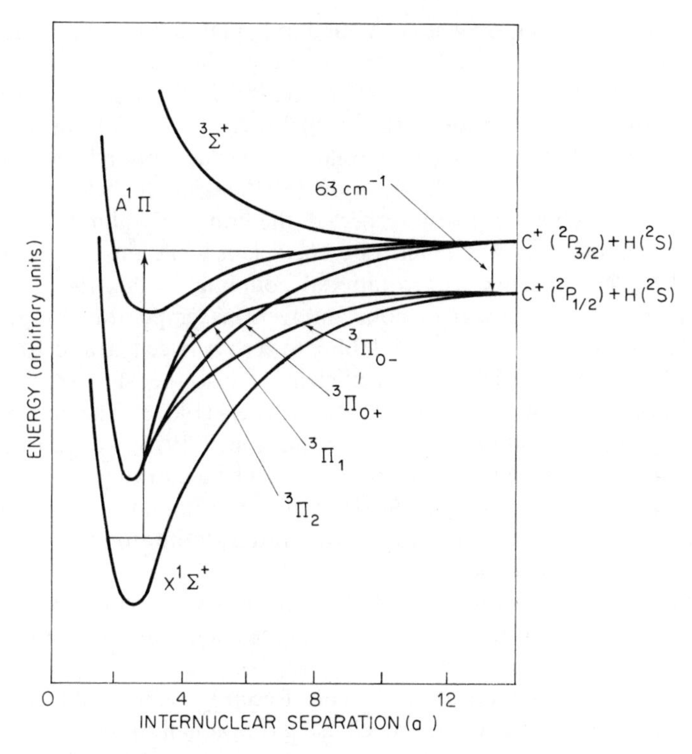

Fig. 16. Schematic of the lower potential curves of CH^+, with the spin-orbit splitting of the C^+ atom greatly exaggerated (Graff *et al.*, 1983b).

sections for the dissociation of diatomic molecules into atoms with non-vanishing electronic angular momenta. For the specific case of interest here, dissociation into $^2P_{3/2,1/2}$ and $^2S_{1/2}$ atoms, they show that non-adiabatic interactions such as those considered by Graff and coworkers (1983b) can have major effects on the angular distributions, and further that quantum interference effects should be observable in the total cross-sections. Experimental measurements to date on CH^+ do not include the direct dissociations and angular distributions necessary for a comparison with this elegant theory. However, such experiments should be possible, and CH^+ photodissociation is probably an excellent case for investigating these predictions experimentally.

The experimental and theoretical results of Graff and coworkers were used to calculate the rate coefficient for the formation of CH^+ by radiative association for the temperature range from 10 to 1000 K (Graff, Moseley and Roueff, 1983). The resonant contribution was determined in an inelastic scattering formalism that included the effects of non-adiabatic couplings for all quasi-bound levels, both the shape resonances above and the predissociated levels below the adiabatic dissociation limit of the $A^1\Pi$ radiating state. The non-resonant contribution was calculated in a quantum treatment that took into account the kinetic and electronic energy distributions predicted by current models of the diffuse interstellar clouds. The result of the calculation for the resonant contribution is shown in Fig. 17. The solid line shows the energy-dependent rate coefficient; the effect of the quasi-bound levels between the two dissociation limits is clearly evident in the structure below zero on the energy scale, which was set with respect to the upper $^2P_{3/2}$ limit. The dashed line is the temperature-dependent rate coefficient while the dot-dashed line shows the effect of neglecting rotational coupling in this rate. The non-resonant contribution is nearly a constant over this temperature range, with a value of about 12×10^{-18} cm^3 s^{-1}, falling to zero between 10 and 100 K. The total radiative association rate coefficient is just the sum of these two contributions and is given in Graff and coworkers (1983c). The best previous calculation, that of Abgrall, Guisti-Suzor and Roueff (1976), was constant with temperature at a value of 18×10^{-18} cm^2 s^{-1}. The major differences between these two calculations are the inclusion of the quasi-bound levels between the two dissociation limits and the proper accounting for the atomic fine structure states and collision energies.

Another significant result of the photofragment spectroscopy study of CH^+ (Helm *et al.*, 1982a) was the observation that the magnitude of the Λ-doubling observed for the rotationally quasi-bound levels could not be explained by the existing standard theoretical treatment. Recently, Helm and coworkers (1982b) have presented a new approach using rotationally adiabatic potentials which include the off-diagonal interactions between Born–Oppenheimer states which arise from the Coriolis Hamiltonian. The resulting calculation is in excellent agreement with the observed Λ-doubling, both at the low J values observed in emission and at the high J values observed in predissociation. It is

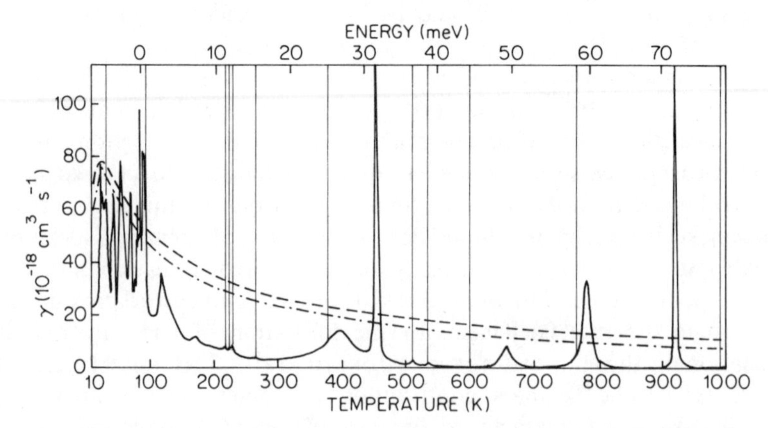

Fig. 17. Calculated resonant contribution to the radiative association rate for $C^+ + H \rightarrow CH^+ + h\nu$. The different curves are described in the text (Graff *et al.*, 1983c).

now possible to include the effect of Λ-doubling in the calculation of collisional processes.

Carrington and coworkers (1982) have observed the dissociation of CH^+ from levels near its dissociation limit using a CO_2 laser and the apparatus shown in Fig. 1. They find a dense spectrum of narrow lines, often appearing as doublets, which are consistent with earlier observation in ultraviolet transitions of Cosby, Helm and Moseley (1980). These transitions are so far unidentified and are very puzzling since they do not appear to fit, even qualitatively, into our present understanding of CH^+ near the lowest atomic threshold. If they in fact arise from levels near this lowest dissociation limit, those levels could further enhance the radiative association rate.

VI. OTHER DIATOMIC IONS

The three diatomic ions discussed in the preceding sections have been the subject of substantially the most detailed studies using techniques of ion photofragment spectroscopy. However, a number of other diatomic ions have been studied, and in some cases these studies illustrate problems not encountered in the above-discussed research. Some of these ions are discussed briefly in this section.

A. Rare Gas Dimer Ions

The first three excited states of the rare gas dimer ions are not bound, at least in the range of observed transitions. They have been extensively studied, both theoretically and experimentally, using techniques of collision spectroscopy. Interest in their optical absorptions arose during the development of excimer and rare gas halogen lasers, in which these ions are formed. Potential curves for Ar_2^+ which were determined using photofragment spectroscopy by Moseley and coworkers (1977) in a manner similar to the previously discussed work on H_2^+ and O_2^+ are shown in Fig. 18. A calculation of the photodissociation cross-section from the ground to the $^2\Pi_g$ state at 300 K is in excellent agreement in shape with the measurement of this cross-section using gas-phase drift tube techniques by Miller and coworkers (1976), but is too small by almost a factor of 3. Angular distributions were not consistent with the perpendicular nature that would be expected from these potential curves, but had both perpendicular and parallel components.

These potential curves do not, however, take into account the effect of the rather large spin-orbit splitting in the Ar^+ ion, which results in splitting each of the Π curves into closely spaced and nearly parallel 3/2 and 1/2 components. The result of this coupling is to both increase the magnitude of the cross-section to the $^2\Pi_g$ state and to impart a parallel component to the angular distribution. The effects of this spin-orbit interaction were even more dramatic

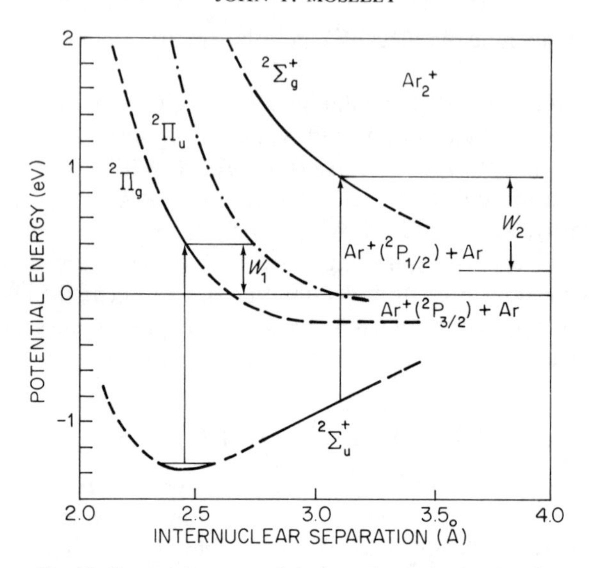

Fig. 18. Potential curves of Ar_2^+ as determined using ion photofragment spectroscopy (Moseley *et al.*, 1977).

in Kr_2^+ (Abouaf *et al.*, 1978), where the 1/2 and 3/2 components of the $^2\Pi_g$ state are well separated and can be viewed individually.

B. Alkali Dimer Ions

Although the photodissociation of Na_2^+ ions has been observed in the context of the two-photon ionization of Na_2 (Nitz *et al.*, 1979; Wagner and Isenor, 1982), this ion has not been studied to the extent of other ions discussed in this review. More recently, Helm and Moller (1983a, 1983b) have used a field ionization source to extract an intense fast beam of Cs_2^+ ions (as well as heavier clusters) from a liquid cesium surface, and begun studies of both direct and predissociative (Helm, Cosby and Huestis, 1983) processes. The lowest potential curves for this ion are similar to those for He_2^+—the Π states are missing for the same reason. However, several states correlating to the low-lying excited Cs atom dissociation limits can also be observed with visible photons, and this makes the spectrum more complicated than for the rare gas dimers.

Since the study of this ion is still in its initial stages, definitive conclusions have not been reached regarding many of the observations. We will therefore not discuss the results to date in detail. However, these results indicate that the study of all of the alkali dimer ions, including the mixed dimers, is likely to be very productive in the future.

C. Hydride Ions

Predissociations have been observed by fast ion beam photofragment spectroscopy in NH^+, PH^+, and SH^+ by Edwards, Maclean and Sarre (1982a). Using a single-mode krypton ion laser coaxial with the ion beam and velocity tuning, it was possible to resolve nuclear hyperfine structure in all of these ions. In all cases, the predissociation appeared to be sufficiently weak that the lines were not significantly lifetime broadened. A typical resolution of 40 MHz was attained.

The ion PH^+ has been studied in more detail using a tunable dye laser in both single-mode and multimode configurations (Edwards *et al.*, 1983). The (0, 1) and (1, 2) bands of the $A^2\Delta$–$X^2\Pi$ system were investigated, and improved molecular constants were obtained. The predissociation was tentatively attributed to the $^2\Sigma^-$ state, which has been predicted to cross the A state near $v' = 0$. At high resolution, most lines showed a fully resolved quartet hyperfine pattern, which arises from nuclear hyperfine interactions between the proton and ^{31}P nucleus, each of which have spin of one-half. A complete set of hyperfine parameters were obtained for both the X and A states.

Bound to rotationally quasi-bound vibration–rotation transitions have been observed in HeH^+ by Carrington and coworkers (1981), using the apparatus depicted in Fig. 1 and a single CO_2 laser. The transition is analogous to the second transition depicted in Fig. 3 for HD^+. As in the HD^+ case, a detailed study of these transitions will provide a severe test of theoretical descriptions of molecular states near dissociation limits.

D. Highly Excited NO^+

Although it plays a major role in atmospheric chemistry and is easily formed in ionized gases, very little spectroscopic information has been obtained on NO^+. Only the $A \rightarrow X$ band system has been rotationally analysed—most available information about higher excited states comes either from photoionization studies or from calculations (Albritton, Schmeltekopf and Zare, 1979). The long-lived and highly excited $a^3\Sigma^+$ state is believed to play a major role in many ionized gas environments, but has never been spectroscopically observed! In spite of attempts by both gas-phase and fast ion beam techniques, this situation remains. However, Cosby and Helm (1981) have observed photodissociation of more highly excited states of this ion, and have studied a predissociation identified as

$$NO^+(b^3\Sigma^-, v'' = 8 \text{ to } 10) \rightarrow NO^+(2^3\Pi, v' = 0 \text{ to } 19) \rightarrow N + O^+ \quad (10)$$

The $2^3\Pi$ state is highly perturbed, and molecular constants were not determined nor were rotational assignments made.

E. Doubly Ionized Nitrogen

The only double-charged ion studied to date by photofragment spectroscopy is N_2^{2+}. Cosby, Moller and Helm (1983) have studied five structured bands between 14 900 and 19 500 cm^{-1} in this molecule, which leads to $2N^+$ photofragments. The bands are attributed to absorptions from the $X^1\Sigma_g^+$ ($v = 0, 1, 2$) levels to unidentified levels of the $1^1\Pi_u$ state. The highest observed levels apparently predissociate by tunnelling through the Coulomb barrier, while the lowest one is predissociated by the $1^1\Sigma_u^-$ state. This is an interesting situation, since all of these states are in the $N^+ + N^+$ continuum, and thus all of the bound states are due to the formation of a barrier in the potential curve. The $X^1\Sigma_g^+$ ($v = 0$) levels is formed from a kinetic energy analysis of the photofragments to lie 4.8 ± 0.2 eV above the $N^+ + N^+$ dissociation limit.

VII. TRIATOMIC IONS

As soon as we move from diatomic to triatomic ions, the problem of understanding in detail the dissociation process becomes substantially more difficult, but, at the same time, more interesting. In addition to the fact that the spectroscopy itself is more complicated, one of the fragments in the dissociation is now a molecule, and therefore has vibrational and rotational modes which may be excited in the dissociation. A simple measurement of the fragment kinetic energy distribution is now unlikely to yield sufficient information to fully determine the system. Furthermore, the potential curves involved now are really surfaces, and are much less well known than the diatomic ion potential curves with which we have dealt so far.

In spite of these difficulties, considerable progress is being made, both experimentally and theoretically, although so far the two have not come together in a unified way to the extent that has been true for some of the diatomic molecules discussed above. Photofragment spectroscopy of triatomics can play a very important role in understanding simple low-energy reactive collisions of the type

$$AB + C \rightarrow (ABC)^* \rightarrow A + BC \tag{11}$$

since the collision complex (ABC)* will involve dissociative states of the type probed in photodissociation experiments.

A. Theory of Triatomic Photodissociation

The major fraction of the theoretical work on the photodissociation of molecules larger than diatomics has been performed with applications to specific neutral molecules in mind. It is far beyond the scope of this paper to review these contributions, since they have not yet been compared with any experimental results from ion photofragment spectroscopy. However, such comparisons will certainly be made in the near future, as more experimental results become available. A brief but up-to-date review of the theoretical situation can be found in a recent volume in this series (Leone, 1983). The work which is most in the spirit of this review is that of Freed, Band and Morse (Band and Freed, 1977; Morse, Band and Freed, 1983; Morse and Freed, 1983; Morse, Freed and Band, 1979). In this section we will discuss theoretical work only as it has been applied to experimental results on ion photofragment spectroscopy.

B. N_2O^+

The triatomic ion which has been the subject of the most experimental and theoretical research on its photodissociation to date is N_2O^+. Orth and Dunbar (1977) studied the cross-section for the channel

$$N_2O^+ + h \rightarrow NO^+ + N \tag{12}$$

between 2300 and 5000 Å, using an ICR cell and a mercury arc lamp. Thomas, Dale and Paulson (1977a) used a fast ion beam crossed with a doubled dye laser to record the spectrum between 3000 and 3420 Å with vibrational resolution, and identified the transitions as vibrational subbands of the

$$\tilde{X}^2\Pi(0, n'', m'') \quad \tilde{A}^2\Sigma^+(1, n', m') \tag{13}$$

transition. Larzilliere and coworkers (1980) and Abed and coworkers (1983) have studied some of these transitions in more detail using coaxial beams, velocity tuning and UV lines from a Kr^+ ion laser.

At the photon energies of these experiments, three dissociation channels are energetically available. These are ground state $NO^+(X^1\Sigma^+)$ with the N atom in either the ^4S state or ^2D state and ground state $O^+ + N_2$. Only NO^+ photofragments are observed; Larzilliere and coworkers conclude that the dissociation is due to spin-orbit coupling with the lowest $4\Sigma^-$ state, which correlates adiabatically with the $N(^4S)$ dissociation limit. The resolution of this study was about the same as that of normal emission spectroscopy, and it turns out that in this case the predissociation lifetime is sufficiently long that emission is also observed from these levels. A comparison of the spectra obtained by both techniques was found to simplify the analysis of the photodissociation work and to bring complementary information on the various unimolecular decay modes of excited N_2O^+.

Frey, Kakoschke and Schlag (1982) have also begun an investigation of the N_2O^+ predissociation. They also use a coaxial beams arrangement, but with a pulsed doubled dye laser, thus limiting the resolution to about 0.075 cm^{-1} in the best cases. However, they obtain a much wider range of tunability and have observed eight bands, some overlapped, in the region from 3175 to 3285 Å. A detailed study and analysis, including measurement of the photofragment kinetic energy, is underway.

Beswick and Horani (1981) have used a distorted wave half-collision model to calculate predissociation linewidths and final vibrational distributions for this predissociation channel. The calculations involve a parameterization chosen to fit experimental data for the (0, 0, 0) and (1, 0, 0) levels, and make predictions for higher levels. In the collinear model, the predissociation rate is given by

$$\gamma n_1 n_3 = \mathrm{Re} \sum_{v} \sum_{v'} \langle n_1, n_3 | v, \varepsilon \rangle F(v, v') \langle n_1, n_3 | v', \varepsilon \rangle \tag{14}$$

and the dissociation linewidth by

$$\Gamma_{n_1 n_3} = 2\Pi |X_{\mathrm{if}}|^2 \gamma_{n_1 n_3} \tag{15}$$

where n_1 and n_3 are the quantum numbers of the vibrational levels of the predissociated state, $|v, \varepsilon\rangle$ is the dissociative wave function immediately after the transition to the dissociative state and $F(v, v')$ is the operator for scattering on the repulsive potential surface, and introduces the effect of vibrational transitions in the BC fragment during the 'half-collision' process. If $\delta_{vv'} = 0$, i.e. all final state interactions are neglected, then Eq. (14) reduces to a simple 'Golden Rule' expression. In Eq. (15), X_{if} is the electronic non-adiabatic coupling between the initial and final states. The predictions of this calculation are in general, but not precise, agreement with fluorescence lifetimes and quantum yields from emission studies, but could be much better tested by linewidth measurements from photofragment spectroscopy. Unfortunately, such results are not yet available.

Abed and coworkers (1983) used a single-mode laser and a wide range of velocity tuning to study transitions in this system at high resolution. A spectrum of the $\tilde{X}^2\Pi_{3/2}(0, 0, 0) \rightarrow \tilde{A}^2\Sigma^+(1, 0, 0)$ transition is shown in Fig. 19, with spectroscopic identification. Note that the velocity tuning was performed by varying the beam energy from 15 to 100 keV. At higher resolution, all lines appear as doublets, with a regular fixed spacing. At still higher resolution, as shown in Fig. 20a, each component of the doublet is still further split, into two and four components each. This suggests that the $\tilde{A}$ state has an unusually large hyperfine structure.

Hyperfine interactions have seldom been studied in the optical spectra of triatomic molecules, let alone ionized ones. Thus a detailed study of this interaction, including experiments on isotopic species, was carried out. In the roughly 50-cm^{-1} region of Fig. 19, over 150 rotational transitions were studied at high resolution, resulting in the observation of about 900 hyperfine components. The Hamiltonian model of Zare and coworkers (1973) with appropriate modification to describe the hyperfine interaction was used to reduce the data to molecular constants. Figure 20b shows a calculated spectrum using the results of this fit.

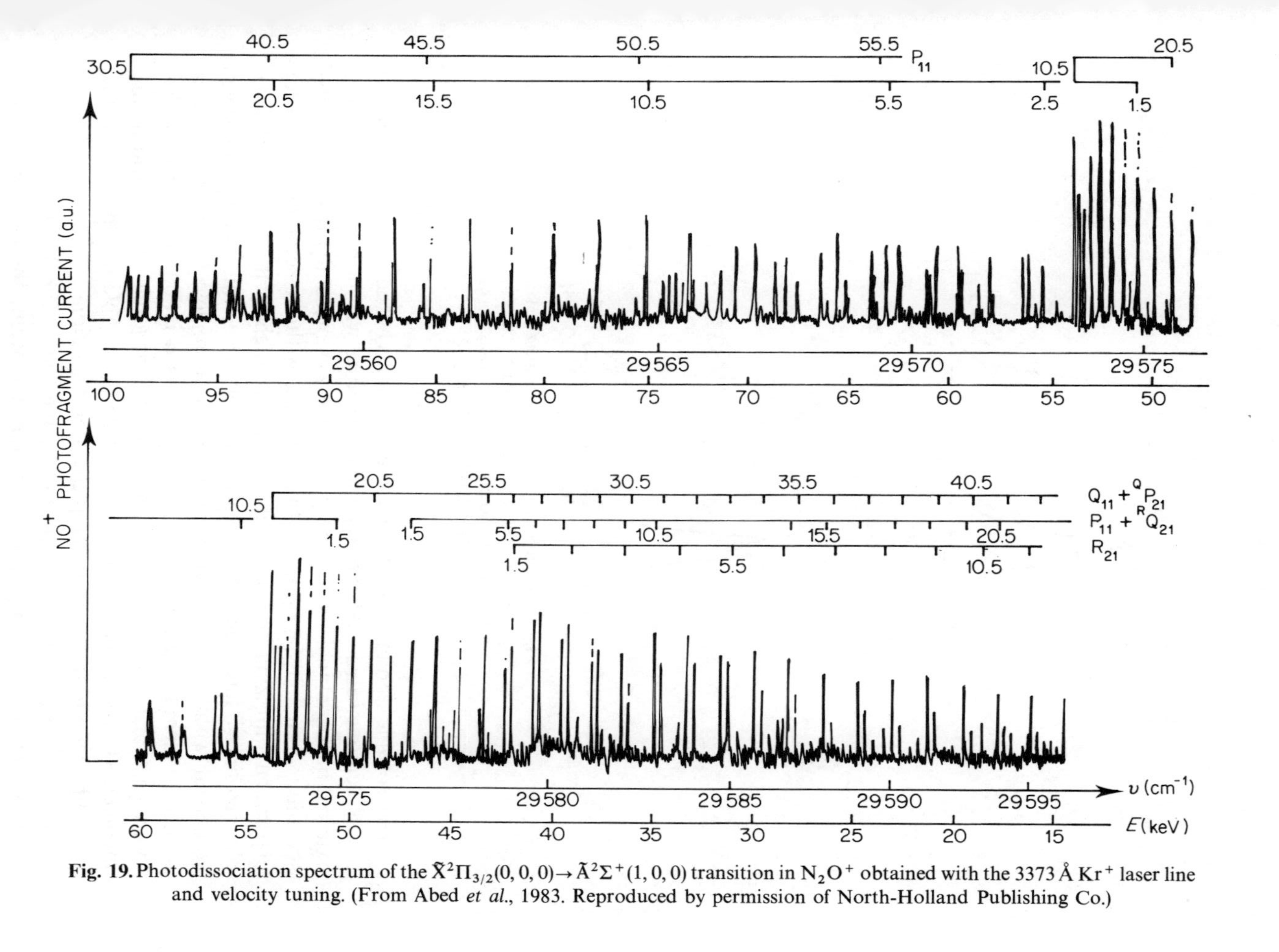

Fig. 19. Photodissociation spectrum of the $\tilde{X}^2\Pi_{3/2}(0,0,0) \rightarrow \tilde{A}^2\Sigma^+(1,0,0)$ transition in N_2O^+ obtained with the 3373 Å Kr^+ laser line and velocity tuning. (From Abed *et al.*, 1983. Reproduced by permission of North-Holland Publishing Co.)

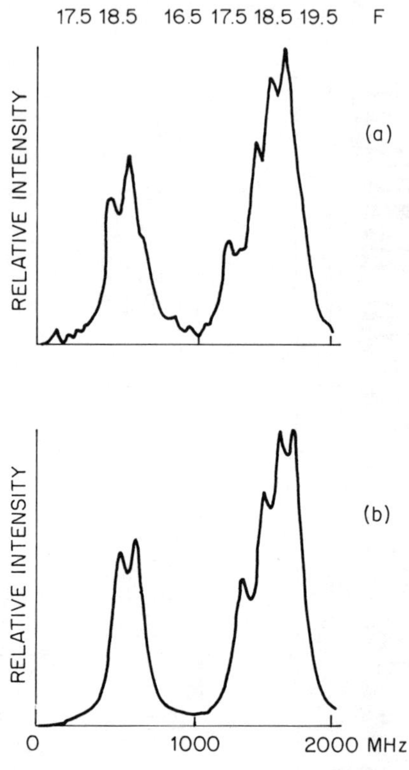

Fig. 20. Hyperfine structure in the spectrum of Fig. 19: (a) observed; (b) calculated. (From Abed *et al.*, 1983. Reproduced by permission of North-Holland Publishing Co.)

The isotopic studies led to two main conclusions (Abed *et al.*, 1982). Firstly, it was demonstrated that the hyperfine structure was primarily due to the external nitrogen nucleus. Secondly, evidence was presented that the predissociation occurs by ejection of this nucleus, without scrambling. Future experiments utilizing both isotopic species and higher vibrational levels are planned since it has been shown (Abed *et al.*, 1982) that this information, when combined with optical spectra, provides the complete set of information needed to identify the exact nature of the predissociation mechanism in this molecule. Furthermore, at the higher vibrational levels, the linewidths are predicted (Beswick and Horani, 1981) to be observable by this experiment, providing an additional valuable piece of information.

More recently, this same group (Lerme *et al.*, 1983) have added an electrostatic energy analyser to their coaxial beam photofragment spectrometer, and have measured the kinetic energy spectrum of NO^+ ions from the predissociation of specific vibrational and rotational levels of the A state. The NO^+ fragment vibrational distributions extracted from these spectra were very close to those observed by Brehm and coworkers (1974) from the photoionization of N_2O at 584 Å, and the shape of the spectrum agrees with that obtained by Thomas, Dale and Paulson (1977a). However, the distributions differ markedly from those found by Nenner and coworkers (1980) from the photoionization of N_2O using synchrotron radiation. The spectra also indicate significant rotational redistribution in the dissociation.

The calculation of Beswick and Horani (1981) predicted that NO^+ would be populated primarily in lower vibrational levels than were observed by Lerme and coworkers. However, more recent work (Komika, 1981) suggests that the dissociation process is more complicated than previously thought. Clearly, the N_2O^+ predissociation is far from well understood, but amenable to a very detailed investigation. Future work on this ion should be very productive.

C. Hydrogen Triatomic Ions

The simplest polyatomic molecule is the H_3^+ ion. It was discovered by J. J. Thomson in 1911 and has been subject to many investigations in the intervening years. However, these investigations were all either theoretical or involved $H^+ + H_2$ collisions, until the very recent and elegant spectroscopic work of Oka (1980) and of Shy and coworkers (1981). Both of these were infrared absorption experiments, and therefore not properly within the purview of this review. At about the same time, Carrington, Buttenshaw and Kennedy (1982) reported a remarkably complex infrared predissociation observed in H_3^+, resulting in $H^+ + H_2$ photofragments, using the apparatus of Fig. 1. The reported density of transitions, about 100 cm^{-1} and our lack of detailed knowledge of $H_3{}^+$ very near the dissociation limit made it difficult to predict whether or not even the states involved could be identified, let alone the individual transitions. However, the presumed relationship between the states involved in this dissociation and the collision complex important in $H^+ + H_2$ collisions made it important to pursue this investigation.

Subsequently, Carrington and Kennedy (1984) have measured the frequencies and intensities of more than 26 000 lines spanning only 222 cm^{-1} of the infrared region. Kinetic energy spectra indicate that both the ground and the predissociating state lie above the lowest $H^+ + H_2$ dissociation limit. The strongest transitions appear to cluster in groups which correlate with the positions of the $J = 3 \rightarrow 5$ transitions for $v = 0, 1, 2, 3$ of the ground state of the H_2 molecule. This has led to the development of a model in which the H_3^+ ions giving rise to the observed predissociation spectrum are considered as H_2–H^+ collision complexes, in which the vibrational and rotational quantum numbers of the H_2 partner are largely conserved. This model leads to reasonably quantitative agreement with the observations for these strongest transitions (which incidentally are only about 7 per cent. of the total observed) and can qualitatively explain the very high density of weak transitions.

Although the basis for this model, i.e. the H_2–H^+ complex, is very likely correct, the model involves many assumptions, and overall the experimental observations can only be said to be qualitatively explained at best. The challenge for theory is clear, particularly when one considers that this is only a tiny window into the infrared predissociation of this molecule, the information from photofragment kinetic energy analysis has been used only in a general way, and is not even yet available over much of the spectrum, and similar observations on the three isotopic species containing deuterium are just beginning to become available. It is exicting to speculate that future study of the infrared predissociation of this molecule will lead us to a detailed understanding of that most basic molecular collision, $H^+ + H_2$, and to a better understanding in general of triatomic molecules near their dissociation limits.

D. The Ozone Positive Ion

Prior to the initiation of photodissociation experiments, our knowledge of the ozonide ion O_3^+ came almost entirely from photoionization studies and from theoretical calculations. Vestal and Mauclaire (1977a) observed two broad structures in the photodissociation of this ion into $O^+ + O_2$, one peaked at about 6000 Å and the other at about 4000 Å. A more detailed study near the low-energy threshold revealed that the onset of the photodissociation was about 1.85 eV. Figure 21 shows an energy level diagram for O_3^+, as well as some related ozone energy levels, as given by Moseley, Ozenne and Cosby (1981). These are all fairly well-established values, obtained prior to any photofragment spectroscopy work, as discussed in the references. From this figure, the expected threshold for dissociation into O^+, in the absence of any excitation in the initial O_3^+, is 2.165 eV. Vestal and Mauclaire (1977a) tentatively attributed their lower energy threshold to the presence of O_3^+ in the 1^2B_2 state in their ion beam.

Moseley, Ozenne and Cosby (1981) used photofragment energy distributions of the O^+ fragments obtained over the wavelength range from 4579 to 7525 Å to study the energetics of this dissociation. They saw no evidence for population of the $^2B_2$ state and were able to interpret their results based on the presence of vibrational excitation of the ozonide ions. They also reported spacings between the first three symmetric stretch vibrational levels of the ion, as well as energy partition between rotation and vibration in the dissociation. The results were entirely consistent with the energy levels of Fig. 21. The energy partition measurements could provide a good test of O_3^+ potential surfaces involved in the dissociation, but such a comparison has not yet been made.

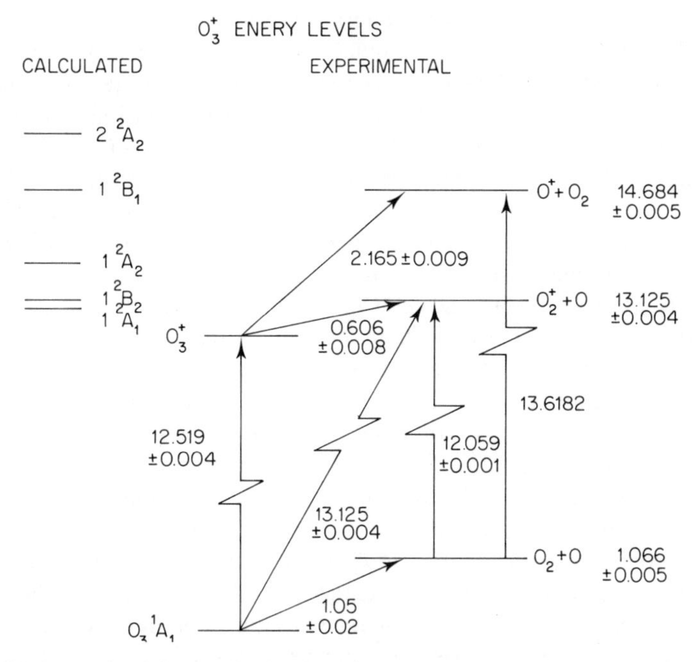

Fig. 21. Energy level diagram for O_3 and O_3^+, using generally accepted values (Moseley, Ozenne and Cosby, 1981).

Hiller and Vestal (1982) have reexamined the threshold region at higher resolution and over a wider wavelength range, and found results similar to, but not identical with, those of Vestal and Mauclaire. The results can be interpreted consistently with those of Moseley, Ozenne and Cosby, and hence with the energies shown in Fig. 21, under the assumption of vibrational excitation in the beam. However, they were interpreted under the assumption of no vibrational excitation, and hence led to an implied upper limit on the bond dissociation energy of neutral ozone of 0.761 eV, about 0.3 eV lower than the accepted value. If this interpretation can be supported, it is an important one, with broad implications in the chemistry and energetics of ozone. Unfortunately, the less exciting interpretation seems the most likely, for the reasons discussed below.

The assumption of virtually no vibrational excitation is almost unprecedented in research using fast molecular ion beams; indeed, evidence is found for such excitation in almost every such experiment equipped to detect it, including the one of Hiller and Vestal (Hiller and Vestal, 1980, 1981; Vestal and Mauclaire, 1977a, 1977b). The statement in Hiller and Vestal (1982) that reconciliation of their results with the accepted bond energy for ozone would require that 'most of our O_3^+ beam be vibrationally excited by more than 0.3 eV is not supported by Fig. 1 of that paper; none of the O_3^+ need be excited above 0.3 eV to allow the alternative interpretation, and indeed only about 10 per cent. of it need be excited above 0.25 eV. Furthermore, in the experiment of Moseley, Ozenne and Cosby, vibrational excitation is observed up to 0.6 eV above the ground level, with no higher excitation detected. This is consistent with the fact that O_3^+ dissociates into $O_2^+ + O$ at this energy (see Fig. 21). Hiller and Vestal's reinterpretation of the results of Moseley, Ozenne and Cosby does not take this observation into account, and further requires that most of the O_3^+ in this experiment has one quanta of excitation in the first bending mode only. This seems unlikely.

Finally, as Moseley, Ozenne and Cosby point out, the accepted value of the bond energy of ozone does not stand only on the calorimetric measurements, but is also given from the combination of the ionization potential of oxygen and the threshold for photoionization of ozone into $O_2^+ + O$. As shown in Fig. 21, these two values are in excellent agreement.

All of this argument does not disprove the interpretation of Hiller and Vestal, and further work on this ion is probably warranted to definitively clarify the situation. One possible experiment would be to measure the photodissociation cross-section of O_3^+ in the threshold region in a gas-phase experiment where the vibrational excitation could be more controlled. Until further independent evidence is presented in support of the lower value for the bond energy of ozone, the best value for use is probably the one given from photoionization, $1.066 + 0.005$ eV.

E. Other Studies of Triatomic Ions

More limited studies have been carried out on a number of other triatomic ions. All of this work indicates that more detailed investigations are likely to be productive in a manner similar to the studies of N_2O^+ and H_3^+ discussed above.

Orth and Dunbar (1980) measured total photodissociation cross-sections for COS^+ and CS_2^+ in the ultraviolet using an ICR spectrometer and an arc lamp. In the case of COS^+ a quite substantial cross-section was observed for dissociation into $S^+ + CO$. For CS_2^+, a much smaller cross-section was found, again with S^+ the observed photofragment. The cross-section for COS^+ shows a peak around 3000 Å, a region which Frey, Kakoschke and Schlag (1982) have shown can be investigated in detail.

Edwards, Maclean and Sarre (1982b) have observed transitions to predissociated levels of $H_2S^+(A^2A_1)$, which result in S^+ photofragments, using Kr^+ ion laser lines around 4100 Å. A high density of very narrow lines were observed, indicating that this will be a very interesting molecule to study using tunable radiation in this region. Edwards and coworkers (1983) have also observed the spectrum of PH_2^+ using a tunable laser over the range from 4200 to 6700 Å. Dissociation into $P^+ + H_2$ was attributed to $X^1A_1 \rightarrow A^1B_1$

transitions, while dissociation into $PH^+ + H$ was attributed to $A^1B_1 \rightarrow B^1A_1$ transitions. An analysis of the observed structure is in progress.

Detailed photofragment studies of triatomic ions are truly in their infancy. The brief studies mentioned above, as well as several other privately reported observations, make it clear that this should be a productive area for many years.

VIII. LARGER POLYATOMIC IONS

A. Cluster Ions

Cluster molecules may be considered as molecules composed of two identifiable partners held together by a weak bond. In the case of ions, clusters form easily, since the neutral partner may be attracted to the ionized one by the polarization force between the two. Cluster ions are important in the atmosphere, and since they are weakly bound, photodissociation by sunlight can be an important loss mechanism. Furthermore, they can often be thought of to a certain approximation as quasi-diatomic molecules, since the original cluster partners largely retain their identity in the cluster.

Based on this description, one might expect two types of absorption in these ions. One type would be characteristic of the absorptions in the partner molecules and might lead to dissociation if a sufficient fraction of the absorbed energy can be transferred to the cluster bond. The other type would be characteristic of the cluster molecule itself–perhaps similar to absorptions in diatomic molecules.

An important difference exists between clusters of the same molecule (homomolecular clusters) and those containing different molecules (heteromolecular clusters). With the homomolecular clusters, electron sharing even in the case of a weak bond will be essentially complete, while with the heteromolecular clusters the partner of lower ionization potential will, in the ground state, be more positively charged than the other partner. Thus homomolecular clusters will be true chemically bonded molecules, while heteromolecular clusters will tend more to be a complex held together by electrostatic attraction due to polarization.

To date, no evidence has been found in positive ion clusters for the first type of dissociation. For example, Miller, Heidrich and Moseley (1975) measured the photodissociation cross-section of N_4^+ over the range of 5700 to 6700 Å, which covers a large number of transitions in the N_2^+ $X \rightarrow A$ system. A smoothly varying photodissociation characteristic of a transition to a repulsive state was observed, with no evidence of $X \rightarrow A$ transitions.

Beyer and Vanderhoff (1976) and Smith and Lee (1978) have measured the photodissociation cross-section for O_4^+ over a total range from 4000 to 8500 Å. This cross-section shows a broad-peak characteristic of a direct dissociation to a repulsive state. Smith and Lee were able to fit this cross-section with reasonable parameters by assuming a diatomic model. Similar results were found for $NONO^+$, although this cross-section did show an apparent double-peaked structure near its maximum, and for $CO_2CO_2^+$. To a first approximation at least, all of these ions have direct dissociations that are similar to those common in homonuclear diatomic ions.

Bowers, Illies and Jarrold (1983) are using a fast ion beam photofragment spectrometer to investigate the dissociation of $NONO^+$ and of the related ion, $N_2O_2^+$, in more detail. In preliminary work, they found, as did Smith and Lee, that the $NONO^+$ ion dissociates facilely at visible wavelengths. However, the $N_2O_2^+$ is not observed to photodissociate at 5145 Å. This is consistent with the suggestion of Beyer and Vanderhoff (1976) that in a cluster molecule AB^+, where the molecule A has the lower ionization potential, one can apparently characterize the lowest photodissociation as a transition between the weak electrostatically bound A^+—B ground state and an excited charge transfer state A—B^+. The upper state will probably be repulsive or only weakly bound. This situation apparently applies to all heteromolecular clusters observed so far: $N_2O_2^+$, $CO_2O_2^+$ and $O_2H_2O^+$ (Smith and Lee, 1978).

B. Methyl Iodide Ions

The photofragmentation of CH_3I^+ was first observed by McGilvery and Morrison (1977), using a triple quadrupole arrangement and coaxial beams. They attributed their spectrum to the following process:

$$CH_3I^+(\tilde{X}^2E_{3/2,1/2}) + h\nu \rightarrow CH_3I^+(\tilde{A}^2E_{1/2}) \rightarrow CH_3^+ + I(^2P^0_{3/2}) \qquad (16)$$

Subsequently, Goss, Morrison and Smith (1981a, 1981b) investigated these transitions in more detail, assigning the major transitions and determining vibrational constants. The spectrum is very congested, however, and the presence of high rotational and vibrational excitation further complicated the analysis.

In order to simplify the spectrum and obtain higher resolution, Chupka and coworkers (1983) have developed a unique new technique for ion photofragmentation studies. With this technique the ions are formed by multiphoton ionization in a molecular beam and subsequently photodissociated by a second tunable laser. The photofragmentation pattern is observed using a time-of-flight mass spectrometer. This technique has been applied to CH_3I^+, but so far only over a very limited part of the wavelength range studied by Goss, Morrison and Smith. A substantial simplification of the spectrum is indeed obtained, and significant differences with both the vibrational constants and numbering of Goss, Morrison and Smith are found. However, more work is required to clarify this situation and to obtain new molecular constants.

A high-resolution study of the photodissociation of this ion was performed using a single-mode coaxial laser by Cosby (1980). A wealth of rotational structure was resolved, which is so far unassigned. It was also determined that the predissociation is slow, with a lifetime of about 4 ns, and that most of the excess energy in the dissociation goes into excitation of the fragments, with typically only a few millielectronvolts of kinetic energy observed.

Recently, Tadjeddine and coworkers (1982) have investigated the angular distribution of the CH_3^+ photofragments at three wavelengths near 5900 Å. At all three wavelengths, an anisotropy parameter $\beta = 0.12 \pm 0.03$ was found, indicating a transition with both parallel and perpendicular components. This is perfectly consistent with the transition given in Eq. (16), but leaves the coupling to the dissociation continuum unexplained. In agreement with Cosby (1980), a kinetic energy release of less than 20 meV was found at all wavelengths studied. This paper presents a detailed discussion of the excited state structure of this ion and the various possible interactions, which should be of use in future, more detailed, research on this photodissociation.

The related ions CH_3Cl^+ and CH_3Br^+ have also been observed to photodissociate between 3000 and 6000 Å by Vestal and Futrell (1974) and by Dunbar (1971). Although only the general shape of the cross-section has been established so far (and there are indications that the lifetimes of the dissociating states are short, leading to diffuse spectra), these ions may be of interest for future studies.

C. Large Molecular Ions

Molecular ions larger than the ones discussed so far in this review have been studied in photodissociation, and results up to 1978 are covered in the review by Dunbar (1979). For obvious reasons, these studies do not yet go into the details of fragmentation dynamics and, indeed, most of them have been performed using ICR techniques which are not well suited for studying the dynamics of the process. Recently, Harris and coworkers (1981) have reported the construction of a large double-focusing spectrometer having both electric and magnetic sectors, which is being devoted to the study of the photodissociation of organic ions. Relative photoabsorption cross-sections at Ar^+ ion laser wavelengths, as well as fragmentation patterns, have been obtained for several ions (see also Kingston *et al.*, 1982). Significant differences are observed between the photofragmentation and the unimolecular fragmentation patterns.

IX. NEGATIVE IONS

Photofragmentation studies of significant detail have been carried out on only two negative ions, O_3^- and CO_3^-. Investigations of both of these ions have proved to be very interesting, and indeed have generated some controversy in the literature which is still not entirely settled.

A. The CO_3^- Ion

The photodissociation of CO_3^- into $O^- + CO_2$ was first reported by Moseley, Bennett and Peterson (1974), with absolute cross-section measurements of Ar^+ ion laser wavelengths, using a drift tube mass spectrometer. Subsequently, Cosby and Moseley (1975) discovered a structured cross-section characteristic of a predissociation around 6000 Å and analysed this band to obtain vibrational constants and to set an upper limit on the bond energy of this ion of about 1.9 eV (Moseley *et al.*, 1976a).

Later, Dotan and coworkers (1977) used ion–molecule reaction rate equilibrium measurements to establish that the bond energy of CO_3^- exceeded that of O_3^- by at least 0.58 eV. Given accepted values of the bond energy of ozone and the electron affinities of ozone and atomic oxygen, this result implies that the bond energy of CO_3^- must exceed 2.27 eV, in clear contradiction to the result stated above. Dotan and coworkers stated: 'If the present and photodissociation limits for $D(CO_2 + O^-)$ are to agree, then $EA(O_3) = 1.70$ eV. The resolution of this discrepancy clearly requires additional information.'

Such information was forthcoming. Vestal and Mauclaire (1977) studied this region of the photodissociation of CO_3^- with the triple quadrupole system and found a similar structure to that reported by Moseley and coworkers but a much smaller absolute cross-section, whose magnitude depended on the ion source pressure. They tentatively attributed the photodissociation around 6000 Å to the presence of an electronically excited state which could serve as the origin for the observed transitions. Smith, Lee and Moseley (1979) were able to show, however, that the CO_3^- ions studied in the drift tube experiment were thoroughly relaxed prior to laser interaction. They also presented evidence that the lifetime of the excited state involved in the transition was relatively long—of the order of 5×10^{-7} s.

Subsequently, Hiller and Vestal (1980) made a more extensive study, covering the range from 4000 to 6350 Å. They found results consistent with those of Moseley and coworkers at wavelengths below 5000 Å, but again found much smaller cross-sections at the longer wavelengths. In addition, they found evidence for a two-photon dissociation process at the longer wavelengths, which saturated at remarkably low laser power. They concluded that 'photodissociation of CO_3^- observed at wavelengths longer than 551 nm (2.25 eV) is probably due to two photon processes in the results of Moseley and coworkers and contains contributions from both two photon and collision induced dissociation in our present and earlier results'. A corollary to this conclusion is that the dissociation energy of this ion may be as large as 2.25 eV, in excellent agreement with the thermochemical evidence of Dotan and coworkers (1977).

Subsequently, Moseley and coworkers (1980) confirmed the presence of a two-photon dissociation in the drift tube experiments, and found that this process saturates at about 10 W of laser power, as shown in Fig. 22. The most likely explanation for the overall process is the excitation to a long-lived excited state just below the dissociation threshold, with a wavelength dependence and magnitude as given by the reported photodissociation cross-section of Moseley and coworkers (1976a), followed by a transition from this state to a dissociative state. This second transition must have a cross-section of the order of $10^{-16}\ cm^2$ in order to saturate at the low laser power observed.

Moseley and coworkers (1980) also observed that the total photodestruction of CO_3^- remained linear with laser power down to about 3 W, as shown in Fig. 22. This additional destruction is almost certainly photodetachment, which must also be a two-photon process since the electron affinity of CO_3 is

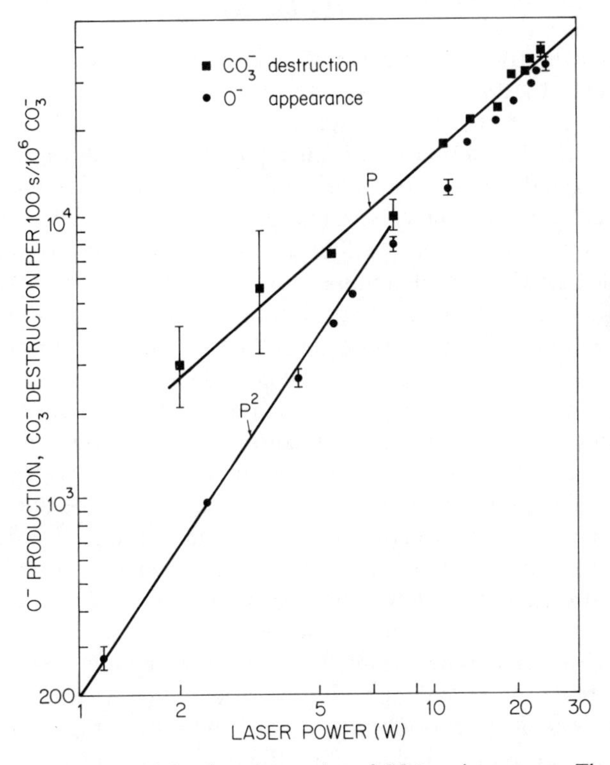

Fig. 22. Dependence of the photodestruction of CO_3^- on laser power. The curve marked O^- represents only photodissociation into this ion; the other curve represents the total photodestruction (Moseley *et al.*, 1980).

about 3 eV. If this conclusion is correct, the second photodetachment step must have an even larger cross-section than the second photodissociation step.

Recently, Hunton and coworkers (1983) have confirmed the existence of this first excited state and found evidence that its lifetime is in the microsecond range. They further clearly observe the dissociation of this state both by the absorption of a second photon and by collisional dissociation, as suggested by Hiller and Vestal. However, photodetachment was apparently not observed. A more detailed manuscript describing these studies as well as studies on the hydrates of CO_3^- is in preparation by these authors.

Although there remains much to be learned about the excited state structure of this ion and the details of its dissociation and detachment processes, at least it can be said that the disagreement about its bond energy has been resolved in favour of the higher value. This is an important result, since that disagreement brought into question the validity of a number of other important thermodynamic properties of O_3, CO_3 and their ions.

B. The O_3^- Ion

The first definitive measurements of the photodissociation of O_3^- were reported by Cosby and coworkers (1975), using a drift tube mass spectrometer. A diffuse but structured cross-section was observed for the dissociation into $O^- + O_2$, which decreased towards an apparent threshold near 2 eV. Similar results were found in an ion beam apparatus by Vestal and Mauclaire (1977).

Subsequently, Cosby and coworkers (1978) made a more detailed study of the photodissociation of this ion, but due to the diffuseness of the structure were unable to definitively determine vibrational constants—two alternative interpretations were proposed. A study of the dependence of the cross-section at various wavelengths on the number of thermalizing collisions between the O_3^- ions and the background O_2 gas indicated strongly that all dissociation at wavelengths of 6050 Å (2.05 eV) and longer originated in vibrationally excited states. The origin of the observed transition was found to be 2.146 or 2.163 eV, depending on which set of molecular constants was chosen.

Although this photodissociation is structured, it may be due to a direct dissociation process. It has all of the characteristics of the process described by Pack (1976) for an excited state potential surface which is bound along its symmetric stretch and bending coordinates but is directly dissociative along the asymmetric stretch coordinate. The dissociation time is sufficiently long that the molecule undergoes symmetric stretch and bending vibrations during the dissociation process, leading to the observed diffuse structure. An ion beam study of the kinetic energy and angular distributions of the photofragments would be helpful in further elucidating the dissociation process.

Later, Hiller and Vestal (1981) made photodissociation measurements on this ion in a beam apparatus which are in excellent agreement with those of Cosby and coworkers (1978) and extend the range of measurements out to

3900 Å. In addition, the dissociation into the $O_2^- + O$ channel was observed, beginning at about 2.4 eV. This is a very extensive set of measurements, but their detailed interpretation is far from clear. Hiller and Vestal (1981) chose as the origin of the observed transition a weak feature near 2.05 eV, a region where Cosby and coworkers (1978) had reported the observation of vibrational excitation. Due to this choice for the origin, they obtain somewhat different values for the vibrational constants than did Cosby and coworkers.

Perhaps more importantly, they chose the observed threshold for the production of O_2^- at 2.41 eV as the true bond energy for dissociation into $O + O_2^-$. This value, when combined with the accepted values for the electron affinities of molecular oxygen and of ozone (Novick *et al.*, 1979), yields a value for the bond energy of ozone of 0.747 eV, in agreement with proposed value for this bond energy energy based on work on O_3^+ (Hiller and Vestal, 1982), but again in disagreement with the accepted value of 1.05 eV or so.

Again, the more prosaic explanation of vibrational excitation in the O_3^- beam could be used to bring all thermodynamic cycles involving ozone and its ions into agreement. If the conclusions of Cosby and coworkers (1978) are correct, the data of Hiller and Vestal show evidence for such excitation. Both sets of authors agree that the magnitude of the cross-section at the peak near 2.15 eV is 1×10^{-18} cm^2, and further agree well on the magnitude of the cross-section at higher energies. However, for the structure at 2.05 eV Hiller and Vestal report a cross-section of 2×10^{-19} cm^2 while Cosby and coworkers report a cross-section which has this value at the minimum drift distance used, 5.1 cm, but which decreases to a value of 1×10^{-19} cm^2 at 30.5 cm drift distance. They interpret this decrease to reflect a decrease in vibrational excitation in the ion due to the increased number of thermalizing collisions. If this conclusion is valid, then the O_3^- beam of Hiller and Vestal has up to 0.3 eV of vibrational excitation, and thus the reported threshold for dissociation into $O_2^- + O$ of 2.41 eV may be 0.3 eV too low. In fact, the data show a significant increase in the O_2^- channel beginning just at 2.7 eV, which could correspond to the true thermodynamic threshold for this dissociation.

Again, it is not possible to rigorously disprove the interpretation of Hiller and Vestal. However, the preponderance of the evidence is that the commonly accepted values for the various energy levels of ozone and its ions, as given in Fig. 21, along with the electron affinity of ozone of 2.1028(25) as given by Novick and coworkers (1979), are much more likely to be the correct ones than those given by Hiller and Vestal (1982, Table I).

X. CONCLUSIONS AND FUTURE DIRECTIONS

The study of ion photofragment spectroscopy is now about a dozen years old and has demonstrated the capability to obtain very high resolution spectroscopic measurements as well as to study the dynamics of the dissociation process in considerable detail. The most complete studies have been made on diatomic ions, and work currently under way in several laboratories indicates that this will continue to be an active and productive area of research. Experimental and theoretical results to date lead to the anticipation that further work will result in a much better understanding of that intermediate region where neither the atomic nor the standard molecular description applies particularly well to the system.

Results of similar detail and specificity on triatomic ions are just beginning to appear, and the theoretical description of the dynamics is much more difficult and less advanced. However, this is a key area of research towards a better understanding of simple reactions. The 'half-collision' approach should prove particularly useful in this regard. The anticipated development of tunable, single-mode lasers in the ultraviolet and of sources of more relaxed ions will open up many important species to study, as well as simplify the experimental spectra.

Larger ions will probably continue to be studied in a more general way for the near future. Measurements of total photodissociation cross-sections, fragmentation channels and thermodynamic properties, as well as the application of simple models (such as the quasi-diatomic model), will lead the way eventually to more detailed studies. However, the complexity of the spectra at high resolution will make interpretation very difficult.

Negative ions have been largely neglected and are deserving of more study. These ions usually do not have bound states below the photodetachment threshold, but many will undergo direct photodissociation processes.

Clearly, the field is vigorous and growing. About half of the references in this review, and most of the key ones, have appeared since 1980. Researchers continue to contribute dramatic new ideas, opening up new areas to study. Only perhaps a half-dozen molecular ions have really been studied in much detail using these techniques, while preliminary results show that a much larger number are amenable to such studies. Researchers who have contributed to this new and exciting way to investigate molecular ions are to be congratulated and encouraged to continue.

Acknowledgements

The author has benefited tremendously from his association over the past eight years with Prof. Jean Durup and Dr. Philip C. Cosby, as well as with others too numerous to list here but who can be found as coauthors on papers in the reference list. Special thanks are also due Prof. Alan Carrington for communicating his H_3^+ results prior to publication, as well as his review article on H_2^+. Much of the research in which the author participated was supported by NSF Chemical Physics and a part of the time required to prepare this article was supported by NSF Grant No. CHE 79-18074.

References

Abed, S., Broyer, M., Carre, M., Gaillard, M. L., Larzilliere, M. (1982). *Phys. Rev. Lett.*, **49**, 120 (1982).

Abed, S., Broyer, M., Carre, M., Gaillard, M. L., Larzilliere, M. (1983). *Chem. Phys.*, **74**, 97.

Abgrall, H., Guisti-Suzor, A., and Roueff, E. (1976). *Astrophys. J.*, **27**, L69.

Abouaf, R., Huber, B. A., Cosby, P. C., Saxon R. P., and Moseley, J. T. (1978). *J. Chem. Phys.*, **68**, 2406.

Adams, N. G., and Smith, D. (1975). *Inst. J. Mass. Spec. Ion. Phys.*, **21**, 349.

Albritton, D. L., Schmeltekopf, A. L., Harrop, W. J., Zare, R. N., and Czarny, J. (1977). *J. Mol. Spectrosc.*, **67**, 157.

Albritton, D. L., Schmeltekopf, A. L., and Zare, R. N. (1979). *J. Chem. Phys.*, **71**, 3271.

Band, Y. B., and Freed, K. F. (1977). *J. Chem. Phys.*, **67**, 1462.

Band, Y. B., and Freed, K. F. (1981). *Chem. Phys. Lett.*, **79**, 238.

Band, Y. B., Freed, K. F., and Kouri, D. J. (1981). *Chem. Phys. Lett.*, **79**, 233.

Beebe, N. H., Thurlstrup, E. W., and Andersen, A. (1976). *J. Chem. Phys.*, **64**, 2080.

Bersohn, R., and Lin, (1969). *Adv. Chem. Phys.*, **16**, 67.

Beswick, J. A. (1979). *Chem. Phys.*, **42**, 191.

Beswick, J. A., and Horani, M. (1981). *Chem. Phys. Lett.*, **78**, 4.

Beyer, R. A., and Vanderhoff, J. A. (1976). *J. Chem. Phys.*, **65**, 2313.

Bowers, M. T., Illies, A. J., and Jarrold, M. F. (1983). Submitted for publication.

Brehm, B., Frey, R., Kustler, A., and Eland, J. H. A. (1974). *Int. J. Mass Spec. Ion Phys.*, **13**, 251.

Brown, J. M., Milton, D. J., Watson, J. K. G., Zare, R. N., Albritton, D. L., Horani, M., and Rostas, J. (1981). *J. Mol. Spectrosc.*, **90**, 139.

Broyer, M., Larzilliere, M., Carre, M., Gaillard, M. L., Velghe, M., and Ozenne, J. B. (1981). *Chem. Phys.*, **63**, 445.

Carre, M., Druetta, M., Gaillard, M. L., Bukow, H. H., Horani, M., Roche, A. L., and Velghe, M. (1980). *Mol. Phys.*, **40**, 1453.

Carrington, A., and Buttenshaw, J. (1981a). *Molec. Phys.*, **44**, 267.

Carrington, A., and Buttenshaw, J. (1981b). *J. Phys. Paris*, **40**, C1.

Carrington, A., Buttenshaw, J., and Kennedy, R. (1982). *Mol. Phys.*, **45**, 753.

Carrington, A., Buttenshaw, J., and Kennedy, A. (1983). *Mol. Phys.*, **48**, 775.

Carrington, A., Buttenshaw, J., Kennedy, R. A., and Softley, T. P. (1981). *Mol. Phys.*, **44**, 1233.

Carrington, A., Buttenshaw, J., Kennedy, R. A., and Softley, T. P. (1982). *Mol. Phys.*, **45**, 747.

Carrington, A., Buttenshaw, J., and Roberts, P. G., (1979). *Mol. Phys.*, **38**, 1711.

Carrington, A., and Kennedy, R. A. (1983). 'Spectroscopy and structure of the hydrogen molecular ion', Review article (in press).

Carrington, A., and Kennedy, R. A. (1984). *J. Chem. Phys.*, **81**, 91.

Carrington, A., Roberts, P. G., and Sarre, P. J. (1977). *Mol. Phys.*, **34**, 291.

Carrington, A., Roberts, P. G., and Sarre, P. J. (1978). *Mol. Phys.*, **35**, 1523.

Carrington, A., and Sarre, P. J. (1979). *J. Phys. Paris*, **40**, C1.

Chupka, W. A., Colson, S. D., Seaver, M. S., and Woodward, A. M. (1983). *Chem. Phys. Lett.*, **95**, 171.

Cosby, P. C. (1980). *35th Symp. Mol. Spec.*, Ohio State University.

Cosby, P. C., Bennett, R. A., Peterson, J. R., and Moseley, J. T. (1975). *J. Chem. Phys.*, **63**, 1612.

Cosby, P. C., and Helm, H. (1981). *J. Chem. Phys.*, **75**, 3882.

Cosby, P. C., Helm, H., and Moseley, J. T. (1980). *Astrophys. J.*, **235**, 52.

Cosby, P. C., Moller, R., and Helm, H. (1983). *Phys. Rev.*, **A28**, 766.

Cosby, P. C., and Moseley, J. T. (1975). *Phys. Rev. Lett.*, **34**, 1603.
Cosby, P. C., Moseley, J. T., Peterson, J. R., and Ling, J. H. (1978). *J. Chem. Phys.*, **69**, 2771.
Cosby, P. C., Ozenne, J. B., Moseley, J. T., and Albritton, D. L. (1980). *J. Mol. Spectros.*, **79**, 203.
Dehmelt, H. G., and Jefferts, K. B. (1962). *Phys. Rev.*, **125**, 1318.
Dotan, I., Davidson, J. A., Streit, G. E., Albritton, D. L., and Fehsenfeld, F. C. (1977). *J. Chem. Phys.*, **67**, 2874.
Dunbar, R. C. (1971). *J. Am. Chem. Soc.*, **93**, 4354.
Dunbar, R. C. (1974). In *Chemical Reactivity and Reaction Paths* (Ed. G. Klopman), pp. 339–366, Wiley-Interscience, New York.
Dunbar, R. C. (1979). In *Gas Phase Ion Chemistry* (Ed. M. T. Bowers), Vol. 2, pp. 181–219, Academic Press, New York.
Durup, J. (1981). *Chem. Phys.*, **59**, 351.
Edwards, C. P., Jackson, P. A., Maclean, C. S., and Sarre, P. J. (1983). *Bull. des Soc. Chimique de Belgique* (In press).
Edwards, C. P., Maclean, C. S., and Sarre, P. J. (1982a). *J. Chem. Phys.*, **76**, 3828.
Edwards, C. P., Maclean C. S., and Sarre, P. J. (1982b). *Chem. Phys. Lett.*, **87**, 11.
Eyler, J. R. (1974). *Rev. Sci. Instru.*, **45**, 1154.
Freiser, B. S., and Beauchamp, J. L. (1974). *J. Am. Chem. Soc.*, **96**, 6260.
Frey, R., Kakoschke R., and Schlag, E. W. (1982). *Chem. Phys. Lett.*, **93**, 227.
Goss, S. P., Morrison, J. D., and Smith, D. L. (1981a). *J. Chem. Phys.*, **75**, 757.
Goss, S. P., Morrison, J. D., and Smith, D. L. (1981b). *J. Chem. Phys.*, **75**, 1820.
Graff, M. M. (1983a). Photodissociation and radiative association of CH^+, PhD thesis, University of Oregon.
Graff, M. M., Moseley, J. T., Durup, J., Roueff, E. (1983b). *J. Chem. Phys.*, **78**, 2355.
Graff, M. M., Moseley J. T., Roueff, E. (1983c). *Astrophys. J.*, **269**, 796.
Greiman, F., Mahan, B. H., and O'Keefe, A. (1980). *J. Chem. Phys.*, **72**, 4246.
Greiman, F. J., Moseley, J. T., Saxon, R. P., and Cosby, P. C. (1980). *Chem. Phys.*, **51**, 169.
Hansen, J. C., Moseley, J. T., Roche, A. L., Cosby, P. C. (1982). *J. Chem. Phys.*, **77**, 1206.
Hansen, J. C., Kuo, C. H., and Moseley, J. T. (1983a). *J. Chem. Phys.*, **79**, 1111.
Hansen, J. C., Moseley, J. T., Cosby, P. C. (1983b). *J. Mol. Spectros.*, **98**, 48.
Harris, F. M., Mukhtar, E. S., Griffiths, I. W., and Beynon, J. H. (1981). *Proc. Roy. Soc. Lond. A*, **374**, 461.
Helm, H., Cosby, P. C., and Huestis, D. L (1980). *J. Chem. Phys.*, **73**, 2629.
Helm, H., Cosby, P. C., Graff, M. M., Moseley, J. T. (1982a). *Phys. Rev.*, **A25**, 304.
Helm, H., Cosby, P. C., Saxon, R. P., and Huestis, D. L. (1982b). *J. Chem. Phys.*, **76**, 2516.
Helm, H., and Moller, R. (1983a). *Rev. Sci. Instrum.*, (in press).
Helm, H., and Moller, R. (1983b). *Phys. Rev.*, **A27**, 2493.
Helm, H., Cosby, P. C., and Huestis, D. L. (1983c). *J. Chem. Phys.*, **78**, 6451.
Hiller, J. F., and Vestal, M. L. (1980). *J. Chem. Phys.*, **72**, 4713.
Hiller, J. F., and Vestal, M. L. (1981). *J. Chem. Phys.*, **74**, 6096.
Hiller, J. F., and Vestal, M. L. (1982). *J. Chem. Phys.*, **77**, 1248.
Huber, B. A., Miller, T. M., Cosby, P. C., Zeman, H. D., Leon, R. L., Moseley, J. T., and Peterson, J. R. (1977). *Rev. Sci. Instrum.*, **48**, 1306.
Hunton, D. E., Hofmann, M., Lindeman, T. G., and Castleman, A. W., Jr. (1983). *Chem. Phys. Lett.*, **96**, 328.
Jefferts, K. B. (1968). *Phys. Rev. Lett.*, **20**, 39.
Jefferts, K. B. (1969). *Phys. Rev. Lett.*, **23**, 1476.

Kingston, E. E., Morgan, T. G., Harris, F. M., and Beynon, J. H. (1982). *Int. J. Mass. Spec. Ion Phys.*, **43**, 261.

Komika, N. (1981). Thesis, Universite de Paris VI, unpublished.

Larzilliere, Michel, Carre, M., Gillard, M. L., Rostas, J., Horani, M., and Velghe, M. (1980). *J. de chimie physique*, **77**, 689.

Leone, S. R. (1983). In *Advances in Chemical Physics* (Ed. K. Lawley) vol. 50.

Lerme, J., Abed, S., Holt, R. A., Larzilliere, M., and Carre, M. (1983). *Chem. Phys. Lett.*, **96**, 403.

McGilvery, D. C., and Morrison, J. D. (1977). *J. Chem. Phys.*, **67**, 368.

Malegat, L. (1980). These de doctorat de 3eme cycle. Universite de Paris—Sud, France.

Marian, C. M., Marian, R., Peyerimhoff, S. D., Hess, B. A., Buenker, R. J., and Seger, G. (1982). *Mol. Phys.*, **46**, 779.

Miller, T. M., Heidrich, J. L., and Moseley, J. T. *IX ICPEAC Abstracts of Papers*, University of Washington, Seattle, p. 7, 1975.

Miller, T. M., Ling, J. H., Saxon, R. P., and Moseley, J. T. (1976). *Phys. Rev.*, **A13**, 2171.

Morse, M. D., Freed, K. F., and Band, Y. B. (1979). *J. Chem. Phys.*, **70**, 3604.

Morse, M. D., and Freed, K. F. (1983). *J. Chem. Phys.*, **78**, 6045.

Morse, M. D., Band, Y. B., and Freed, K. F. (1983). *J. Chem. Phys.*, **78**, 6066.

Moseley, J. T. (1982). *J. Phys. Chem.*, **86**, 3282.

Moseley, J. T. (1983). In *Appl. Atomic Coll. Phys.* (Ed. E. W. McDaniel), Vol. 5, pp. 269–283, Academic Press, New York.

Moseley, J. T., Bennett, R. A., and Peterson, J. R. (1974). *Chem. Phys. Lett.*, **26**, 288.

Moseley, J. T., Cosby, P. C., Bennett, R. A. and Peterson, J. R. (1975). *J. Chem. Phys.*, **62**, 4826.

Moseley, J. T., Tadjeddine, M., Durup, J., Ozenne, J. B., Pernot, C., and Tabche-Fouhaille, A. (1976). *Phys. Rev. Lett.*, **37**, 891.

Moseley, J. T., Cosby, P. C., Ozenne, J. B., and Durup, J. (1979). *J. Chem. Phys.*, **70**, 1474.

Moseley, J. T., and Durup, J. (1981). In *Ann. Rev. Phys. Chem.* (Ed. B. S. Rabinovitch), Vol. 32, pp. 53–76, Annual Reviews, Inc., Palo Alto.

Moseley, J. T., Hansen, J. C., Graff, M. M., and Grieman, F. G. (1980). *Bull. Am. Phys. Soc.*, **25**, 1139.

Moseley, J. T., Ozenne, J. B., and Cosby, P. C. (1981). *J. Chem. Phys.*, **74**, 337.

Moseley, J. T., Saxon, R. P., Huber, B. A., Cosby, P. C., Abouaf, R., and Tadjeddine, M. (1977). *J. Chem. Phys.*, **67**, 1659.

Nenner, I., Guyon, P. M., Baer, T., and Govers, T. R. (1980). *J. Chem. Phys.*, **72**, 6587.

Nitz, D. E., Hogan, P. B., Schearer, L. D., and Smith, S. J. (1979). *J. Phys.*, **B12**, L103.

Novick, S. E., Engelking, P. C., Jones, P. L., Futrell, J. H., and Lineberger, W. C. (1979). *J. Chem. Phys.*, **70**, 2652.

Oka, T. (1980). *Phys. Rev. Lett.*, **45**, 531.

Orth, R. G., and Dunbar, R. C. (1977). *J. Chem. Phys.*, **66**, 1616.

Orth, R. G., and Dunbar, R. C. (1980). *Chem. Phys.*, **45**, 195.

Ozenne, J. B., Durup, M., Odom, R. W., Pernot, C., Tabche-Fouhaille, A., and Tadjeddine M. (1976). *Chem. Phys.*, **16**, 75.

Ozenne, J. B., Pham, D., and Durup, J. (1972). *Chem. Phys. Lett.*, **17**, 422.

Pack, R. T. (1976). *J. Chem. Phys.*, **65**, 4765.

Pernot, C., Durup, J., Ozenne, J. B., Beswick, J. A., Cosby, P. C., and Moseley, J. T. (1979). *J. Chem. Phys.*, **71**, 2787.

Pernot, C., Ozenne, J. B., Panczel, M., and Durup, J. (1977). *10th ICPEAC Abstracts Papers*, **1**, 94 (Paris Commissarat a l'Engergie Atomique).

Ray, R. D., and Certain, P. R. (1977). *Phys. Rev. Lett.*, **38**, 824.

Richardson, C. B., Jefferts, K. B., and Dehmelt, H. G. (1968). *Phys. Rev.*, **65**, 80.
Richardson, J. H., Stephenson, L. M., and Brauman, J. I. (1974). *J. Am. Chem. Soc.*, **96**, 3671.
Shy, J. T., Farley, J. W., Lamb, Jr., W. E., and Wing, W. H. (1980). *Phys. Rev. Lett.*, **45**, 535.
Smith, G. P., and Lee, L. C., and Moseley, J. T. (1979). *J. Chem. Phys.*, **71**, 4034.
Tabché-Fouhaillé, A., Durup, J., Moseley, J. T., Ozenne, J. B., Pernot, C., and Tadjeddine, M. (1976). *Chem. Phys.*, **17**, 81.
Tadjeddine, M. (1979). *J. Chem. Phys.*, **71**, 3891.
Tadjeddine, M., Bouchoux, G., Malegat, L., Durup, J., Pernot, C., and Weiner, J. (1982). *Chem. Phys.*, **69**, 229.
Tadjeddine, M., Huber, B. A., Abouaf, R., Cosby, P. C., and Moseley, J. T. (1978). *J. Chem. Phys.*, **69**, 710.
Thomas, T. F., Dale, F., and Paulson, J. F. (1977a). *J. Chem. Phys.*, **67**, 793.
van Asselt, N. P. F. B., Maas, J. G., and Los, J. (1974a). *Chem. Phys. Lett.*, **24**, 555.
van Asselt, N. P. F. B., Maas, J. G., and Los, J. (1974b). *Chem. Phys.*, **5**, 429.
van Asselt, N. P. F. B., Maas, J. G., and Los, J. (1975). *Chem. Phys.*, **11**, 253.
Vestal, M. L., and Futrell, J. H. (1974). *Chem. Phys. Lett.*, **28**, 559.
Vestal, M. L., and Mauclaire, G. H. (1977a). *J. Chem. Phys.*, **67**, 3758.
Vestal, M. L., and Mauclaire, G. H. (1977b). *J. Chem. Phys.*, **67**, 3767.
von Busch, F., and Dunn, G. H. (1972). *Phys. Rev.*, **A5**, 419.
Wagner, G., and Isenor, N. R. (1982). *Can. J. Phys.*, **61**, 40.
Wolniewicz, L., and Poll, J. D. (1980). *J. Chem. Phys.*, **73**, 6225.
Zare, R. N. (1964). Ph.D. Thesis, Harvard University.
Zare, R. N. (1974). *Mol. Photochem.*, **4**, 1.
Zare, R. N., Schmeltekopf, A. L., Harrop, W. J., and Albritton, D. J. (1973). *J. Mol. Spectrosc.*, **46**, 37.
Zeman, H. D. (1978). *Rev. Sci. Instrum.*, **48**, 1079.

Photodissociation and Photoionization
Edited by K. P. Lawley

THE FRANCK–CONDON PRINCIPLE IN BOUND-FREE TRANSITIONS

JOEL TELLINGHUISEN

Department of Chemistry, Vanderbilt University, Nashville, Tennessee 37235, USA

CONTENTS

I. INTRODUCTION

In the spectroscopic analysis of discrete (or bound–bound) molecular transitions, one typically measures and assigns various sharp features—rotational lines and band edges. The correct interpretation of these features leads to estimates of the energy levels involved in the transitions and to rotational and vibrational constants for the electronic state or states involved. From these energy levels or spectroscopic constants one can deduce potential curves or surfaces which are capable of reproducing the energy levels more or less reliably in a quantum mechanical calculation. For diatomic molecules, which are my primary concern in this paper, the methods of inverting the spectroscopic data to obtain potential curves are based on the Rydberg–Klein–Rees or RKR method (Rydberg, 1931; Klein, 1932; Rees, 1947). Although this method is a first-order semiclassical scheme, numerical tests have shown that it is often quantum mechanically reliable within $\sim 0.1\ cm^{-1}$ (Mantz *et al.*, 1971), which is comparable to experimental error in many cases. Moreover, there are methods for correcting a first-order RKR potential to render it 'exact'—through inclusion of higher-order semiclassical terms (Gouedard and Vigue, 1983; Kirschner and Watson, 1974; Le Roy, 1980) or via an inverted potential approach (Kosman and Hinze, 1975; Vidal and Scheingraber, 1977). Thus for diatomic molecules the analysis of discrete transitions can lead to potential curves whose reliability is limited solely by the range and quality of the spectroscopic data.

Having a reliable potential curve, one can calculate radial wavefunctions for any desired rovibrational level $E(v, J)$. The numerical methods commonly used for this calculation have been around for two decades (Cashion, 1963; Cooley, 1961; Zare and Cashion, 1963; Zare, 1964) and may also be considered to be exact. These wavefunctions may then be used to calculate certain important radial matrix elements, the squares of which are proportional to the intensities. For electronic transitions these intensity factors can often be well approximated as the product of a Franck–Condon factor (the square of a radial overlap integral) and a quantity which depends only weakly on the 'average' internuclear distance for the transition (Fraser, 1954). Thus, starting from the measurement of spectral *frequencies*, one can predict *intensities*, apart from a scaling factor which often varies slowly with frequency. A comparison between observed and predicted intensities can yield this scaling factor, which is proportional to the transition strength function.

When one or both of the levels involved in an electronic transition lie in the vibrational continuum, the transition is termed bound-free, free-bound or free-free; and the spectrum is *diffuse*, i.e. it no longer displays the wealth of structure characteristic of the discrete transition. Indeed, the spectrum may still display considerable structure, but now that structure is entirely Franck–Condon structure, i.e. is directly relatable to the aforementioned squared radial matrix elements. In other words the intensity information is no longer a *redundancy* of the analysis; rather it now represents the primary data and hence is the *only* source of information about the potentials from the spectrum in question. Such spectra can still be analysed quantitatively, through trial-and-error spectral simulations, or in special cases, by direct inversion methods. The potentials resulting from such analyses may be quite reliable—often much better than those obtained from molecular scattering data, for example. However, in most cases they are still crude by comparison with the potentials obtained from discrete spectra via the RKR method or one of its improvements. In fact, the accuracy and precision of potentials determined from analysis of diffuse spectra depend strongly on how much is already known about one or both potentials from other sources.

In this paper I discuss the role of the Franck–Condon principle in the interpretation and analysis of bound-free transitions in diatomic molecules. My emphasis is on radiative transitions, both absorption and emission, but I include also a brief discussion of predissociation. The paper is organized as follows. First, I summarize radiation relations of relevance to diatomic transitions. Then I discuss theoretical aspects of the Franck–Condon principle, including the classical and semiclassical implementations. The main points are illustrated with many examples, mostly from my own repertoire, since these are most familiar to me. I conclude with a review of the recent literature of this field.

II. RADIATION RELATIONS

A. Exact Treatment

Expressions of relevance to the quantitative analysis of radiative transitions in diatomic molecules appear frequently in the literature. Here I summarize relations particularly pertinent to the analysis of low-resolution and diffuse spectra, following closely a similar previous treatment (Tellinghuisen and Moeller, 1980). The latter work in turn drew heavily on the books by Mitchell and Zemansky (1934), Condon and Shortley (1951) and Herzberg (1950), and also on notes by Brewer (1967; see also Brewer and Hagan, 1979) which have been reproduced by Steinfeld (1974).

The absorption coefficient k_ν (in units cm^{-1}) for a transition from lower level B having degeneracy g_B to upper level A having degeneracy g_A is related to the radiative decay coefficient A_AB (in units s^{-1}) for spontaneous emission from A to B by

$$k_\nu \delta\nu = (8\pi c \nu^2)^{-1}\left(\frac{g_\mathrm{A}}{g_\mathrm{B}}\right) A_\mathrm{AB} \delta N_\mathrm{B} \tag{1}$$

where c is the speed of light, δN_B is the number density of particles in the lower level capable of absorbing radiation in an interval $\delta\nu$ about the wavenumber ν (in cm^{-1}), and stimulated emission is negligible. Equation (1) is valid in the low-power limit, where the concentration of molecules in the lower level is altered negligibly by the radiation field. This condition is easy to violate with pulsed lasers, but hard to violate with conventional light sources. For narrow-line absorption ν varies negligibly across the breadth of the line, and Eq. (1) can be integrated to give

$$\int k_\nu \, \mathrm{d}\nu = (8\pi c \nu_0^2)^{-1}\left(\frac{g_\mathrm{A}}{g_\mathrm{B}}\right) A_\mathrm{AB} N_\mathrm{B} \tag{2}$$

where ν_0 is the wavenumber at the centre of the absorption line and N_B is the total concentration of absorbers in level B.

For electric dipole transitions the spontaneous emission coefficient A_{AB} is proportional to the product of ν_0^3 and the absolute squared matrix element $|\langle a|\boldsymbol{\mu}|b\rangle|^2$ of the dipole moment operator $\boldsymbol{\mu}$ between two states a and b of levels A and B respectively. For atoms Condon and Shortley (1951) defined a line strength S_{AB} as the sum of this squared matrix element over all states a and b. This S_{AB} is symmetric with respect to upper and lower levels, such that absorption and emission are proportional to S_{AB}/g_B and S_{AB}/g_A respectively. In particular the radiative decay coefficient is

$$A_{AB} = \left(\frac{64\pi^4}{3h}\nu_0^3\right)\left(\frac{S_{AB}}{g_A}\right) \tag{3}$$

where h is Planck's constant. The line strength can be expressed further as a product of a reduced matrix element and a factor depending on the angular momenta alone. For diatomic transitions between upper and lower levels $(v'J')$ and $(v''J'')$, this reduction yields

$$S_{AB}(v'J', v''J'') = s_{J'J''}|\langle v''_{J''}|\mu_e(R)|v'_{J'}\rangle|^2 \tag{4}$$

The matrix element in Eq. (4) is an integral of the product of upper- and lower-state vibrational wavefunctions and the dipole strength function $\mu_e(R)$. The latter quantity is the electronic matrix element $\langle B|\boldsymbol{\mu}|A\rangle$ and is a function of the internuclear separation R. The vibrational wavefunctions are solutions of the one-dimensional Schrödinger equation for the effective potentials $U_{AJ'}(R)$ and $U_{BJ''}(R)$, where, for example,

$$U_{AJ'}(R) = U_{A0}(R) + \kappa/R^2 \tag{5}$$

The quantities on the right-hand side of Eq. (5) are the rotationless potential and the centrifugal term. The latter is strongly dependent on the rotational quantum number; e.g. κ is proportional to $J(J+1)$ for $^1\Sigma$ states. Through this term the vibrational wavefunction is an implicit function of the rotational quantum number, as indicated in the notation of Eq. (4).

In writing Eq. (4) I have made the usual assumption that the molecular wavefunctions for both electronic states are expressible in the product form. $\psi_{evr} = \psi_e(R)\psi_{vr}(R, \theta, \varphi)$, where ψ_e represents the electronic wavefunction and ψ_{vr} the rotation–vibration wavefunction. In this approximation the solution to the electronic Schrödinger equation yields the rotationless potential curves and electronic wavefunctions. The subsequent solution of the Schrödinger equation for the nuclear motion (see below) employs these potentials and yields the eigenvalues $E(v, J)$ as functions of the vibrational and rotational quantum numbers. From an empirical standpoint the validity of the separability assumption is supported by the observation that for the vast majority of molecular electronic states, most of the rovibrational energy levels can indeed be accounted for in terms of single rotationless potentials. In practice spectroscopists start with the observed energy levels and work back to the potentials through the RKR method, discussed below. Thus it is possible for spectroscopists to employ many of the expressions derived in this section in a very precise fashion, but still in almost total ignorance of the electronic wavefunctions, the only quantity explicitly requiring the latter being the transition strength function.

The rotational line strength $s_{J'J''}$ in Eq. (4) is obtained from the angular momentum properties of the rotational wavefunctions for the upper and lower states. The $s_{J'J''}$ are normalized such that summation over all allowed J'' or J' for a given J' or J'' yields a quantity proportional to $(2J'+1)$ or $(2J''+1)$ respectively (Herzberg, 1950). The choice of the proportionality constant has been a source of some confusion in the literature and effectively determines the definition of $\mu_e(R)$. To deal with this problem, Whiting and coworkers (1980) have suggested a convention whereby the $s_{J'J''}$'s sum to $(2-\delta_{0,\Lambda'+\Lambda''})(2S+1)(2J+1)$ for all transitions connecting with all states having a given J' $(=J)$ or J'' $(=J)$. In other words, the line strengths are normalized to $(2S+1)(2J+1)$ for spin-allowed Σ–Σ transitions, or twice that for all others. When Hund's case c is more appropriate, the line strengths sum to $(2J+1)$ for 0–0 transitions and $2(2J+1)$ for all others (Tellinghuisen, 1982a).

With the incorporation of Eq. (4) for diatomics, Eqs. (2) and (3) become

$$\int k_\nu \, d\nu = \left(\frac{8\pi^3\nu}{3hc}\right)\left(\frac{s_{J'J''}}{2J''+1}\right) \cdot G_{ab}|\langle v'_{J'}|\mu_e(R)|v''_{J''}\rangle|^2 N(v'',J'') \tag{6}$$

and

$$A(v'J',v''J'') = \left(\frac{64\pi^4}{3h}\right)\nu^3\left(\frac{s_{J'J''}}{2J'+1}\right) \cdot G_{em}|\langle v''_{J''}|\mu_e(R)|v'_{J'}\rangle|^2 \tag{7}$$

The quantities G_{ab} and G_{em} account for electronic degeneracy. Like the line strengths, electronic degeneracies have been a source of confusion, in part because they must be defined with respect to experimental parameters. If the experimental resolution is such that electronically nondegenerate rotational lines (i.e. single components of Λ or Ω doublets) are resolved, then G_{ab} and G_{em} are unity. Accordingly the concentration $N(v'',J'')$ and line strength $s_{J'J''}$ are the quantities relevant to this specific Λ or Ω component. When Λ or Ω doubling is not resolved experimentally, G_{ab} and G_{em} become $1/g''$ and $1/g'$, respectively, where $g=1$ for $\Omega=0$ states and 2 for all others. In this case $s_{J'J''}$ must include both overlapped transitions and $N(v'',J'')$ refers to the total concentration in the (v'',J'') level, i.e. both Ω components if $\Omega''\neq 0$. For still lower resolution with more extensive overlap, these definitions must be further modified in similar fashion. (For example results are given below for unresolved rotational structure in diffuse spectra.) When other conventions are used for G_{ab} and G_{em} (as they have been frequently in the literature), the effect is a change in the definition of $\mu_e(R)$ from the recommendation of Whiting and coworkers (1980). In any event the relation $G_{ab}/G_{em} = g'/g''$ must hold for consistency with the fundamental radiation relations. Finally, there is one further type of degeneracy—that associated with nuclear spin statistics in homonuclear diatomics. The effects of nuclear spin degeneracy can usually be accommodated by simply altering the concentration appropriately (Tellinghuisen and Moeller, 1980).

For diffuse spectra it is the differential form of Eq. (6) which is of interest, since now the 'line' is the entire spectrum; and it is the *structure* in this spectrum as well as its integrated intensity that we are interested in. For bound-free absorption the appropriate expression is

$$k_\nu(v''J'') = \left(\frac{8\pi^3\nu}{3hc}\right)\left(\frac{S_{J'J''}}{2J''+1}\right) \cdot G_{ab}|\langle \varepsilon'_{J'}|\mu_e(R)|v''_{J''}\rangle|^2 N(v'', J'') \tag{8}$$

in which the continuum wavefunctions $\langle \varepsilon'_{J'}|R\rangle$ are energy normalized per unit wavenumber. From now on I will drop the J subscripts in the matrix elements for brevity, but it must be remembered that these are always implicitly dependent on rotation and that in general this dependence cannot be neglected. In diffuse transitions the Λ (or Ω) doubling is never resolved and the individual rotational branches (P, Q and R) often differ negligibly. (There are exceptions to the latter statement, as noted below. Note that although the *branch* dependence for a given initial J' or J'' may be negligible, the dependence on the *initial* J over the range of J occurring in typical experiments is not, as noted above.) In that case it is appropriate to sum Eq. (8) over J' to obtain the total continuum absorption by the molecules in a specific Ω substate of level J''. The absorption cross-section $\sigma_{\nu,ab}$ $[= k_\nu/N(v'', J'')]$ then becomes

$$\sigma_{\nu,ab}(v''J'') = \left(\frac{8\pi^3\nu}{3hc}\right)\left(\frac{d}{g''}\right)|\langle \varepsilon'|\mu_e(R)|v''\rangle|^2 \tag{9}$$

where d is 1 for Σ–Σ transitions (0–0 transitions in case c), 2 otherwise. In the simulation of typical bound-free absorption spectra, this expression must be averaged over v'' and J'', with appropriate population weighting factors. For molecules in thermal equilibrium, the populations are simply proportional to rotational degeneracies and Boltzmann factors. When the spectrum is not highly structured, the J'' averaging can be contracted to a few properly chosen levels, or even to the single level $J'' = J''_{av}(T)$ (the average J'' at temperature T), with negligible error (LeRoy, Macdonald and Burns, 1976; Tellinghuisen *et al.*, 1976).

For bound-free stimulated emission the counterpart to Eq. (9) is

$$\sigma_{\nu,\mathrm{se}}(v'J') = \left(\frac{8\pi^3\nu}{3hc}\right)\left(\frac{d}{g'}\right)|\langle \varepsilon''|\mu_e(R)|v'\rangle|^2 \tag{10}$$

Similarly, the Einstein coefficient for spontaneous bound-free emission is

$$A(v'J', \varepsilon'') = \left(\frac{64\pi^4}{3h}\right)\nu^3\left(\frac{d}{g'}\right)|\langle \varepsilon''|\mu_e(R)|v'\rangle|^2 \tag{11}$$

Thus the total spontaneous emission coefficient for level (v', J') is

$$A_T(v'J') = \left(\frac{64\pi^4}{3h}\right)\left(\frac{d}{g'}\right)\langle \nu^3\mu_e^2\rangle \tag{12}$$

where the last factor is the average of $\nu^3\mu_e^2$, obtained by summing Eq. (7) over v'' and integrating Eq. (11) over ε'' (see below).

When the continuum wavefunctions are energy normalized in wavenumber units, Eq. (11) gives the bound-free emission coefficient in units $(\mathrm{cm}^{-1}\mathrm{s})^{-1}$. Nowadays spectra are usually recorded on grating spectrometers, which in first approximation are constant wavelength interval devices. Thus it is customary to intensity calibrate such instruments in units proportional to quantum flux per unit wavelength interval or power per unit wavelength interval. To convert Eq. (11) to compatible units, one uses the appropriate conversion relation for equivalent distribution functions in different independent variables. For example, the expression (Mies and Smith, 1966)

$$A_\nu\,\mathrm{d}\nu = A_\lambda\,\mathrm{d}\lambda \tag{13}$$

can be used to convert to units proportional to quantum flux per Å. We see that in this case A_λ is obtained from Eq. (11) by multiplying by $10^{-8}\nu^2$ (Tellinghuisen, 1974a). For spectra calibrated in power per Å, another factor of $hc\nu$ is required.

B. Broad-Band Expressions

Starting with potential curves for upper and lower electronic states, one can use the equations given in the preceding section to calculate intensities of discrete transitions and bound-free spectra of any type—absorption or emission, for single or multiple bound levels in the initial state. At this point I want to develop approximate expressions which are appropriate for the analysis of low-resolution or broad-band absorption spectra. This treatment is useful for several reasons. (1) It permits estimation of several important parameters—particularly the average transition strength $\langle\mu_e^2\rangle$ and the radiative decay coefficient A_T—from low-resolution extinction coefficients, which are often very easy to obtain. (2) It reestablishes the connection with atomic spectra, showing that the entire electronic band for the diatomic is equivalent to a single atomic line or multiplet. (3) It leads directly to expressions which can be used for quantitative calculations via the classical method, discussed below. Equations equivalent to some of those given below were obtained over 20 years ago by Stafford (1960) and Strickler and Berg (1962), but from a different standpoint. The present treatment follows that of Tellinguisen and Moeller (1980), with modifications to bring it into consistency with the different degeneracy convention of Whiting and coworkers (1980).

We start with Eq. (9) for the bound-free absorption cross section and its counterpart for the discrete spectrum, obtained by dividing both sides of Eq. (6) by $N(v'', J'')$ and summing over J'. To obtain the total integrated spectrum, we sum over v' and integrate over ε' on the right-hand side, then thermally average over v'' and J''. For the matrix elements of Eqs. (8) and (9), the sum over v' and integral over ε' yields $\langle v'J'|\mu_e^2(R)|v'J'\rangle$ (Mies and Julienne, 1979). The thermal averaging then yields a 'system' average $\langle \mu_e^2 \rangle$. In evaluating the integral of σ_ν on the left-hand side of the equation, we make one minor change from the procedure used to obtain Eq.(2): We recognize that ν can vary significantly ($\sim 10\%$) across a molecular band system and include this factor under the integral. The result is

$$\int \sigma_{\nu,\mathrm{ab}} d \ln \nu = \left(\frac{8\pi^3}{3hc}\right)\left(\frac{d}{g''}\right)\langle \mu_e^2 \rangle \tag{14}$$

Using Eq. (12) we can relate the A coefficient to $\sigma_{\nu,\mathrm{ab}}$:

$$A_T \approx 8\pi c \langle \nu_{\mathrm{fl}}^3 \rangle \left(\frac{g''}{g'}\right) \int \sigma_{\nu,\mathrm{ab}} d \ln \nu \tag{15}$$

which is the equivalent of Eq. (2) for atoms. To use Eq. (15) one must be able to estimate the average cubed frequency for fluorescence, which generally differs from that for absorption. From an experimental standpoint the use of these equations for discrete spectra is only appropriate when saturation effects from sharp-line absorption can be eliminated, e.g. by pressure broadening the lines (Rabinowitch and Wood, 1936) or extrapolating to the limit of zero absorption (Tellinghuisen, 1973a). Also, Eq. (15) assumes that $\langle \mu_e^2 \rangle$ is the same for fluorescence as for absorption, which can be a poor approximation, because absorption and fluorescence may sample different regions of R.

A number of other quantities are often used to characterize the radiative properties of molecules. The molar extinction coefficient ε_ν is related to the absorption cross-section $\sigma_{\nu,\mathrm{ab}}$ by

$$\sigma_{\nu,\mathrm{ab}} N'' l = \ln(10) \varepsilon_\nu C l \tag{16}$$

where C is the concentration in mol/l. Thus one can convert from cross section (units Å^2) to extinction coefficient (units $\mathrm{l\ mol^{-1}\ cm^{-1}}$) using

$\varepsilon_\nu = 2.6153 \times 10^4 \sigma_\nu$. Another occasionally used quantity is the dimensionless absorption oscillator strength, which can be obtained from

$$f \approx 4.319 \times 10^{-9} \int \varepsilon_\nu \, d\nu \tag{17}$$

where ε_ν is in the units given above. In terms of $\langle \mu_e^2 \rangle$, the oscillator strength is

$$f \approx 4.702 \times 10^{-7} \left(\frac{d}{g''} \right) \langle \nu \rangle \langle \mu_e^2 \rangle \tag{18}$$

where μ_e is in debyes. In fact Eq. (18) is equivalent to Eq. (14), but with the integral in the latter approximated as $\int \sigma_\nu \, d\nu / \langle \nu \rangle$. In practice it makes little difference which expression is used. For example, for a band shape of form $\nu e^{-b(\nu - \nu_0)}$, the two methods agree within 1 per cent when the full width at half height is taken to be 20 per cent of the peak frequency, ν_0.

III. FRANCK–CONDON CALCULATIONS

A. Potentials

At the heart of most of the relations given in the previous section is a squared matrix element of the form $|\langle v' | \mu_e(R) | v'' \rangle|^2$ or $|\langle \varepsilon' | \mu_e(R) | v'' \rangle|^2$, which can be approximated as the product of a Franck–Condon factor $|\langle v' | v'' \rangle|^2$ or Franck–Condon density $|\langle \varepsilon' | v'' \rangle|^2$ and an 'average' transition strength. In this section I will discuss the evaluation of these matrix elements by 'exact' numerical methods, as well as by approximate classical and semiclassical methods. Although the exact calculations are straightforward, the classical and semiclassical methods yield more insight into the Franck–Condon principle and in many cases are close enough to exact to become the method of choice. This statement is particularly true of the classical method, which can provide a very simple and quick estimation of spectra for thermal distributions of molecules.

All methods of evluating the intensity factors start with potential curves for the electronic states. Bound potentials for which there are extensive rotational and vibrational data from discrete spectra are usually calculated by the RKR method, as was noted earlier. Although the fundamental relations for this calculation were given by Klein in 1932, they were not utilized extensively until about 1960, when high-speed computers became available. Initially various approximation methods were used to treat or avoid the integrable singularities in the Klein f and g integrals (Jarmain, 1960; Rees, 1947; Vanderslice *et al.*, 1959; Weissman, Vanderslice and Battino, 1963). Then Kasper (1963) and Zare (1964) demonstrated that Gaussian quadrature could be used to evaluate these integrals directly. Later I showed that a Gauss–Mehler quadrature with a suitable weight function was much more efficient (Tellinghuisen, 1972a). The weight function effectively neutralizes the singularities in the integrals and permits part-per-million accuracy for levels up to within a few per cent of the dissociation limit, with only four quadrature points (Tellinghuisen, 1974b). It is interesting to note that this method could even have been used in pre-computer days; however, it took computers to permit the discovery and verification of its accuracy. Since ~1970 several other methods of evaluating Klein's f and g integrals have been proposed, which in some cases are comparable in accuracy and efficiency to the Gauss–Mehler scheme (Dickinson, 1972; Fleming and Rao, 1972, and see also Zeleznik, 1965; Kaiser, 1970; Kirschner and Watson, 1973; Mantz *et al.*, 1971; Telle and Telle, 1981). Thus this problem can be considered 'solved'. As I mentioned in Section I, current efforts in this area are aimed at removing the usually quite small errors inherent in the first-order WKB approximation behind Klein's equations for the f and g integrals.

For bound states which are less well characterized experimentally, the RKR method may not be appropriate. Thus when only the first few bound v levels are known, simple closed-form potentials like the Morse or Hulburt–Hirschfelder curves (Herzberg, 1950) may be adequate. In other cases the RKR method may be used in part. For example, heavy diatomics are often difficult to analyse rotationally but easy to analyse vibrationally. When only vibrational constants are known, the RKR f integral can still be used to calculate the *width* of the potential, $R_+ - R_-$, as a function of the vibrational energy. Then if one branch of the potential can be approximated by a suitable expression, the other can be calculated from these turning-point differences. In one such scheme (Tellinghuisen and Henderson, 1982) the Morse curve was found to be a surprisingly reliable approximation of the repulsive branches for a number of well-known potentials.

In RKR calculations the classical turning points on the potential are generated as functions of v or E. The RKR method is not restricted to integer v values, so it is customary to produce curves over a v-grid of step size $\Delta v = 0.1$ to 0.5. The extra points are particularly useful near the curve minimum, because a relatively large range of R is encompassed by the turning points for $v = 0$. Since the RKR calculations are very fast, the only minor drawback to the use of a dense v-grid is the need for a larger array to store the potential. In the computational methods described below, the potential is needed at specified (usually equidistant) values of R. For this purpose interpolation procedures are needed. Numerical tests have shown that for the calculation of quantum wavefunctions, eight-point Lagrangian interpolations are about as good as any other technique (LeRoy and Bernstein, 1968; Zare, 1964). For levels near the dissociation limit, the repulsive branches of RKR curves nearly always flare in or out in an unphysical manner. It has long been considered that such behaviour is mainly attributable to inadequacies in the rotational constants, often within experimental error. Recently Wells, Smith and Zare (1983) have used a simple Morse-like model in a quantitative verification of this dependency. It is generally desirable to smooth the repulsive branch to remove such anomalies, and to extend both branches to smaller and larger R than spanned by the highest level included in the RKR calculations. This can be done in various ways, e.g. by using Morse curves (Jarmain, 1971), curves of form R^{-n} (Zare, 1964) or Hulburt–Hirschfelder potentials (Albritton, Schmeltekopf and Zare, 1973). When the repulsive branch is smoothed in this manner, it is wise to adjust the attractive branch to preserve the RKR widths, $R_{+}-R_{-}$, as mentioned above.

Unbound or weakly bound potentials (and the vibrational continua of strongly bound potentials) are generally much more poorly determined than are bound potentials, as was mentioned earlier. In some cases such potentials have been estimated from scattering data; for tractability these curves are usually represented as simple closed-form expressions containing several adjustable patameters. The latter approach is often desirable or necessary also in the analysis of bound-free spectra, although direct inversion methods can be used in certain cases (see below). When a bound-free spectrum accesses the vibrational continuum of a deeply bound state which is already well characterized within the well, the determination of the continuum region of this potential may be quite precise, since it is usually reasonable to assume a 'smooth' connection to the repulsive branch within the well.

Although my emphasis in this paper is on the determination of potentials from experimental spectra, it should be noted that *ab initio* and semiempirical theoretical methods can now produce very accurate potential curves. This is especially true for light diatomics, but is increasingly true for heavy diatomics as well (see, for example, Hay, Wadt and Dunning, 1979; Mies, Stevens and Krauss, 1978). For bound wells for which ample discrete spectral data are available, theoretical methods can match the RKR method only for selected light hydrides (notably H_2). However, in many cases spectral data are not yet available for much of the well, and theoretical results can complement the experimental data to yield a more extensive potential (see, for example, Konowalow and Olson, 1979). Repulsive potentials and the repulsive walls of bound potentials are usually much less well determined experimentally, as noted above. Thus a theoretical potential can at least represent a good starting approximation in the analysis of a diffuse spectrum. For the experimentally more elusive quantity $\mu_e(R)$, theoretical methods may prove to be an even greater asset (Hefferlin, 1976; Julienne, Neumann and Krauss, 1976; Klemsdal, 1973).

B. Quantum Method

The solution to the three-dimensional Schrödinger equation for the rotation and vibration of a diatomic molecule can be written in the product form, $\psi_{vr} = \psi_r(\theta, \varphi)\psi_v(R)/R$, with the rotational wavefunction ψ_r obtainable in closed form. The vibrational wavefunction then becomes the solution of the one-dimensional Schrödinger equation, which can be written in the form,

$$\frac{d^2\psi_v(R)}{dR^2} + H[E - U_J(R)]\psi_v(R) = 0, \qquad (19)$$

where $U_J(R)$ is the effective potential of Eq. (5) and the eigenvalues E are functions of the quantum numbers v and J (magnetic interactions are neglected). If the energies are in units cm^{-1} and R is in Å, the constant H is

$$H = 10^{-16} \times \frac{8\pi^2 c\mu}{h} = 0.05932\,\mu_A \qquad (20)$$

where μ_A is in amu in the second expression. In the evaluation of the matrix elements of the preceding section, the volume element contains a factor of R^2, which cancels the similar factors in ψ_{vr}. Thus these matrix elements are, for example,

$$\langle v'_{J'}|\mu_e(R)|v''_{J''}\rangle = \int \psi_{v'J'}(R)\psi_{v''J''}(R)\mu_e(R)dR \qquad (21)$$

in which there is no need to show complex conjugates, since the wavefunctions can always be made real. (Expressions analogous to Eq. (21) result for all operators which involve R^n or combinations thereof; however, for operators which involve differentiation with respect to R, one must be careful to derive the correct expression from the full wavefunction.)

In a few special cases (e.g. the harmonic and Morse oscillators), the solutions to Eq. (19) are available in closed form. However, for realistic potentials, especially those generated by the RKR method, the solutions must be obtained numerically. Although some authors have used expansions over harmonic oscillator or Morse basis sets for this purpose, most use methods which produce ψ_v in tabular form at preselected values of R. The method of Cooley (1961), Cashion (1963) and Zare (1963, 1964), which utilizes the Numerov finite difference method with a predictor–corrector formula based on Löwdin (1963), is probably the most widespread procedure. However, other techniques have been described (Bernstein, 1960; Gordon, 1969; Lawley and Wheeler, 1981; LeRoy and Bernstein, 1968; Tobin and Hinze, 1975).

Equation (19) is appropriate for both bound and unbound (continuum) levels. For the former the finite difference solution is carried out iteratively, starting with a trial eigenvalue and converging on the nearest true eigenvalue. In the method of Zare and Cashion (1963), the solution is obtained in two segments, by inward and outward integration; and the predictor–corrector formula adjusts the eigenvalue to achieve continuity in the first derivative at the connection point. At the end ψ_v is normalized to unity. A similar procedure can be used used for continuum wavefunctions—called box normalization. In this method the wavefunction is forced to have a node at some suitably large R (the 'box' distance), which effectively makes the continuum discrete. As for bound functions, this method involves iterative solution for the eigenvalues and functions (unless the box size is made adjustable, in which case any continuum E value can be accommodated). Then in the evaluation of bound-free matrix elements, one must specifically include a density of states. It is usually more convenient to use energy-normalized wavefunctions, which effectively combine both steps. In this procedure one simply generates the outward-going function (using, for example, the same Numerov procedure used for the bound levels) until the amplitude stabilizes. Then the function is normalized to an amplitude P (Buckingham, 1961; Jarmain, 1971),

$$P = \left(\frac{8\mu}{h^3 c\varepsilon}\right)^{1/4} \tag{22}$$

where ε is the energy above the dissociation asymptote in cm^{-1}. Use of Eq. (22) gives the wavefunction in cgs units, $(\mathrm{erg\,cm})^{-1/2}$, which can be converted to $(\mathrm{cm}^{-1}\,\text{Å})^{-1/2}$:

$$P' = 10^{-4} \times \left(\frac{8c\mu}{h\varepsilon}\right)^{1/4} = 0.27844\left(\frac{\mu_A}{\varepsilon}\right)^{1/4} \tag{23}$$

In fact it is not even necessary to extend the wavefunction to very large R if one normalizes to the *local* kinetic energy (Roche and Tellinghuisen, 1979),

$$\varepsilon^\dagger = \varepsilon - U_J(R^\dagger) \tag{24}$$

where $R^\dagger$ is the maximum R for the integration. Note that continuum functions can be calculated by this method quicker and more easily than bound functions, because no iteration is required. Note further that provided the step size for the integration grid is small enough, this ψ_ε must always be correct in a relative sense. That is, the only error can be in the form of a constant scale factor resulting from inaccurate normalization due to not extending ψ_ε to large enough R for the WKB-type approximation of Eq. (24) to be valid. Regarding the step size, it is a good rule of thumb to include at least ten points per node. This translates into

$$\Delta R \approx 5 \times 10^6 \left(\frac{h}{2\mu c E_{\max}}\right)^{1/2} \tag{25}$$

where $E_{\max}$ is the maximum kinetic energy in cm^{-1} $[= E - U_{J,\min}(R)]$ and ΔR is in Å. Of course, the effect of changing the range and density of the grid for the calculation of the wavefunctions, both bound and unbound, can and should be checked.

The method for calculating continuum wavefunctions outlined above can also be used to characterize shape resonances, or levels which are 'bound' behind a barrier—e.g. a rotational barrier or a barrier in the rotationless potential. In this case one simply generates outward and normalizes as usual in the region beyond the barrier. In penetrating the barrier the wavefunction will usually increase its amplitude many orders of magnitude, so that the energy-normalized ψ_ε will appear flat behind the barrier. However, near a resonance the inner portion of ψ_ε will expand in amplitude, possibly becoming several orders of magnitude larger than the outer portion. One can determine the width of the resonance (which is relatable to a predissociation rate) from the dependence of this inner amplitude on ε. The position of the resonance can be determined from an examination of the outer segment of ψ_ε as a function of ε: As the resonance is traversed, the phase changes by π, and the sign of the first loop changes from positive to negative or vice versa. Actually the resonance can be located very accurately and much more easily using the first-order semiclassical quantization condition that is at the heart of the RKR method (see, for example, Tellinghuisen and Exton, 1980). It is often appropriate to treat the resonance levels as truly bound (which for most purposes they are, except for levels very near the top of the barrier and for very light molecules). In this case one can use the usual bound function routines, being careful to start the inward integration within the nonclassical region of the barrier rather than the classical region beyond. The R value of the barrier maximum is a reasonable choice for this outer limit to the integration grid (Tellinghuisen and Exton, 1980).

The Franck–Condon overlap integrals are generally evaluated by pointwise summation of the product of the two wavefunctions, which are available in tabular form from the finite-difference calculations. For this purpose it is convenient to determine the wavefunctions over common grid points, although interpolation can be used if the grids for the two wavefunctions are not identical. An R-dependent $\mu_e(R)$ can be incorporated in these calculations in an obvious way. Of course the range of the integration can be restricted to the region where both wavefunctions have appreciable amplitude, which is slightly more than the overlap of the two classical regions. This means that in bound-free calculations, the free wavefunctions need be *stored* only out to the maximum R value of the bound wavefunction, even though it may be desirable to *generate* out to larger R for accurate normalization. In free-free calculations both wavefunctions extend in principle to $R = \infty$, and it is usually necessary to 'discretize' the spectrum by means of a large-R wall. (This is equivalent to box normalization, although one can still energy normalize such box functions by means of Eq. (22) or Eq. (23); see Herman and Sando, 1978, for an alternative procedure.) An exception is the case where $\mu_e(R)$ dies off at large R, effectively terminating the range of integration. In practice it makes little difference whether the integrals are evaluated by the trapezoidal rule or by an extended Simpson's formula, because the wavefunctions are usually tabulated on a sufficiently fine grid. Here again the magnitude of grid errors is easily checked by changing the grid.

Franck–Condon factors and densities satisfy the important closure relation

$$\sum_{v''} |\langle v''|v'\rangle|^2 + \int |\langle \varepsilon''|v'\rangle|^2 \, d\varepsilon'' = 1 \tag{26}$$

and an analogous expression for summation over factors connecting with a given v'' level. Note that the units of normalization dictate the units of $d\varepsilon''$ in Eq. (26), so when the continuum functions are normalized in cm, $d\varepsilon''$ is in cm^{-1}. It is Eq. (26) that permitted the total radiative decay rate in Eq. (12) to be expressed in terms of an average value of $v^3\mu_e^2$. Closure can similarly be used to show that (Mies and Julienne, 1979)

$$\sum_{v''} |\langle v''|\mu_e(R)|v'\rangle|^2 + \int |\langle \varepsilon''|\mu_e(R)|v'\rangle|^2 \, d\varepsilon'' = \langle v'|\mu_e^2(R)|v'\rangle \, . \tag{27}$$

which I used in the derivation of Eq. (14). I have recently (Tellinghuisen, 1984a) shown that a similar result holds to a very good approximation ($\sim 0.2\%$) for the sum of $v^3|\langle v''|\mu_e(R)|v'\rangle|^2$, which occurs in the expression for $A_T(v'J')$ (Eq. 12). In the relevant expectation value, analogous to that of Eq. (27), v^3 is replaced by its classical equivalent, the cubed difference potential (see below). The classical analogue of the latter sum rule has been used previously by Julienne (1978) and Ehrlich and Osgood (1979a).

In the R-centroid approximation (Fraser, 1954),

$$|\langle v'|\mu_e(R)|v''\rangle|^2 \approx |\mu_e(\bar{R})|^2 |\langle v'|v''\rangle|^2 \tag{28}$$

and similarly for bound-free matrix elements, where $\bar{R}$ is the R-centroid.

$$\bar{R} = \frac{\langle v'|R|v''\rangle}{\langle v'|v''\rangle} \tag{29}$$

It is easy to show that the approximation becomes rigorous when $\mu_e(R)$ is a linear function of R. If we further neglect the linear term, we obtain the Condon approximation (Condon, 1928), in which the matrix element is simply proportional to the Franck–Condon factor. When $\mu_e(R)$ contains terms in R^n beyond the linear term, the R-centroid approximation is equivalent to replacing $\overline{R^n}$ by $\bar{R}^n$. The R-centroid method is quite reliable for cases where there is a single classical transition point, i.e. a single R value R^* which solves $v = U'(R^*) - U''(R^*)$, in which case $R = R^*$ to a very good approximation (Noda and Zare, 1982; Smith, 1968a). When there are multiple transition points, the R-centroid method is unreliable (Tellinghuisen *et al.*, 1980); Noda and Zare (1982) have shown that in this case the range of integration can be segmented to yield a modifed R-centroid method, which agrees much better with exact calculations. The latter are anyway advisable in the final stage of an analysis of $\mu_e(R)$ from intensity data, unless only a weak dependence on R is indicated.

All of the calculations discussed in this section start with the solution of the Schrödinger equation (19) for the bound and/or free energy levels in the potentials $U'(R)$ and $U''(R)$. If the transition strength function is also known, then the quantum calculation of discrete and bound-free spectra is straightforward. In the usual case one or both of these potentials may not be well known, and the job of calculating the spectrum becomes a trial-and-error procedure of adjusting the unknown potential(s) and transition strength function until the calculated spectrum agrees satisfactorily with experiment. This task is somewhat open-ended, and one may need to adopt appropriate models for the unknowns to keep the trial-and-error simulation tractable and to assess the reliability of the analysis. This point is clarified in the discussion of examples below. In special cases exact expressions can be used to extract an unknown potential from a bound-free spectrum (Smith, 1968b). In others, approximate methods are effective (Child, Essen and LeRoy, 1983; Chow and Smith, 1971; Smith, 1968b). However, much of the time the trial-and-error approach is needed, at least in part. The calculations can usually be made to 'converge' in 5 to 20 cycles, with typically much less than a minute of CPU time needed for each spectrum. Success is facilitated by a direct graphical comparison of experimental and calculated spectra (Tellinghuisen *et al.*, 1979). Thus the trial-and-error approach, though indirect, is nonetheless practicable. Actually, trial and error needs no special justification, since it is already the *modus operandi* of much, if not most, of science.

C. Classical Methods

The Franck–Condon principle and the Born–Oppenheimer approximation both stem from the realization that the nuclei in a molecule are massive compared with the electrons, and that therefore they alter their motions sluggishly in reaction to changes in the electronic structure. For electronic transitions this means that to a very good approximation the nuclear positions and momenta are conserved during the transition (Franck, 1925; Condon, 1926, 1928, 1947). Thus for diatomics such transitions can be represented on potential diagrams as vertical lines from the initial energy E_{i} and position R to an energy E_{f} in the final state for which the kinetic energy, $E_{\mathrm{f}} - U_{\mathrm{f},J}(R)$, is the same as in the initial state, $E_i - U_{i,J}(R)$. The inset to Fig. 1 illustrates this point for an emission transition which will be discussed more fully below. The dashed curve $X(R)$ represents the locus of points which classically conserve nuclear position and momentum for the indicated E' level, i.e.

$$X(R) = U''(R) + E' - U'(R) \tag{30}$$

This curve is usually called the *Mulliken difference potential* (Tamagake and Setser, 1977), since Mulliken (1971b) first used such curves to interpret bound-free spectra. This curve may also be said to represent all of the *classical*

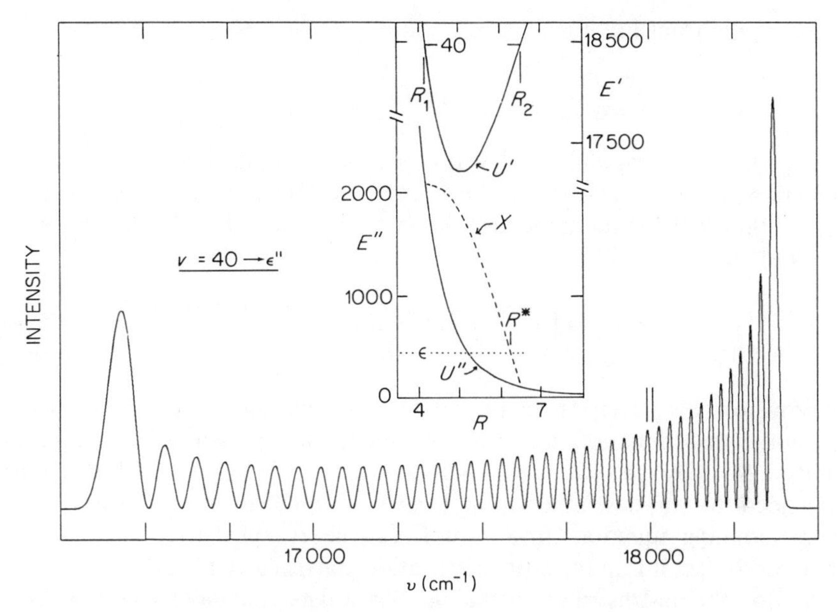

Fig. 1. Potential curves (inset, units cm^{-1} and Å) and calculated bound-free emission spectrum illustrating reflection structure. The calculations for this figure and Figs. 2 to 11 were carried out for the Cs_2 molecule, using potentials which have been described in more detail in Tellinghuisen *et al.*, (1980) and Tellinghuisen (1984c). (Reproduced from Tellinghuisen, 1984c, by permission of Academic Press, Inc.)

transition points, or, in semiclassical parlance, the *points of stationary phase* (see below). Equation (30) can be rewritten to yield a very simple expression for the transition wavenumber as a function of R:

$$\nu(R) = E' - E'' = E' - X(R) = U'(R) - U''(R) \equiv V(R) \tag{31}$$

where $V(R)$ is usually referred to as simply the *difference potential* (Hedges, Drummond and Gallagher, 1972, and references cited therein). The classical method employs Eq. (31) to relate spectral wavenumber to the distance or distances R_ν which solve $\nu = V(R)$. The problem of calculating spectra by this method then focuses on finding a suitable way to weight the contributions from different R values.

The classical formula for the absorption cross-section can be obtained directly from expressions in Section II by noting that by virtue of closure, (Eq. 26), the Franck–Condon density can be viewed as a probability distribution in ε'', or equivalently in the spectral wavenumber ν. We need only use the equivalent of Eq. (13) to convert to a distribution in R:

$$P_\nu \, d\nu = P(R) \, dR \tag{32}$$

whereupon, from Eq. (9)

$$\sigma_{v,\mathrm{cl}} = \left(\frac{8\pi^3 v}{3hc}\right)\left(\frac{d}{g''}\right)\mu_\mathrm{e}^2(R_v)P(R_v)\left|\frac{\mathrm{d}V(R)}{\mathrm{d}R}\right|_{R_v}^{-1} \tag{33}$$

An alternative viewpoint which leads to the same result starts with Eq. (1) in differential form and uses $\mathrm{d}N''_v = N''P(R)\,\mathrm{d}R$ (Tellinghuisen and Moeller, 1980). Similarly the spontaneous emission spectrum A_λ has the classical form (Tellinghuisen, 1983b),

$$A_{\lambda,\mathrm{cl}} = \left(\frac{64\pi^4}{3h}\right)v^5\mu_\mathrm{e}^2(R_v)P(R_v)\left|\frac{\mathrm{d}V(R)}{\mathrm{d}R}\right|_{R_v}^{-1} \tag{34}$$

where now $P(R)$ refers to a probability distribution in the excited state. Although these equations have been obtained from expressions for absorption and emission by specific (v, J) levels, it is clear that the meaning of $P(R)$ can be extended to an ensemble of molecules in different levels, just as the corresponding equations in Section II can be averaged over initial distributions. In fact it is in the latter application, particularly the case where $P(R)$ refers to a thermal distribution, that the classical method is of most value. This is because the classical method averages the fluctuation structure which occurs in quantum and semiclassical calculations; but in many cases of thermal emission or absorption, such structure is anyway smoothed by the averaging over initial states, so nothing is lost in the classical calculation. And much can be gained in ease of computation.

Equations (33) and (34) contain a factor of $|\mathrm{d}V(R)/\mathrm{d}R|$ in the denominator, so the expressions 'blow up' when this derivative is zero, or equivalently, when the upper and lower potentials are parallel. This situation is that which is associated with 'satellite bands' in atomic line-broadening theory and with a 'head-of-heads' in classical spectroscopy (Gallagher, 1975). The singularity is related to a minimum or maximum wavelength for the spectrum, and can be formally removed through semiclassical modifications (Lam, Gallagher and Hessel, 1977; Sando and Wormhoudt, 1973) or by convoluting Eq. (33) or Eq. (34) with an appropriate 'instrumental' resolution function, while limiting the total integrated spectrum to the values given in Section II.B. However, since the spectra calculated from these equations are usually reliable except very near the singularity, such steps are often not necessary.

Equations very similar to (33) and (34) are the basis for quantitative spectral calculations by the reflection method (Bayliss, 1937; Condon, 1928; Coolidge, James and Present, 1936; Finkelnburg, 1938; Gislason, 1973; Herzberg, 1950; Mies and Julienne, 1979; Tellinghuisen, 1973a; Winans and Stueckelberg, 1928). In this case $|\mathrm{d}V/\mathrm{d}R|$ is replaced by $|\mathrm{d}U_\mathrm{f}/\mathrm{d}R|$, and $P(R)$ is replaced by a properly weighted sum of quantum probability densities, e.g. for absorption

(Tellinghuisen and Moeller, 1980)

$$P(R)\left|\frac{\mathrm{d}V}{\mathrm{d}R}\right|^{-1} \leftrightarrow \sum_{v''J''} P(v''J'')\left|\frac{\mathrm{d}U'}{\mathrm{d}R}\right|_{R_v}^{-1} \psi^2_{v''J''}(R_v) \tag{35}$$

where $P(v''J'')$ is the weight for level (v'', J''), normalized to sum to unity, and R_v is, for each term in the sum, the solution of $v = U'(R_v) - E(v'', J'')$. In thermal absorption $P(v''J'')$ is a product of a degeneracy and a Boltzmann factor, divided by the rotation–vibration partition function. Through Eq. (35) the probability densities in the initial state are 'reflected' onto the final potential. The reflection method is appropriate when U_f is sufficiently steep in the transition region (Child, 1980a; Mies, 1973; Mies and Julienne, 1979) and when only small vibrational quantum numbers are involved in the initial state. In fact, under conditions where levels $v_i > 0$ make insignificant contributions to the spectrum, the reflection method is probably the method of choice for simple estimation of unstructured spectra. However, as contributions from $v_i > 0$ become important, the classical method becomes more reliable and much easier to use (see below).

What I have here called the classical method has its origins in atomic line-broadening theory, where it is usually called the quasi-static theory (Allard and Kielkopf, 1982; Ch'en and Takeo, 1957; Cooper, 1967; Jablonski, 1945). The molecular interpretation of line broadening through this theory has been developed and applied most extensively by Gallagher and coworkers in the study of metal–rare gas interactions (Carrington and Gallagher, 1974; Cheron, Scheps and Gallagher, 1976, 1977; Drummond and Gallagher, 1974; Gallagher, 1975; Gallagher, 1982; Gallagher and Holstein, 1977; Hedges, Drummond and Gallagher, 1972; Lam, Gallagher and Drullinger, 1978; Scheps *et al.*, 1975; West and Gallagher, 1978; West, Shuker and Gallagher, 1978; York, Scheps and Gallagher, 1975). These authors and others have also shown that for thermal emission or absorption (Hedges, Drummond and Gallagher, 1972),

$$P(R) \propto R^2 \mathrm{e}^{-hcU_i(R)/kT} \tag{36}$$

where the dissociation asymptote is the zero of energy for the potential $U_i(R)$. If the difference potential is monotonic in the region where $P(R)$ is appreciable, the upper and lower potentials can be extracted from the temperature dependence of the spectrum (apart from an inherent uncertainty in the absolute R_e values). In at least one case, NaAr, subsequent high-resolution studies (Smalley *et al.*, 1977; Tellinghuisen *et al.*, 1979) have confirmed the validity of this approach. The classical method has also been applied to the alkali dimers (Benedict, Drummond and Schlie, 1977, 1979), and in the case of Na_2 and Li_2, Lam, Gallagher and Hessel (1977) have shown that quantum, classical and semiclassical spectra for thermal distributions are all in good agreement (a few per cent). I have recently (Tellinghuisen, 1983b, 1984b) given quantum *vs.* classical comparisons for simpler cases of bound-free absorption and emission in I_2. In the latter work I found that the classical method agreed with the quantum spectrum within 1 per cent of peak absorption when $hc\omega_e < kT/2$ in the initial state. In addition to these works there has been some use of the classical method in the simulation of diatomic spectra, but usually in cases which might be considered to be extreme line broadening (e.g. Balling, Wright and Havey, 1982; Castex, 1981; Castex *et al.*, 1981; Düren *et al.*, 1982; Exton and Snow, 1978; Gadea *et al.*, 1983; Gerardo and Johnson, 1974; Pfaff and Stock, 1982; Sayer *et al.*, 1978, 1979, 1980; Tam *et al.*, 1978). In view of its demonstrated reliability for the calculation of unstructured spectra for heavy molecules and/or high T, it is perhaps deserving of more attention from molecular spectroscopists.

It is worth noting that another school of classical methodology has arisen in the last few years (Bergsma *et al.*, 1984; Heller, 1981; Noid, Koszykowski and Marcus, 1977; Simons, 1982). As in the method I have outlined, the focus is on the relation of spectral frequency to difference potentials and surfaces via Eq. (31) and its analogue for polyatomics. However, the spectra are calculated in the time domain using trajectory calculations to sample the initial potential surface. In effect the trajectory calculations determine $P(R)$ in Eqs. (33) and (34) by time averaging, whereas Gallagher's method utilizes ensemble averaging. In fact, Bergsma and coworkers (1984) have recently demonstrated the formal equivalence of these methods. The new classical methods look particularly promising for the interpretation of a range of spectral phenomena in polyatomic molecules, where they offer considerable savings in computational effort over quantum methods, in effect permitting calculations that would otherwise be impossible or prohibitively expensive. However, in light of the non-negligible errors that can occur in the classical calculation of spectra for light diatomics at low T ($hc\omega_e > 2kT$; see Tellinghuisen, 1984b), there is a need for further benchmarking of these new methods against quantum results for selected cases where the latter can be obtained.

D. Semiclassical Methods

1. *Reflection Structure*

In most semiclassical treatments of intensities in diatomic transitions, the focus is on various closed-form approximations to the Franck–Condon matrix elements which appear in the equations in Section II. These treatments range from the 'primitive' approximation, in which the overlap integrals are estimated by the stationary-phase method (valid only in the classical region), to various 'uniform' treatments, which yield more accurate estimates of the integrals, valid in the non-classical as well as the classical region (Child,

1980a; Miller, 1975). The stationary-phase method (Jablonski, 1945; Landau and Lifshitz, 1958) has been utilized by a number of workers, including Mies (1968; also Mies and Julienne, 1979), Smith (1968a), Miller (1970, 1975), Sando (1974; also Sando and Wormhoudt, 1973), Golde (1975), Szudy and Baylis (1975), Adler and Wiesenfeld (1977), and myself (Tellinghuisen, 1975; also Tellinghuisen *et al.* (1980). Similarly, the more reliable (but more complicated) uniform methods have been investigated by, among others, Miller (1968, 1975), Connor and Marcus (1971), Berry and Mount (1972), Szudy and Baylis (1975), Bieniek (1977), Krüger (1979), Golde and Kvaran (1980a, 1980b), and especially Child (1974), 1975, 1978, 1980a; also Hunt and Child, 1978; Uzer and Child, 1980; Child and Shapiro, 1983). In the present work I will confine my attention to the primitive method, which may not be reliable for numerical calculations (Child, 1980a) but gives a fundamentally correct and easily seen *qualitative* picture of the important dependences in the Franck–Condon matrix elements. The present discussion is an abbreviated version of results given previously (Tellinghuisen, 1975; Tellinghuisen *et al.*, 1980).

To be specific I will consider bound-free emission from a single bound upper-state level v' ($\equiv v$) to all levels $\varepsilon''(\equiv \varepsilon)$ in the vibrational continuum of the lower state. The primitive semiclassical wavefunctions are (omitting normalization constants)

$$\chi_v = k_v^{-1/2}\cos\phi_v; \qquad \chi_\varepsilon = k_\varepsilon^{-1/2}\cos\phi_\varepsilon \tag{37}$$

where k is the local wavenumber and ϕ is the phase, e.g.

$$k_\varepsilon^2 = \frac{8\pi^2\mu c}{h}[\varepsilon - U''(R)]$$

$$\phi_\varepsilon = \int_{R''}^{R^*} k_\varepsilon \,\mathrm{d}R - \frac{\pi}{4} \tag{38}$$

Here R'' is the classical turning point on the potential $U''(R)$ for energy ε. Both of the latter quantities are in units cm^{-1}, and the reference of energy for ε is the dissociation limit of the lower state, $U''(\infty) = 0$. The functional part of the Franck–Condon integrals is

$$\langle \varepsilon | \mu_e(R) | v \rangle \approx \tfrac{1}{2}\int (k_\varepsilon k_v)^{-1/2}\mu_e(R)(\cos\phi_+ + \cos\phi_-)\mathrm{d}R \tag{39}$$

where $\phi_\pm = \phi_\varepsilon \pm \phi_v$. The stationary-phase method (Landau and Lifschitz, 1958) recognizes that the integral of the product of two sinusoidal functions can only accumulate appreciably in regions where the two functions have the same periodicity, or equivalently where the phases ($\phi_\pm$) in Eq. (39) are changing slowly with R. Provided μ_e varies slowly with R (compared with the periods of the wavefunctions), the condition for a point of stationary phase is

$$\frac{\partial \phi_{\pm}}{\partial R}(=k_{\varepsilon} \pm k_{v}) = 0 \tag{40}$$

No solution to Eq. (40) exists for ϕ_+, so we neglect the contribution from $\cos\phi_+$ in Eq. (39). For ϕ_- the condition becomes $k_\varepsilon(R^*) = k_v(R^*)$, which is precisely the Franck–Condon principle; and the points of stationary phase lie on the curve $X(R)$ defined in Eq. (30).

If there is a single point of stantionary phase for continuum energy ε, the Franck–Condon integral becomes

$$\langle \varepsilon | \mu_e(R) | v \rangle \propto A \cos\left(W \pm \frac{\pi}{4} \right) \tag{41}$$

where

$$A = \mu_e(R^*)[k(R^*)|V'(R^*)|]^{-1/2} \tag{42}$$

and

$$W = \int_{R''}^{R^*} k_\varepsilon(R)\,\mathrm{d}R - \int_{R_1}^{R^*} k_v(R)\,\mathrm{d}R \tag{43}$$

In Eq. (43) R_1 is the inner turning point on $U'(R)$ for level v (see Fig. 1), and the plus or minus sign in the argument of Eq. (41) is the same as the sign of $V'(R^*)$ $[=(\mathrm{d}V/\mathrm{d}R)_{R=R^*}]$. I have shown no ε subscript on R^* and R'', but it should be remembered that both quantities are functions of ε.

When there is a single point of stationary phase for all classically accessible R, the function $X(R)$ increases or decreases monotonically with R, and the primitive method predicts that the spectrum will display *reflection* structure—one spectral peak for each peak in the radial probability density for the initial level v_i (Bergeman and Liao, 1980; Child, 1980a; Hunt and Child, 1978; Tellinghuisen *et al.*, 1980). Although this result is based on an admittedly crude treatment of the Franck–Condon integral, which is grossly invalid near the classical turning points, quantum test calculations verify that it is essentially correct. For example, the spectrum in Fig. 1 was calculated for $v' = 40$ and shows the anticipated 41 peaks. By increasing the mass to 700 amu, I pushed this test to $v' = 100$ for the same regions of these same potentials, with similar results. Thus it appears that a monotonically sampled difference potential can indeed be expected to yield reflection structure. This point is pursued further below.

To calculate the approximate separation between spectral peaks, we differentiate W with respect to ε and set $\Delta W = \pi$ to obtain

$$\Delta\varepsilon = (h\varepsilon/2\mu c)^{1/2}\left(\int_{R''}^{R^*} \left[\frac{\varepsilon - U''(R)}{\varepsilon} \right]^{-1/2} \mathrm{d}R \right)^{-1} \tag{44}$$

The quantity in the large parentheses is a length. For a simple hard sphere potential ($U'' = 0$ for $R > R''$, $U'' = \infty$ for $R \leq R''$), it is just $L_\varepsilon = R^* - R''$. In other cases this length can be expressed as $(L_\varepsilon + x)$, where $x > 0$ for a completely repulsive $U''(R)$ and < 0 for an attractive $U''(R)$. Thus the fluctuation interval is proportional to $\varepsilon^{1/2}/(L_\varepsilon + x)$, which is equivalent to a result obtained by Mulliken (1971b) using a different approach. If we approximate $U''(R)$ as linear in the region $R'' < R < R^*$, we obtain

$$\Delta\varepsilon = \left(\frac{h}{8\mu c}\right)^{1/2}\left(\frac{a}{L_\varepsilon}\right)^{1/2} \tag{45}$$

where a is the slope of U'' in the stated region. Thus the fluctuation interval increases where the lower potential becomes steep and decreases where U'' flattens out, in agreement with the reflection principle. These results are borne out qualitatively in the spectrum of Fig. 1. The fluctuation interval decreases with increasing v (decreasing ε), then increases again very slightly for the last few high-frequency peaks, where $(L_\varepsilon + x)$ goes to zero.

Although the results of this section have been obtained for the specific case of a bound-free transition, they apply also to bound-bound transitions, because the expression for W is the same. Of course, in a bound-bound transition only certain values of 'ε' (the eigenvalues) are allowed in the final state. Thus if the fluctuation interval $\Delta\varepsilon$ is comparable to the vibrational interval in the final state, the 'discretization' of the latter will obscure the inherent reflection pattern and the classification becomes moot.

The occurrence of reflection *structure* should not be taken as justification for using the reflection *method* (described above) to analyse the spectrum. In the reflection method the bound initial wavefunction is 'reflected' onto the final state potential, whereas the stationary-phase approximation places the emphasis on the Mulliken difference potential. Thus it is reasonable to expect the reflection method and the stationary-phase approximation to come into agreement as the Mulliken difference potential converges on the real final potential, which means that the reflection method can be trustworthy for sufficiently steep final potentials and/or suitably low initial quantum numbers v_i, as was noted earlier. (Actually the primitive semiclassical method is *not* very good in the limit $v_\mathrm{i} = 0$; see Child, 1980a. However, the reflection method *is*, because the conservation of nuclear momentum is not a significant feature of the transition.)

For the analysis of bound-free spectra displaying reflection structure, several direct inversion procedures have been described (Child, Essen and LeRoy, 1983; Chow and Smith, 1971; Smith, 1968b.) All of these require prior knowledge of one potential. Smith (1968b) used the stationary-phase method to derive expressions for the determination of an unknown bound initial potential involved in free-free transitions (modelled as bound-free) to a steeply repulsive, known final potential. Later Chow and Smith (1971) described a least-squares procedure for the solution of the companion problem—transitions from a known bound initial level to an unknown repulsive state. In both cases the repulsive state was modelled as an exponential for convenience. The latter problem (i.e. known bound to unknown repulsive) has also been treated very recently by Child, Essen and LeRoy (1983). These authors employed the more reliable uniform harmonic approximation of Hunt and Child (1978) and made no assumption about the form of the repulsive curve. All of these methods appear capable of moderate-to-good performance. However, the particular examples treated—namely, well-resolved reflection structure from a single initial bound level—are also very easy to analyse by exact quantum trial-and-error procedures (see, for example, Tellinghuisen *et al.*, 1979). Since a quantum calculation is anyway desirable as a consistency check at the end of such an analysis, there may be little gained from such inversion methods. However, they are esthetically satisfying and add convincing evidence in support of the reliability of the semiclassical treatments of Franck–Condon factors.

The stationary-phase method itself has been tested against exact calculations on several occasions (Adler and Wiesenfeld, 1977; Child, 1980a; Mies, 1968; Mies and Julienne, 1979). As noted earlier, the method is not very good for low v_i (especially $v_i = 0$), where a large fraction of the quantum probability density lies in the non-classical region. However, for large v_i it can perform quite well, both for reflection structure (Mies, 1968) and, with suitable modifications (Mies and Julienne, 1979), for interference structure (see below). It is of course still limited to the transitions which can be ascribed to the classical regions of both potentials, but for large v_i these account for most of the total transition probability. Furthermore, in contradiction to the statements of some authors, its validity is not very dependent on reduced mass, except insofar as μ determines v_i for a given energy in a given potential. For example, Mies's comparisons were carried out for $v_i \sim 15$ in He_2 and H_2.

2. *Interference Structure*

Next I turn to the case where there is an extremum in the difference potential, as shown for emission in Fig. 2. Here there are two points of stationary phase for all classically allowed ε. There are, of course, other cases where $X(R)$ may be monotonic for some ε, 'diatonic' for other; or more than two points may occur for a given ε. But I will treat just the two-point case of Fig. 2.

For ε sufficiently below the maximum in $X(R)$ we simply add the contributions from the two points of stationary phase, obtaining

$$\langle \varepsilon | \mu_e(R) | v \rangle \propto A_1 \cos\left(W_1 - \frac{\pi}{4} \right) + A_2 \cos\left(W_2 + \frac{\pi}{4} \right) \qquad (46)$$

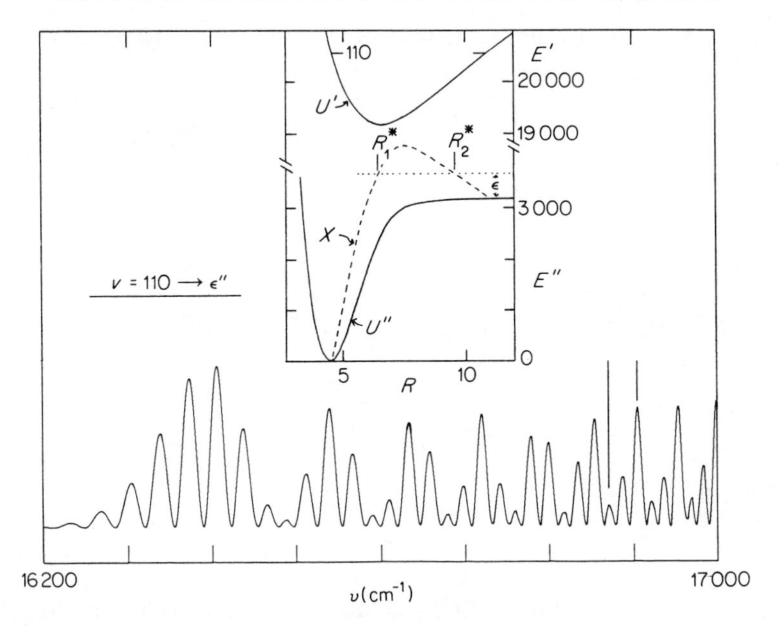

Fig. 2. Potential curves and bound-free emission spectra illustrating interference structure for Cs_2. (Reproduced from Tellinghuisen, 1984c, by permission of Academic Press, Inc.)

where A_i and W_i are as defined in Eqs. (42) and (43), with the subscripts referring to the two points of stationary phase, R_1^* and R_2^*. To illustrate clearly the origin of the interference structure that arises in this case, we rewrite Eq. (46) as

$$\langle \varepsilon | \mu_e(R) | v \rangle \propto 2A_1 \cos W_+ \cos W_- + (A_2 - A_1) \cos\left(W_2 + \frac{\pi}{4} \right) \tag{47}$$

where $W_+ = (W_1 + W_2)/2$ and $W_- = (W_1 - W_2)/2 - \pi/4$. Both W_+ and W_- vary monotonically with ε, so that when the second term in (47) is small, the spectrum has the appearance of the modulated sinusoidal function, $\cos^2 W_+ \cos^2 W_-$, with the second factor providing the envelope under which the first factor oscillates. Where R_1^* and R_2^* coalesce at the maximum in $X(R)$, the present treatment must be modified (see below). Nevertheless, the modulated sinusoidal remains a qualitatively correct description of the spectrum in this case, as is evident in Fig. 2. I have suggested (Tellinghuisen, 1977) that such structure be designated *interference* structure, in recognition of the manner in which the contributions to the overlap integral at R_1^* and R_2^* can add constructively or destructively.

In the interference-type spectrum of Fig. 2 there is what we might call fine structure and coarse structure. We can evaluate the fluctuation intervals in both as before. For the fine structure interval we obtain (Tellinghuisen *et al.*, 1980)

$$\Delta\varepsilon_{\mathrm{f}} = \left(\frac{h\varepsilon}{2\mu c}\right)^{1/2} (L_{\mathrm{f}} + x)^{-1} \tag{48}$$

where

$$L_{\mathrm{f}} + x = \int_{R''}^{R_{\mathrm{av}}} \left[\frac{\varepsilon - U''(R)}{\varepsilon}\right]^{-1/2} \mathrm{d}R \tag{49}$$

with $R^*_{\mathrm{av}} = (R^*_1 + R^*_2)/2$ and $L_{\mathrm{f}} = R^*_{\mathrm{av}} - R''$. Similarly, for the coarse structure interval

$$\Delta\varepsilon_{\mathrm{c}} = \left(\frac{2h\varepsilon}{\mu c}\right)^{1/2} (L_{\mathrm{c}} + y)^{-1} \tag{50}$$

where

$$L_{\mathrm{c}} + y = \int_{R^*_1}^{R^*_2} \left[\frac{\varepsilon - U''(R)}{\varepsilon}\right]^{-1/2} \mathrm{d}R \tag{51}$$

with $L_{\mathrm{c}} = R^*_2 - R^*_1$. Thus the fine and coarse structure both show the same dependence on ε and μ but different dependence on $X(R)$. In a crude sense the coarse structure depends only on the *shape* of $X(R)$, whereas the fine structure depends on the location of $X(R)$ relative to the repulsive branch of $U''(R)$. For example, a finer fine structure interval requires a larger displacement of $X(R)$ from the repulsive wall of $U''(R)$, and a finer coarse structure requires a 'fatter' $X(R)$.

The spectra displayed in Fig. 3 show that these dependences are again qualitatively correct. These spectra were calculated by quantum methods using a single potential for the ground state and two different upper potentials, as shown in the inset. For spectrum a the upper curve was a Morse curve, and for spectrum b the potential was obtained by simply translating the Mulliken difference potential $X(R)$ 1.2 Å to larger R (Tellinghuisen *et al.*, 1980). The two spectra were calculated for vibrational levels having nearly the same energy, so that a given spectral frequency corresponds to the same ε value in both cases. Note that the coarse structure is roughly the same in the two spectra, but the fine structure is considerably finer in b than in a, in keeping with the longer distance the lower-state wavefunction must 'travel' before it reaches the region of stationary phase (i.e. the larger L_{f} value in Eq. (48).) The slightly smaller coarse structure interval in b is attributable to the changing shape of U'' in the regions sampled in the two calculations: The flatter U'' in b yields larger (less negative) values of y in Eq. (51). Note that although the difference potentials are identical in shape, by construction, the *real* upper potentials differ considerably. It follows that a translation of $U'(R)$ along the R axis alters

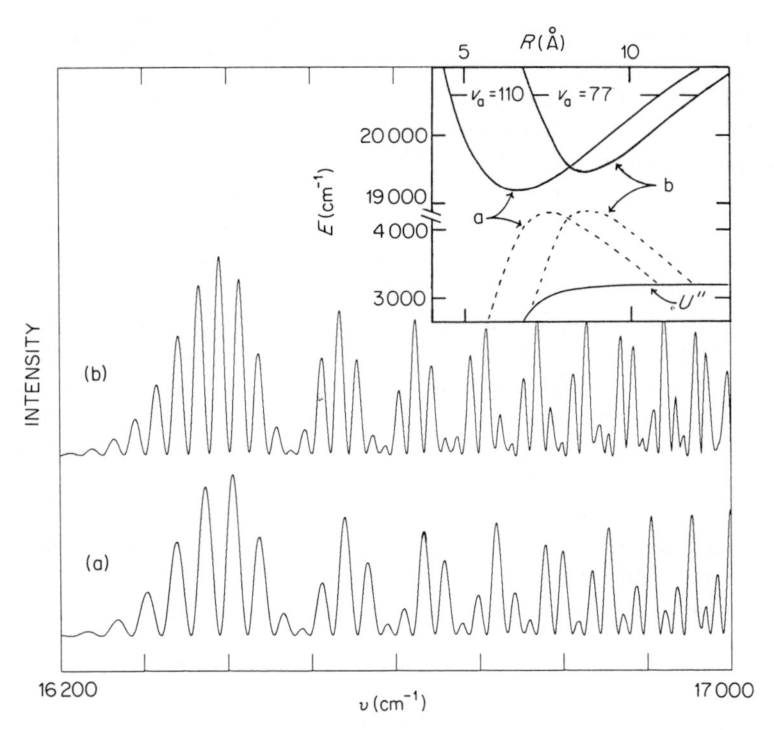

Fig. 3. Bound-free emission spectra of Cs_2 showing interference structure, calculated for a fixed lower potential and two different upper potentials, as illustrated. Potential (b) was obtained from (a) by simply translating the Mulliken difference potential 1.2 Å to larger *R*. (Reproduced from Tellinghuisen *et al.*, 1980, by permission of North-Holland Publishing Company.)

the spectrum in a more complex manner than the present shift of $X(R)$, as may in fact be seen in the sample calculations of Hunt and Child (1978). Thus in trial-and-error simulations it is often wise to work directly with the difference potential in adjusting the unknown potentials. One further point is noteworthy: The more highly structured spectrum of case b is associated with a *smaller* vibrational quantum number in the initial level, in contradiction with naive intuition.

As mentioned earlier, the primitive semiclassical treatment diverges in the region where the two points of stationary phase coalesce, which is the region associated with the long-wavelength end of the spectra in Figs. 2 and 3. As also noted earlier, this is the region responsible for the satellites of line-broadening theory and the heads-of-heads in bound-bound molecular transitions. The theoretical treatment of this region, and of the two-point problem in general, have inspired a number of contributions. The problem in the usual stationary-phase approach can be overcome by retaining the next higher (cubic) term in the expansion of ϕ_- about R^* (Carrier, 1966; Miller, 1968, 1970; Sando and

Wormhoudt, 1973). To obtain a more workable final result, Sando and Wormhoudt (1973) further approximated the difference potential as quadratic near the extremum; numerical comparisons with exact quantum results for free-bound absorption by colliding H atoms showed good qualitative and fair quantitative agreement. Golde (1975) employed a semiclassical treatment of the simple model—harmonic oscillator initial state to hard sphere final state—in a semiquantitative analysis of rare gas halide emission spectra. The same model was later treated quantum mechanically by Alder and Wiesenfeld (1977) and Tamagake and Setser (1977). The latter authors also pointed out that because of the simple sinusoidal nature of the final state wavefunction, the coarse structure in the spectrum could be expressed in terms of the momentum representation of the bound initial wavefunction. For this particular model there is also a simple relationship between the number of broad oscillations (i.e. oscillations in the modulation envelope) and the quantum number in the initial state (Adler and Wiesenfeld, 1977; Mies and Julienne, 1979), namely $(v + 2)/2$ peaks for even v and $(v + 1)/2$ for odd. Hunt and Child (1978) and Mies and Julienne (1979) have discussed the role of Fourier transforms (i.e. momentum representations) in more complex cases. In both works the approximate expressions for the coarse structure are shown to agree well with exact numerical results.

3. *Transition Moment Function*

In writing Eq. (41) I assumed that $\mu_e(R)$ varies slowly (or linearly) near R^*. If this condition holds, $\mu_e(R^*)$ serves as a slowly varying scale factor in Eqs. (42) and (46). For reflection structure the scaling is a single-value function of the spectral frequency, and the spectrum retains its reflection character. In interference spectra the different scaling at the multiple points of stationary phase can lead to distortion of the basic modulated sinusoidal pattern. If $\mu_e(R)$ varies strongly and rapidly in the region of concern, the predictions of the primitive semiclassical treatment may not be reliable for either of the two simple cases considered here.

As was noted earlier, when there is a single point of stationary phase, that point is, to a good approximation, the R centroid. This result was intimated by Jarmain and Fraser (Fraser, 1954) and was proved semiclassically by Smith (1968a), but only recently has been clearly stated from a quantum mechanical approach by Noda and Zare (1982). When there are multiple points of stationary phase, the R^n-centroids are weighted averages of the contributions from the different R^* values, so that in general $\overline{R^n} \neq \bar{R}^n$ and the R-centroid method is not reliable. As mentioned earlier, Noda and Zare (1982) have shown that in this case a fairly reliable modified R-centroid method can be regained by appropriately subdividing the range of integration.

IV. FRANCK–CONDON DISTRIBUTIONS

A. Quantum Integrals and Stationary Phase

In the preceding section I noted that Franck–Condon distributions can be characterized as displaying reflection or interference structure, depending on whether the initial wavefunction samples a monotonic or polytonic region of the difference potential, $V(R)$. In this section I pursue this point further, illustrating the conceptual validity of the stationary-phase approximation by displaying the quantum wavefunctions, their product, and the accumulated overlap integral at selected frequencies in the spectra of Figs. 1 and 2. The last quantity is particularly illustrative, since it shows that the integral indeed accumulates near the points of stationary phase when such points exist (Tellinghuisen, 1984c).

The specification of the sampled region of $V(R)$ is important, since it determines the range and nature of the Franck–Condon distribution. For example, in many electronic transitions the low initial levels v_i (for which ψ_{vi} extends over a small range of R) sample a monotonic region of $V(R)$, whereas the high levels sample a polytonic region. Thus the spectrum spans a small

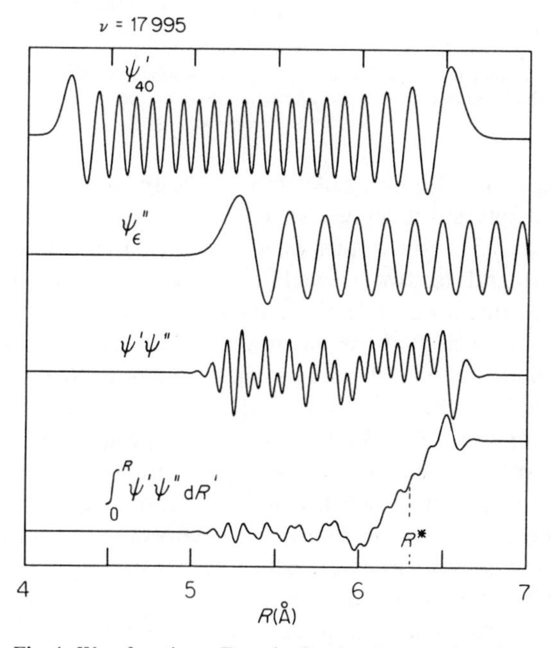

Fig. 4. Wavefunctions, Franck–Condon integrand and accumulated integral at 17995 cm^{-1} in the spectrum of Fig. 1. (Reproduced from Tellinghuisen, 1984c, by permission of Academic Press, Inc.)

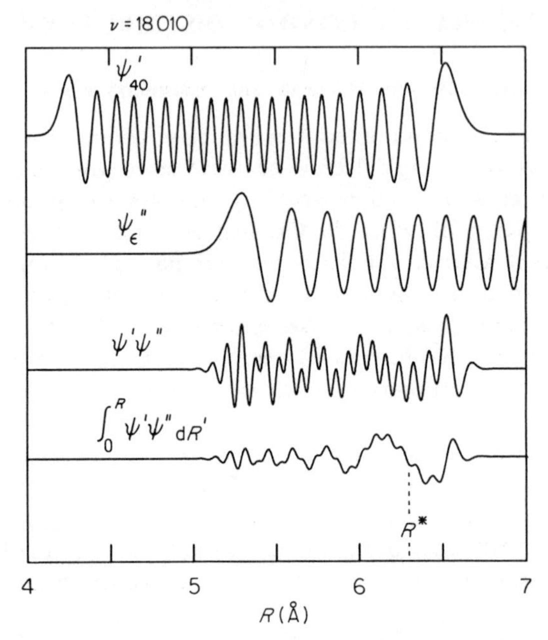

Fig. 5. Wavefunctions, Franck–Condon integrand and accumulated integral at 18 010 cm^{-1} in the spectrum of Fig. 1. (Reproduced from Tellinghuisen, 1984c, by permission of Academic Press, Inc.)

frequency range and displays reflection structure for low v_i, but is more extended and displays interference structure for high v_i.

In Figs. 4 and 5 I show the wavefunctions, their products and the accumulated integrals for two selected frequencies in the spectrum of Fig. 1—one a peak, the other a valley. In the first case the wavefunctions are in phase near R^*, and it is clear that the integral accumulates in this region. In the second, most of the 'action' still occurs near R^*, but now the wavefunctions are out of phase here and there is no net overlap as $R \to \infty$. Figures 6 and 7 show similar results for selected peaks in the spectrum of Fig. 2. Again the integrals accumulate near the points of stationary phase (now two in number). In the first case the contributions add constructively, producing a strong peak in the spectrum. In Fig. 7 the two contributions partially cancel, giving a weak peak.

In Fig. 8 I have pushed this point further, by 'perturbing' the upper potential of Fig. 2 in such a way as to create *four* points of stationary phase in a part of the continuum. The spectrum is now quite complex in the region $\nu >$ 16 900 cm^{-1}, and there is no longer any simple, easily recognized pattern. A small region of the spectrum is shown at expanded scale in Fig. 9, with selected frequencies chosen to display the wavefunctions and integrals in Figs. 10 and 11. Again the points of stationary phase are clearly important in the

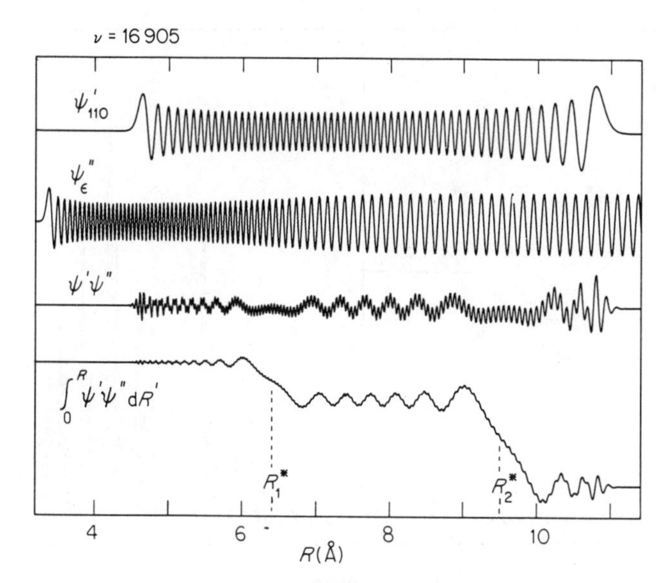

Fig. 6. Wavefunctions, Franck–Condon integrand and accumulated integral at $16\,905\,\text{cm}^{-1}$ in the spectrum of Fig. 2. (Reproduced from Tellinghuisen, 1984c, by permission of Academic Press, Inc.)

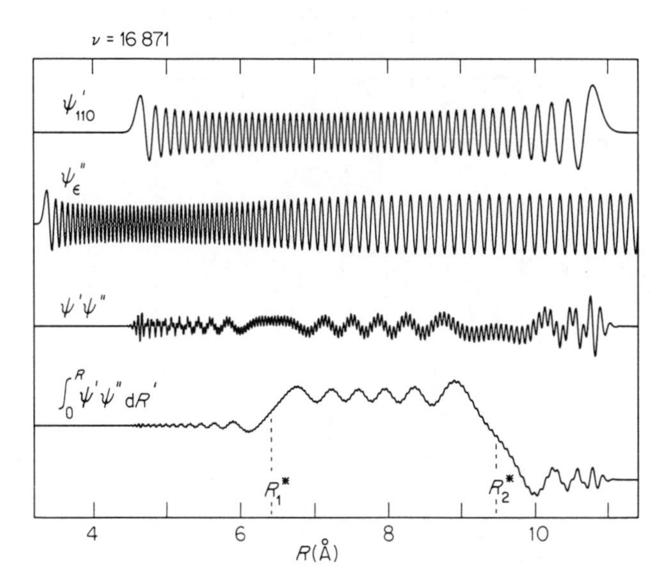

Fig. 7. Wavefunctions, Franck–Condon integrand and accumulated integral at $16\,871\,\text{cm}^{-1}$ in the spectrum of Fig. 2. (Reproduced from Tellinghuisen, 1984c, by permission of Academic Press, Inc.)

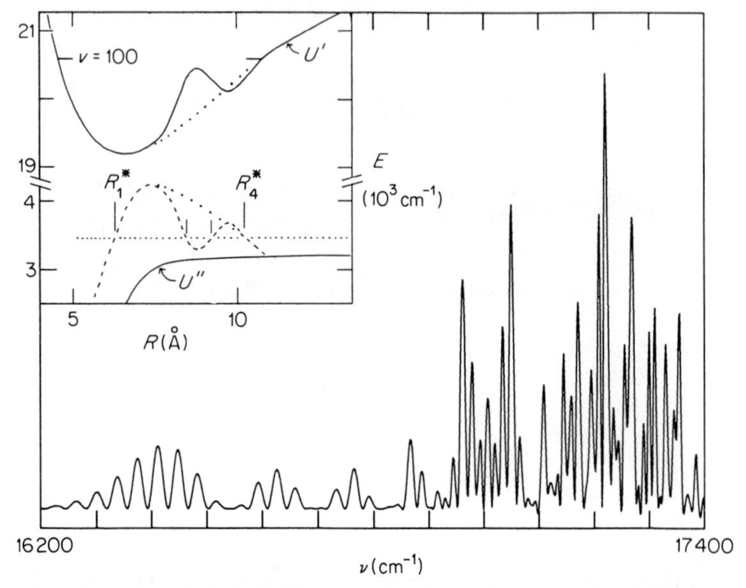

Fig. 8. Spectrum calculated for the potentials of Fig. 2 after introducing the indicated 'perturbation' (inset) of the upper potential, which is designed to create a region where there are four points of stationary phase. The dotted curves represent the unperturbed quantities. The intensity scale remains arbitrary but is roughly the same as in Fig. 2, i.e. the decrease by a factor of two in the intensity of the peaks below 16 900 cm^{-1} is real. (Reproduced from Tellinghuisen, 1984c, by permission of Academic Press, Inc.)

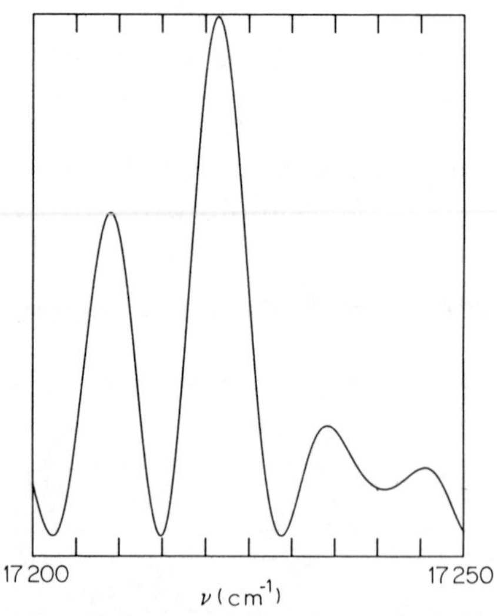

Fig. 9. Region 17 200 to 17 250 cm^{-1} of Fig. 8 at expanded wavenumber scale. (Reproduced from Tellinghuisen, 1984c, by permission of Academic Press, Inc.)

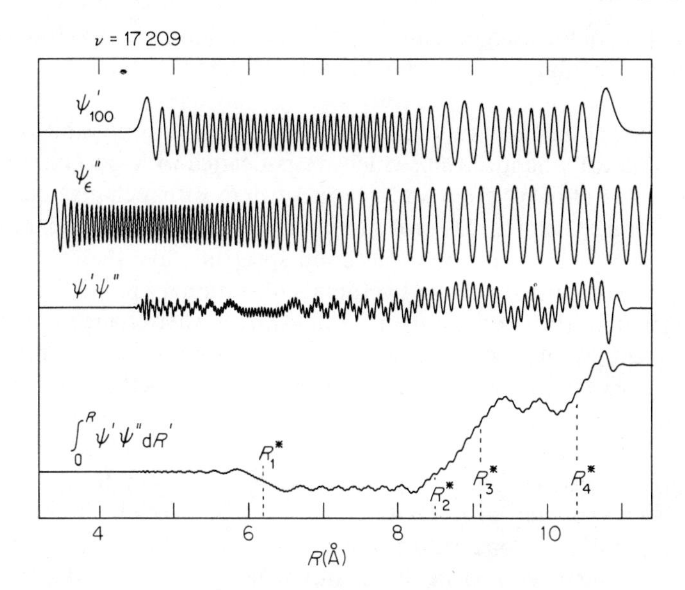

Fig. 10. Wavefunctions, integrand and accumulated overlap integral at 17 209 cm^{-1} in the spectrum of Fig. 9. (Reproduced from Tellinghuisen, 1984c, by permission of Academic Press, Inc.)

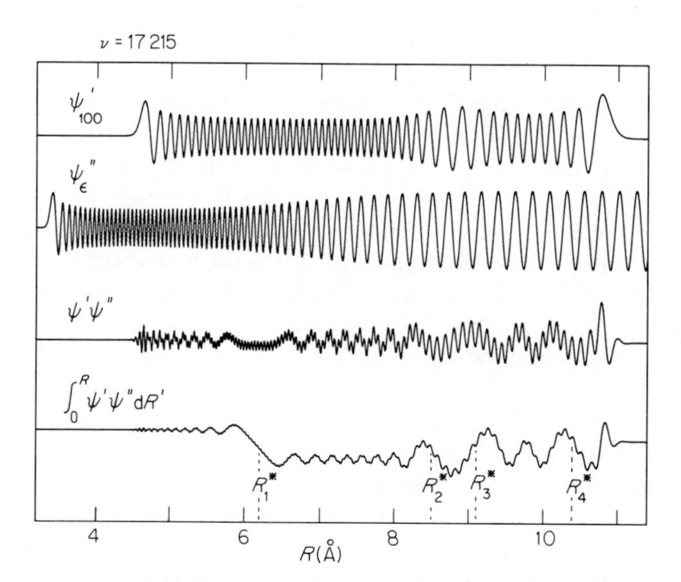

Fig. 11. Wavefunctions, integrand and accumulated overlap integral at 17 215 cm^{-1} in the spectrum of Fig. 9. (Reproduced from Tellinghuisen, 1984c, by permission of Academic Press, Inc.)

accumulation of the integrals, although R_2^* and R_3^* are now so close together that their contributions are not clearly distinguishable at the chosen frequencies.

There is one other new result in Figs. 8 and 9. It is now possible to observe nonzero valleys in the spectrum, which are associated with local minima in the absolute value of the overlap integral rather than with a change of sign. The primitive semiclassical theory predicts that there is a sign change in the overlap integral for adjacent peaks in a reflection spectrum, like that of Fig. 1. For interference structure involving two points of stationary phase, Eqs. (46) and (47) suggest that the overlap integral will *nearly* always change sign between peaks, except possibly where $\cos(W_-) \sim 0$, i.e. where the spectrum is weak. With more than two points of stationary phase, it appears that nonzero valleys can occur more frequently. Of course, if $\mu_e(R)$ is made to vary sufficiently rapidly with R, these conditions can be violated and even the characteristic reflection or interference structure can be altered. For example, in simulations of the $E \rightarrow B$ spectrum of I_2 (Tellinghuisen, 1975), I observed nonvanishing valleys separating strong peaks when a strongly R-dependent transition strength was incorporated in the quantum integrals. (Note that in experimental spectra the intensity in the valleys will never vanish completely, even for a single v_i, because of averaging of the P/R or P/Q/R rotational branch structure.)

In Fig. 12 I have reproduced a figure from Tamagake and Setser's 1977 paper, which shows the Franck–Condon structure calculated for KrF, for $v' = 21$ of a fixed upper potential and a range of different lower potentials. In the first case the difference potential is monotonic and the spectrum is a pure reflection spectrum. In examples II to V the lower potential becomes progressively flatter, producing an extremum in the difference potential. As the two-point region of $X(R)$ expands, the characteristic interference structure develops at the red end and spreads to the blue. By example V virtually the entire spectrum shows this structure, but in II through IV there is still considerable reflection-type structure at the blue end, where only a single point of stationary phase is involved. Example VI is the hypothetical case of a flat lower potential (no wall), for which the lower wavefunction is a travelling wave. This is the case where the spectrum can be expressed in terms of the momentum representation of the initial state. In principle it is also the case for a pure interference spectrum, but without the wall to produce standing waves in the lower state, the spectrum displays just the coarse structure. Note that in IV-VI the difference potential has roughly constant shape near its extremum, and the coarse structure is nearly invariant. On the other hand the fine structure interval decreases in III through V, as the repulsive wall 'backs away' from the Franck–Condon region. All of these results are consistent with the predictions of the primitive semiclassical treatment of Section III.D.

The examples that I have presented show that the qualitative picture of

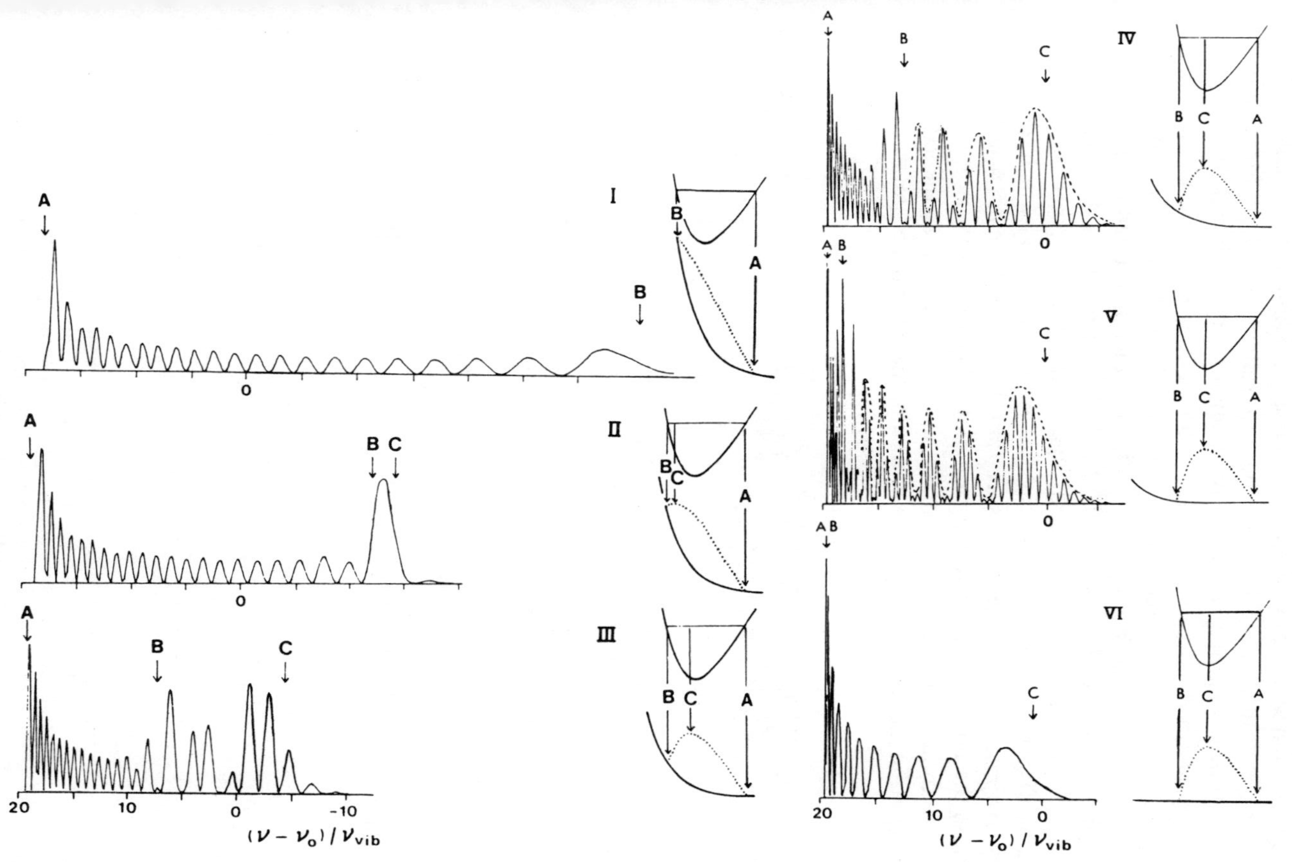

Fig. 12. Bound-free Frank–Condon densities calculated for $v' = 21$ of the B state of KrF to various assumed X potentials, which grow progressively 'flatter' in I to VI. The spectrum changes concomitantly from purely reflection to purely interference structure. Contributions to each spectrum from the regions of the two classical turning points are indicated by A and B, from the extremum in the difference potential by C. (Reproduced from Tamagake and Setser, 1977, by permission of the authors and the American Institute of Physics.)

Franck–Condon distributions which emerges from the primitive semiclassical treatment is fundamentally correct. Of course the primitive treatment covers only the classically accessible region, but that suffices to account for most of the total transition probability anyway, except for very low initial vibrational quantum numbers. Because the definition of a spectral 'peak' and the 'sampling region' must remain ambiguous, the condition for reflection structure should be taken as a strong guideline rather than a rigorous rule. For example, in the spectrum of Fig. 1, the difference potential does have an extremum just outside the classical region, and the computed spectrum actually displays subsidiary peaks on the low-frequency end which are too weak to appear on the scale of this spectrum. These peaks are not discernible because the bound initial wavefunction dies off rapidly in the non-classical region and damps them out. This must generally be true near the left-hand turning point, because the repulsive branch of the potential is almost always

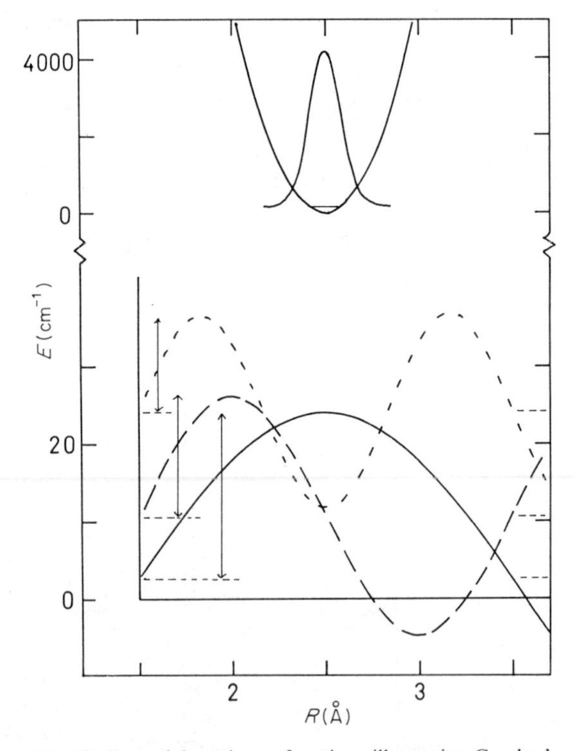

Fig. 13. Potentials and wavefunctions illustrating Condon's model for 'internal diffraction' for KrF ($\omega_e = 310$ cm^{-1}). The continuum wavefunctions in the lower state are for the first (highest frequency) spectral peak (bottom), first valley (middle) and second peak (top). (Reproduced from Tellinghuisen, 1984c, by permission of Academic Press, Inc.)

quite steep. However, there may be cases similar to this, but where the initial state function dies off slowly in the non-classical region (e.g. near the right-hand turning point for levels near dissociation), such that the additional peaks may be more prominent. Conversely, there are cases (e.g. the 2880-Å system of I_2; (see Viswanathan and Tellinghuisen, 1983, and below) where the extremum lies within the classical region but the spectrum still displays *primarily* reflection character.

As one final example in this section, I show in Fig. 13 the quantum wavefunctions for the harmonic oscillator→hard sphere model which has served as a prototype for the study of interference structure by many authors. This is in fact the case discussed by Condon (1928) when he coined the term 'internal diffraction'. The occurrence of structure in the spectrum is evident from simple symmetry considerations. As ε increases in the final state, the sinusoidal wavefunctions contract in amplitude and period, producing peaks and valleys in the spectrum as the peaks and nodes alternately come into coincidence with the symmetrical Gaussian ψ'_0. If the hard wall in the final state is displaced sufficiently to small R from ψ'_0, the structure can be both rich and sharp.

B. Illustrations

1. *Radiative Transitions*

In this section I want to amplify the points made previously, through examples of experimental spectra which illustrate the Franck–Condon principle in its many facets. The examples cover reflection structure, interference structure, and intermediate cases; and they include cases where extensive vibrational and rotational averaging occurs in the initial state, as well as the 'pure' case of a single initial level.

The simplest structured continuum, and the easiest to analyse (by either trial-and-error simulations or inversion) is the case of reflection structure from a single initial (v, J) level. A good example is the $A\,^2\Pi \rightarrow X\,^2\Sigma^+$ transition of NaAr, shown in Fig. 14. These spectra were recorded for selected v' levels of both fine structure components of the A state, using tunable laser excitation of NaAr in a nozzle beam (Smalley *et al.*, 1977; Tellinghuisen *et al.*, 1979). The analysis was by trial-and-error simulation and involved adjusting the repulsive branch of the X state, with the bound wells of both states already well determined from discrete spectra. (However, the bound-free fluorescence was used to determine the v' numbering, through the reflection principle). The lack of quantitative agreement at long wavelengths in Fig. 14 was attributed to a decline in the sensitivity of the spectrometer. Both this effect and a variation of $\mu_e(R)$ with R could yield such a discrepancy. However, the *positions* and *shapes* of the spectral peaks and nodes are insensitive to such effects, so the unknown

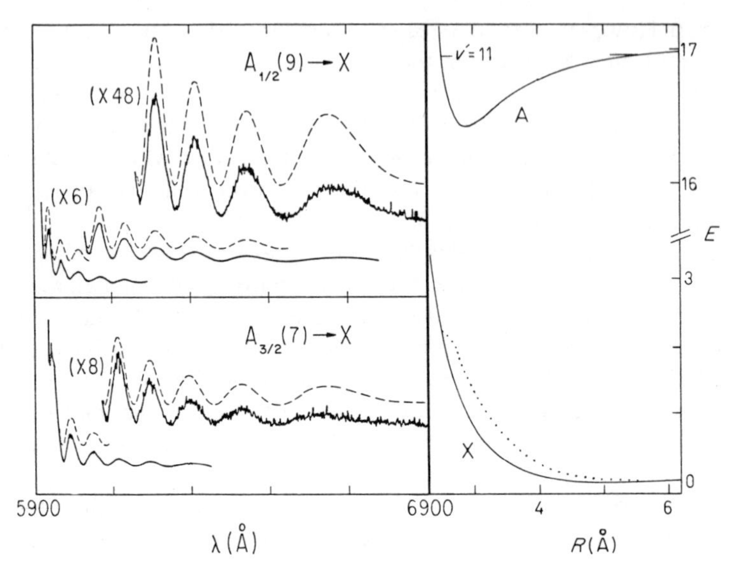

Fig. 14. Observed and calculated (dashed curves) fluorescence spectra of NaAr. The potentials (energies in $10^3\ \mathrm{cm}^{-1}$; note scale change) and Mulliken difference potential for $v' = 11$ are shown to the right. (Reproduced, with modifications, from Tellinghuisen *et al.*, 1979, by permission of the American Institute of Physics.)

potential can be determined without knowledge of $\mu_e(R)$ (which in this case was thought to be nearly constant); the latter can then be obtained as a wavelength-dependent scaling factor. This result is true in a more restricted sense for interference structure as well; however, it will not hold in either case if $\mu_e(R)$ varies too radically over the sampled region of R.

This same NaAr example serves to illustrate other points. The short-wavelength end of the spectrum is actually discrete, as the large-R transitions terminate within the shallow (40-cm^{-1}) bound well of the X state. However, for the purpose of the spectral simulations, this spectrum was treated as completely bound-free and compared with low-resolution experimental spectra, in which the sharp-line structure was not resolved. To make the calculated spectrum bound-free, we lopped off the attractive branch of the X potential at $R = R_e''$, setting $U'' = 0$ for $R > R_e''$. This step is permissable, since the Mulliken difference potential $X(R)$ spans essentially just the region $R < R_e$ for all of the simulated spectra. For this region the wavefunctions determined for the modified X curve must agree *exactly* (apart from a constant scale factor related to different normalization) with those calculated for the true (bound) X curve at the same energy. Hence if contributions to the overlap integrals are negligible beyond $R = R_e''$, the Frank–Condon densities (FCDs) calculated for the pseudocontinuum model must be identical to those for the true X curve.

The verity of this statement is illustrated in Fig. 15, where the pseudo-continuum FCDs are compared with the FCDs calculated from the discrete Franck–Condon factors (FCFs) using (Tellinghuisen, 1973b)

$$\begin{aligned} \mathrm{FCD}(v' \to v'') &= |\langle v''|v'\rangle|^2 (\mathrm{d}G''_v/\mathrm{d}v'')^{-1} \\ &\approx 2|\langle v''|v'\rangle|^2/(G''_{v+1} - G''_{v-1}) \end{aligned} \tag{52}$$

in which G_v represents the vibrational energy. It should be apparent from this example that the FCD does not 'care' about the asymptotic behaviour of the wavefunction, so similar results would have been obtained if, for example, the X potential had been made to extrapolate smoothly to an asymptote far below $U'' = 0$. In other words the *amplitude and shape* of the wavefunction in the Franck–Condon region are determined entirely by the potential in the region, $R \leq R^*$. If the attractive branch of the X curve is present, it simply serves to 'discretize' the spectrum. (An exception to the above occurs for a very narrow region just above the dissociation limit of a bound state; see Allison and Dalgarno, 1971.)

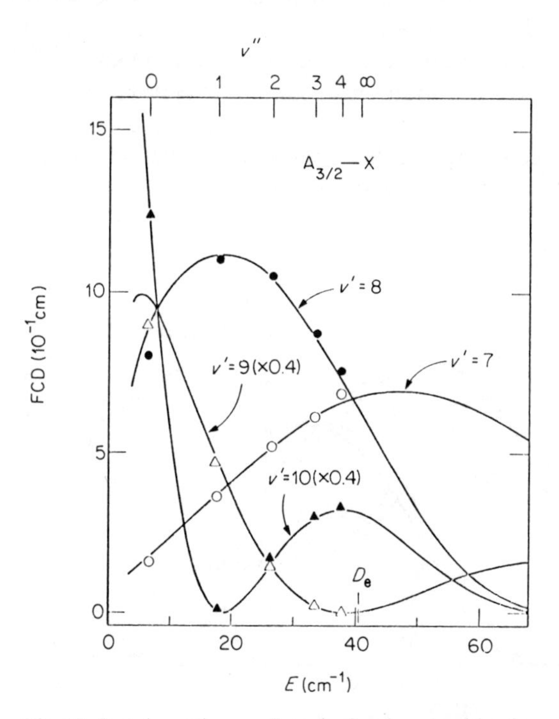

Fig. 15. Pseudocontinuum Franck–Condon densities for $v' = 7$ to 10 of NaAr A → X system, compared with results calculated from discrete FCFs using Eq. (52) (points). (Reproduced from Tellinghuisen *et al.*, 1979, by permission of the American Institute of Physics.)

Another result evident in Fig. 15 is the fact that the FCD is continuous across the dissociation limit. This point was first discussed by Allison and Dalgarno (1971) and Smith (1971) and is illustrated more graphically in Fig. 16, which shows $B \leftarrow X$ absorption in I_2 near the B-state dissociation limit (Tellinghuisen, 1973b). The spectra show only a portion of the inherent reflection structure, namely the high-frequency peaks for $v'' = 1$ and 2 and the tail of the single peak for $v'' = 0$. The rest of the distributions lie fully in the discrete spectrum. Note also the relatively weak dependence on J, which is as expected when the spectrum is not very highly structured. (For reference, the average J'' in I_2 at room temperature is 67.) One other result emerges from these calculations: Eq. (52) can be used to estimate discrete FCFs for levels very near dissociation. The calculation of bound wavefunctions for such levels can be very tedious, since the range of integration must extend to large R, with a dense enough grid to properly capture the loops over the deepest part of the well. The bound-free FCDs above dissociation usually do not demand such a large integration range, so are easier to obtain (Tellinghuisen, 1978).

The spectra in Fig. 16 were calculated for room temperature. In heavy diatomics like I_2 the excited v'' levels are appreciably populated at 300 K. Although the absorption cross-sections for individual v'' levels display reflection structure, the additional peaks for $v'' > 0$ are washed out by thermal averaging when the final potential is steep, as it is here. Another example of this case is illustrated in Fig. 17, which shows the contributions from the first four v' levels to the $D'(2_g) \rightarrow 2_u{}^3\Delta$ transition in I_2, at $T = 360$ K (Tellinghuisen,

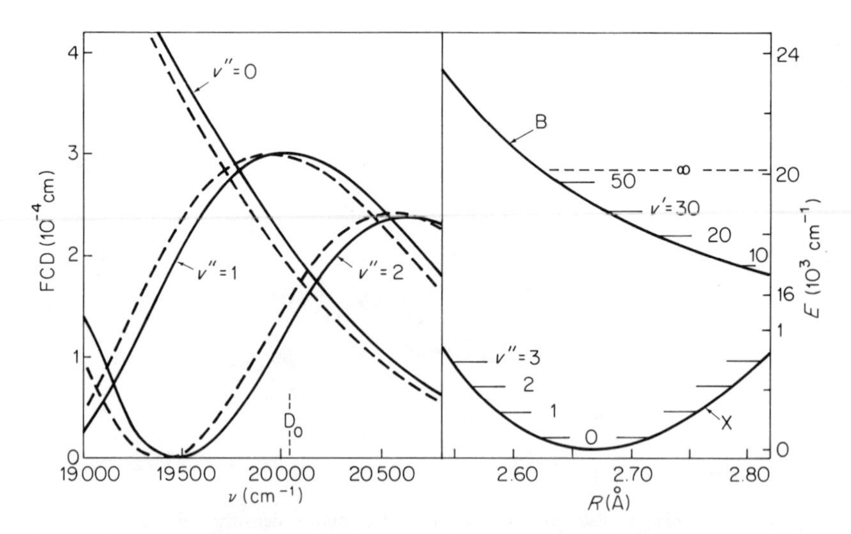

Fig. 16. The Franck–Condon density for $v'' = 0$ to 2 of I_2 B←X, in the region of the B-state dissociation limit. For the discrete regions the FCD was obtained using Eq. (52). Potentials are shown to the right (note different energy scales). (Reproduced from Tellinghuisen, 1973b, by permission of the American Institute of Physics.)

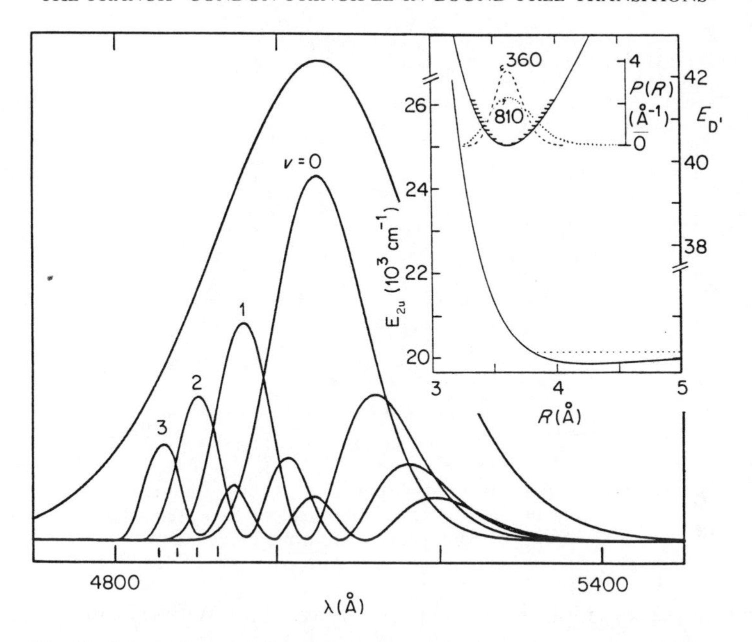

Fig. 17. Calculated $D' \rightarrow 2_u{}^3\Delta$ spectrum of I_2 at 360 K, with contributions from the first four v' levels indicated. The potentials are shown in the inset, together with the classical radial distribution functions at two temperatures. (Reprinted with permission from Tellinghuisen, 1983b. Copyright © 1983 American Chemical Society.)

1983b). Here $\omega'_e = 104\,\text{cm}^{-1}$ and $v' = 0$ accounts for only one-third of the total population. The spectrum has the simple bell shape that is particularly easy to simulate by the classical method, which in fact was used to deduce the lower potential here. Thermal absorption or emission spectra like this one broaden with increasing T (Sulzer and Wieland, 1952), as the higher v levels in the initial state make relatively larger contributions. In the classical treatment this dependence appears as a simple broadening of the radial distribution function $P(R)$ of Eq. (36) and Fig. 17.

If the difference potential is only gently sloping over a region of R near the repulsive potentials of both states (e.g. as in Fig. 1), the inherent reflection structure may persist with thermal averaging. This situation occurs for the $B \rightarrow X$ and $D \rightarrow X$ transitions in most of the rare gas halides at high buffer gas pressures (as employed in rare gas halide lasers). For example, the $B \rightarrow X$ spectrum of XeI is shown in Fig. 18, together with the contributions from the first four v' levels (Tellinghuisen *et al.*, 1976; Tellinghuisen, 1982b). It is noteworthy that the peaks in the overall spectrum do *not* correspond in a simple way to those in the individual v' components. Thus it is necessary to analyse such spectra by simulation. For these particular transitions in XeBr, XeI, KrF, and KrCl, we adopted the model of a Rittner excited state potential

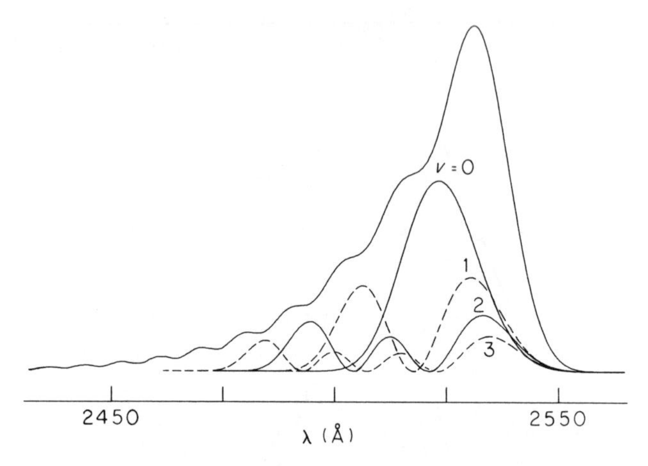

Fig. 18. Computed B → X spectrum of XeI at 360 K, showing contributions from first four v' levels. (Reproduced from Tellinghuisen, 1982b, by permission of Academic Press, Inc.)

and a simple R^{-n} repulsive curve for the lower state. With this model we were able to estimate ω_e' and the slope of $U''(R)$ in the Franck–Condon region within ~ 10 per cent and the curvature of $U''(R)$ within ~ 30 per cent. In the case of KrCl we have observed enough structure in the spectrum to make the simulation a factor of 3 more precise and to obtain a crude estimate of the R dependence of $\mu_e(R)$ in the Franck–Condon region (McKeever, Moeller and Tellinghuisen, 1984).

In spectra like that of Fig. 18, rotational averaging has a negligible effect on the peak positions and only a minor effect (some blurring) on the qualitative structure (Tellinghuisen *et al.*, 1976). However total neglect of rotation will lead to systematic errors, since the spectrum for $J = 0$ is appreciably different from that for $J_{av}(T)$. I have found that such spectra can be simulated satisfactorily using just the latter single J value, with anticipation of some further smoothing of the structure on complete rotational averaging. For the latter it usually suffices to use 3–5 representative J values, obtained by subdividing the rotational population distribution into equal segments (LeRoy, Macdonald and Burns, 1976; Tellinghuisen *et al.*, 1976).

Another case similar to that of Fig. 18 is the 'yellow' band of CsXe (Tellinghuisen and Exton, 1980), which is associated with the forbidden 7s–6s atomic Cs transition (Fig. 19). Here the structure which persists in thermal averaging is attributable to the higher v' levels, which are significantly populated in this case. Figure 20 shows that the peaks in the high-v' components nearly coincide over a large range of v'. Again this result is attributable to a nearly flat difference potential. It is also noteworthy that a spectrum very similar to that of Fig. 19 has been observed in absorption (Moe,

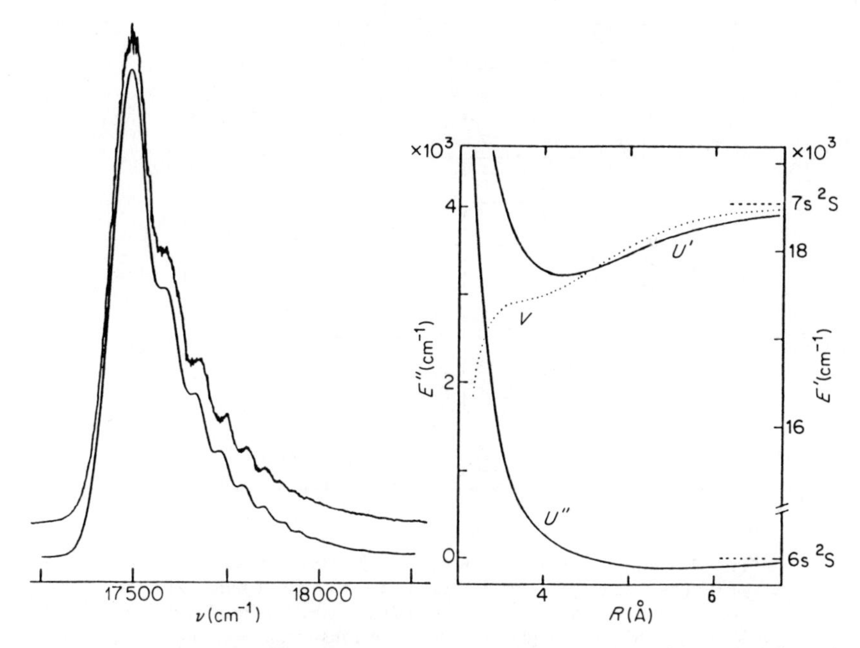

Fig. 19. Observed and calculated (lower) CsXe emission spectra, with potentials and difference potential (dotted) shown to right. (Reproduced, with modifications, from Tellinghuisen and Exton, 1980, by permission of the Institute of Physics (UK).)

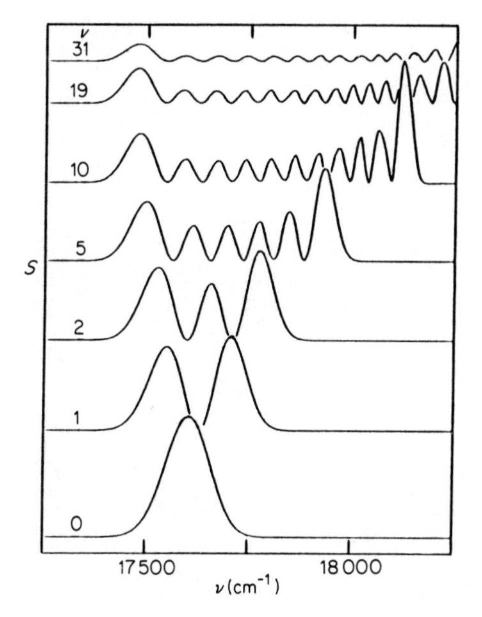

Fig. 20. Contributions to CsXe spectrum in Fig. 19 from selected v' levels. (Reproduced from Tellinghuisen and Exton, 1980, by permission of the Institute of Physics (UK).)

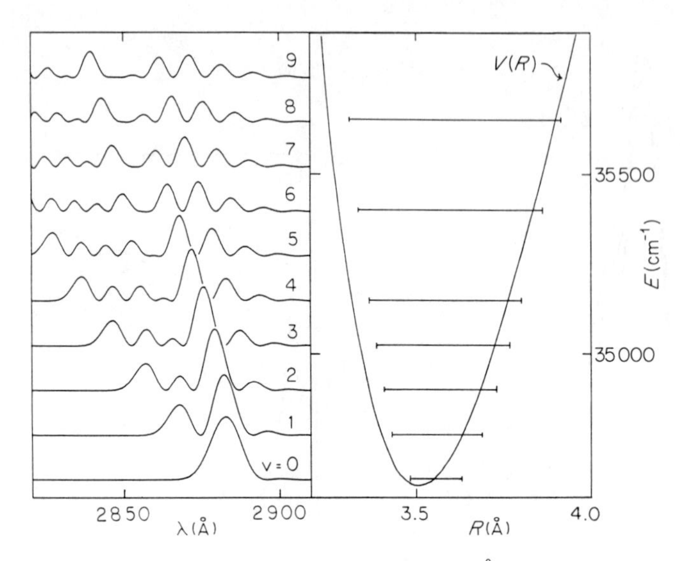

Fig. 21. Contributions of v' levels 0 to 9 to the 2880-Å emission spectrum of I_2. The difference potential is shown to the right, together with the classically accessible regions for selected v' levels. (Reproduced from Viswanathan and Tellinghuisen, 1983, by permission of Academic Press, Inc.)

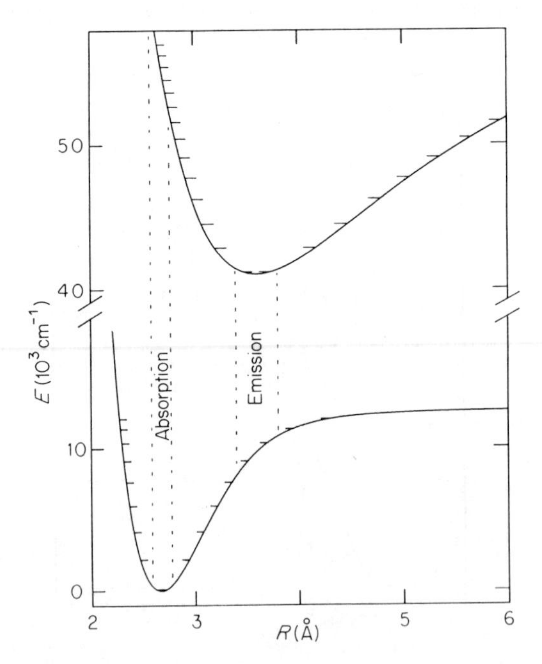

Fig. 22. Potentials for D and X states of I_2, showing regions sampled in D–X absorption and emission under conditions of thermal equilibrium in the initial state. (Reproduced from Tellinghuisen, 1983a, by permission of North-Holland Publishing Company.)

Tam and Happer, 1976; Sayer, Ferray and Lozinget, 1979), from which we can conclude that thermal emission and absorption both sample essentially the same region of the difference potential.

The 2880-Å band of I_2 (Viswanathan and Tellinghuisen, 1983) is another case where the thermal emission spectrum displays the characteristic undulatory pattern of Figs. 18 and 19. Here as in NaAr the transition terminates in a weakly bound lower state, and most of the emission from $v' = 0$ is discrete. However, the discrete structure is experimentally hard to resolve, and the low-resolution spectrum can be modelled conveniently using the same pseudocontinuum treatment that was applied to NaAr. In this case it appears that all v' levels sample an extremum in $V(R)$ (see Fig. 21), so in principle one might expect interference structure for all v'. In practice, however, the first four v' levels still show mainly reflection structure, with interference structure clearly evident for $v' \approx 4$ and higher.

The D–X system of I_2 (Figs. 22 and 23) is a good example of a case where thermal absorption and emission both display reflection structure but sample considerably different regions of the difference potential (Tellinghuisen, 1983a). As a consequence, absorption occurs in the vacuum ultraviolet, while

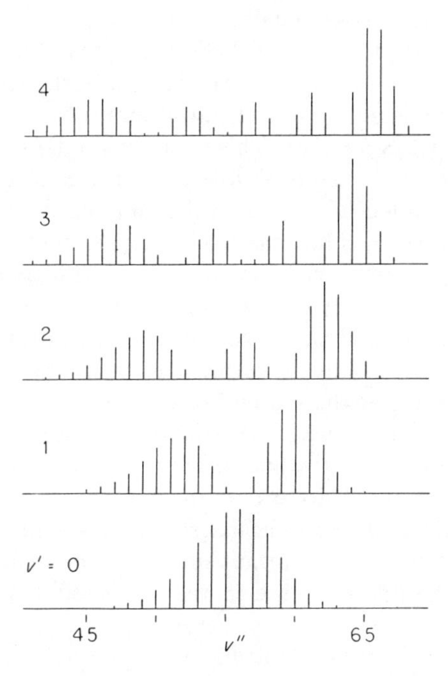

Fig. 23. Franck–Condon factors for D → X emission in I_2 from first five v' levels, plotted as functions of v''. (Reproduced from Tellinghuisen, 1983a, by permission of North-Holland Publishing Company.)

emission from low v' levels occurs near 3250 Å. On the other hand, fluorescence from high v' (e.g. the levels produced by excitation in the VUV) accesses a much larger region of $V(R)$, including a minimum, and the resulting spectrum (Fig. 24) is a beautiful example of interference structure. Indeed this system is the archetype of the highly structured continuum (Mulliken, 1970, 1971a, 1971b; Tellinghuisen, 1974a).

Figure 23 displays reflection structure in the discrete spectrum for low v_i levels. An example for higher v_i is the $B \rightarrow X$ fluorescence of I_2, shown for $v' = 21$ in Fig. 25 (Tellinghuisen, 1978, 1984c). It is interesting to note that in his 1947 paper Condon discussed this system as an example of internal diffraction, which he used in a more general sense than in his 1928 paper, to refer to features off the main Franck–Condon parabola. It is clear from Fig. 25 that this is an example of reflection structure—a fact which the experimental and computational methods of 1947 could not have revealed. Reflection structure is present in this transition to $v' \sim 32$, beyond which ψ'_v begins to sample a nearly flat region of $V(R)$, producing one very large FCF for a single v'–v'' band in each progression (e.g. an FCF of 0.45 for the 43–83 band; see Tellinghuisen, 1978, and Koffend, Bacis and Field, 1979). Note that the reflection pattern is just discernible in Fig. 25; if the fluctuation interval were much finer, the pattern would be obscured.

The long-wavelength limits of the 2880-Å spectra in Fig. 21 and the $D \rightarrow X$ spectra in Fig. 24 are both fixed by the extrema in the respective difference potentials, and there is no significant transition probability to the red of the red-most group of peaks *for any* v'. This same situation occurs for high v' in the rare gas halides, where emission at low pressures involves a large range of initial (v, J) levels (Golde, 1975; Tamagake and Setser, 1977; Tellinghuisen, 1982b). (As was noted in connection with Fig. 18, the low v' levels in these transitions display mainly reflection structure.) With rotational and vibrational averaging (particularly the latter) the fine structure is obliterated, but the coarse structure or modulation envelope persists. From a computational standpoint the fine structure is surprisingly resistant to such averaging, especially in the peaks at the red end of the spectrum. A similar potential configuration is associated with the 6100-Å fluorescence band of Cs_2 (see Figs. 2 and 3), and here too the fine structure is largely washed out unless a single-mode laser is used to excite the spectrum (Exton, Pichler and Tellinghuisen, 1981). In this case the fine structure is so fine that even the P/R branch structure can be an important smoothing mechanism for large J.

I believe that the fine structure in the Cs_2 6100-Å system is as sharp ($\Delta\varepsilon_f \sim 25\ \mathrm{cm}^{-1}$) as any bound-free structure reported to date. However, it is possible for bound-free structure to be much sharper still, as was shown in Figs. 8 and 9. Under some circumstances such structure might even be mistaken for discrete structure. The most highly structured bound-free spectra will always be associated with heavy rather than light molecules, and generally

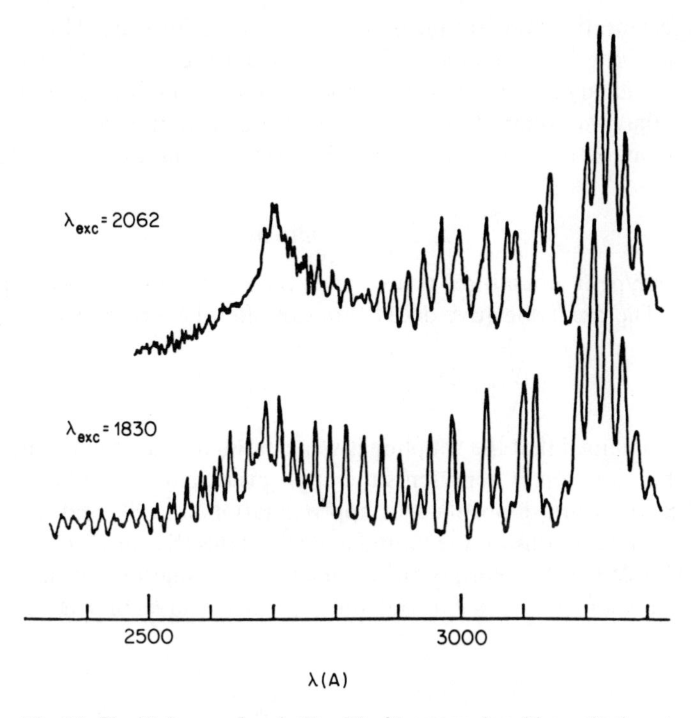

Fig. 24. The McLennan bands (D → X) of I_2, as produced by excitation at 1830 and 2062 Å. The approximate v' levels are 200 and 100 respectively. The increase in intensity near 2700 Å is attributed to another transition, produced by collisional transfer out of the D state. (Reproduced from Tellinghuisen, 1974a, by permission of North-Holland Publishing Company.)

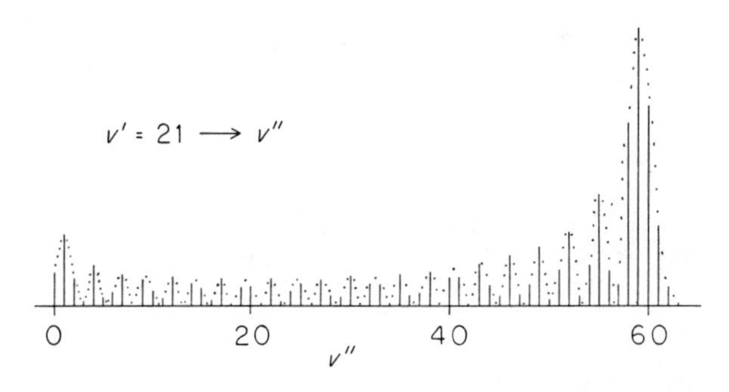

Fig. 25. Franck–Condon factors for $v' = 21 \rightarrow v''$ in the B → X spectrum of I_2. The dotted curve shows the approximate reflection envelope of the structure. (Reproduced from Tellinghuisen, 1984c, by permission of Academic Press, Inc.)

with large vibrational quantum numbers in the initial state. However, the dependence on v_i is not simple, as is evident from Figs. 3 and 24. Of course, large mass and large quantum numbers are the conditions for the approach to classical behaviour. Interestingly it is in this approach to the classical limit that Franck–Condon distributions exhibit their richest quantum structure.

2. *Predissociation*

Another type of bound-free transition is the radiationless process, predissociation. The predissociative decay rate can often be written as

$$k_p = \left(\frac{4\pi^2}{h}\right)|\langle\Psi_c|H'|\Psi_b\rangle|^2 \tag{53}$$

where it is assumed that the coupling is weak enough to use the Fermi golden rule. In this equation H' represents the appropriate part of the Hamiltonian omitted in the Born–Oppenheimer approximation, and Ψ_b and Ψ_c are the *complete* wavefunctions for the bound and free states (Kronig, 1928; Wentzel, 1927, 1928). When the Born–Oppenheimer approximations of these wavefunctions are substituted in Eq. (53), the dependence of k_p on v in the bound

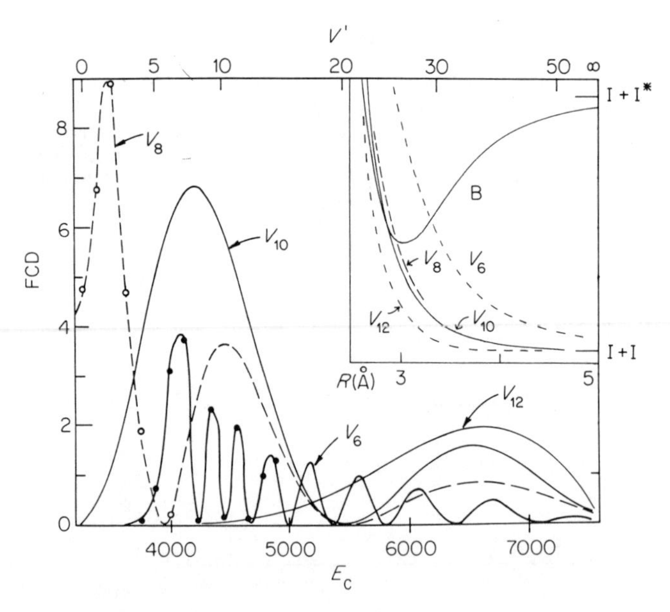

Fig. 26. Franck–Condon-type factors (units $10^{26}\,cm^{-3}$—the Franck–Condon integral contains a factor of R^{-2}) for $B \rightarrow {}^1\Pi_u$ predissociation in I_2, for various assumed ${}^1\Pi_u$ potentials (inset). (Reproduced from Tellinghuisen, 1972b, by permission of the American Institute of Physics.)

state is usually proportional to $|\langle \varepsilon|W_e(R)|v\rangle|^2$, where $W_e(R)$ represents an electronic coupling strength, analogous to $\mu_e(R)$ for radiative transitions. This matrix element is analogous to that in the expressions for radiative transitions, and predissociation is similarly governed by the Franck–Condon principle (Herzberg, 1950; Mulliken, 1960). Like radiative transitions, predissociation may be a strongly structured function of the vibrational quantum number in the initial state. However, that structure now represents a special cut in the Franck–Condon distributions for the various v levels, namely, the energy-conserving transitions. Thus the structure in k_p as a function of v cannot be simply categorized as reflection or interference structure, although it can often be interpreted in terms of one or two points of stationary phase.

As an example of the kind of structure which can occur in predissociation, I have shown in Fig. 26 the FCD calculated for $B(O_u^+\,{}^3\Pi) \rightarrow {}^1\Pi_u$ predissociation in I_2 for several hypothetical ${}^1\Pi_u$ potentials (Tellinghuisen, 1972b). The most highly structured FCD occurs for the right-branch crossing of the B curve by ${}^1\Pi_u$, for which the phase relation between the bound and free wavefunctions changes rapidly with v (and E) at the point of stationary phase—the curve crossing point. For the left-branch crossings the two potentials are nearly parallel, and the relative phases change more slowly with v, giving less structure. For the case where the bound curve is embedded in the continuum of the repulsive state, there is no point of stationary phase within the bound well, and much of the overlap occurs in the non-classical regions of one or both states. In this case the FCD increases as the two curves come closer together. However, in all cases there is a decrease near the dissociation limit, which is

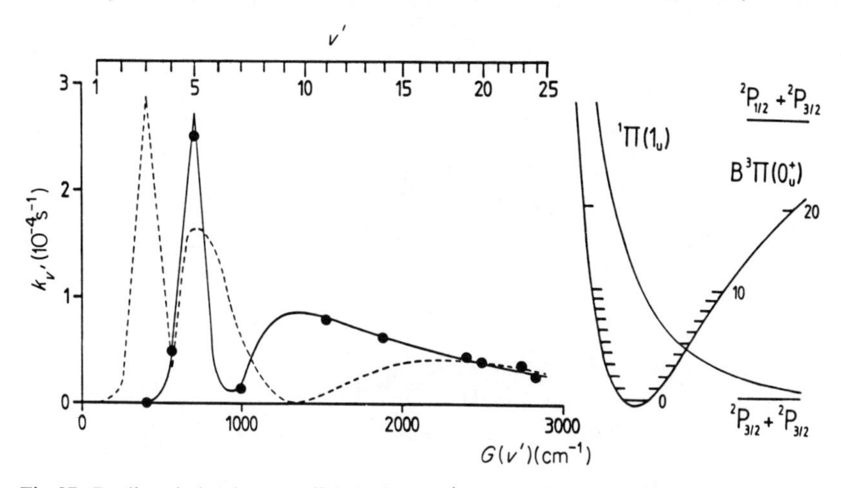

Fig. 27. Predissociative decay coefficients for $B \rightarrow {}^1\Pi_u$ predissociation in Br_2. Solid curve was calculated using potentials shown to right. Solid points are experimental values. Dashed curve was calculated for ${}^1\Pi_u$ potential of LeRoy, Macdonald and Burns (1976), which crosses the B curve between $v' = 2$ and 3. (Reproduced from Clyne, Heaven and Tellinghuisen, 1982, by permission of the American Institute of Physics.)

due to the decreasing amplitude of ψ_v in the small-R overlap region at high v. Note that the latter is an *initial* state effect, so the increasing density of bound levels near dissociation does not play any role.

Figure 27 shows k_p for the analogous predissociation in Br_2 (Clyne and Heaven, 1978). Here the right-branch crossing is forced by energetic considerations, as shown in the accompanying potential diagram, but k_p still shows only a relatively slow variation with v. Detailed analysis (Child, 1980b; Clyne, Heaven and Tellinghuisen, 1982) shows that the ${}^1\Pi_u$ curve lies roughly parallel to the B curve above $v \approx 10$. In Fig. 28 I have shown the $B \rightarrow {}^1\Pi_u$ FCD for selected v levels. The reference of energy in each case is $E(v)$, so the values at the origin are those which are pertinent to the predissociation. This figure illustrates how the predissociative FCD is equivalent to a constant-frequency cut in the ensemble of bound-free spectra. It is interesting to note that Child (1980a; also Hunt and Child, 1978) has taken just the opposite view and treated bound-free radiative transitions as a generalization of predissociation.

Another interesting example of vibrational predissociation is the $C({}^2\Sigma_u^+) \rightarrow B({}^2\Sigma_u^+)$ predissociation in N_2^+ (Roche and Tellinghuisen, 1979; Tellinghuisen and Albritton, 1975). The experimentally observed predissociation of the C state of N_2^+ was a source of puzzlement for several years, but it now appears that the $C \rightarrow B$ process can account quantitatively for the observed rates, including their monotonic dependence on v and surprisingly strong isotope dependence. For $C \rightarrow B$ predissociation the

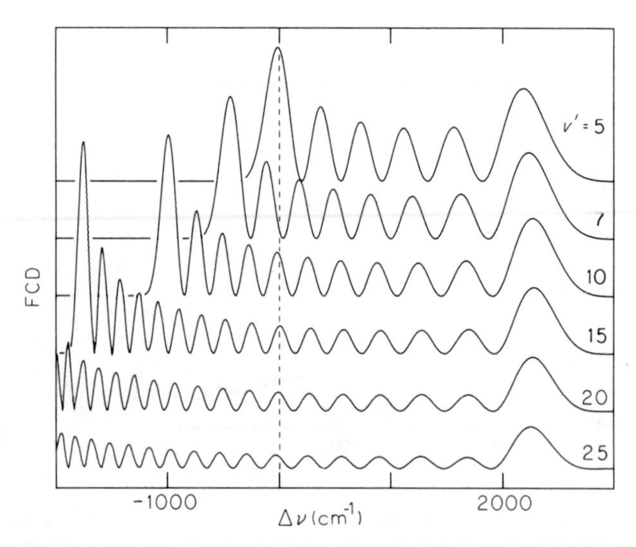

Fig. 28. Franck–Condon Densities for $B \rightarrow {}^1\Pi_u$ in Br_2. The dashed line marks the values at $\Delta\nu = 0$, which are appropriate for the predissociative coupling.

coupling is by means of the nuclear kinetic energy operator. To a good approximation the matrix element in Eq. (53) can be expressed as a sum of two matrix elements in the Born–Oppenheimer representation,

$$\langle \Psi_c | H' | \Psi_b \rangle \propto 2\langle \varepsilon | W_e(R) \partial/\partial R | v \rangle + \langle \varepsilon | W'_e(R) | v \rangle \tag{54}$$

where

$$W_e(R) = \langle \psi_{eC} | \partial/\partial R | \psi_{eB} \rangle \tag{55}$$

and $W'_e(R)$ represents the derivative of $W_e(R)$ with respect to R. As shown in Fig. 29, the C state is embedded in the continuum of the B state, so there are no points of stationary phase (i.e. no curve crossings) and the primitive semiclassical treatment cannot apply. Accordingly, the vibrational overlap in the normal sense is almost vanishingly small. However, two effects mediate to make C→B predissociation feasible: (1) the matrix element $\langle \varepsilon | \partial/\partial R | v \rangle$ is an 'amplified' FCD through the derivative operator, as was pointed out long ago by Van Vleck (1936); and (2) $W_e(R)$ is a strongly peaked function of R, as a consequence of the configuration change associated with an avoided crossing between the B and C states. The latter effect in particular serves to amplify a small portion of the Franck–Condon integral, giving an appreciable matrix element for two otherwise nearly orthogonal wavefunctions. The end result is a k_p value as large as $10^9\,\mathrm{s}^{-1}$ (for $v = 7$ of $^{14}N_2^+$) for two states which on naive inspection might not be expected to interact. The interesting isotope dependence is explained by the sensitivity of the matrix elements to phase relations between the bound and free wavefunctions, which change with

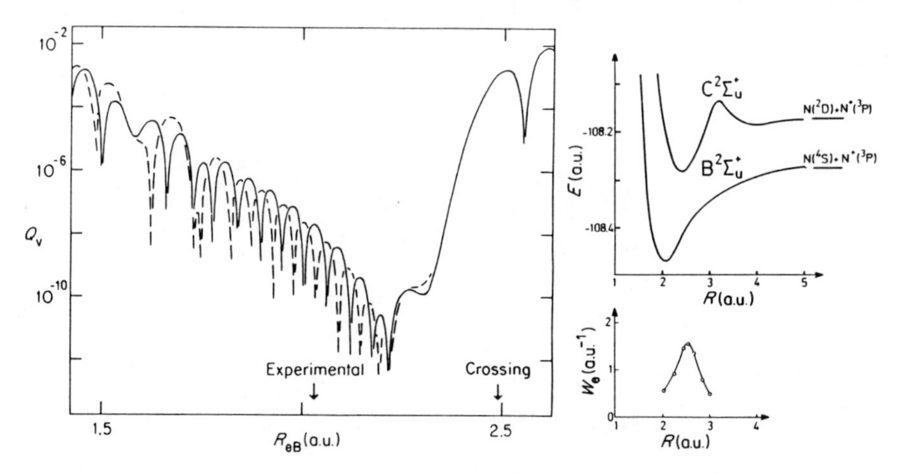

Fig. 29. Franck–Condon-type factors for $C(v' = 5) \rightarrow B$ predissociation in N_2^+, as a function of internuclear distance for the B state, for $^{14}N_2^+$ (solid) and $^{15}N_2^+$ (dashed). The arrows mark the experimental R_e value and the value which produces a curve crossing with the C state at $v' = 5$. The potentials and electronic coupling element are shown to the right. (Reproduced from Roche and Tellinghuisen, 1979, by permission of Taylor and Francis, Ltd.)

reduced mass. As an indication of this sensitivity, I have shown in Fig. 29 the results of a computational experiment in which the B state was translated along the R axis with respect to C (Roche and Tellinghuisen, 1979). The quantity plotted is the square of the usually dominant first matrix element on the right side of Eq. (54). Near the right side of the plot, the repulsive branch of the B state crosses that of C (see arrow). As Ψ_v is shifted to smaller R, the plotted quantity first drops by ten orders of magnitude, then rises again in an oscillatory fashion. Near the experimental R_{eB} value the results for $^{14}N_2^+$ and $^{15}N_2^+$ are almost exactly out of phase.

As a final example I illustrate in Fig. 30 the use of the computational methods described in Section III.B to characterize shape resonances. While this is not a Franck–Condon phenomenon, it is still relevant to a bound-free process—namely, predissociation by tunnelling (e.g. rotational predissociation). Figure 30 shows the standing-wave solutions to the Schrödinger equation near and slightly above and below the highest resonance for a hypothetical excited state potential of HeAr. The wavefunctions were obtained as described earlier, by simply generating outwards and normalizing in the region beyond the barrier. Note that the amplitude inside the barrier is maximal at or near the resonance, and the phase of ψ outside the barrier changes by π as the resonance is traversed. This particular resonance is broad, as it lies near the top of the barrier. Consequently, the amplitude of ψ_{in} remains appreciable several cm^{-1} above and below the resonance. For this same potential there are two lower resonances, both of which are so narrow that it

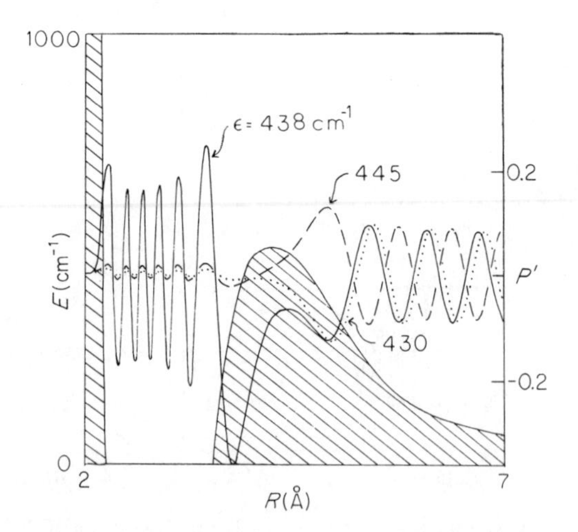

Fig. 30. Energy-normalized free wavefunctions in the vicinity of the highest shape resonance for a hypothetical state of HeAr having a potential well 3000 cm^{-1} deep with a barrier (shaded) 500 cm^{-1} high.

takes some trial-and-error effort to zero in on them. Correspondingly the peak amplitude of ψ_{in} is orders of magnitude greater than that of ψ_{out}. It is worth noting that the *positions* of all three resonances were determined within 1 cm^{-1} using the first-order WKB quantization condition

$$(v + \tfrac{1}{2}) = \left(\frac{8c\mu}{h}\right)^{1/2} \int_{R_1}^{R_2} [E - U_J(R)]^{1/2} \, dR \tag{56}$$

and solving for integer v (see, for example, Tellinghuisen, 1973c, 1983c).

V. RECENT LITERATURE

Diffuse spectra have been a subject of interest and curiosity to spectroscopists since the dawn of modern spectroscopy. For example, in his 1938 monograph, Finkelnburg devoted four chapters to diffuse spectra in diatomics and cited 525 references. In fact it is interesting to note that his Fig. 45, which depicts various possible upper and lower potential curves and relates them to the resulting spectra, is sufficient to account for virtually all possible discrete and diffuse transitions in diatomics—including phenomena being investigated in the currently faddish areas of 'radiative collisions' and 'radiation-assisted chemistry' (George, 1982; Goodzenko and Yakovlenko, 1974; Harris and Lidow, 1974; Ku, Inoue and Setser 1983; Weiner, 1980). The former term is usually meant to designate processes such as

$$A^* + B + hc\nu \rightarrow A + B^* \tag{57}$$

where A and B are atoms and the equation is interpreted mechanistically rather than phenomenologically (i.e. $B + hc\nu \rightarrow B^{**}$ followed by $B^{**} + A \rightarrow B^* + A$ is precluded). In radiation-assisted chemistry one or both of A and B are molecules, and a change in composition occurs in the net reaction. The study by Falcone and coworkers (1977), in which A = Sr and B = Ca, is often cited as the first experimental example of a radiative collision. However, at least in the weak-field limit, equivalent processes have been known to molecular spectroscopists for decades as free–free molecular transitions, and they have been treated quantitatively from a theoretical standpoint as early as the mid-1960s by Mies and Smith (1966) and Doyle (1968a, 1968b). It is true that the emphasis is on the fate of the heavy particles in the study of radiative collisions, while spectroscopists focus their attention on the photons. However, there is usually a simple mapping between the spectrum (absorption or emission) and the collision dynamics. The other distinctions are mainly matters of viewpoint. For example, the collision dynamicist speaks of process (57) as occurring off-resonance by virtue of the presence of the radiation field, whereas the molecular spectroscopist thinks of the collision as tuning the transition *into* resonance (i.e. altering the difference potential until $V(R) = \nu$ for the given AB *molecular* transition—or, equivalently, 'bending' the atomic energy levels).

Many experimental examples of free-free and free-bound diatomic transitions are now known (see below).

Since about 1940 it appears to me that the study of diffuse diatomic spectra has proceeded along two more-or-less separate paths, with relatively few convergences: (1) atomic line-broadening studies and (2) 'purely' molecular studies. I will make no attempt to review work on line broadening, since there are already recent comprehensive reviews of this subject (see especially Allard and Kielkopf, 1982). In any event, quantum structure *per se* has seldom been a salient feature of the line-broadening work, because such structure is generally lost to a combination of inadequate experimental resolution and extensive averaging over initial state energies. Regarding diffuse molecular spectra, much of the early work dealt with essentially unstructured spectra, e.g. the bell-shaped bands one typically observes for bound-free absorption by molecules in thermal equilibrium (see, for example, Sulzer and Wieland, 1952). In the 1960s there developed an interest in 'structured continua', particularly through the works of Mies and Smith (1966), Smith (1967, 1968a, 1968b, 1968c) and Mies (1968), which dealt mainly with partially resolved reflection structure in bound-free and free-free transitions in rare gas dimers ('excimers'). About 1970 two interesting cases of bound-free interference structure were noted—in H_2 by Dalgarno, Herzberg and Stephens (1970) and the previously mentioned McLennan bands of I_2 by Mulliken (1970, 1971a, 1971b). Prior to this time there was a tendency for spectroscopists to interpret the Franck–Condon principle in a too-limited fashion, in terms of vertical transitions from the classical turning points. These examples reawakened interest in the principle, as they pointed up the importance of conservation of nuclear kinetic energy as well as position; and they led eventually to an appreciation of the role of the difference potential in the Franck–Condon principle.

There have been numerous examples of structured continua cited since 1970. In Table I, I have attempted to summarize by molecule those cases which have been observed experimentally in that time, together with information about the nature of the structure and the extent to which it has been analysed. The reader might bear in mind that a number of these examples represent 'rediscoveries' of systems known already by the time of Finkelnburg's monograph, redone now with better equipment and better interpretation. By the same token there remain many examples which were first noted decades ago but still have not been correctly interpreted or quantitatively analysed. I should emphasize that the summary in Table I by no means exhausts all studies of diffuse diatomic spectra in the last 15 years. However, I *have* included all cases I could find where quantum structure was a significant feature of the observations or their interpretation. The list includes over 120 papers dealing with more than 50 different molecular transitions, of the bound-free, free-bound or free-free type. Spectroscopists are often inclined to concentrate on their own molecules to the extent that they neglect equivalent

TABLE I

Recent work involving observation and/or analysis of structured continua in diatomic molecules. Works are listed by molecule and transition. The structure is categorized as reflection (R) or interference (I), with a few cases which can be considered as intermediate (R–I). Under 'type', the symbols have the following significance: E—experimental, A—analysis (Q—quantum mechanical, S—semiclassical, RM—reflection method). (The designation 'A' is used only where some manner of quantitative interpretation is included.)

Transition	Structure	Reference	Type	Comments
Halogens				
I_2 $D(O_u^+) \rightarrow X(^1\Sigma_g^+)$	I	Mulliken (1970, 1971a, 1971b)	A(S)	First correct interpretation of diffuse bands originally observed by McLennan (1913, 1914)
		Tellinghuisen (1974a)	E, A(Q)	Analysis of fluorescence from $v' \simeq 100$ and 200, excited by I atomic radiation at 2062 and 1830 Å
		Hemmati and Collins (1980)	E	Excitation by ArF laser, at high and low pressures; several other transitions observed
		Martin *et al.* (1981)	E	Survey of emission produced by ArF laser excitation, at high and low buffer gas pressure
I_2 $E(O_g^+) \rightarrow B(O_u^+ {}^3\Pi)$	I	Rousseau and Williams (1974) Rousseau (1975)	E	Excitation to $v' = 51$ and 45 (see Brand *et al.*, 1982) by optical–optical double resonance, $E \leftarrow B \leftarrow X$
		Tellinghuisen (1975)	A(S, Q)	Analysis of spectrum of Rousseau and Williams (1974)
		Brand *et al.* (1982)	A(Q)	Refined analysis of spectra of Rousseau and Williams (1974) and Rousseau (1975), using precise E-state constants
I_2 $B \rightarrow X$	I	Tellinghuisen (1978)	A(Q)	Complex interference structure calculated for emission from $v' = 62$ to continuum just above dissociation limit

TABLE I (Contd.)

Transition	Structure	Reference	Type	Comments
I_2 D → a′(O_g^+)	R	Tellinghuisen (1974a)	E	Diffuse McLennan bands to red of 3300 Å assigned to this transition but not analysed
		Lawley *et al.* (1982)	A(S, Q)	Analysis of spectra reported by Martin *et al.* (1981)
I_2 O_g^-($^3P_1$) → O_u^-	R–I	Viswanathan and Tellinghuisen (1983)	E, A(Q)	Discrete transition with characteristic 'excimer' profile, modelled as bound-free; see Fig. 21
Br_2 ???	R, I	Tellinghuisen *et al.* (1981)	E	Many diffuse bands, including structured continua at 2000–2200 and 3150–3350 Å; assignment uncertain
Br_2 D → X(?)	?	MacDonald, Donovan and Gower (1983)	E	Three different fluorescence transitions excited by F_2 laser. Upper state labelled 'K' in literature but probably same as D in I_2
Br_2 ???	?	Kvaran and Bragason (1983)	E	Spectra similar to low-pressure spectra of Tellinghuisen *et al.* (1981), from hollow cathode discharge
IBr D → X(?)	?	MacDonald *et al.* (1983)	E	Two fluorescence transitions excited by ArF laser
Metal dimers				
Cs_2 E′(O_u^+) → X($^1\Sigma_g^+$)	I	McClintock and Balling (1969)	E	Diffuse fluorescence bands reported near 6100 Å; E′ → X assignment from Tellinghuisen *et al.* (1980)—see below
		Exton, Snow and Hillard (1978)	E	6100-Å bands observed for excitation at several wavelengths using Ar^+ laser
		Kato and Yoshihara (1979)	E	Fine structure observed in 6100-Å bands, confirming interference character of spectrum

		Höning *et al.* (1979)	E	6100-Å bands observed for excitation of Cs_2 in a heated cell, but not in nozzle-beam experiments; ${}^3\Sigma_u^+$ suggested as terminus, but symmetry-forbidden
		Tellinghuisen *et al.* (1980)	E, A(S, Q)	Extensive study of effects of initial-state vibrational and rotational averaging. $O_u^+ - X$ assignment strongly indicated, with upper state being lowest ion-pair state
		Kato (1980)	E, A(Q)	Excited-state potential similar to that obtained by Tellinghuisen *et al.* (1980), but assigned as $E({}^1\Pi_u)$. Latter unlikely, for reasons given by Tellinghuisen *et al.* (1980)
$Cs_2\ D({}^1\Sigma_u^+) \rightarrow X$	I	Raab, Weickenmeier and Demtröder (1982)	E	Dye laser excitation of diffuse fluorescence to continuum and discrete levels near dissociation limit
Cs_2 ???	?	McClintock and Balling (1969), Kato and Yoshihara (1979) and Tellinghuisen *et al.*(1980)	E	Diffuse bands reported near 6400, 7200 and 7600 Å. The 7200-Å system probably $D \rightarrow X$ (Raab, Weickenmeier and Demtröder, 1982)
NaK $d({}^3\Pi_1) \rightarrow a({}^3\Sigma^+)$	R	Breford and Engelke (1978, 1979)	E, A(RM)	Well-developed reflection spectrum excited by Ar^+ 4765 Å; called '$D\,{}^1\Pi \rightarrow a$' but later attributed to $d \rightarrow a$
		Eisel, Zevgolis and Demtröder (1979)	E	Laser excitation at 4765 and 4880 Å; spectra similar to those of Breford and Engelke (1978)
		Kato and Noda(1980)	A(Q)	Spectra of Breford and Engelke (1978) assigned to $v' = 13$ of d state, but assumptions of analysis make results for the d state approximate
		Child, Essen and Le Roy (1983)	A(S, Q)	Analysis of spectrum of Breford and Engelke (1978) by direct inversion. D state ruled out as origin, but character of d remains uncertain

TABLE I (Contd.)

Transition	Structure	Reference	Type	Comments
NaK $C(^1\Sigma^+) \rightarrow X(^1\Sigma^+)$	I	Eisel, Zevgolis and Demtröder (1979)	E	Discrete spectrum with adjoining continuum following laser excitation to high v'
		Noda and Kato (1982)	A(Q)	Calculated spectrum for $v' \simeq 50$ agrees with observations of Eisel, Zevgolis and Demtröder (1979)
Na_2 $A(^1\Sigma_u^+) \rightarrow X(^1\Sigma_g^+)$	I	Verma, Bahns and Stwalley (1981) and Verma *et al.* (1983)	E, A(Q)	Laser excitation to $v' = 34$, fluorescence to discrete and continuum levels near dissociation limit
Na_2 $C^* \rightarrow X$, $a(^3\Sigma_u^+)$	?	Allegrini *et al.* (1977) and Allegrini and Moi (1980)	E	Diffuse bands, 4160-4570 Å, produced by excitation of $Na(3^2P)$. Interpreted as free-bound $B \rightarrow X$ emission but later altered (see below, also Lyyra and Bunker, 1979, and Kopystynska and Kowalczyk, 1979)
		Woerdman (1978)	E	Violet diffuse bands attributed to at least two transitions '$C^* \rightarrow X$', with C^* produced by collisional transfer from C $(^1\Sigma_u^+)$
Na_2 ???	R	Gole *et al.* (1982)	E	Fluctuation bands in absorption attributed to Na_3 but could possibly involve Na–Na free-free absorption
Rb_2 ???	?	Brom and Broida (1974)	E	Structured 'quasi-continuum' fluorescence in 5400-6100-Å region produced by laser excitation; interpretation uncertain
Ca_2 $A(^1\Sigma_u^+) \rightarrow X(^1\Sigma_g^+)$	R	Sakurai and Broida (1976)	E, A(RM)	Excitation by Ar^+ laser; assigned as '$B \rightarrow A$' but later reassigned to $A \rightarrow X$ (Vidal, 1980)
		Wyss (1978)	E	More extensive data, still assigned to '$B \rightarrow A$'
		Vidal (1980)	E	Spectra excited at 9 Ar^+ laser wavelengths, 4545-5145 Å, producing v' levels 2-35 (usually several v' levels simultaneously)

Mg_2 $A(^1\Sigma_u^+) \rightarrow X(^1\Sigma_g^+)$	R	Scheingraber and Vidal (1977)	E, A(Q)	Laser-induced bound-bound-free structured continuum, possibly overlapped by less structured free-bound-free component; detailed analysis to obtain X potential
K_2 ???	?	Rebbeck and Vaughan (1971)	E	Diffuse bands, 5600-5725 Å, from low-pressure discharges through K vapour
RbCs $^1\Pi \rightarrow X(^1\Sigma^+)$	I	Kato and Kobayashi (1983)	E, A(Q)	Observation and approximate analysis of fluorescence near 7200 Å
Sr_2 $A(^1\Sigma_u^+) \rightarrow X(^1\Sigma_g^+)$	R	Bergeman and Liao (1980)	E, A(Q)	Fluorescence attributed to both bound-bound-free and free-bound-free processes from Ar^+ and dye laser excitation. Structure present in both
Hg_2 $O_u^+ \rightarrow X(O_g^+)$	R	Ehrlich and Osgood (1978, 1979a)	E, A(S)	ArF laser excitation to $v' \simeq 57$ in free-bound process, followed by highly structured bound-free fluorescence
Hg_2 $1_u \rightarrow X$	R	Ehrlich and Osgood (1979b)	E, A(RM)	1_u state pumped from lower reservoir states by infrared lasers (HF, DF, HBr)
Rare gas excimers				
He_2 $A(^1\Sigma_u^+) \leftrightarrow X(^1\Sigma_g^+)$	R	Mies and Smith (1966), Smith (1967) and Mies (1968)	E, A(Q, S, RM)	Structure in diffuse bands of Tanaka and Yoshino (1963) interpreted as free-free spectrum
		Smith (1968b,c)	A(S)	Determination of A potential by direct inversion
		Chow and Smith (1971)	A(S)	Determination of X potential by direct inversion
		Chow, Smith and Waggoner (1971),	E	Structure in free-bound absorption
		Sando and Dalgarno (1971), Sando (1971), Sando (1972), Mukamel and Kaldor (1973) and Matsuura and Fukuda (1979b)	A(Q)	Quantum treatment of free-free and free-bound absorption and emission

TABLE I (Contd.)

Transition	Structure	Reference	Type	Comments
Ne_2 $Ne_2^* \leftrightarrow X(O_g^+)$	R, I(?)	Tanaka and Yoshino (1972), Tanaka and Walker (1981), and references cited therein	E	Undulatory structure in diffuse and discrete bands assigned to $Ne_2^* = 1_u(^3P_2)$, $O_u^+(^3P_1)$, $O_u^+(^1P_1)$; structure similar to that in He_2 and Ar_2
Ar_2 $Ar_2^* \leftrightarrow X(O_g^+)$	R, I(?)	Michaelson and Smith (1970, 1974)	E, A(S)	Undulatory diffuse band studied in emission by time-resolved detection of pulsed discharges; O_g^+ upper state obtained by direct inversion
		Tanaka, Walker and Yoshino (1979) and references cited therein; also Castex *et al.* (1981)	E	Several undulatory band systems, diffuse and discrete; latter analysed by Freeman, Yoshino and Tanaka (1979)
		Matsuura and Fakuda (1979a, 1981) and references cited therein	E, A(S)	Structured continua assigned to $1_u(^3P_2)$ states in Ne_2, Ar_2, Kr_2; 1_u potential obtained by direct inversion
Kr_2 $Kr_2^* \leftrightarrow X(O_g^+)$	R, I(?)	Tanaka, Yoshino and Freeman (1973)	E	Diffuse structure in absorption by Kr_2 and ArKr
		Gadea *et al.* (1983) and references cited therein	E, A(S, Q)	Absorption spectra analysed by classical and quantul calculations
		Matsuura and Fukuda (1981)	E, A(S)	(See above)

Xe_2 Undulatory bands like those observed in the lighter Rg_2 molecules seem not to have been observed yet in Xe_2. Dutuit *et al.* (1980) and Castex (1981) provide recent entries to the literature on this much-studied molecule. See also Mies (1973) for a computational study.

Transition	Structure	Reference	Type	Comments
RgRg′ $RgRg'^* \leftrightarrow X(O^+)$	?	Castex (1977) and Freeman, Yoshino and Tanaka (1977) and references cited therein	E	Many diffuse bands, some structured, reported for Xe/Rg and Kr/Rg mixtures

Metal–rare gas excimers				
ARgA(np$^2\Pi$) – X(ns$^2\Sigma^+$)	R	Carrington *et al.* (1973)	E, A(Q)	Quantum interpretation of undulations in CsRg and RbRg spectra, absorption and emission, observed earlier by Jefimenko and Chen (1957) and Chen and Phelps (1973)
		Carrington and Gallagher (1974)	E	Quantum undulations in both bound-free and free-free emission from RbXe
		Scheps *et al.*(1975)	E	Quantum undulations observed for all LiRg except LiXe
		Tellinghuisen *et al.* (1979)	E, A(Q)	Analysis of reflection spectra from selected v' levels of $A_{1/2}$ and $A_{3/2}$
$AR_g(n+1)s\,^2\Sigma^+ - X$	R	Tam *et al.* (1975) Marek and Niemax (1976) and Exton, Snow and Hillard (1978)	E	Observation of structured CsRg 'yellow' bands following laser excitation of Cs atoms in Rg = Ne, Ar, Kr, Xe. Similar bands noted for RbRg
		Sayer *et al.* (1976) and Tam and Moe (1976)	E	CsRg yellow bands observed in discharges
		Moe, Tam and Happer (1976) and Sayer *et al.* (1979)	E	CsRg yellow bands in absorption
		Tellinghuisen and Exton (1980)	E, A(Q)	Analysis of yellow emission band for CsXe
$AR_g\,B(n\mathrm{p}^2\Sigma^+) - X$	I	Sando (1974)	A(S)	Analysis of CsAr blue satellite of Chen and Phelps (1973)
		Herman and Sando (1978)	A(Q)	Analysis of B–X and A–X quantum structure, free-free and free-bound, for LiHe and LiAr
HgAr ???	?	Kielkopf and Miller (1974)	E	Structured continua near 2537 line attributed to free-free transitions in HgAr, HgKr, HgXe (see also Fuke, Saito and Kaya, 1983)
		Grycuk (1977) and Grycuk and Czerwosz (1981)	E	Quantitative study for HgKr; some features attributed to Hg_2

TABLE I (Contd.)

Transition	Structure	Reference	Type	Comments
Rare gas halides (*RgX*)				
$B(^2\Sigma^+) \leftrightarrow X(^2\Sigma^+)$,	R–I	Golde and Thrush (1974)	E	ArCl $B \rightarrow X$ emission from reactions of Ar* with Cl_2 and CCl_4; also ArO and KrO emission from analogous reactions
$B \rightarrow A(^2\Pi_{1/2})$ and	R			
$C(^2\Pi_{3/2}) \rightarrow A(^2\Pi_{3/2})$	R			
		Velazco and Setser (1975)	E	XeX $B \rightarrow X$ and $C \rightarrow A$ emission from reactions of Xe* with F-, Cl-, Br-, I-bearing reagents
		Ewing and Brau (1975) and Brau and Ewing (1975)	E	Observation and qualitative analysis of XeX (all X) and KrF spectra produced by electron beam excitation at high pressures
		Tellinghuisen *et al.* (1976)	E, A(Q)	Analysis of high-pressure emission spectra of XeBr, XeI, KrF; includes summary of earlier work by laser developmental laboratories
		Tamagake and Setser (1977)	E, A(Q)	Analysis of low-pressure KrF spectra to obtain v' population distributions
		Tamagake *et al.* (1979, 1981), Kolts, Velazco and Setser (1979) and Dreiling and Setser (1981)	E, A(Q)	Initial state vibrational distributions for B and C states of XeCl, XeBr, XeI, from simulation of low-pressure spectra
		Casassa, Golde and Kvaran (1978)	E	$B \rightarrow A$ spectra reported for ArX, KrX (X = Br, I) from low-pressure reactions of Rg*
		Setser *et al.* (1979)	E	XeX and KrX spectra from reactions of Xe* and Kr*; review of earlier work
		Adler and Wiesenfeld (1979)	A(S)	Analysis of KrF v' distributions
		Golde and Kvaran (1981a, 1980b)	E, A(S)	ArBr spectra from low-pressure reactions of Ar*
		Golde and Poletti (1981)	E	ArCl and ArBr $B \rightarrow X$ and $D \rightarrow X$ emission from Ar* reactions
		Tellinghuisen (1982b)	E, A(Q)	Review of earlier work

		Wren, Setser and Ku (1982)	E	Excited state distribution from tesla discharge spectra of XeCl, XeF
		Inoue, Ku and Setser (1983)	E, A(Q)	Bound-bound and bound-free XeCl $B \rightarrow X$ emission from free-bound laser excitation
Others				
H_2^+ $2p\sigma_u \leftarrow 1s\sigma_g$	R	Dunn (1968)	A(Q, RM)	Calculation of photodissociation cross-section for H_2^+ and D_2^+; comparison with reflection method
H_2 $^3\Sigma_g^+ \leftarrow {}^3\Sigma_u^+$	R	Doyle (1968a, 1968b)	A(Q)	Calculation of free-bound absorption coefficient
H_2 $B(^1\Sigma_u) \leftrightarrow X(^1\Sigma_g^+)$	I	Dalgarno, Herzberg and Stephens (1970)	E, A(Q)	First clear identification of interference structure in experimental spectra
		Stephens and Dalgarno (1972)	A(Q)	Calculation of $B \rightarrow X$ radiative dissociation for H_2, HD, D_2; also $C(^1\Pi_u) \rightarrow X$
		Sando and Wormhoudt (1973) and Sando (1974)	A(S, Q)	Comparison of classical, semiclassical and quantal band shapes for free-bound $B \leftarrow X$ absorption
		Schmoranzer and Zietz (1978)	E	$B \rightarrow X$ fluorescence from single v' levels 7 to 9, following excitation by synchrotron radiation
O_2 $B(^3\Sigma_u^-) \leftarrow X(^3\Sigma_g^-)$	R	Bixon, Raz and Jortner (1969)	E, A(Q)	Quantum oscillations reported for bound-free absorption from $v''=0$, but apparently a computational artifact (see Allison and Dalgarno, 1971, and Allison, Dalgarno and Pasachoff, 1971)
HgBr $B(^2\Sigma^+) \rightarrow A(^2\Pi)$	R	Lapatovich Gibbs and Proud (1982)	E	Undulations on blue side of broad, diffuse emission band
HgX $B \rightarrow X(^2\Sigma^+)$	I	Dreiling and Setser (1982)	E	Flowing afterglow emission for X = I, Br, Cl, F, including first observation of $B \rightarrow X$ for HgF
NaI $NaI^* \leftrightarrow X(^1\Sigma^+)$	?	Foth, Polyani and Telle (1982)	E	Structure in free-bound emission; includes survey of bound-free-bound resonance Raman work (such spectra are usually discrete)

phenomena in other molecules. I hope that Table I will serve as a reminder that what may appear new to molecule XY is probably not new to diatomic spectroscopy. Accordingly, structured continua have by now 'come of age' and should be accepted as a standard feature of spectroscopy rather than a novelty.

VI. CONCLUSION

The Franck–Condon principle was first stated by Franck (1925) and Condon (1926) in connection with the interpretation of a bound-free phenomenon—the photodissociation of I_2 by visible light. Thus it is fitting that the Franck–Condon principle has reached its zenith in the interpretation of the wealth of structure which can arise in bound-free transitions. It is also interesting that the same molecule which served as the initial example has remained through the years as probably the most instructive single source of information on the principle.

The primitive semiclassical method serves to unify the classical and quantum mechanical versions of the Franck–Condon principle. The classical version states that transitions terminate at the point or points (R^*, E^*) which conserve nuclear position and momentum. Quantum mechanics says that particles cannot be localized in this manner, and that therefore one must evaluate transition probabilities through overlap integrals containing as integrand the product of two wavefunctions which extend over the entire classical region and into the non-classical region. The semiclassical stationary-phase approximation tells us that these integrals nevertheless accumulate mostly in the vicinity of the classical transition points, which are the points of stationary phase. The primitive semiclassical method leads further to the prediction of reflection structure in the case of a monotonically sampled difference potential, interference structure otherwise. The examples that I have presented show that this picture is fundamentally correct.

It has been fashionable in recent years to quote Condon's term, 'internal diffraction', when referring to any Franck–Condon distribution displaying more than the simplest type of reflection structure (e.g. a single peak). Condon originally (1928) used this term with reference to a special case of what I now call interference structure, shown in Fig. 13, which he advanced to explain diffuse bands observed by Rayleigh (1928) in the spectrum of Hg_2. Later (1947) he used the term in a more general way to refer to cases where spectral peaks are associated with inner loops of the overlapping wavefunctions. Some of his later examples can now be shown to be cases of reflection structure, as was noted in connection with Fig. 25. Interestingly, in the same 1928 paper in which he coined 'internal diffraction', Condon also introduced the reflection method. Thus it is clear that by 1928 Condon had an appreciation of both reflection and interference structure in diatomic intensity distributions. In the intervening 55 years we have learned a great deal about the application of the principle, largely through computational methods made possible by the advent of the computer. But perhaps the most significant new *qualitative* feature we have learned in this time is the paramount role played by the difference potential in determining the range and nature of Franck–Condon distributions.

Acknowledgements

I would like to thank Mark Child for helpful comments on an early draft of this manuscript. I also thank the University of Canterbury, where a portion of the work was completed, for the use of facilities and the assistance of technical staff. This work was supported in part by the Air Force Office of Scientific Research under Grant AFOSR-83-0110.

References

Adler, S. M., and Wiesenfeld, J. R. (1977). *J. Mol. Spectrosc.*, **66**, 357.
Adler, S. M., and Wiesenfeld, J. R. (1979). *Chem. Phys.*, **43**, 21.
Albritton, D. L., Schmeltekopf, A. L., and Zare, R. N. (1973). Unpublished.
Allard, N., and Kielkopf, J. (1982). *Rev. Mod. Phys.*, **54**, 1103.
Allegrini, M., Alzetta, G., Kopystynska, A., Moi, L., and Orriols, G. (1977). *Opt. Commun.*, **22**, 329.
Allegrini, M., and Moi, L. (1980). *Opt. Commun.*, **32**, 91.
Allison, A. C., and Dalgarno, A. (1971). *J. Chem. Phys.*, **55**, 4342.
Allison, A. C., Dalgarno, A., and Pasachoff, N. W. (1971). *Planet. Space Sci.*, **19**, 1463.
Balling, L. C., Wright, J. J., and Havey, M. D. (1982). *Phys. Rev.*, **A26**, 1426.
Bayliss, N. S. (1937). *Proc. Roy. Soc. London*, **158**, 551.
Benedict, R. P., Drummond, D. L., and Schile, L. A. (1977). *J. Chem. Phys.*, **66**, 4600.
Benedict, R. P., Drummond, D. L., and Schile, L. A. (1979). *J. Chem. Phys.*, **70**, 3155.
Bergeman, T., and Liao, P. F. (1980). *J. Chem. Phys.*, **72**, 886.
Bergsma, J. P., Berens, P. H., Wilson, K. R., Fredkin, D. R., and Heller, E. J. (1984). *J. Phys. Chem.*, **88**, 612.
Bernstein, R. B. (1960). *J. Chem. Phys.*, **33**, 795.
Berry, M. V., and Mount, K. E. (1972). *Rep. Prog. Phys.*, **35**, 315.
Bieniek, R. J. (1977). *Phys. Rev.*, **A15**, 1513.
Bixon, M., Raz, B., and Jortner, J. (1969). *Mol. Phys.*, **17**, 593.
Brand, J. C. D., Hoy, A. R., Kalkar, A. K., and Yamashita, A. B. (1982). *J. Mol. Spectrosc.*, **95**, 350.
Brau, C. A., and Ewing, J. J. (1975). *J. Chem. Phys.*, **63**, 4640.
Breford, E. J., Engelke, F. (1978). *Chem. Phys. Lett.*, **53**, 282.
Breford, E. J., and Engelke, F. (1979). *J. Chem. Phys.*, **71**, 1994.
Brewer, L. (1967). Unpublished.
Brewer, L., and Hagan, L. (1979). *High Temp. Sci.*, **11**, 233.
Brom, J. M., and Broida, H. P. (1974). *J. Chem. Phys.*, **61**, 982.
Buckingham, R. A. (1961). In *Quantum Theory I, Elements* (Ed. D. R. Bates), Academic Press, New York.
Carrier, G. F. (1966). *J. Fluid Mech.*, **24**, 641.
Carrington, C. G., Drummond, D., Gallagher, A., and Phelps, A. V. (1973). *Chem. Phys. Lett.*, **22**, 511.

Carrington, C. G., and Gallagher, A. (1974). *J. Chem. Phys.*, **60**, 3436.
Casassa, M. P., Golde, M. F., and Kvaran, A. (1978). *Chem. Phys. Lett.*, **59**, 51.
Cashion, J. K. (1963). *J. Chem. Phys.*, **39**, 1872.
Castex, M. C. (1977). *J. Chem. Phys.*, **66**, 3854.
Castex, M. C. (1981). *J. Chem. Phys.*, **74**, 759.
Castex, M. C., Morlais, M., Spiegelmann, F., and Malrieu, J. P. (1981). *J. Chem. Phys.*, **75**, 5006.
Chen, C. L., and Phelps, A. V. (1973). *Phys. Rev.*, **A7**, 470.
Ch'en, S.-Y., and Takeo, M. (1957). *Rev. Mod. Phys.*, **29**, 20.
Cheron, B., Scheps, R., and Gallagher, A. (1976). *J. Chem. Phys.*, **65**, 326.
Cheron, B., Scheps, R., and Gallagher, A. (1977). *Phys. Rev.*, **A15**, 651.
Child, M. S. (1974). In *Molecular Spectroscopy* (Eds. R. F. Barrow, D. A. Long and D. J. Millen) Vol. 2, Chemical Society Specialist Periodical Report.
Child, M. S. (1975). *Mol. Phys.*, **29**, 1421.
Child, M. S. (1978). *Mol. Phys.*, **35**, 759.
Child, M. S. (1980a). In *Semiclassical Methods in Molecular Scattering and Spectroscopy* (Ed. M. S. Child), p. 127, Reidel, London.
Child, M. S. (1980b). *J. Phys. B: At. Mol. Phys.*, **13**, 2557.
Child, M. S., Essen, H., and Le Roy, R. J. (1983). *J. Chem. Phys.*, **78**, 6732.
Child, M. S., and Shapiro, M. (1983). *Mol. Phys.*, **48**, 111.
Chow, K.-W., and Smith, A. L. (1971). *J. Chem. Phys.*, **54**, 1556.
Chow, K.-W., Smith, A. L., and Waggoner, M. G. (1971). *J. Chem. Phys.*, **55**, 4208.
Clyne, M. A. A., and Heaven, M. C. (1978). *J. Chem. Soc. Faraday Trans. II*, **74**, 1992.
Clyne, M. A. A., Heaven, M. C., and Tellinghuisen, J. (1982). *J. Chem. Phys.*, **76**, 5341.
Condon, E. U. (1926). *Phys. Rev.*, **28**, 1182.
Condon, E. U. (1928). *Phys. Rev.*, **32**, 858.
Condon, E. U. (1947). *Am. J. Phys.*, **15**, 365.
Condon, E. U., and Shortley, G. H. (1951). *The Theory of Atomic Spectra*, Cambridge Univ. Press, London.
Connor, J. N. L., and Marcus, R. A. (1971). *J. Chem. Phys.*, **55**, 5636.
Cooley, J. W. (1961). *Math. Comp.*, **15**, 363.
Coolidge, A. S., James, H. M., and Present, R. D. (1936). *J. Chem. Phys.*, **4**, 193.
Cooper, J. (1967). *Rev. Mod. Phys.*, **39**, 167.
Dalgarno, A., Herzberg, G., and Stephens, T. L. (1970). *Astrophys. J.*, **162**, L49.
Dickinson, A. S. (1972). *J. Mol. Spectrosc.*, **44**, 183.
Doyle, R. O. (1968a). *J. Quant. Spectrosc. Radiat. Transfer*, **8**, 1555.
Doyle, R. O. (1968b). *Astrophys. J.*, **153**, 987.
Dreiling, T. D., and Setser, D. W. (1981). *J. Chem. Phys.*, **75**, 4360.
Dreiling, T. D., and Setser, D. W. (1982). *J. Phys. Chem.*, **86**, 2276.
Drummond, D. L., and Gallagher (1974). *J. Chem. Phys.*, **60**, 3426.
Dunn, G. H. (1968). *Phys. Rev.*, **172**, 1.
Düren, R., Hasselbrink, E., Tischen, H., Milosević, S., and Pichler, G. (1982). *Chem. Phys. Lett.*, **89**, 218.
Dutuit, O., Castex, M. C., La Calvé, J., and Lavollée, M. (1980). *J. Chem. Phys.*, **73**, 3107.
Ehrlich, D. J., and Osgood, Jr., R. M. (1978). *Phys. Rev. Lett.*, **41**, 547.
Ehrlich, D. J., and Osgood, Jr., R. M. (1979a). *Chem. Phys. Lett.*, **61**, 150.
Ehrlich, D. J., and Osgood, Jr., R. M. (1979b). *IEEE J. Quant. Elect.*, **QE-15**, 301.
Eisel, D., Zevgolis, D., and Demtröder, W. (1979). *J. Chem. Phys.*, **71**, 2005.
Ewing, J. J., and Brau, C. A. (1975). *Phys. Rev.*, **A12**, 129.
Exton, R. J., Pichler, G., and Tellinghuisen, J. (1981). In *Spectral Line Shapes* (Ed. B. Wende), p. 983, de Gruyter and Co., Berlin.

Exton, R. J., and Snow, W. L. (1978). *J. Quant. Spectrosc. Radiat. Transfer*, **20**, 1.
Exton, R. J., Snow, W. L., and Hillard, M. E. (1978). *J. Quant. Spectrosc. Radiat. Transfer*, **20**, 235.
Falcone, R. W., Green, W. R., White, J. C., Young, J. F., and Harris, S. E. (1977). *Phys. Rev.*, **A15**, 1333.
Finkelnburg, W. (1938). *Kontinuierliche Spektren*, Springer, Berlin.
Fleming, H. E., and Rao, K. N. (1972). *J. Mol. Spectrosc.*, **44**, 189.
Foth, H. J., Polyani, J. C., and Telle, H. H. (1982). *J. Phys. Chem.*, **86**, 5027.
Franck, J. (1925). *Trans. Faraday Soc.*, **21**, 536.
Fraser, P. A. (1954). *Can. J. Phys.*, **32**, 515.
Freeman, D. E., Yoshino, K., and Tanaka, Y. (1977). *J. Chem. Phys.*, **67**, 3462.
Freeman, D. E., Yoshino, K., and Tanaka, Y. (1979). *J. Chem. Phys.*, **71**, 1780.
Fuke, K., Saito, T., and Kaya, K. (1983). *J. Chem. Phys.*, **79**, 2487.
Gadea, F. X., Spiegelmann, F., Castex, M. C., and Morlais, M. (1983). *J. Chem. Phys.*, **78**, 7270.
Gallagher, A. (1975). In *Atomic Physics* (Eds. G. zu Putlitz, E. W. Weber and A. Winnacker), Vol. 4, p. 559, Plenum Press, New York.
Gallagher, A. (1982). In *Physics of Electronic and Atomic Collisions*, (Ed. S. Datz), p. 403, North-Holland, Amsterdam.
Gallagher, A., and Holstein, T. (1977). *Phys. Rev.*, **A16**, 2413.
George, T. F. (1982). *J. Phys. Chem.*, **86**, 10.
Gerardo, J. B., and Johnson, A. W. (1974). *Phys. Rev.*, **A10**, 1204.
Gislason, E. A. (1973). *J. Chem. Phys.*, **58**, 3702.
Golde, M. F. (1975). *J. Mol. Spectrosc.*, **58**, 261.
Golde, M. F., and Kvaran, A. (1980a). *J. Chem. Phys.*, **72**, 434.
Golde, M. F., and Kvaran, A. (1980b). *J. Chem Phys.*, **72**, 442.
Golde, M. F., and Poletti, R. A. (1981). *Chem. Phys. Lett.*, **80**. 23.
Golde, M. F., and Thrush, B. A. (1974). *Chem. Phys. Lett.*, **29**, 486.
Gole, J. L., Green, G. J., Pace, S. A., and Preuss, D. R. (1982). *J. Chem. Phys.*, **76**, 2247.
Goodzenko, L. I., and Yakovlenko, S. I. (1974). *Phys. Lett.*, **46A**, 475.
Gordon, R. (1969). *J. Chem. Phys.*, **51**, 14.
Gouedard, G., and Vigue, J. (1983). *Chem. Phys. Lett.*, **96**, 293.
Grycuk, T. (1977). *Chem. Phys. Lett.*, **50**, 309.
Grycuk, T., and Czerwosz, E. (1981). *Physica*, **106C**, 431.
Harris, S. E., and Lidow, D. B. (1974). *Phys. Rev. Lett.*, **33**, 674.
Hay, P. J., Wadt, W. R., and Dunning, Jr., T. H. (1979). *Ann. Rev. Phys. Chem.*, **30**, 311.
Hedges, R. E. M., Drummond, D. L., and Gallagher, A. (1972). *Phys. Rev.*, **A6**, 1519.
Hefferlin, R. (1976). *J. Quant. Spectrosc. Radiat. Transfer*, **16**, 1101.
Heller, E. J. (1981). In *Potential Energy Surfaces and Dynamics Calculations* (Ed. D. G. Truhlar), p. 103, Plenum Press, New York.
Hemmati, H., and Collins, G. J. (1980). *Chem. Phys. Lett.*, **75**, 488.
Herman, P. S., and Sando, K. M. (1978). *J. Chem. Phys.*, **68**, 1153.
Herzberg, G. (1950). *Spectra of Diatomic Molecules*, Van Nostrand, Princeton.
Höning, G., Czajkowski, M., Stock, M., and Demtröder, W. (1979). *J. Chem. Phys.*, **71**, 2138.
Hunt, P. M., and Child, M. S. (1978). *Chem. Phys. Lett.*, **58**, 202.
Inoue, G., Ku, J. K., and Setser, D. W. (1983). *J. Chem. Phys.*, **76**, 733.
Jablonski, W. (1945). *Phys. Rev.*, **68**, 78.
Jarmain, W. R. (1960). *Can. J. Phys.*, **38**, 217.
Jarmain, W. R. (1971). *J. Quant. Spectrosc. Radiat. Transfer*, **11**, 421.
Jefimenko, O., and Chen, S.-Y. (1957). *J. Chem. Phys.*, **26**, 913.

Julienne, P. S. (1978). *J. Chem. Phys.*, **68**, 32.
Julienne, P. S., Neumann, D., and Krauss, M. (1976). *J. Chem. Phys.*, **64**, 2990.
Kaiser, E. W. (1970). *J. Chem. Phys.*, **53**, 1686.
Kasper, J. V. V. (1963). Unpublished work cited by Zare (1964).
Kato, H. (1980). *Int. J. Quant. Chem.*, **18**, 287.
Kato, H., and Kobayashi, H. (1983). *J. Chem. Phys.*, **79**, 123.
Kato, H., and Noda, C. (1980). *J. Chem. Phys.*, **73**, 4940.
Kato, H., and Yoshihara, K. (1979). *J. Chem. Phys.*, **71**, 1585.
Kielkopf, J. F., and Miller, R. A. (1974). *J. Chem. Phys.*, **61**, 3304.
Kirschner, S. M., and Watson, J. K. G. (1973). *J. Mol. Spectrosc.*, **47**, 234.
Kirschner, S. M., and Watson, J. K. G. (1974). *J. Mol. Spectrosc.*, **51**, 321.
Klein, O. (1932). *Z. Phys.*, **76**, 226.
Klemsdal, H. (1973). *J. Quant. Spectrosc. Radiat. Transfer*, **13**, 517.
Koffend, J. B., Bacis, R., and Field, R. W. (1979). *J. Chem. Phys.*, **70**, 2366.
Kolts, J. H., Velazco, J. E., and Setser, D. W. (1979). *J. Chem. Phys.*, **71**, 1247.
Konowalow, D. D., and Olson, M. L. (1979). *J. Chem. Phys.*, **71**, 450.
Kopystynska, A., and Kowalczyk, P. (1979). *Opt. Commun.*, **28**, 78.
Kosman, W. M., and Hinze, J. (1975). *J. Mol. Spectrosc.*, **56**, 93.
Kronig, R. (1928). *Z. Physik*, **50**, 347.
Krüger, H. (1979). *Theoret. Chim. Acta*, **51**, 311.
Ku, J. K., Inoue, G., and Setser, D. W. (1983). *J. Phys. Chem.*, **87**, 2989.
Kvaran, A., and Bragason, H. (1983). *J. Chem. Soc. Faraday II*, **78**, 2131.
Lam, L. K., Gallagher, A., and Drullinger, R. (1978). *J. Chem. Phys.*, **68**, 4411.
Lam, L. K., Gallagher, A., and Hessel, M. M. (1977). *J. Chem. Phys.*, **66**, 3550.
Landau, L. D., and Lifshitz, E. M. (1958). *Quantum Mechanics–Nonrelativistic Theory*, Pergamon, London.
Lapatovich, W. P., Gibbs, G. R., and Proud, J. M. (1982). *Appl. Phys. Lett.*, **41**, 786.
Lawley, K. P., MacDonald, M. A., Donovan, R. J., and Kvaran, A. (1982). *Chem. Phys. Lett.*, **92**, 322.
Lawley, K. P., and Wheeler, R. (1981). *J. Chem. Soc. Faraday II*, **77**, 1133.
Le Roy, R. J. (1980). In *Semiclassical Methods in Molecular Scattering and Spectroscopy* (Ed. M. S. Child), p. 109, Reidel, London.
Le Roy, R. J., and Bernstein, R. B. (1968). *J. Chem. Phys.*, **49**, 4312.
Le Roy, R. J., Macdonald, R. G., and Burns, G. (1976). *J. Chem. Phys.*, **65**, 1485.
Löwdin, P.-O. (1963). *J. Mol. Spectrosc.*, **10**, 12.
Lyyra, M., and Bunker, P. R. (1979). *Chem. Phys. Lett.*, **61**, 67.
McClintock, M., and Balling, L. C. (1969). *J. Quant. Spectrosc. Radiat. Transfer*, **9**, 1209.
MacDonald, M., Donovan, R. J., and Gower, M. C. (1983). *Chem Phys. Lett.*, **97**, 72.
MacDonald, M., Wilkinson, J. P. T., Fotakis, C., Martin, M., and Donovan, R. J. (1983). *Chem. Phys. Lett.*, **99**, 250.
McKeever, M. R., Moeller, M. B., and Tellinghuisen, J. (1984). To be published.
McLennan, J. C. (1913). *Proc. Roy. Soc.*, **A88**, 289.
McLennan, J. C. (1914). *Proc. Roy. Soc.*, **A91**, 23.
Mantz, A. W., Watson, J. K. G., Narahari Rao, K., Albritton, D. L., Schmeltekopf, A. L., and Zare, R. N. (1971). *J. Mol. Spectrosc.*, **39**, 180.
Marek, J., and Niemax, K. (1976). *Phys. Lett.*, **57A**, 414.
Martin, M., Fotakis, C., Donovan, R. J., and Shaw, M. J. (1981). *Nuovo Cim.*, **63B**, 300.
Matsuura, Y., and Fukuda, K. (1979a). *J. Phys. Soc. Japan*, **46**, 1397.
Matsuura, Y., and Fukuda, K. (1979b). *J. Phys. Soc. Japan*, **47**, 1033.
Matsuura, Y., and Fukuda, K. (1981). *J. Phys Soc. Japan.*, **50**, 933.

Michaelson, R. C., and Smith, A. L. (1970). *Chem. Phys. Lett.*, **6**, 1.
Michaelson, R. C., and Smith, A. L. (1974). *J. Chem. Phys.*, **61**, 2566.
Mies, F. H. (1968). *J. Chem. Phys.*, **48**, 482.
Mies, F. H. (1973). *Mol. Phys.*, **26**, 1233.
Mies, F. H., and Julienne, P. S. (1979). *IEEE J. Quant. Elect.*, **QE-15**, 272.
Mies, F. H., and Smith, A. L. (1966). *J. Chem. Phys.*, **45**, 994.
Mies, F. H., Stevens,W. J., and Krauss, M. (1978). *J. Mol. Spectros.*, **72**, 303.
Miller, W. H. (1968). *J. Chem. Phys.*, **48**, 464.
Miller, W. H. (1970). *J. Chem. Phys.*, **52**, 3563.
Miller, W. H. (1975). *Adv. Chem. Phys.*, **30**, 77.
Mitchell, A. C. G., and Zemansky, M. W. (1934). *Resonance Radiation and Excited Atoms*, Cambridge University Press, London.
Moe, G., Tam, A. C., and Happer, W. (1976). *Phys. Rev.*, **A14**, 349.
Mukamel, S., and Kaldor, U. (1973). *Mol. Phys.*, **26**, 291.
Mulliken, R. S. (1960). *J. Chem. Phys.*, **33**, 247.
Mulliken, R. S. (1970). *Chem. Phys. Lett*, **7**, 11.
Mulliken, R. S. (1971a). *J. Chem. Phys.*, **55**, 288.
Mulliken, R. S. (1971b). *J. Cem. Phys.*, **55**, 309.
Noda, C., and Kato, H.(1982). *Chem. Phys. Lett.*, **86**, 415.
Noda, C., and Zare R. N. (1982). *J. Mol. Spectrosc.*, **95**, 254.
Noid, D. W., Koszykowski, M. L., and Marcus, R. A. (1977). *J. Chem. Phys.*, **67**, 404.
Pfaff, J., and Stock, M. (1982). *J. Chem. Phys.*, **77**, 2928.
Raab, M., Weickenmeier, H., and Demtröder, W. (1982). *Chem. Phys. Lett.*, **88**, 377.
Rabinowitch, E., and Wood, W. C. (1936). *Trans. Farada Soc.*, **32**, 540.
Rayleigh, L. (1928). *Proc. Roy. Soc.*, **A119**, 349.
Rebbeck, M. M., and Vaughan, J. M. (1971). *J. Phys. B: At. Mol. Phys.*, **4**, 258.
Rees, A. L. G. (1947). *Proc. Phys. Soc.*, **59**, 998.
Roche, A. L., and Tellinghuisen, J. (1979). *Molec. Phys.*, **38**, 129.
Rousseau, D. L. (1975). *J. Mol. Spectrosc.*, **58**, 481.
Rousseau, D. L., and Williams, P. F. (1974). *Phys. Rev. Lett.*, **33**, 1368.
Rydberg, R. (1931). *Z. Phys.*, **73**, 376.
Sakurai, K., and Broida, H. P. (1976). *J. Chem. Phys.*, **65**, 1138.
Sando, K. M. (1971). *Mol. Phys.*, **21**, 439.
Sando, K. M. (1972). *Mol. Phys.*, **23**, 413.
Sando, K. M. (1974). *Phys. Rev.*, **A9**, 1103.
Sando, K. M., and Dalgarno, A. (1971). *Mol. Phys.*, **20**, 103.
Sando, K. M., and Wormhoudt, J. C. (1973). *Phys. Rev.*, **A7**, 1889.
Sayer, B., Ferray, M., and Lozingot, J. (1979). *J. Phys. B: At. Mol. Phys.*, **12**, 227.
Sayer, B., Ferray, M., Visticot, J. P., and Lozingot, J. (1980). *J. Phys. B: At. Mol. Phys.*, **13**, 177.
Sayer, B., Ferray, M., Lozingot, J., and Berlande, J. (1976). *J. Phys. B: At. Mol. Phys.*, **9**, L293.
Sayer, B., Visticot, J. P., and Pascale, J. (1978). *J. Physique*, **39**, 361.
Scheingraber, H., and Vidal, C. R. (1977). *J. Chem. Phys.*, **66**, 3694.
Scheps, R., Ottinger, C., York, G., and Gallagher, A. (1975). *J. Chem. Phys.*, **63**, 3581.
Schmoranzer, H., and Zietz, R. (1978). *Phys. Rev.*, **A18**, 1472.
Setser, D. W., Dreiling, T. D., Brashears, Jr., H. C., and Kolts, J. H. (1979). *Disc. Faraday Soc.*, **67**, 255.
Simons, J. (1982). *J. Phys. Chem.*, **86**, 3615.
Smalley, R. E., Auerbach, D. A., Fitch, P. S. H., Levy, D. H., and Wharton, L. (1977). *J. Chem. Phys.*, **66**, 3778.

Smith, A. L. (1967). *J. Chem. Phys.*, **47**, 1561.
Smith, A. L., (1968a). *J. Mol. Spectrosc.*, **28**, 269.
Smith, A. L. (1968b). *J. Chem. Phys.*, **49**, 4813.
Smith, A. L. (1968c). *J. Chem. Phys.*, **49**, 4817.
Smith, A. L. (1971). *J. Chem. Phys.*, **55**, 4344.
Stafford, F. E. (1960). Ph.D. Thesis, University of California, Berkeley, Lawrence Radiation Lab. Report UCRL-8854.
Steinfeld, J. I. (1974). *Molecules and Radiation*, MIT Press, Cambridge, Mass.
Stephens, T. L., and Dalgarno, A. (1972). *J. Quant. Spectrosc. Radiat. Transfer*, **12**, 569.
Strickler, S. J., and Berg, R. A. (1962). *J. Chem. Phys.*, **37**, 814.
Sulzer, P., and Wieland, K. (1952). *Helv.Phys. Acta*, **25**, 653.
Szudy, J., and Baylis, W. E. (1975). *J. Quant. Spectrosc. Radiat. Transfer*, **15**, 641.
Tam, A. C., and Moe, G. W. (1976). *Phys. Rev.*, **A14**, 528.
Tam, A. C., Moe, G., Park, W., and Happer, W. (1975). *Phys. Rev. Lett.*, **35**, 85.
Tam, A. C., Yabuzaki, T., Curry, S. M., and Happer, W. (1978). *Phys. Rev.*, **A18**, 196.
Tamagake, K., Kolts, J. H., and Setser, D. W. (1979). *J. Chem. Phys.*, **71**, 1264.
Tamagake, K., and Setser, D. W. (1977). *J. Chem. Phys.*, **67**, 4370.
Tamagake, K., Setser, D. W., and Kolts, J. H. (1981). *J. Chem. Phys.*, **74**, 4286.
Tanaka, Y., and Walker, W. C. (1981). *J. Chem. Phys.*, **74**, 2760.
Tanaka, Y., Walker, W. C., and Yoshino, K. (1979). *J. Chem. Phys.*, **70**, 380.
Tanaka, Y., and Yoshino, K. (1963). *J. Chem. Phys.*, **39**, 3081.
Tanaka, Y., and Yoshino, K. (1972). *J. Chem. Phys.*, **57**, 2964.
Tanaka, Y., Yoshino, K., and Freeman, D. E. (1973). *J. Chem. Phys.*, **59**, 5160.
Telle, H., and Telle, U. (1981). *J. Mol. Spectrosc.*, **85**, 248.
Tellinghuisen, J. (1972a). *J. Mol. Spectrosc.*, **44**, 194.
Tellinghuisen, J. (1972b). *J. Chem. Phys.*, **57**, 2397.
Tellinghuisen, J. (1973a). *J. Chem. Phys.*, **58**, 2821.
Tellinghuisen, J. (1973b). *J. Chem. Phys.*, **59**, 849.
Tellinghuisen, J. (1973c). *Chem. Phys. Lett.*, **18**, 544.
Tellinghuisen, J. (1974a). *Chem. Phys. Lett.*, **29**, 359.
Tellinghuisen, J. (1974b). *Computer Phys. Commun.*, **6**, 221.
Tellinghuisen, J. (1975). *Phys. Rev. Lett.*, **34**, 1137.
Tellinghuisen, J. B. (1977). In *Etats Atomiques et Moleculaires Couples a un Continuum: Atomes et Molecules Hautement Excites*, Colloquium No. 273 of the CNRS, p. 317.
Tellinghuisen, J. (1978). *J. Quant. Spectrosc. Radiat. Transfer*, **19**, 149.
Tellinghuisen, J. (1982a). *J. Chem. Phys.*, **76**, 4736.
Tellinghuisen, J. (1982b). In *Applied Atomic Collision Physics*: Vol. 3, *Gas Lasers* (Eds, E. W. McDaniel and W. L. Nighan), Pure and Applied Physics, Vol. 43-3, p. 251, Academic Press, New York.
Tellinghuisen, J. (1983a). *Chem. Phys. Lett.*, **99**, 373.
Tellinghuisen, J. (1983b). *J. Phys. Chem.*, **87**, 5136.
Tellinghuisen, J. (1983c). *Chem. Phys. Lett.*, **102**, 4.
Tellinghuisen, J. (1984a). *Chem. Phys. Lett.*, **105**, 241.
Tellinghuisen, J. (1984b). *J. Chem. Phys.*, **80**, 5472.
Tellinghuisen, J. (1984c). *J. Mol. Spectrosc.*, **103**, 455.
Tellinghuisen, J., and Albritton, D. L. (1975). *Chem. Phys. Lett.*, **31**, 91.
Tellinghuisen, J., Berwanger, P., Ashmore, J. G., and Viswanathan, K. S. (1981). *Chem. Phys. Lett.*, **84**, 528.
Tellinghuisen, J., and Exton, R. J. (1980). *J. Phys. B: At. Mol. Phys.*, **13**, 4781.
Tellinghuisen, J., Hays, A. K., Hoffman, J. M., and Tisone, G. C. (1976). *J. Chem. Phys.*, **65**, 4473.

Tellinghuisen, J., and Henderson, S. D. (1982). *Chem. Phys. Lett.*, **91**, 447.
Tellinghuisen, J., and Moeller, M. B. (1980). *Chem. Phys.*, **50**, 301.
Tellinghuisen, J., Pichler, G., Snow, W. L., Hillard, M. E., and Exton, R. J. (1980). *Chem. Phys.*, **50**, 313.
Tellinghuisen, J., Ragone. A., Kim, M. S., Auerbach, D. J., Smalley, R. E., Wharton, L., and Levy, D. H. (1979). *J. Chem. Phys.*, **71**, 1283.
Tobin, F. L., and Hinze, J. (1975). *J. Chem. Phys.*, **63**, 1034.
Uzer, T., and Child, M. S. (1980). *Molec. Phys.*, **41**, 1177.
Vanderslice, J. T., Mason, E. A., Maisch, W. G., and Lippincott, E. R. (1959). *J. Mol. Spectrosc.*, **3**, 17.
Van Vleck, J. H. (1936). *J. Chem. Phys.*, **4**, 327.
Velazco, J. E., and Setser, D. W. (1975). *J. Chem. Phys.*, **62**, 1990.
Verma, K. K., Bahns, J. T., Rajaei–Rizi, A. R., Stwalley, W. C., and Zemke, W. T. (1983). *J. Chem. Phys.*, **78**, 3599.
Verma, K. K., Bahns, J., and Stwalley, W. C. (1981). *J. Phys. Chem.*, **85**, 2884.
Vidal, C. R. (1980). *J. Chem. Phys.*, **72**, 1864.
Vidal, C. R., and Scheingraber, H. (1977). *J. Mol. Spectrosc.*, **65**, 46.
Viswanathan, K. S., and Tellinghuisen, J. (1983). *J. Mol. Spectrosc.*, **101**, 285.
Weiner, J. (1980). *Chem. Phys. Lett.*, **76**, 241.
Weissman, S., Vanderslice, J. T., and Battino, R. (1963). *J. Chem. Phys.*, **39**, 2226.
Wells, B. H., Smith, E. B., and Zare. R. N. (1983). *Chem. Phys. Lett.*, **99**, 244.
Wentzel, G. (1927). *Z. Physik.*, **43**, 524.
Wentzel, G. (1928). *Physik. Z.*, **29**, 321.
West, W. P., and Gallagher, A. (1978). *Phys. Rev.*, **A17**, 1431.
West, W. P., Shuker, P., and Gallagher, A. (1978). *J. Chem. Phys.*, **68**, 3864.
Whiting, E E., Schadee, A., Tatum, J. B., Hougen, J. T., and Nichols, R. W. (1980). *J. Mol. Spectrosc.*, **80**, 249.
Winans, J. G., and Stueckelberg, E. C. G. (1928). *Proc. Nat. Acad. Sci.*, **14**, 867.
Woerdman, J. P. (1978). *Opt. Commun.*, **26**, 216.
Wren, D. J., Setser, D. W., and Ku, J. K. (1982). *J. Phys. Chem.*, **86**, 284.
Wyss, J. (1978). *Photoluminescence of Calcium Molecules*, Ph.D. Thesis, University of California, Santa Barbara.
York, G., Scheps, R., and Gallagher, A. (1975). *J. Chem. Phys.*, **63**, 1052.
Zare, R. N. (1963).Lawrence Berkeley Laboratory Report UCRL-10925.
Zare, R. N. (1964). *J. Chem. Phys.*, **40**, 1934.
Zare, R. N., and Cashion, J. K. (1963). Lawrence Berkeley Laboratory Report UCRL-10881.
Zeleznik, F. J. (1965). *J. Chem. Phys.*, **42**, 2836.

Photodissociation and Photoionization
Edited by K. P. Lawley

THEORETICAL ASPECTS OF PHOTODISSOCIATION AND INTRAMOLECULAR DYNAMICS

PAUL BRUMER

Department of Chemistry, University of Toronto, Toronto, Ontario, Canada

and

MOSHE SHAPIRO

Department of Chemical Physics, Weizmann Institute of Science, Rehovot 766100, Israel

CONTENTS

I. INTRODUCTION

The nature of intramolecular dynamics in isolated molecules remains, after almost 50 years of study, a topic of active interest. Both formal and experimental examination of the nature of intramolecular dynamics involves preparation of the molecule in a non-stationary state, by any of a variety of means, followed by subsequent interrogation of the molecular time evolution. Two broad classes of prepared states are readily identified. In the first, the molecule is excited to an energy regime where the system remains bound. Under these circumstances the integrity of the molecule is preserved and traditional chemistry (i.e. formation of new molecular species) does not occur. The molecular dynamics is then generally measurable via coupling to the radiation field—e.g. via spontaneous emission. Many experiments, especially in radiationless transitions, fall into this category. In the second class of excitation processes, excitation is to unbound stationary or non-stationary states allowing for molecular rearrangement. In such instances measurement of product properties provides one convenient tool for studying the nature of the prepared state and its dynamical behaviour from preparation to measurement. Included in this category are such processes as predissociation and unimolecular decay via collisional activation or photoexcitation. The latter, photodissociation and its interrelationship with intramolecular dynamics, is the subject of this review. Of specific interest is the proper quantum formulation of both intramolecular dynamics and photodissociation and the extent to which the theory of photodissociation has, and can, provide information about the dynamics of molecules prepared with sufficient energy to dissociate. Only topics specific to the issue of the interconnection between photodissociation and intramolecular dynamics are included. No effort is made to discuss other aspects of these topics, reviewed in other recent articles.[1]

This paper is organized as follows. Section II contains a general introduction to preparation, time evolution and measurement in isolated molecular systems. Section III specializes the preparation to that occurring in photodissociation and provides a completely general treatment of photoexcitation and subsequent time evolution in small molecules. These dynamical discussions are supplemented, in Section IV, by a discussion of quantum statistical intramolecular dynamics in general and its relationship to statisticality in photodissociation. Detailed calculations and their ability to shed light on intramolecular dynamics are the subject of Section V. Finally, in Section VI we briefly note, via two sample systems, the status of photodissociation experiments on larger molecules and the extent to which they provide insight into intramolecular processes.

II. ISOLATED MOLECULE DYNAMICS

Consider a molecule described by the (matter) Hamiltonian H_M with eigenvalues E_j and eigenstates $|\phi_M(E_j)\rangle$ (also denoted $|\phi_j\rangle$ below), i.e.

$$H_M|\phi_M(E_j)\rangle = E_j|\phi_M(E_j)\rangle \tag{1}$$

Here H_M is the full Hamiltonian describing both nuclei and electrons and $|\phi_M(E_j)\rangle$ are either bound or scattering states. We introduce, for complete generality, states

$$|\chi_i\rangle = \sum_j C_j^i |\phi_M(E_j)\rangle$$

whose nature is intimately linked to the specific preparation of interest.

Examination of the dynamics of a molecule in isolation implies the possibility of identifying a time, here conveniently chosen as $t = 0$, when the preparation device ceases to influence the molecule. The subsequent dynamics of the molecule is fully described by the density matrix $\boldsymbol{\rho}(t)$:

$$\begin{aligned} \boldsymbol{\rho}(0) &= \sum_i W_i |\chi_i\rangle\langle\chi_i| = \sum_{i,j,k} W_i C_j^i C_k^{i*} |\phi_j\rangle\langle\phi_k| \\ \boldsymbol{\rho}(t) &= \sum_{i,j,k} W_i C_j^i C_k^{i*} |\phi_j\rangle\langle\phi_k| \exp[-i(E_j - E_k)t/\hbar] \end{aligned} \tag{2}$$

Here W_i are the weights of the pure states $|\chi_i\rangle$ determined in the initial preparation of the molecule. We restrict attention to small or intermediate size molecules defined as systems where the natural linewidths of the individual molecular eigenstates do not overlap. Equation (2) excludes spontaneous emission; if this process is of interest it may be readily incorporated to give

$$\boldsymbol{\rho}(t) = \sum_{ijk} W_i C_j^i C_k^{i*} |\phi_j\rangle\langle\phi_k| \exp[-i(E_j - E_k)t/\hbar - (\Gamma_j^r + \Gamma_k^r)t/2]$$

where Γ_j^r is the radiative width of the jth state.

Measurement of the evolving system corresponds to determining

$$\bar{A}(t) = Tr[\boldsymbol{\rho}(t)\boldsymbol{A}] \tag{3}$$

for properties $\boldsymbol{A}$ of interest. For a general $\boldsymbol{A} = \sum_{l,m} a_{lm} |\phi_l\rangle\langle\phi_m|$:

$$\bar{A}(t) = \sum_{ijk} a_{kj} W_i C_j^i C_k^{i*} \exp[-i(E_j - E_k)t/\hbar - (\Gamma_j^r + \Gamma_k^r)t/2]$$

Several features of these equations are worthy of emphasis. Firstly, note that the time evolution of the system is intimately linked to, indeed heavily determined by, the initial preparation. This dependence is explicit in the values of W_i and the choice of $|\chi_i\rangle$, both being affected by the specific nature of the preparation device and by the system response in the preparation step. The ability to experimentally disentangle preparation effects from molecular properties requires *detailed* knowledge of the nature of the preparation device, a requirement not met, for example, in a host of recent laser experiments. Secondly, note that, barring spontaneous emission, excitation of the system to a precise energy level yields a system which displays no time evolution. A similar remark applies to completely incoherent excitation to a mixture $\rho(0) = \sum W_i |\phi_i\rangle\langle\phi_i|$. Thirdly, again disregarding spontaneous emission, intramolecular dynamics consists solely of interference between exact molecular eigenstates. Measurement of this time dependence consists of projecting $\boldsymbol{\rho}(t)$

onto a specific operator, in accordance with Eq. (3). The resultant observation depends intimately on the nature of A, i.e. on what is being measured. Thus, for example, $\bar{A}(t)$ is constant for any A which commutes with H_M, i.e. for $A = \sum d_{jj}|\phi_j\rangle\langle\phi_j|$. The most obvious of such properties are the populations of the states $|\phi_i\rangle$ where $A = |\phi_i\rangle\langle\phi_i|$ and $\bar{A}(t) = \sum W_i|C_i|^2$. Alternatively, A may not commute with H_M, an example being the measurement of the population of a zeroth-order mode (e.g. bond oscillator) in the molecule. Under these circumstances $\bar{A}(t)$ is time dependent in accord with Eq. (3). Fourthly, note that Eq. (3), as applied to the purely bound molecule, cannot show relaxation to a constant long time limit, the picture normally adopted in the classical limit. Rather, Eq. (3) will typically display an initial rapid dephasing (whose timescale is given by the inverse of the frequency width of the initial preparation) followed by weak ragged oscillations. In the long time limit $\bar{A}(t)$ will show recurrences. Any anticipated agreement between classical and quantal calculations of $\bar{A}(t)$ must be limited to timescales shorter than the time required to discern, in the finite-time Fourier transform, discrete frequencies.[2]

Finally, note that this approach properly includes the possibility that extensive intramolecular dynamics is observed in systems wherein the $|\phi_i\rangle$ are separable, and conversely no dynamics in instances wherein the $|\phi_i\rangle$ are highly coupled. Specifically, the former will occur when preparation produces a superposition of unequal energy separable $|\phi_i\rangle$ and the latter occurs in the preparation of the system at a fixed total energy.

There is little ambiguous about the above formulation of intramolecular processes, where the term *dynamics* is reserved for molecular time evolution. It is, nonetheless, quite distinct from two approaches whose language has dominated the interpretation of a host of experiments. In the first,[3] an essentially spectroscopic approach, interpretation is in terms of zero-order modes. For electronic excitation these are usually singlet and triplet states whereas normal modes are most common for vibrational analysis. Observation of extensive coupling between such zeroth-order modes is termed intramolecular dynamics. In our view such a description is inappropriate to experiments only probing the nature of bound stationary eigenstates. It may be applicable in a consistent way, however, to experiments capable of preparing the zero-order states, in which case the remaining coupling serves to induce time-dependent dynamics in the prepared system. Note, however, that the detailed relationship between the nature of the stationary states and features of the time evolution is just beginning to be understood. The second approach[4] considers intramolecular dynamics from a time-dependent viewpoint with the initial state comprised of a Gaussian wavepacket localized in phase space. Here again the observed dynamics must be understood to relate directly to a particular preparation which leaves the system in this rather unique type of state, if indeed this is at all realistic (see Section III).

The explicit nature of the state $\psi(t)$ arising in photodissociation is derived in the next section in its most general form. Formulae appropriate to several specific types of experiment are also provided.

III. QUANTUM DYNAMICS OF PHOTODISSOCIATION

A qualitative picture of the photodissociation of a polyatomic molecule separates the process into three steps:

1. *Preparation*, in which a molecule is energized by a photon:

$$A - B + h\nu \rightarrow (A - B)^*$$

2. *Intramolecular dynamics*, in which energy is envisioned to migrate from one mode to another:

$$(A - B)^* \rightarrow (A - B)^\dagger$$

3. *Dissociation*, in which the energy now exists in the 'right' mode and causes a chemical bond to break:

$$(A - B)^\dagger \rightarrow A + B$$

In the above, A and B are (in general) two clusters of atoms which may or may not be internally stable, i.e. in the latter case subsequent dissociation may result.

In spite of the intuitive appeal of the above model it is no more than a useful descriptive framework. This is so because the model implicitly assumes a separation between the three processes. Clearly, dissociation (step 3) can occur while the preparation (step 1) is still in progress. Moreover, the preparation can energize the dissociating mode directly, and this direct dissociation will interfere with the indirect route, described by steps 1 to 3. In addition, the three-stage model may involve virtual excitation to zero-order states which need not conserve energy.

In order to give a consistent description of the interference between these virtual processes, and in order to set the scene for an exact treatment of photodissociation dynamics, we adopt a different approach. In this approach no artificial separation between preparation and decay is assumed and we formulate the problem using the fully interacting stationary scattering states. These are the 'true molecular eigenstates' in the continuum. Note, of importance later below, that contrary to bound state problems, each continuum state has N-fold degeneracy, where N is the number of energetically accessible (open) eigenstates of the separate fragments (A or B of Step 1). This degeneracy is in addition to that imposed by the strictly conserved quantities, such as the total angular momentum.

With the asymptotic states in mind, we define state-to-state photodissociation as a process in which the system, while absorbing one or more photons, undergoes a transition from $\phi_M(E_i)$, a bound state of the matter Hamiltonian

H_{M}:

$$(E_i - H_{\mathrm{M}})|\phi_{\mathrm{M}}(E_i)\rangle = 0 \tag{4}$$

to $\phi_{\mathrm{M}}(E, \mathbf{m}^-)$, an 'incoming' continuum state of the same Hamiltonian:

$$(E - H_{\mathrm{M}})|\phi_{\mathrm{M}}(E, \mathbf{m}^-)\rangle = 0 \tag{5}$$

The index $\mathbf{m}^-$ is a reminder of the possible asymptotic states, labelled by $\mathbf{m}$ of the separable asymptotic Hamiltonian H_0, where

$$H_0 \equiv \lim_{R \to \infty} H_{\mathrm{M}} \tag{6}$$

and $\mathbf{R}$ is the A − B displacement vector.

We next consider the effect of the radiation field on this material system. The problem is conceptually very simple if viewed using the fully interacting (matter + radiation) stationary states. These are the solution of the full Schrödinger equation written schematically as

$$(E + \sum_k M_k \hbar\omega_k - H)|\psi(E, \mathbf{n}^-, \mathbf{M}^-)\rangle = 0 \tag{7}$$

where H, the total Hamiltonian, is given as

$$H = H_{\mathrm{M}} + H_{\mathrm{R}} + H_{\mathrm{I}} \equiv H_{\mathrm{f}} + H_{\mathrm{I}} \tag{8}$$

where H_{M} is the matter Hamiltonian, H_{R} the free radiation Hamiltonian and H_{I} the matter radiation interaction. $\mathbf{M} = (M_1, \ldots, M_k, \ldots)$ is a collection of photon occupation numbers and the $\mathbf{M}^-$ notation is a reminder of the fact that in the long time limit the interaction is switched off at an infinitely slow rate (adiabatic switching), i.e.

$$H_{\mathrm{I}} = \lim_{\varepsilon \to +0} H_{\mathrm{I}} \exp(-\varepsilon t) \tag{9}$$

and $\mathbf{M}$ becomes a set of good quantum numbers. In this limit $|\psi(E, \mathbf{m}^-, \mathbf{M}^-)\rangle$ goes over to $|\phi_{\mathrm{M}}(E, \mathbf{m}^-), \mathbf{M}\rangle$, a stationary solution of the non-interacting H_{f}:

$$(E + \sum_k M_k \hbar\omega_k - H_{\mathrm{f}})|\phi_{\mathrm{M}}(E, \mathbf{m}^-), \mathbf{M}\rangle = 0 \tag{10}$$

The use of the fully interacting stationary states reduces the description of all time-dependence phenomena to the specification of initial conditions and the computation of *time-independent* amplitudes. The initial boundary conditions we wish to consider are such that

$$|\psi(t = 0)\rangle = |\phi_{\mathrm{M}}(E_i), \boldsymbol{\alpha}^i\rangle \tag{11}$$

where $\boldsymbol{\alpha}^i \equiv (\alpha_1, \ldots, \alpha_k, \ldots), |\alpha_k\rangle$ are coherent states of the k mode, i.e.[5]

$$|\alpha_k\rangle = \exp(-\tfrac{1}{2}|\alpha_k|^2) \sum_{N_k} \frac{\alpha_k^{N_k}|N_k\rangle}{(N_k!)^{1/2}} \tag{12}$$

We choose α_k to be a smooth function of ω_k so as to generate a smooth pulse in the coordinate and time.

The initial state (Eq. 11) can be written with the aid of Eq. (12) as a linear superposition of eigenstates of H_f:

$$|\psi(t=0)\rangle = \exp(-\tfrac{1}{2}|\alpha^i|^2)|\phi_M(E_i)\rangle \sum_{\mathbf{N}} a(\boldsymbol{\alpha}^i, \mathbf{N})|\mathbf{N}\rangle \tag{13}$$

where

$$|\alpha^i|^2 = \sum_k |\alpha_k|^2 \tag{14}$$

and

$$a(\boldsymbol{\alpha}^i, \mathbf{N}) = \prod_k \alpha_k^{N_k}/(N_k!)^{1/2} \tag{15}$$

The notation $\sum_{\mathbf{N}}$ implies the multiple summation $\sum_{N_1}\sum_{N_2}\cdots\sum_{N_k}\cdots$. The subsequent dynamics is obtained by allowing the evolution operator

$$U(t) = \exp\left(\frac{-iHt}{\hbar}\right) \tag{16}$$

to operate on $|\psi(t=0)\rangle$:

$$|\psi(t)\rangle = \exp\left(\frac{iHt}{\hbar}\right)|\psi(t=0)\rangle \tag{17}$$

We now choose to expand $U(t)$ in terms of the fully interacting *incoming* states of Eq. (7). In this way we pin down the temporal evolution at both the $t=0$ and $t=\infty$ ends. Using Eqs. (7), (13) and (17) we have that

$$|\psi(t)\rangle = \exp(-\tfrac{1}{2}|\alpha^i|^2)\sum_{\mathbf{N}} a(\boldsymbol{\alpha}^i, \mathbf{N}) \sum_{\mathbf{Mm}} \int dE' |\psi(E', \mathbf{m}^-, \mathbf{M}^-)\rangle A(E', \mathbf{m}, \mathbf{M}|i, \mathbf{N}|t) \tag{18}$$

where $A(E', \mathbf{m}, \mathbf{M}|i, \mathbf{N}|t)$, the photodissociation probability amplitudes, are defined as

$$A(E', \mathbf{m}, \mathbf{M}|i, \mathbf{N}|t) \equiv \exp\left[-i\left(\frac{E'}{\hbar} + \sum_k M_k\omega_k\right)t\right]\langle\psi(E', \mathbf{m}^-, \mathbf{M}^-)|\phi_M(E_i), \mathbf{N}\rangle \tag{19}$$

Using adiabatic switching Eq. (9), it can be shown[1,5] that

$$A(E', \mathbf{m}, \mathbf{M}|i, \mathbf{N}|t) = \lim_{\varepsilon\to+0} \frac{\langle\psi(E', \mathbf{m}^-, \mathbf{M}^-|H_I|\phi_M(E_i), \mathbf{N}\rangle \exp\left[-i\left(E'/\hbar + \sum_k M_k\omega_k\right)t\right]}{\hbar(\omega_{E_i} + i\varepsilon + \sum_k (M_k - N_k)\omega_k} \tag{20}$$

where

$$\omega_{E_i} = \frac{E' - E_i}{\hbar} \tag{21}$$

The integral in Eq. (18) can be performed using the residue theorem, to yield

$$\lim_{\varepsilon \to +0} \int_{\omega_{th}}^{\infty} \frac{d\omega f(\omega) \exp(-i\omega t)}{\omega + i\varepsilon} = -2\pi i f(0) \tag{22}$$

This result holds, assuming $\omega_{th} = -\infty$ and $f(\omega)$ has *no poles* in the lower half-plane. Under these circumstances we can write that

$$|\psi(t)\rangle = -2\pi i \exp(-\tfrac{1}{2}|\alpha^i|^2) \sum_{\mathbf{M},\mathbf{N}} a(\boldsymbol{\alpha}^i, \mathbf{N}) |\psi(E, \mathbf{m}^-, \mathbf{M}^-)\rangle$$

$$\cdot \langle \psi(E, \mathbf{m}^-, \mathbf{M}^-)| H_1 |\phi_{\mathbf{M}}(E_i), \mathbf{N}\rangle \exp\left[-i\left(E_i/\hbar + \sum_k N_k \omega_k \right) t \right] \tag{23}$$

where E, the (asymptotic) energy of the material system, is constrained by Eq. (22) to a discrete set f values:

$$E = E_i + \hbar \sum_k (N_k - M_k)\omega_k \tag{24}$$

This is the *on-resonance* condition. Although each of the initial $|\phi_{\mathbf{M}}(E_i), \mathbf{N}\rangle$ components is not an energy eigenstate, only projections onto $\langle \psi(E, \mathbf{m}^-, \mathbf{M}^-)|$ with the E of Eq.(24) survive.

It is important to note that the on-resonance condition, Eq. (24), holds true only when $\psi(E, \mathbf{m}^-, \mathbf{M}^-)$ has no poles in the lower energy half-plane and ω_{th} is not bounded from below. When such poles exist, e.g. in the case of scattering resonances, transient 'off-resonance' components which vanish asymptotically arise. Under these circumstances Eq. (22) is replaced by

$$\lim_{\varepsilon \to +0} \frac{f(\omega) \exp(-i\omega t)}{\omega + i\varepsilon} = \frac{P_v f(\omega) \exp(-i\omega t)}{\omega} - i\pi f(0)\delta(\omega) \tag{25}$$

If $f(\omega)$ has no poles in the lower half-plane and $\omega_{\text{th}} = -\infty$, the contribution of the principal value integral (the first term) is identical to the second term and Eq. (22) is obtained. If any one of these two conditions is not met, all values of ω contribute to the first term and an additional 'off-resonance' transient results. However, as $t \to \infty$ we can perform the contour integration over ever-decreasing semicircles in the lower complex half-plane and thus avoid the poles of $f(\omega)$. As a result, the integral of Eq. (25) approaches, as $t \to \infty$, the 'on-resonance' value of Eq. (22), namely

$$\lim_{t \to \infty, \varepsilon \to +0} \frac{f(\omega) \exp(-i\omega t)}{\omega + i\varepsilon} \to -2i\pi f(0)\delta(\omega) \tag{26}$$

The result embodied in Eqs. (22) to (24) is very surprising because it shows that, under the conditions of validity of Eq. (22), the absorption (or scattering) of radiation to a *continuum* of final molecular states is on resonance *at all times*. This would be a trivial statement had we started from an eigenstate of the matter plus radiation Hamiltonian. However, each component $|\phi_M(E_i), \mathbf{N}\rangle$ of our initial state is manifestly *not* an eigenstate of the total Hamiltonian, it being an eigenstate of H_f the matter plus *free* radiation (Eq. 8). The on-resonance condition is due to a complete cancellation of all off-resonance components where a true continuum of molecular state exists, and H goes over to H_f at infinitely long times.

In practice, due to the finite threshold, Eq. (22) is never rigorously true and some off-resonance components remain at finite times. Deviations from on-resonance behaviour of greater significance occur if scattering resonances exist, since as pointed out above these resonances correspond to poles in the lower ω half-plane.

In addition to the introduction of an 'off-resonance' transient the existence of resonances substantially modifies the temporal behaviour of the 'on-resonance' components, due to the sharp energy dependence of the transition amplitude for physical energies. This last effect is usually termed 'predissociation', although, strictly speaking, the predissociation phenomenon incorporates both the 'on-resonance' and 'off-resonance' terms which are different manifestations of the same poles in the complex energy plane.

In order to further analyse the 'on-resonance' components we proceed by specializing the treatment to the case of one-photon dissociation. This means that the final state has one photon less than the initial state in one of the radiation modes.

In the dipole approximation H_I is given by[5]

$$H_I = i\sum_l \left(\frac{\hbar\omega_l}{2\varepsilon_0 V}\right)^{1/2} \boldsymbol{\varepsilon}_l\cdot\boldsymbol{\mu}(a_l - a_l^+) \tag{27}$$

where V is the cavity volume, ε_0 the permitivity of free space, $\boldsymbol{\varepsilon}_l$ the polarization direction of the l mode, $\boldsymbol{\mu}$ the electric dipole function and a_l^+, a_l are, respectively, the creation and annihilation operators of photons in the l mode. For the on-resonance term we have, by Eq. (24), that for each radiation mode only one choice of final set of occupation numbers $\mathbf{M}$ is possible for the case of *single-photon* dissociation. This choice,

$$\begin{aligned} M_K &= N_K \qquad K \neq l \\ M_l &= N_l - 1 \qquad K = l \end{aligned} \tag{28}$$

for the l-radiation mode is denoted $\mathbf{M} = \mathbf{N}_l$. The on-resonance condition implies, by Eqs. (21), (24) and (28), that the molecular energy, post-absorption, is

$$E = E_l \equiv E_i + \hbar\omega_l \tag{29}$$

The transition amplitude is given in the dipole approximation as (see

Eqs. 22 and 27)

$$A(E,\mathbf{m},\mathbf{M}|i,\mathbf{N}|t)=i\left(\frac{\hbar}{2\varepsilon_0 V}\right)^{1/2}\sum_l \omega_l^{1/2}\langle\psi(E,\mathbf{m}^-,\mathbf{M}^-|$$
$$\times[|N_1,\ldots,N_{l-1},\ldots\rangle N_l^{1/2}$$
$$-|N_1,\ldots,N_{l+1},\ldots\rangle(N_l+1)^{1/2}]\ \boldsymbol{\varepsilon}_l\cdot\boldsymbol{\mu}|\phi_\mathrm{M}(E_i)\rangle \quad (30)$$

and is greatly simplified by assuming condition (28) for each of the terms in the sum over radiation modes. Using the rotating wave approximation,[5] which is equivalent to neglecting the second term in the square brackets above, we have, by Eq. (28), that

$$A(E,\mathbf{m},\mathbf{M}|i,\mathbf{N}|t)=i\left(\frac{\hbar}{2\varepsilon_0 V}\right)^{1/2}\sum_l (N_l\omega_l)^{1/2}\delta_{\mathbf{M}\mathbf{N}_l}$$
$$\times\langle\psi(E_l,\mathbf{m}^-,\mathbf{N}_l^-|\mathbf{N}_l\rangle\boldsymbol{\varepsilon}_l\cdot\boldsymbol{\mu}|\phi_\mathrm{M}(E_i)\rangle \quad (31)$$

It follows from Eq. (23) that the wavefunction, at time t, is of the form

$$|\psi(t)\rangle=\pi\left(\frac{2h}{\varepsilon_0 V}\right)^{1/2}\exp(-\tfrac{1}{2}|\alpha^i|^2)\sum_{l,m}\omega_l^{1/2}\sum_{\mathbf{N}}N_l^{1/2}(\boldsymbol{\alpha}^i,\mathbf{N})|\mathbf{N}_l\rangle$$
$$\times|\psi(E_i,\mathbf{m}^-,\mathbf{N}_l)\rangle\langle\psi(E_l,\mathbf{m}^-,\mathbf{N}_l)|\boldsymbol{\varepsilon}_l\cdot\boldsymbol{\mu}|\phi_\mathrm{M}(E_i)\rangle$$
$$\times\exp\left[-i\left(\frac{E_i}{\hbar}+\sum N_K\omega_K\right)t\right] \quad (32)$$

where $\psi(E_l,\mathbf{m}^-,\mathbf{N}_l)$ is the probability amplitude that a photon number state ($|\mathbf{N}_l\rangle$), observed with unit probability when the matter-radiation term is switched off, is also observed when it is present, i.e.,

$$|\psi(E_l,\mathbf{m}^-,\mathbf{N}_l)\rangle\equiv\langle\mathbf{N}_l|\psi(E_l,\mathbf{m}^-,\mathbf{N}_l^-)\rangle \quad (33)$$

In the weak-field limit this quantity is independent of $\mathbf{N}_l$ since

$$|\psi(E,\mathbf{m}^-,\mathbf{N}^-)\rangle\approx|\phi_\mathrm{M}(E,\mathbf{m}^-)\rangle|\mathbf{N}\rangle \quad (34)$$

We then have, using Eq. (29) and the definition of the coherent states (Eq. 12) that

$$|\psi(t)\rangle=\pi\left(\frac{2\hbar}{\varepsilon_0 V}\right)^{1/2}\sum_{l,m}|\alpha_l|\omega_l^{1/2}\exp\left[\frac{-iE_lt}{\hbar}+\theta_l\right]$$
$$\times|\phi_\mathrm{M}(E_l,\mathbf{m}^-)\rangle\langle\phi_\mathrm{M}(E_l,\mathbf{m}^-)|\boldsymbol{\varepsilon}_l\cdot\boldsymbol{\mu}|\phi_\mathrm{M}(E_i)\rangle|\boldsymbol{\alpha}^i(t)\rangle \quad (35)$$

The state $|\boldsymbol{\alpha}^i(t)\rangle$ is a multimode coherent state (see Eq. 11), evolving freely in time:

$$|\boldsymbol{\alpha}^i(t)\rangle=\prod_K\sum_{N_K}|N_K\rangle\exp(-\tfrac{1}{2}|\alpha_K|^2)[|\alpha_K|\exp(i\omega_Kt+i\theta_K)]^{N_K}/(N_K!)^{1/2} \quad (36)$$

where θ_K, the phases of the α_K members (i.e. $\alpha_K \equiv |\alpha_K| \exp(i\theta_K)$ determine the degree of coherence of the $|\boldsymbol{\alpha}^i(t)\rangle$ pulse. As shown in Eq. (35), this in turn determines the properties of the wavepacket created as a result of photon absorption. If θ_l are a set of random numbers we expect, by Eq. (35), a cancellation of many terms in the summation over the l radiation modes. This will not affect the rate of photodissociation to one of the final free states of the fragments[1a,6] but will change the transient properties, such as fluorescence, of $\psi(t)$.

Equation (35) may be viewed as the basis for all 'radiationless transitions' type of phenomena. We see that the photodissociation process creates a superposition state whose energetic width, as embodied by the choice of α_l of the initial pulse, is carried over directly from the energetic width of the incident pulse $|\boldsymbol{\alpha}^i(t)\rangle$. It is impossible to observe any energy levels other than those embodied in the l summation of Eq. (35). Thus the picture that 'intramolecular vibrational relaxation' (IVR) leads to distribution of the total material energy between modes at finite times is misleading.

The above comments hold true in the strong field case (Eq. 32) as well, except that, in that case, it is no longer possible to separate the energy into its material and radiative components.

In order to further clarify the nature of the superposition state created by photon absorption to a continuum of material states, we proceed by exploring the limit in which the radiation field is treated classically. To do so we identify[5]

$$F_l = \left(\frac{2\hbar\omega_l}{\varepsilon_0 V} \right)^{1/2} |\alpha_l| \tag{37}$$

as the expectation value of the electric field amplitude of the l mode. In order to consider a continuous set of modes, the summation over l is replaced by an integration over ω according to

$$\sum_l \rightarrow \int \rho_\omega \, \mathrm{d}\omega \tag{38}$$

where

$$\rho_\omega = \frac{V^{1/3}}{\pi c} \tag{39a}$$

when the incident radiation propagates along one direction with a well-defined polarization vector, and

$$\rho_\omega = \frac{V\omega^2}{\pi^2 c^3} \tag{39b}$$

when the incident radiation has all propagation directions and two polarizations. Defining the field amplitude per wavenumber $F(\omega)$ as

$$F(\omega) = \rho_\omega F_l \tag{40}$$

we have, from Eqs. (35) and (37), that the matter part of the wavefunction is given as

$$|\psi_M(t)\rangle = \left(\frac{\pi}{\hbar}\right)\sum_{\mathbf{m}} \int dE\, F\left[\frac{(E-E_i)}{\hbar}\right] \exp\left[-iEt/\hbar + i\theta(E)\right]$$
$$\times\, |\phi_M(E,\mathbf{m}^-)\rangle\langle\phi_M(E,\mathbf{m}^-)|\boldsymbol{\varepsilon}\cdot\boldsymbol{\mu}|\phi_M(E_i)\rangle \tag{41}$$

This is the well-known[6] semiclassical result for photodissociation by a weak field. Given the bound-free $\langle\phi_M(E,\mathbf{m}^-)|\boldsymbol{\varepsilon}\cdot\boldsymbol{\mu}|\phi_M(E_i)\rangle$ matrix elements we can follow the time development of $\psi_M(t)$. Methods for calculating these integrals are outlined elsewhere.[1a,7] It follows from Eq. (41) that the integration over E is limited by the laser spectral function $F[(E-E_i)/\hbar]$, which, for a typical nanosecond laser pulse, is of the order of 0.1 cm^{-1}. Under these conditions the superposition state for direct dissociation becomes an almost stationary, highly delocalized wavepacket. The picture often advocated[8,9] that in photodissociation one generates a highly localized wavepacket is not accurate. Only when the spectral function is much wider than the lineshape function, given by

$$A_i(E) = \sum_{\mathbf{m}} |\langle\phi_M(E,\mathbf{m}^-)|\boldsymbol{\varepsilon}\cdot\boldsymbol{\mu}|\phi_M(E_i)\rangle|^2 \tag{42}$$

can the photodissociation process be described as generating a well-localized state. This is so because in this case we can treat $F[(E-E_i)/\hbar]$ as a constant function

$$F(\omega) = F \tag{43}$$

Assuming complete coherence, i.e.

$$\theta(E) = \theta \tag{44}$$

we obtain

$$|\psi_M(t)\rangle = \frac{\pi}{\hbar} F \exp(i\theta) \exp(-iH_M t/\hbar)\boldsymbol{\varepsilon}\cdot\boldsymbol{\mu}|\phi_M(E_i)\rangle \tag{45}$$

where we have used the spectral resolution of the matter evolution operator. Under these conditions it is legitimate to consider the time-dependent wavefunction as being given by the matter evolution operator acting on an initial state given by[8] $\boldsymbol{\varepsilon}\cdot\boldsymbol{\mu}|\phi_M(E_i)\rangle$. For direct dissociation the conditions of Eqs. (43) and (44) are fulfilled by a completely coherent pulse whose width is greater than 4000 cm^{-1}, which is a typical linewidth for such systems as CH_3I[10] and ICN.[11] Under these conditions and because of Eq. (44) the pulse duration must be 10^{-15} s. This is unattainable by present techniques and does not represent any current experiment. Thus, for example, one cannot interpret the fluorescence data[9] from the dissociating CH_3I as being governed by Eq. (45). Instead, the more correct form (Eq. 41) must be used.

If dissociation is slower than 1 ps, we may find experimental situations for which Eq. (45) is valid. This is the case when $\phi_M(E, \mathbf{m}^-)$ is dominated by a single resonance. However, when this occurs it is easier to use Eq. (41) directly. This is best done by partitioning the matter Hamiltonian, via two projection operators P, Q such that

$$P + Q = I \tag{46}$$

where Q projects out the quasi-bound manifold and P the direct component. By defining the 'distorted waves' solutions, for the quasi-bound manifold

$$(E_j - QH_M Q)|\chi_j\rangle = 0 \tag{47a}$$

and for the direct component

$$(E \pm i\varepsilon - PH_M P)|\chi(E, \mathbf{m}^{\pm})\rangle = 0 \tag{47b}$$

we can write $\phi_M(E, \mathbf{m}^-)$ in the isolated resonance approximation[7] as

$$|\phi_M(E, \mathbf{m}^-)\rangle = |\chi(E, \mathbf{m}^-)\rangle + \frac{(I + PH_M Q)|\chi_j\rangle \gamma_{j\mathbf{m}}(E)}{E - \hat{E}_j(E) + i\Gamma_j(E)/2} \tag{48}$$

where the 'width amplitudes' $\gamma_{j\mathbf{m}}$ are given as

$$\gamma_{j\mathbf{m}}(E) = \langle \chi_j | QH_{\mathrm{m}} P | \chi(E, \mathbf{m}^+)\rangle \tag{49}$$

the 'total width' Γ_j as

$$\Gamma_j(E) = 2\pi \sum_{\mathbf{m}} |\gamma_{j\mathbf{m}}|^2 \tag{50}$$

and the resonance energy $\hat{E}_j$ as

$$\hat{E}_j(E) = E_j + \frac{1}{2\pi} P_{\mathrm{v}} \int \frac{\mathrm{d}E' \; \Gamma_j(E')}{E - E'} \tag{51}$$

Substituting Eq. (48) into Eq. (41) we obtain for $Q\psi_M(t)$, the resonant component of $\psi_M(t)$, that

$$|Q\psi_M(t)\rangle = \frac{\pi}{\hbar} \sum_{\mathbf{m}} \int \mathrm{d}E \, F\left(\frac{E - E_i}{\hbar}\right) \exp\left[\frac{-iEt}{\hbar} + i\theta(E)\right] |\chi_j\rangle$$
$$\times \frac{\gamma_{j\mathbf{m}}(E)\mu_{\mathbf{m}i}(E)}{\hat{E} - E_j(E) + i\Gamma_j(E)/2} \tag{52}$$

where

$$\mu_{\mathbf{m}i}(E) = \langle \phi_M(E, \mathbf{m}^-) | \boldsymbol{\varepsilon} \cdot \boldsymbol{\mu} | \phi_M(E_i)\rangle \tag{53}$$

If the excitation is faster than the decay, $F(\omega)$ must vary more slowly than the denominator of Eq. (52). Assuming, for simplicity, that $\gamma_{j\mathbf{m}}(E)$, $\hat{E}_j(E)$ and $\mu_{\mathbf{m}i}(E)$ are also slowly varying, we obtain that

$$|Q\psi_M(t)\rangle = \frac{\pi}{\hbar} F \exp(i\theta) |\chi_j\rangle \sum_{\mathbf{m}} \gamma_{j\mathbf{m}} \mu_{\mathbf{m}i} \exp\left[\frac{-i(E_j + \frac{1}{2}\Gamma_j)t}{\hbar}\right] \tag{54}$$

i.e. the resonant component decays exponentially at a rate given by $\Gamma_j/\hbar$. If $|\chi_j\rangle$ is the only state capable of emitting to the ground electronic state this will result in the well-known decay of fluorescence due to photodissociation.

It is important to note that only when the pulse duration is shorter than $\hbar/\Gamma_j$ can we say that $\hbar/\Gamma_j$ is indeed the lifetime of our state. Excitation with laser light whose monochromaticity is better than Γ_j will result in a state which decays at a slower rate than $\Gamma_j/\hbar$.

In summary, this section formulates the exact quantum dynamics of the photodissociation process for arbitrary field and molecular interaction strengths. It was pointed out that the resonance approximation may be much more accurate for bound-free than for bound-bound (photo)transitions due to coherent interference between the various continuum components. The conditions for complete cancellation of 'off-resonance' components were discussed. Finally, the semiclassical weak-field limit of the process was derived. It was pointed out that present-day experiments generally generate a highly delocalized superposition of continuum states and that the lifetime of a quasi-bound level generated by laser excitation is a function of the laser pulse characteristics.[12]

IV. STATISTICAL BEHAVIOUR

The approach described above is purely dynamic in nature. That is, the interpretation of any given experiment requires detailed knowledge of both system properties, i.e. E_i and $|\phi_i\rangle$, and system preparation. Difficulties associated with such calculations, as well as the desire to qualitatively interpret and predict experimental results, have for a long time been the source of interest in identifying systems wherein statistical assumptions regarding the dynamics suffice. Progress towards linking the exact eigenstate description of dynamics to such concepts of statisticality are reviewed below.

A. Bound State Dynamics

It is within the framework of classical mechanics that concepts of statisticality are best formulated and understood.[13] Thus, an introduction via classical mechanics wherein ideas such as relaxation and ergodicity are well defined is necessary. We focus first on ergodicity, the lowest level in the hierarchy of statistical systems. By definition, ergodicity on an energy hypersurface implies that the conservative system, defined by the Hamiltonian H, posesses only one stationary distribution at energy E, i.e. the classical microcanonical distribution $\rho = \delta(E - H)$. As a direct consequence, knowledge of the Hamiltonian alone suffices to completely characterize the stationary properties of the system. Classical ergodicity does not, however, imply relaxation to a statistical limit so that further levels in the hierarchy of statisticality are required in order to make contact with requirements of relaxation in bound molecules.[14] Systems of major interest in this regard are mixing systems which satisfy the conditions (among others) that

$$\lim_{t\to\infty} \langle g(q(t), p(t)), f(q(0), p(0))\rangle = \langle g\rangle\langle f\rangle$$

$$\lim_{t\to\infty} w_{\mathrm{if}} = V_i V_{\mathrm{f}} \tag{55}$$

Here $\langle\rangle$ denotes a microcanonical average, (q, p) are conjugate coordinates and momenta, and w_{if} is the probability of a transition from cell i to cell f in phase space, of volume V_i, V_{f} respectively. The first condition implies a decoupling of correlations with increasing time and the second shows that w_{if} is dynamics-free in the long-time limit. These conditions on a Hamiltonian system ensure relaxation of an initial distribution to a weighted sum of microcanonical distributions, the weighting determined by the initial preparation over energy hypersurfaces.

The number of physically interesting systems shown to be mixing is small. One highly useful example, considered below, is the stadium system in which the particle is confined, by infinite potential walls, to a region defined by two parallel lines capped by two semicircles.

The situation in the quantum mechanics of finite systems is far less satisfactory, even in terms of the definition of such ideal systems. Finite quantum systems are known not to relax, hence excluding a direct extension of classical mixing. Efforts to produce satisfactory theories of quantum ergodicity have not led to definitions easily connected to the definition in the classical limit.[16] Typical among these is the von Neumann ergodicity requirement[17] of non-degenerate energy eigenstates. In principle such a condition is reasonable, implying that knowledge of the Hamiltonian only suffices to determined all stationary properties at a given energy. However, many classically regular systems have non-degenerate quantum spectra, a simple example being a particle in a rectangle with two sides of incommensurate length.

One currently popular view[18] is that quantum chaos is manifest, in stationary state quantum mechanics, by non-degenerate wavefunctions with 'complicated and irregular' nodal patterns which are 'similar' to their nearby (in energy) neighbours. Further, the distribution $P(s)$ of adjacent level spacings $P(s)$ is expected to peak away from $s = 0$, one proposed form[19] being the Wigner distribution:

$$P(s) = 2\alpha^2 s\,\mathrm{e}^{-\alpha^2 s^2} \qquad \alpha = \pi^{1/2}/2\bar{s} \tag{56}$$

where $\bar{s}$ is the average level spacing. A connection between this $P(s)$ distribution and nodal properties of ψ has recently been proposed[20] but the relation to formal aspects of quantum ergodicity has yet to be established. Note, furthermore, that the criterion of 'complicated and irregular' nodal patterns, although observed for several oscillator systems[21] as well as for the stadium,[22] is qualitative and may well be coordinate system dependent.

Two issues must be addressed in order to ascertain the role in dynamics of this proposed criteria of quantum chaos. Firstly, are these proposed properties always observed for systems displaying classically chaotic behaviour? Secondly, what are the observable *dynamic consequences* of the proposed $P(s)$ distribution and associated irregular nodal patterns? Several studies designed to address these issues have been carried out. Specific applications have compared simple two-degrees-of-freedom systems, i.e. a particle in a rectangle or stadium, which display regular and mixing motion, respectively, in classical mechanics. Several studies of this kind are summarized briefly below.

Initial studies[22] of the exact stadium wavefunctions at high energies showed the existence of highly erratic nodal patterns, as in Fig. 1b, in addition to a $P(s)$ distribution peaking away from $s = 0$. Subsequent studies[23], however, have shown the presence of nearby well-ordered stadium wavefunctions, as shown in Fig. 1a. Thus, highly ordered wavefunctions appear even in energy regimes where these manifestations of quantum chaos are observed. Furthermore, the classical stadium is mixing at all energies whereas the low-lying quantum states of the stadium are highly regular, as shown in Fig. 2a to c. Indeed, the

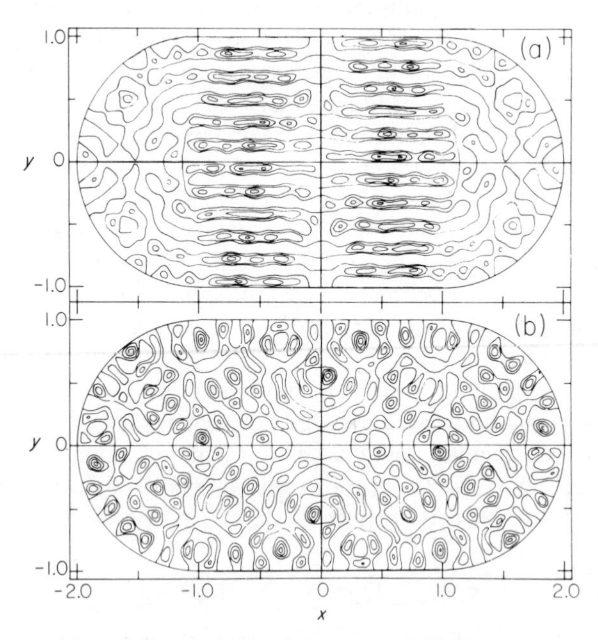

Fig. 1. Stadium wavefunctions at (a) $K = 52.254702$ and (b) $K = 52.382607$ where $E = \hbar^2 k^2/2m$. Stadium area is π; axes are labelled in units of $[\pi/(\pi + 4)]^{1/2}$. Only non-negative contours at 0, 0.5, 1.0 and 1.5 are shown. (From Ref. 23. Reproduced by permission of The Royal Society of Chemistry.)

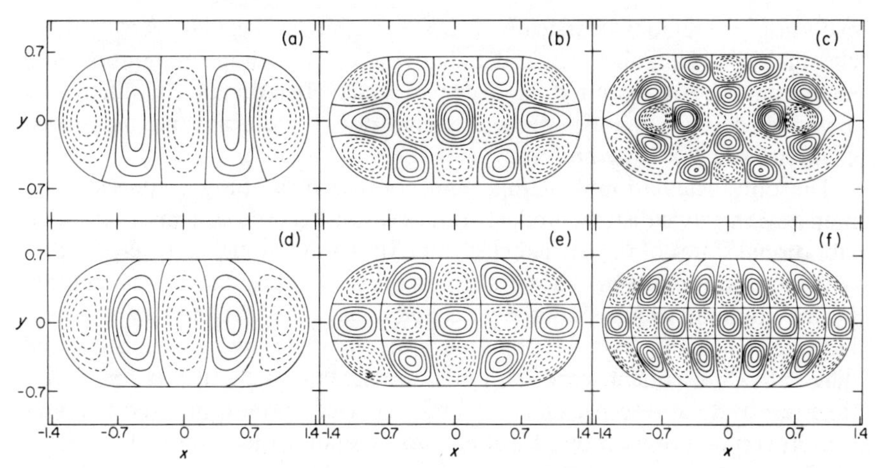

Fig. 2. Sample low-lying stadium wavefunctions (top row) and their adiabatic counterparts (bottom row). (From Ref. 24. Reproduced by permission of North-Holland Publishing Co.)

low-lying wavefunctions have been shown[24] to be *quantitatively* describable by an adiabatic separable Hamiltonian, yielding the wavefunctions shown in Fig. 2d to f. Thus, observations of classical chaos at a given energy do not imply the occurrence of erratic nodal patterns in quantum mechanics.

Do the Wigner $P(s)$ distribution and erratic nodal patterns have observable dynamic consequences? Several calculations addressing this question are of interest. In the first,[25] the autocorrelation function $P(t) = \langle \psi(0) | \psi(t) \rangle$ with $\psi(0) = \sum c_i \phi_i$ was compared for two cases, one where ϕ_i were eigenstates of a separable oscillator system and the second where ϕ_i were eigenstates of the stadium. With $|c_i|^2$ chosen from a similar distribution, corresponding to the assumption that the systems respond similarly to the preparation device, $P(t)$ for both cases was found to be quite similar. Several considerations suggest that this must be the case. Specifically, the initial $P(t)$ fall-off is characterized by the gross frequency width of the Fourier transform of $P(t)$ chosen similarly in each case. The long-time $P(t)$ behaviour, although intimately related to the detailed system spectrum, is generally small and found to be qualitatively similar in the two cases. The conclusion is then clear: if both the regular and irregular systems respond similarly to preparation, yielding similar $|c_i|^2$ distributions, then properties such as $P(t)$ are unable to provide qualitative differences between the two systems. This does not, however, imply that in a given experiment regular and irregular systems will behave similarly. Rather, if they are *observed*, in a similar experiment, to behave differently then this difference must be due to different responses of the two systems to preparation and/or measurement. Further evidence for the fundamental role of the system preparation are computations[26] of the evolution of initial distributions localized in phase space which do display qualitative differences between evolution in regions classically characterizable as regular versus those classifiable as irregular. Thus, detailed studies, including preparation and measurement, are required in order to assess the dynamical differences expected due to chaotic versus regular nodal patterns and associated characteristic $P(s)$ distributions.

The only calculation[27] of this kind consists of a study of pulsed laser preparation of two distinct model two-degrees-of-freedom systems. In the first, vibrational states of the excited electronic state were modelled by eigenstates of the regular particle-in-a-box system. In the second this system was replaced by the stadium which shows erratic nodal patterns. Preliminary results[27] indicated that the initially prepared state in these two systems evolved differently. Subsequent work[28] on measurements in such an experiment, i.e. time-resolved fluorescence emission and dispersed fluorescence, shows qualitatively similar observations. Further work extending these studies to realistic molecular systems is in progress.

B. Photodissociation and Statistical Behaviour

The ability to monitor fragment properties in photodissociation and the inherent chemical interest in such features of dissociation have placed emphasis on the statistical nature of the product distributions.[29] Recent proposals on quantum chaos have, however, focused on bound state systems. The fundamental difference between photodissociation and bound state dynamics lies in the participation of the degenerate continuum in the former process. The general consequences are dramatic: both classical theories of ergodicity, which assume a compact phase space, and quantum mechanical ergodic theories, which require non-degenerate energy eigenstates, are inapplicable. Nonetheless, some connection[30] can be made for the case of decay through an isolated resonance.

Consider statistical behaviour in an isolated molecule at fixed energy E. We denote the molecular state, after the excitation source has ceased to influence the molecule, as the prepared state $|s\rangle$ and expand it as

$$|s\rangle = \int \mathrm{d}E' \sum_{\mathbf{m}} |\phi_{\mathrm{M}}(E', \mathbf{m}^-)\rangle\langle\phi_{\mathrm{M}}(E', \mathbf{m}^-)|s\rangle \tag{57}$$

The probability of observing the asymptotic product state $|E, \mathbf{m}^{\circ}\rangle$, an eigenstate of H_0 (see Eq. 6), is given by $|\langle s|E, \mathbf{m}^-\rangle|^2$. Statistical behaviour in the observed product distribution implies that the observed product distributions are solely a function of E and the total angular momentum (not explicitly included herein). This is rigorously the case only if the degenerate $|\phi_{\mathrm{M}}(E, \mathbf{m}^-)\rangle$ are sufficiently similar to one another so that $\langle s|\phi_{\mathrm{M}}(E, \mathbf{m}^-)\rangle$ is $\mathbf{m}$-independent. Thus statistical behaviour in photodissociation at fixed energy is equal representation of all degenerate eigenstates in the prepared state $|s\rangle$.

The possibility of this occurring in the case of an isolated resonance is readily examined using Eqs. (46) to (54). Specifically, the state $|\phi_{\mathrm{M}}(E,\mathbf{m}^-)\rangle$ at energies near the resonance is of the form given in Eq. (48). Thus

$$\langle s|\phi_{\mathrm{M}}(E,\mathbf{m}^-)\rangle = \langle s|\chi(E,\mathbf{m}^-)\rangle + \frac{\gamma'_{sj}\gamma_{jm}}{E - \hat{E}_j(E) + i\Gamma_j(E)/2} \tag{58}$$

with

$$\gamma'_{sj} = \langle s|(I + PH_{\mathrm{M}}Q)|\chi_j\rangle$$

At these near resonance energies the transition amplitude $\langle s|\phi_{\mathrm{M}}(E,\mathbf{m}^-)\rangle$ is dominated by the second term in Eq. (58), and the product distribution depends solely upon the variation of γ_{jm} with $\mathbf{m}$. If $|\chi_j\rangle$ (see Eq. 47a) is characterized by sufficiently erratic nodal patterns then γ_{jm} may be relatively independent of $\mathbf{m}$, yielding a statistical product distribution. Thus it is through the nature of $|\chi_j\rangle$, the bound state 'responsible' for the resonance, that contact can be made with current theories of quantum ergodicity as manifest in isolated bound eigenfunctions. Note that although this condition is sufficient for statistical behaviour at fixed energy it is not a necessary condition, particularly for alternate approaches[7] which incorporate averaging over energy intervals.

V. CASE STUDIES OF TRIATOMIC AND PSEUDOTRIATOMIC PHOTODISSOCIATION

Triatomic molecules are complex enough to display all the essentials of intramolecular dynamics, while at the same time simple enough to be analysed in some detail. In the examples below we display some of the manifestations of the formalism of preparation and decay developed in Section II. We treat four different cases:

1. Photodissociation of CH_3I which exemplifies a *fast probe* of the initial bound vibrational state.
2. Predissociation of ArN_2 which demonstrates the interference between *direct and resonant* processes.
3. Predissociation of the HeI_2 molecule which serves as a probe of nearly overlapping vibrational resonances.
4. The VUV photodissociation of H_2O in which we demonstrate a connection between resonance widths and the degree of statistically.

A. CH_3I—Fast Probes of Ground State Intramolecular Dynamics

In photofragment spectroscopy,[31,32] a thermal or supersonically cooled beam of molecules is crossed and partly dissociated by a narrow-band laser.

The fragments resulting from the photodissociation process

$$AB(n) + h\nu \rightarrow A(\mathbf{v}) + B$$

are probed by a variety of techniques [32–37] to determine their internal state distributions. Using Eq. (41) it can be shown[1a,6] that the photodissociation cross-section is given by

$$\sigma_n(h\nu, \mathbf{v}) = \frac{3\pi^3 \nu}{c} |\langle \phi(E, \mathbf{v}^-)| \boldsymbol{\mu}\cdot\boldsymbol{\varepsilon} | \phi(E_n)\rangle|^2 \tag{59}$$

where we have dropped the M subscript since we are dealing exclusively with matter states. As in Eq. (29), by energy conservation we have

$$E = E_n + h\nu \tag{60}$$

In the case of the UV photodissociation of CH_3I,

$$CH_3I(n) + h\nu(33\,000\text{–}43\,000\,\text{cm}^{-1}) \rightarrow CH_3(v) + I(^2P_{1/2}) \tag{61}$$

This probe is termed fast because if a localized wavepacket were formed in the excited state then the CH_3 would depart from $I(^2P_{1/2})$ in 10^{-14} to 10^{-15} s. On the basis of time-of-flight and fluorescence data [38–40] it appears that only two modes participate strongly in the process. In the ground state these are the C—I stretch and the C—H_3 'umbrella' bend. In the excited state, as the molecule breaks apart, the C—I stretch becomes an unbounded motion and the C—H_3 bend undergoes considerable vibrational excitation. All other modes seem to remain cold and not affect the process.

In Fig. 3 we show the photodissociation cross-section (see Eq. 59) as a function of $h\nu$ and v—the umbrella bend quantum number, for different initial states $n(=n_1 n_2)$ of CH_3I.[41] Also shown are the nuclear probability densities of the corresponding initial states. The cross-sections were calculated, using realistic potential surfaces, by the artificial channel method.[42,43]

Quite clearly, the topology of the nuclear densities in coordinate space (composed of the C—I and CH_3 distances) is the same as that of the cross-sections in energy space (composed of the total energy, given by $h\nu$, and the fragment energy, given by v). Since all intramolecular phenomena are very much a function of the shape of the nuclear wavefunctions, the fast dissociation in the experiment, if feasible, acts as the ideal probe to intramolecular dynamics. In particular, information regarding the nuclear densities of highly excited molecular can be obtained directly. Transient effects can also be probed by taking 'snapshots' of non-stationary initial states at different times. Naturally the feasibility of such an experiment depends on our ability to dissociate the molecule rapidly. Our ability to do that depends on the existence of an allowed transition to a highly repulsive state in an energy range where visible or UV lasers exist.

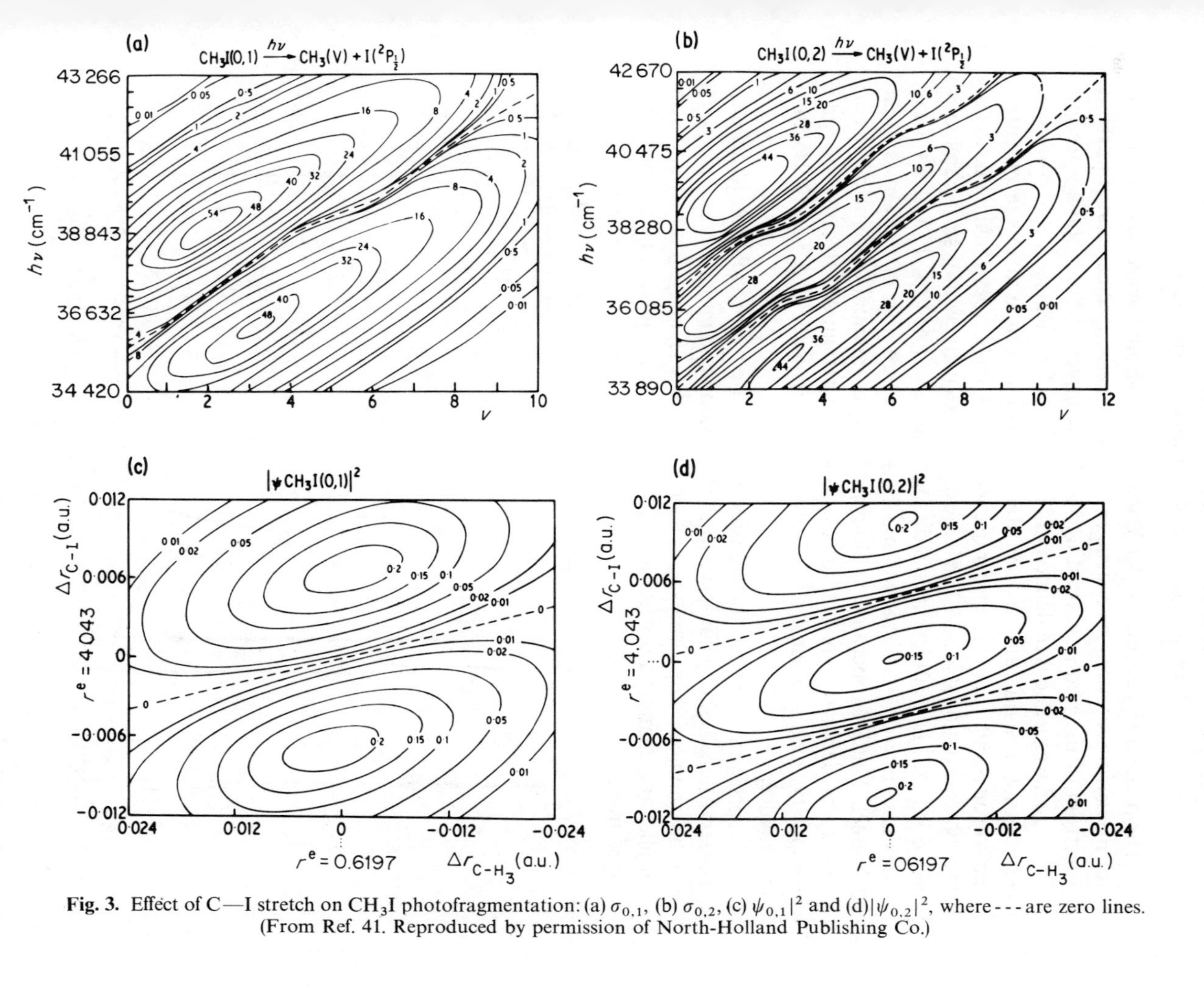

Fig. 3. Effect of C—I stretch on CH_3I photofragmentation: (a) $\sigma_{0,1}$, (b) $\sigma_{0,2}$, (c) $\psi_{0,1}|^2$ and (d)$|\psi_{0,2}|^2$, where - - - are zero lines. (From Ref. 41. Reproduced by permission of North-Holland Publishing Co.)

B. Predissociation of the ArN_2 Complex – The Interference between Direct and Resonance Processes

The predissociation of van der Waals molecules may serve as an ideal tool for measuring interference effects. In Fig. 4 we depict the calculated infrared spectrum of the photodissociating Ar–N_2 complex.[44] Since the absorption lines are narrow and well defined as N_2 rotational modes (see Table I) one can excite such a mode and the resultant dissociation is due to the potential anisotropy which causes transfer of energy from the N_2 rotational mode to the Ar–N_2 translational mode. The large differences in linewidths shown in Fig. 4 are due to the large differences in dissociation rates (see Eqs. 49 and 50). Thus the lines are wide (i.e. dissociation is fast) at the threshold of a new rotational channel. This is so because at threshold only a minimal amount of kinetic energy is deposited in the Ar–N_2 translational mode. Because of their simplicity each linewidth is a direct measurement of a nearly isolated resonance.

Of special interest are the transitions associated with violation of helicity conservation. Helicity conservation[45,46] in non-rigid molecules is analogous to the symmetric-top situation in rigid molecules. Helicity non-conservation implies and change in orientation of the N_2 angular momentum in the body-fixed frame.

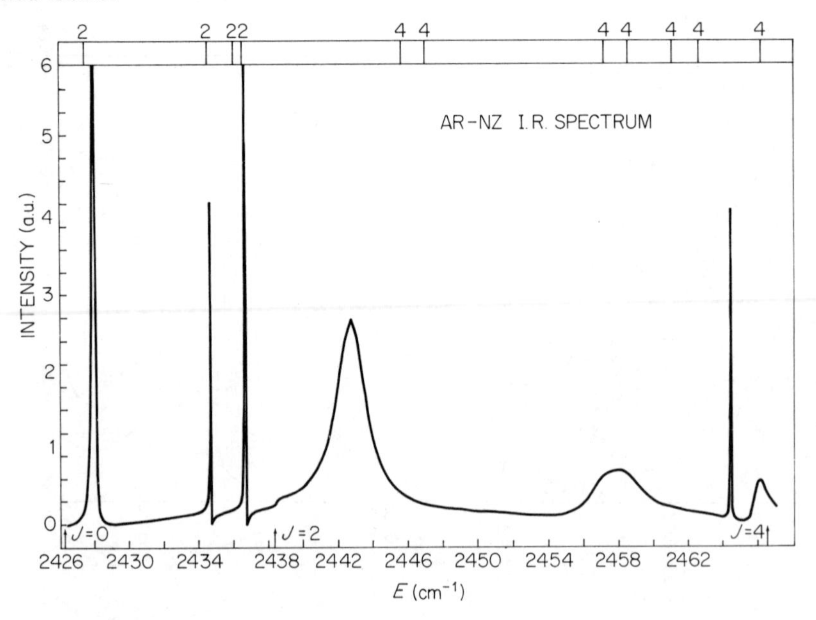

Fig. 4. Ar—N_2 infrared absorption cross-sections in the 2426 to 2468 cm^{-1} range in arbitrary units. The transition corresponds to $J=0$, $v=0 \rightarrow J=1$, $v=1$. Indicated as arrows are the openings of the $j=0$, 2 and 4 channels. (From Ref. 44. Reproduced by permission of North-Holland Publishing Co.)

TABLE I

Positions of $J = 1$, $v = 1$ quasi-bound states and resonances of ArN_2. (From Ref. 44. Reproduced by permission of North-Holland Publishing Co.)

Diabatic				Bound states				Resonances
Assignment			Position[b]	Bound manifold j values	Helicity decoupling Ω^c	Helicity decoupling position[b]	Full close-coupling position,[b] even parity, $\Omega = 0, 1$ $j = 2, 4, 6, 8$	Full close-coupling position,[b] even parity, $\Omega = 0, 1$ $j = 0, 2, 4, 6, 8$
j	n	Ω						
4	1	0	2432.93	2,4,6,8	0	2427.43	2427.49	2427.94
4	1	1	2432.80	2,4,6,8	1	2434.62	2434.49	2434.65
2	3	0	2434.31	2,4,6,8	0	2436.04	2435.97	[e]
2	3	1	2434.21	2,4,6,8	1	2436.48	2436.49	2436.64
4	2	0	2451.72	4,6,8	1	2445.97	2445.57	2442.70
4	2	1	2451.61	4,6,8	0	2446.04	2446.94	[e]
6	0	0	2452.52	4,6,8	1	2457.47	2457.09	2457.82
6	0	1	2452.40	4,6,8	0	2458.02	2458.44	[e]
4	3	0	2463.20	4,6,8	1	2461.34	2460.98	[e]
4	3	1	2463.10	4,6,8	0	2462.14	2462.53	2464.36
6	1	0	2478.51	4,6,8		[d]	2466.11	2466.10

[a]Energy in cm^{-1} relative to the $j = 0$, $v = 0$ dissociation limit.

[b]$h\nu$ in cm^{-1} i.e. position relative to the $J = 0$ ground vibrational–rotational level (at -66.59 relative to the $j = 0$, $v = 0$ dissociation limit).

[c]$\Omega = 1$ eigenvalues are also the exact odd parity (under reflection) levels (forbidden to an optical transition from $J = 0$).

[d]Bound state does not exist for $j = 4, 6, 8$ manifold.

[e]A separate resonance not detected; may either be of width smaller than 0.1 cm^{-1} or more likely has been merged into a neighbouring resonance.

As shown in Table I, some of the lines of Fig. 4 are entirely due to helicity non-conservation (i.e. a transition from $\Omega = 1$ to $\Omega = 0$) an example being the 2434.65 cm^{-1} line which is most likely a result of mixing of the ($j = 4, n = 1, \Omega = 1$) bound state with the ($j = 0, \Omega = 0$) continuum state. Its very narrowness is a result of the near-conservation of the helicity (Ω) quantum number. Yet the transition is easily detectable because the predissociation linewidth is still greater than that due to Doppler broadening.

The most interesting phenomenon, shown in Fig. 4, is the possibility of observing interferences between two levels due to coupling to the continuum. Thus the Fano lineshape[47] at 2436.64-cm^{-1} is apparently due to interference with the wide 2442.70-cm^{-1} resonance. The process is envisioned as a transition from the 2436.49-cm^{-1} bound level to the continuum, followed by a transition to the 2436.49-cm^{-1} bound level. A situation of this kind will exist even if only a very dense set of levels exists in the molecule.

C. HeI_2—Predissociation and Overlapping Resonances

The HeI_2 serves as an excellent example for a slow-probe photodissociation. Schematically, the experiment[48] consists of three steps:

1. Complex formation:

$$I_2(X) + He \xrightarrow{M} He—I_2(X, v = 0) \tag{62}$$

2. Optical excitation:

$$He—I_2(X, v = 0) \xrightarrow{h\nu} He—I_2(B, v) \tag{63}$$

3. Predissociation:

$$He—I_2(B, v) \rightarrow I_2(B, v - n, j) + He \tag{64}$$

In the above X stands for the $X'\sum_{0g}^{+}$ ground electronic state of I_2 and B for the $I_2 - B^3\Pi_{0u}^{+}$ excited state. The optical excitation and predissociation is usually monitored by measuring the I_2 fluorescence.

Contrary to the ArN_2 case, the excitation and decay are usually well separated: even the slow vibrational mode has a period of ~ 1 ps whereas the dissociation lifetime may be as long as 220 ps[48] (for the $v = 15$ case). The excitation fluorescence spectrum[49] contains two resonances at very close proximity ($0.3\,cm^{-1}$) to one another. The origin of these resonances was explained[50] as being due to the existence of two nearly degenerate HeI_2

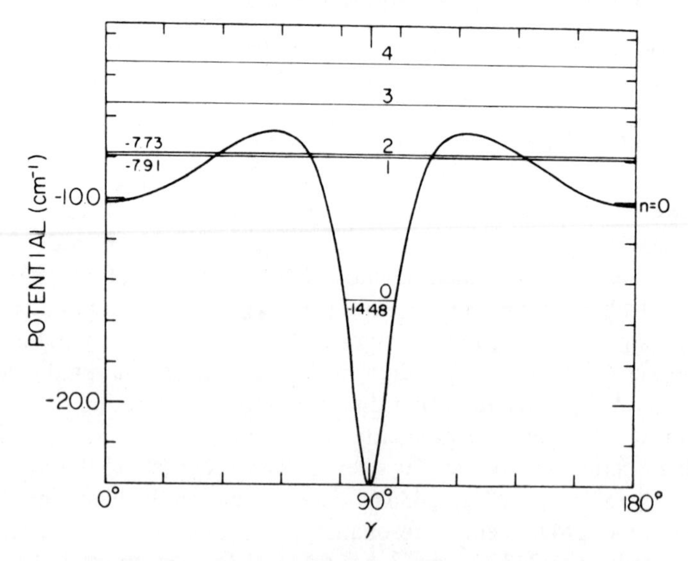

Fig. 5. Effective librational potential and energy levels for the lowest stretching eigenvalue of this HeI_2 complex. (From Ref. 50. Reproduced by permission of the American Institute of Physics.)

librational levels. One level is assigned as the first excited (asymmetric) state in the T-shape configuration; the other as the ground state in a shallow well at the collinear configuration.

The existence of a well in the collinear configuration is partly a result of *intramolecular* dynamics: it results from the lowering of the angle-dependent vibrational ground state at the collinear configuration. Due to the separation of timescales the 'fast' He—I_2 and I—I vibrations can be separated from the slow libration of He about the I_2 bond. As a result, it is possible to solve for the fast motion at a given value of γ[50]-the libration angle. The resulting (γ-dependent) ground vibrational eigenvalue, depicted in Fig. 5, serves as an effective potential for the librational motion. As shown in Fig. 5, this potential supports several librational bound states, among them the two nearly degenerate ones mentioned above.

The strong intramolecular coupling between the He—I_2 and the I_2 vibration thus induces an extra well in the collinear configuration and hence an extra bound state. When the molecule dissociates these bound states become resonances (see Eq. 48) and, as such, are detected in the excitation fluorescence spectrum.[49] A three-dimensional calculation[50] of the photopredissociation of He—I_2 indeed shows (see Fig. 6) absorption spectrum composed of two nearly overlapping lines.

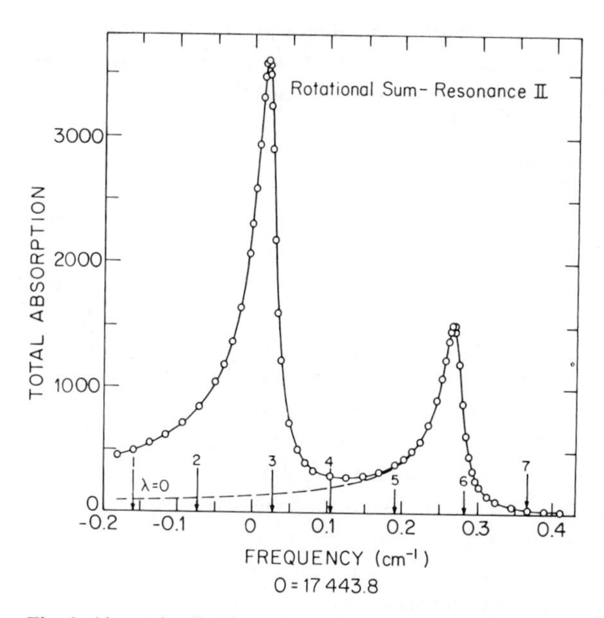

Fig. 6. Absorption lineshape due to $\lambda = 0$ and $\lambda = 1$ components in the resonance II region using the rotational sum formula. (From Ref. 50. Reproduced by permission of the American Institute of Physics.)

The phenomenon outlined here has a direct bearing on the existence of 'two phases' in larger clusters. These two phases have been described as rigid and liquid-like. In order to explain their existence it seems that two different Hamiltonians have to be assumed. In this case we see that dynamical effects, and in particular separation of fast and slow motions, can give rise to the appearance of one effective Hamiltonian displaying rigidity for some configurations and non-rigidity for others.

D. The VUV Photodissociation of H_2O and the 'Rate of Intramolecular Energy Transfer'

A question of fundamental importance to the understanding of intramolecular phenomena is the rate of energy transfer from one mode to another. In view of the discussion of Section II it is legitimate to phrase the problem in this way only when the spectral function of Eq. (40) is wider than Γ_i, the total width (see Eq. 50) of the mode in question.

The proper interpretation of the VUV photodissociation of H_2O allows us to give a quantitative estimate to this rate. The first experimental datum of importance is the photo absorption cross-section. In the 1411 to 1256 Å region it is known to consist of a series of diffuse bands.[51,52]

In Fig. 7 we show the calculated spectrum obtained by performing a complete three-dimensional calculation of the photodissociation of water.[53,54] The calculated spectrum shows a series of resonances. As shown in Fig. 7, their positions agree—considering the experimental uncertainties—very well with the observed diffuse band positions. Also shown in Fig. 7 are the OH rotational state distributions at selected wavelengths.

A close inspection of Fig. 7 reveals an interesting correlation between the time available for intramolecular energy transfer and the rotational distributions. We see that at 'off-peak' energies, e.g. points '2', '4' and '6' in Fig. 7, the rotational state distribution in inverted, i.e. very hot; at energies 'on-peak', e.g. points '1', '3', '5' and '7', a substantially colder, thermal-like, component is added. The instantaneous rotational distribution, following the optical transition from the ground state, may be said to be highly inverted, since the molecules is formed in the bent configuration, whereas the excited B^1A' surface is linear. Thus the degree of relaxation of the OH rotation is seen to be a direct measure of the duration of intramolecular energy transfer. We see that if the molecule lives as much as 0.1 ps ('slow' probe), the approximate lifetimes being deduced from the widths of the resonances shown in Fig. 7a, substantial relaxation occurs (see points '3' and '5'). If the lifetime is shorter ('fast' probe), e.g. points '4' and '6', relaxation cannot take place.

The above example illustrates the roles of fast and slow probes. The fast probe measures the instaneous nuclear density. As the probe, namely dissociation, becomes slower we see relaxation setting in. A combined

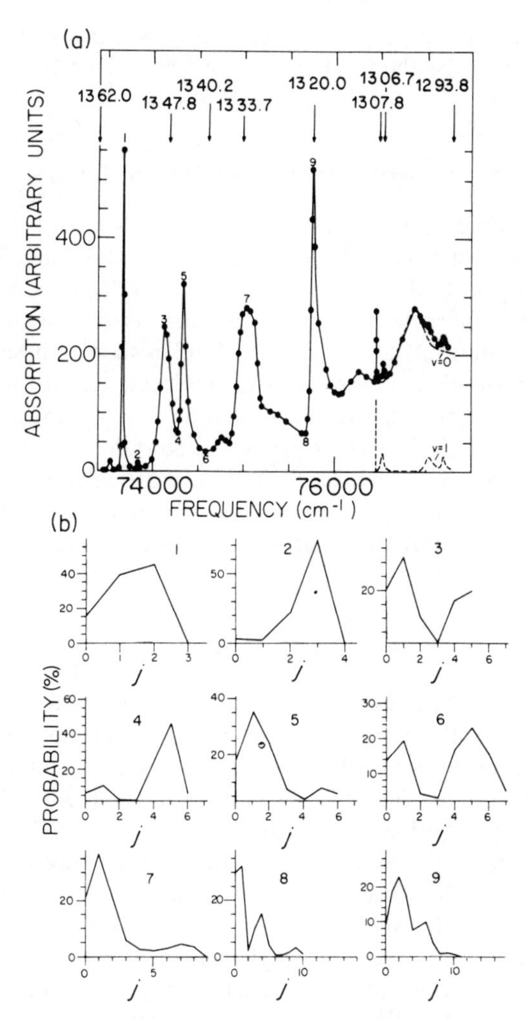

Fig. 7. Absorption spectrum (a) and OH rotational distribution (b) at photon energies marked in (a). The arrows and wavelengths (in angstroms) correspond to the experimental positions of the diffuse bands. (From Ref. 54. Reproduced by permission of the American Institute of Physics.)

measurement of resonance width and rotational state distributions yields the duration of intramolecular energy transfer in water.

In this section we have discussed four examples for the role of dissociation as a probe of intramolecular dynamics. By being able to provide a realistic theoretical treatment of small polyatomics it is possible to enhance the scope of currently available experimental techniques. With the aid of the theoretical calculations one can reliably investigate various aspects of intramolecular dynamics and describe the fundamental process which is being monitored in the experiment.

VI. SAMPLE POLYATOMIC STUDIES

Rapid developments in laser technology are currently being directed towards photodissociation studies of a wide variety of systems. The theoretical ideal, consistent with the discussion in Section II, involves a thorough knowledge of the characteristics of the exciting laser as well as of the initial state (prior to excitation) of the molecules. Studies such as these have not been reported for photodissociation of large molecules although recent spectroscopic studies[55] (both frequency as well as time dependent) on pyrazine approach the theoretical ideal. Nonetheless, considerable insight into gross features of photodissociation in larger molecules are emerging. To indicate the current state of affairs we briefly discuss two experimental examples below; no accurate theoretical photodissociation studies are available on these larger systems.

A. CF_3NO

Laser-induced photodissociation of CF_3NO to yield $CF_3 + NO(^2\Pi_{1/2}, {}^2\Pi_{3/2})$ with excitation wavelengths of $600\,nm < \lambda < 670\,nm$ has been the subject of several studies.[56,57] Initial work was confined to thermal CF_3NO samples excited via a non-transform limited pulsed nanosecond laser; a more recent note[57] utilizes an expansion cooled CF_3NO sample. In both cases high-resolution detection of NO product state resulted from use of a two-photon laser excited fluorescence technique. The results show NO vibrational distributions with predominant population in the ground state ($v = 0$) but with $v = 2$, 3 populations in excess of simple statistical predictions. Further, the $v = 1$ to $v = 0$ population ratio, as well as the population of the upper $^2\Pi_{3/2}$ electronic state, were found to increase with decreasing excitation wavelength.

Results obtained about the mechanism are, thus far, rather qualitative with considerably greater insight expected from ongoing studies with expansion cooled samples. Specifically, the rise time of the NO fluorescence indicates a predissociation mechanism with no long-lived intermediate. Detailed studies by Haas, Houston and coworkers[56,57] as well as by Spears and Hoffland[58] suggest an initial transition to an S_0 state prior to dissociation, i.e.

$$CF_3NO[S_0(^1A')] \overset{h\nu}{\rightarrow} CF_3NO[S_1(^1A'')] \rightarrow CF_3NO[S_0] \rightarrow CF_3 + NO \qquad (65)$$

with rates of flow out of S_0 similar in magnitude to CF_3NO fluorescence decay rates. From the time-dependent exact molecular eigenstate viewpoint Eq. (65)

suggests that the initially prepared state contains superposition states of exact states which individually contain both the S_1 and S_0 character. Initially phases are such that the S_1 character dominates. Under time evolution the S_0 character becomes evident, followed by the asymptotically dissociated state.

In addition to the nature of the intermediate electronic state in photodissociation the experimental results have suggested[56,58] a link between the appearance of rotationally excited NO and the change in CF_3NO geometry upon excitation. Specifically, ground state CF_3NO shows an eclipsed conformer, the excited state being staggered.[59] Thus electronic excitation involves a change in the torsional angle about the CN axis, imparting rotational motion to the departing NO. This proposal will probably undergo reexamination in the light of recent results[57] which display increasing NO rotation with decreasing γ, independent of whether the torsional or skeletal bend mode is initially excited. Additional experimental studies would also allow detailed examination of several predictions,[58] from simple radiationless transitions theory of the relationship between product distributions and modes of excitation.

B. Aryl and Aryl–Alkyl Halides

Molecules containing aromatic chromophores with substituted halogens have been the subject of a series of experiments by Bersohn and coworkers.[32b,60,61] Initial experiments[32b] focused on determining the fragment flux anisotropy parameter β, defined via the angular distribution of the photofragments measured relative to the polarization direction of the incident light. Subsequent studies[60,61] included pulsed excitation sources and time-of-flight detection of product-relative translational energy. Of the many molecules studied we note the interesting results on the series (X = Cl, Br, I):

$$\phi - X \xrightarrow{h\nu} \phi + X \tag{66}$$

$$\phi - CH_2 - X \xrightarrow{h\nu} \phi - CH_2 + X$$

$$\phi - C_2H_4 - X \xrightarrow{h\nu} \phi - C_2H_4 + X \tag{67}$$

Substantial mechanistic changes are found to occur with variations in X in reaction (66) and in the addition of CH_2 groups in (67).

Cumulative experimental results on reaction (66) with λ in the 265 to 320-nm range indicate that the photodissociation is indirect, involving a variety of electronic levels. Specifically supporting this view are the observations that the transition dipole is not related to the C—X axis and that the anisotropy parameter is reduced, a consequence of the longevity—relative to rotation—of the excited molecule. Additional experimental results on related aryl halides (e.g. 4, 4′-diiodobiphenyl) show product translational energy distributions independent of the wavelength of the exciting light (265 to 299 nm), further indicating an indirect photodissociation dynamics. The lifetime of excited iodobenzene is ~ 1 ps and is substantially longer for aryl bromides. The overall suggested mechanism involves initial excitation to a bound excited singlet state delocalized over the molecule followed by crossing to a triplet state with strong C—X repulsion. Differences in the magnitude of spin-orbit coupling in Br and I are deemed responsible for the different ϕ-Br and ϕ-I lifetimes. Excitation at higher energy ($\lambda = 193$ nm) introduces participation by additional electronic states prior to dissociation and several different pathways to dissociation.

Reducing the proximity, and hence the coupling, of the halogen to the ring by introducing CH_2 groups (see reaction 67) leads to significant mechanistic changes. In these cases there is far less dependence of observables on the particular halogen and translational energy distributions peak at lower values than for ϕ-X. The limited data are consistent with conversion from the initially excited singlet state to the ground singlet state, with subsequent dissociation being ordinary unimolecular decay.

These examples illustrate the current status of photodissociation studies of polyatomic systems.[62] The need for further detailed experimental studies as well as detailed theoretical calculations is apparent.

Acknowledgement

Partial support of the Petroleum Research Fund, administered by the American Chemical Society and NSERC, Canada, is gratefully acknowledged.

References

1. (a) Balint-Kurti, G. G., and Shapiro, M., *Adv. Chem. Phys.* (this volume); (b) Leone, S., *Adv. Chem. Phys.* (Vol. 50); Simons, J. P. in *Gas Kinetics and Energy Transfer* (Eds. P. G. Ashmore and J. Donovan, senior reporter), Vol. 2, p. 58, Burlington House, London, 1977; Freed, K. F., and Band, Y. B., in *Excited States* (Ed. E. C. Lim), Vol. 3, Academic Press, New York, 1977).
2. Heller, E. J., *J. Chem. Phys.*, **72**, 1337 (1980).
3. For example, Smalley, R. E., *Ann. Rev. Phys. Chem.*, **34**, 129 (1983).
4. Heller, E. J., *Faraday Disc. Chem. Soc.*, **75**, 141 (1983), and references therein.
5. Loudon, R., *The Quantum Theory of Light*, Clarendon Press, Oxford, 1973.
6. Shapiro, M., and Bersohn, R., *Ann. Rev. Phys. Chem.*, **33**, 409 (1982).
7. Levine, R. D., *Quantum Mechanics of Molecular Rate Processes*, Oxford University Press, London, 1969.
8. Lee, S. Y., and Heller, E. J., *J. Chem. Phys.*, **76**, 3035 (1982).
9. Imre, D., Kinsey, J. L., A. Sinha and Krenos J., *J. Phys. Chem.* (in press).
10. Shapiro, M., and Bersohn, R., *J. Chem. Phys.*, **73**, 3810 (1980).
11. Tipps, W. M., Baronowski, A. T., *Chem. Phys. Lett.*, **71**, 395 (1980).
12. For related discussions in radiationless transitions see Rhodes, W., in *Radiationless Transitions* (Ed. E. C. Jim), Vol. 4, Academic Press, New York, 1980.

13. For a qualitative introduction see Brumer, P., *Adv. Chem. Phys.*, **47**, 201 (1981), and Wightman, A. S., in *Statistical Physics at the Turn of the Decade* (Ed. E. D. G. Cohen), North-Holland, Amsterdam, 1975.
14. The common assumption that regular quasi-periodic systems show no relaxation is incorrect. See Jaffe, C., and Brumer, P., *J. Phys. Chem.*, **88**, 4829 (1984) for a general description.
15. Golden, S., and Longuet-Higgins, H. C., *J. Chem. Phys.*, **33**, 1479 (1960); Kosloff, R., and Rice, S. A., *J. Chem. Phys.*, **74**, 1340 (1981).
16. For an extensive treatment aimed at developing a semiclassical theory directly linked to classical theory see Kay, K., *J. Chem. Phys.*, **79**, 3026 (1983); for a general discussion of quantum chaos see Pechukas, P., *J. Phys. Chem.*, **88**, 4823 (1984).
17. von Neumann, J., *Z. Phys.*, **57**, 30 (1929).
18. Berry, M. V., *Phil. Trans. Roy. Soc. London*, **A287**, 237 (1977); Percival, I. C., *Adv. Chem. Phys.*, **36**, 1 (1977); Berry, M. V., *J. Phys.*, **A60**, 2083 (1977).
19. Porter, C. E., *Statistical Theories of Spectra*, Academic Press, New York, 1965.
20. Pechukas, P., *Phys. Rev. Lett.*, **51**, 943 (1983).
21. For example, Hutchinson, S., and Wyatt, R. E., *Chem. Phys. Lett.*, **72**, 384 (1980).
22. McDonald, S. W., and Kaufmann, A. N., *Phys. Rev. Lett.*, **42**, 1189 (1979).
23. Taylor, R. D., and Brumer, P., *Far. Disc. Chem. Soc.*, **75**, 170 (1983); Gruner, D., Taylor, R. D., and Brumer, P. (to be published).
24. Shapiro, M., Taylor, R. D., and Brumer, P., *Chem. Phys. Lett.*, **106**, 325 (1984).
25. Shapiro, M., and Brumer, P., *Chem. Phys. Lett.*, **72**, 528 (1980); Brumer, P., and Shapiro, M., *Chem. Phys. Lett.*, **90**, 481 (1982).
26. Davis, M., Stechel, E., and Heller, E. J., *Chem. Phys. Lett.*, **76**, 21 (1980).
27. Taylor, R. D., and Brumer, P., *Far. Disc. Chem. Soc.*, **75**, 117 (1983).
28. Gruner, D., Taylor, R. D., and Brumer, P. (to be published).
29. For related issues in statistical behaviour in chemical reactions see Duff, J. W., and Brumer, P., *J. Chem. Phys.*, **67**, 4898 (1977), **71**, 2693 (1979), **71**, 3895 (1979); Brumer, P., *Disc. Far. Soc.*, **75**, 271 (1983); Hamilton, I., and Brumer, P., *J. Chem. Phys.* (in press).
30. Brumer, P., and Shapiro, M., *J. Chem. Phys.*, **80**, 4567 (1984).
31. Busch, G. E., and Wilson, K. R., *J. Chem. Phys.*, **56**, 3638 (1972); Wilson, K. R., in *Chemistry of the Excited State* (Ed. J. N. Pitts, Jr.). Gordon and Breach, New York, 1970.
32. (a) Solomon, J., Jonah, C., Chandra, P., and Bersohn, R., *J. Chem. Phys.*, **55**, 1908 (1971); (b) Dzvonik, M. J., Yang, S. C., and Bersohn, R., *J. Chem. Phys.*, **61**, 4408 (1974).
33. Yang, S. C., Freedman, A., Kawasaki, M., and Bersohn, R., *J. Chem. Phys.*, **72**, 4058 (1980).
34. Butler, J. E., Drozdoski, W. S., and McDonald, J. R., *Chem. Phys.*, **50**, 413 (1980).
35. Zare, R. N., and Dagdigian, P. J., *Science*, **185**, 739 (1974).
36. MacPherson, M. T., and Simons, J. P., *Chem. Phys. Lett.*, **51**, 261 (1977).
37. Kinsey, J. L., *J. Chem. Phys.*, **66**, 2560 (1977); Schmiedl, R., Bottner, R., Zacharias, H., Meirer, U., and Welge, K. H., *Opt. Commun.*, **31**, 329 (1979).
38. Sparks, R. K., Shobatake, K., Carlson, L. R., and Lee, Y. T., *J. Chem. Phys.*, **75**, 3838 (1981).
39. Hermann, H. W., and Leone, S. R., *J. Chem. Phys.*, **76**, 4759 (1982).
40. van Veen, G. N. A., Baller, T., deVries, A. E., and van Veen, N. J. A. (to be published).
41. Shapiro, M., *Chem. Phys. Lett.*, **81**, 521 (1981).
42. Shapiro, M., *J. Chem. Phys.*, **56**, 2582 (1972).
43. Shapiro, M., *Chem. Phys. Lett.*, **46**, 442 (1977); Beswick, J. A., Shapiro, M., and Sharon, R., *J. Chem. Phys.*, **67**, 4045 (1977).

44. Beswick, J. A. and Shapiro, M., *Chem. Phys.*, **64**, 333 (1982).
45. McGuire, P., and Kouri, D. J., *J. Chem. Phys.*, **60**, 2488 (1974).
46. Tamir, M., and Shapiro, M., *Chem. Phys. Lett.*, **31**, 166 (1975).
47. Fano, U., *Phys. Rev.*, **124**, 1866 (1961).
48. Levy, D. H., in *Photoselective Chemistry, Adv. Chem. Phys.* (Eds. J. Jortner, R. Levine and S. A. Rice), Vol. 47, p. 323, New York, 1981.
49. Sharfin, W., Johnson, K. E., Wharton, L., and Levy, D. H., *J. Chem. Phys.*, **71**, 1292 (1979).
50. Segev, E., and Shapiro, M., *J. Chem. Phys.*, **79**, 4969 (1983).
51. Wang, H., Felps, W. S., and McGlynn, S. P., *J. Chem. Phys.*, **67**, 2614 (1977).
52. Watanabe, K., and Zelikoff, M., *J. Opt. Soc. Am.*, **43**, 753 (1953).
53. Segev, E., and Shapiro, M., *J. Chem. Phys.*, **73**, 2001 (1980).
54. Segev, E., and Shapiro, M., *J. Chem. Phys.*, **77**, 5604 (1982).
55. Smith, D. D., Rice, S. A., and Struve, W., *Far Disc. Chem. Soc.*, **75**, 173 (1983); Van der Meer, B. H., Jonkmann, H. Th., Kommandeur, J., Meets, W. L., and Majewski, W. A., *Chem Phys. Lett.*, **92**, 565 (1982).
56. Roellig, M. P., Houston, P. L., Asscher, M., and Hass, Y., *J. Chem. Phys.*, **73**, 5081 (1980).
57. Jones, R. W., Bower, R. T., and Houston, P. L., *J. Chem. Phys.*, **76**, 3339 (1982).
58. Spears, K. G., and Hoffland, L. D., *J. Chem. Phys.*, **74**, 4765 (1981).
59. DeKoven, B. M., Fung, K. H., Levy. D. H., Hoffland, L. D., and Spears, K. G., *J. Chem. Phys.*, **74**, 4755 (1981).
60. Kawasaki, M., Lee, S. J., and Bersohn, R., *J. Chem. Phys.*, **66**, 2647 (1977).
61. Freeman, A., Yang, S. C., Kawasaki, M., and Bersohn, R., *J. Chem. Phys.*, **72**, 1028 (1980).
62. (added in proof) see also recent results on NCNO in Nadler, I., Pfab, J., Reisler, H., and Wittig, C., *J. Chem. Phys.*, **81**, 653 (1984). Also of interest are coherent excitations in bound polyatomic molecules designed to probe intramolecular dynamics via quantum beats – e.g. Muhlbach, J., Dubs, M., Bitto, H., and Huber, J. R. (in press); Lambert, W. R., Felker, P. M., and Zewail, A. H., *J. Chem. Phys.*, **81**, 2217 (1984) and references therein; Sharfin, W., Ivanco, M., and Wallace, S. C., *J. Chem. Phys.*, **76**, 2095 (1982); Chaiken, J., Gurnick, M., and McDonald, J. D., *J. Chem. Phys.*, **74**, 106, 117 (1981); Okajina, S., Saigusa, H., and Lim, E. C., *J. Chem. Phys.*, **76**, 2096 (1982); Moore, C. B., and Weisshaar, *Ann. Rev. Phys. Chem.*, **34**, 525 (1983); Bitto, H., Stafast, H., Russegger, P., and Huber, J. R., *Chem. Phys.*, **84**, 249 (1984).

Photodissociation and Photoionization
Edited by K. P. Lawley

QUANTUM THEORY OF MOLECULAR PHOTODISSOCIATION

G. G. BALINT-KURTI

School of Chemistry, Bristol University, Bristol BS8 1TS

and

M. SHAPIRO

Weizmann Institute of Science, Rehovot, Israel

CONTENTS

I. INTRODUCTION

In this article we describe the fundamental theory of photodissociation processes. This theory constitutes the essential tool needed to analyse the highly resolved experimental data now becoming available from photofragmentation and photodissociation experiments.[1–15] It is only through the use of the theory that it is possible to extract the physically significant, and interesting, molecular quantities from the experimental data. Several reviews of both experimental[1–7] and theoretical[16–20] aspects of photodissociation processes have recently been published. The purpose of the present article is not to duplicate this material, but rather to concentrate on the development of the fundamental theory needed for the analysis and planning of both currently feasible and future experiments.

Photodissociation experiments provide data which contain information about three types of fundamental molecular properties: (1) the lower (or ground) state potential energy surface; (2) the upper excited state potential energy surface; and (3) the transition dipole moment function. The transition dipole moment function depends on the relative positions of the nuclei within the molecule, and it is through the mediation of this function that the light interacts with the molecule. The theoretical framework, to be developed below, has two roles. The first is to provide the apparatus for extracting the fundamental molecular quantities discussed above from the experimental measurements. Its second role is to provide the basis through which we can understand the mechanisms of the many intriguing ways in which light interacts with and leads to the dissociation of molecules.

In Section II the basic molecule–radiation interaction process is considered, in a manner which is valid for both the weak and strong field situation. The very definition of a photofragmentation cross-section is problematic in the strong-field limit. This problem is addressed and a careful analysis is presented of the rate of transition between completely specified initial quantum states of the bound molecule and final states of the fragments. This analysis leads to an unambiguous definition of the photofragmentation cross-section for both weak and strong field regimes. The formal structure of the differential equations and boundary conditions which govern the wavefunctions required in the calculation of the photofragmentation cross-section are presented, and the weak field limit of the theory is derived.

Section III discusses in more explicit detail the calculation of the photofragmentation cross-sections. The section is divided into three subsections. Subsection III.A discusses photodissociation by a strong laser field and presents, in detail, the differential equations and boundary conditions needed for the calculation of the wavefunctions and photofragmentation cross-sections in this limit. The theory presented in Sections II and III.A lays the foundation for the computation of multiphoton dissociation and photofragmentation cross-sections. There are several other related processes to which very similar theories are also applicable. These include:

1. Laser-induced inelastic and reactive collisional processes;
2. Raman spectroscopy and fluorescence, especially of transient species; and
3. Energy level shifts and spectral features in strong laser fields.

In Subsection III.B the differential equations and boundary conditions are further specialized to the specific problem of the photodissociation of a triatomic molecule by a weak radiation field. In this subsection explicit use is made of the conservation of total angular momentum to simplify, as far as possible, the set of coupled differential equations which must be solved for both the bound and continuum state wavefunctions. All the wavefunctions and differential equations are expressed in a body fixed coordinate system appropriate to the photodissociation problem. The last of the three subsections (III.C) discusses the artificial channel method. This method permits the *direct* evaluation of the pivotally important photofragmentation **T** matrix elements, which embody the intrinsic dynamics of the photofragmentation process. The relationship of the artificial channel method to other methods which have been proposed for the evaluation of photofragmentation cross-sections is reviewed.

Section IV discusses the various different types of photofragmentation cross-sections which may be measured. Exact theoretical formulae are given for each type of experimentally measurable cross-section. The theoretical expressions permit an analysis to be made of the different types of intrinsic information content embodied in each of the measurable types of cross-section. These theoretical expressions provide the essential link between the experimentally measurable attributes of a photodissociation process and the underlying molecular properties (i.e. potential energy surfaces and transition dipole moments). An example is given of the extreme sensitivity of the photofragment angular distribution to the nature of the transition dipole moment function for the photofragmentation of a triatomic molecule.

Section V provides a brief summary of applications which have been made of photofragmentation theory to specific systems.

II. GENERAL THEORY OF MOLECULAR PHOTODISSOCIATION

In this section we consider the general problem of a molecule, composed of two clusters A and B, breaking apart under the influence of a radiation field. The process is written schematically as

$$\mathrm{A-B} + h\nu \rightarrow \mathrm{A} + \mathrm{B} \tag{1}$$

We treat the problem in a fairly general way, in that we do not constrain ourselves to weak fields or weak molecular interactions. Therefore, our approach, which is a non-perturbative one, is to recast the problem as a set of coupled differential equations. The results hold for all orders in field intensity or molecular interactions.

We start the derivation by noting that most experiments interrogate the molecular states in the absence of (the strong) radiation. We therefore set the scene by exploring the states of $\hat{H}_{\mathrm{M}}$—the molecular Hamiltonian (in the absence of radiation). For our present discussion it is convenient to write $\hat{H}_{\mathrm{M}}$ as a sum of nuclear kinetic energy terms, $\hat{K}(\mathbf{R})$ and $\hat{K}(\mathbf{r})$, and an electronic term $\hat{H}_{el}(\mathbf{q}|\mathbf{r},\mathbf{R})$:

$$\hat{H}_M = \hat{K}(\mathbf{R}) + \hat{K}(\mathbf{r}) + \hat{H}_{el}(\mathbf{q}|\mathbf{r},\mathbf{R}) \tag{2}$$

where $\mathbf{R}$ is the A to B displacement vector, $\mathbf{r}$ denotes the internal nuclear coordinates of both clusters and $\mathbf{q}$ designates the collection of all electronic coordinates.

The process of photodissociation is defined as a transition from $\phi_M(E_i)$—a bound state of $\hat{H}_M$:

$$(E_i - \hat{H}_M)\phi_M(E_i) = 0 \tag{3}$$

to $\phi_M(E,\mathbf{n}^-)$ and 'incoming' continuum state[21–23] of the same Hamiltonian:

$$(E - \hat{H}_M)\phi_M(E,\mathbf{n}^-) = 0 \tag{4}$$

after absorbing a photon. The index $\mathbf{n}^-$ is a reminder of the possible asymptotic states of $\phi_M(E,\mathbf{n}^-)$, in the following sense: as $R\to\infty$, $\hat{H}_M$ goes over to a separable form ('free Hamiltonian'), $\hat{H}_0$:

$$\hat{H}_M \underset{R\to\infty}{\to} \hat{H}_0 \tag{5}$$

where

$$\hat{H}_0 = \hat{K}(\mathbf{R}) + \hat{K}(\mathbf{r}) + \hat{h}_{el}(\mathbf{q}|\mathbf{r}) \tag{6}$$

with $\hat{h}_{el}(\mathbf{q}|\mathbf{r})$ being the asymptotic form of $\hat{H}_{el}(\mathbf{q}|\mathbf{r},\mathbf{R})$:

$$\hat{H}_{el}(\mathbf{q}|\mathbf{r},\mathbf{R}) \underset{R\to\infty}{\to} \hat{h}_{el}(\mathbf{q}|\mathbf{r}) \tag{7}$$

Because of its form (Eq. 6), the eigenstates of $\hat{H}_0$:

$$(E - \hat{H}_0)\phi_0(E,\mathbf{n}) = 0 \tag{8}$$

are given as simple products:

$$\phi_0(E,\mathbf{n}) = |\mathbf{n}\rangle|\mathbf{k}_\mathbf{n}\rangle, \tag{9}$$

where $|\mathbf{n}\rangle$ are the internal states of the A and B clusters:

$$[\varepsilon_\mathbf{n} - \hat{K}(\mathbf{r}) - \hat{h}_{el}(\mathbf{q}|\mathbf{r})]|\mathbf{n}\rangle = 0 \tag{10}$$

and $|k_\mathbf{n}\rangle$ are the (free) translational states, given by

$$[E - \varepsilon_\mathbf{n} - \hat{K}(\mathbf{R})]|k_\mathbf{n}\rangle = 0 \tag{11}$$

It can be shown[21] that $\phi_M(E,\mathbf{n}^-)$ go over, as t (the time) $\to\infty$, to a well-defined free state $\phi_0(E,\mathbf{n})$. More precisely, we have that[21]

$$\exp[i(E-E')t/\hbar]\langle\phi_0(E,\mathbf{n})|\phi_M(E',\mathbf{n}'^-)\rangle \underset{t\to\infty}{\to} \delta(E-E')\delta_{\mathbf{nn}'} \tag{12}$$

The $\mathbf{n}^-$ notation thus serves as a reminder of the long-time limit of $\phi_M(E,\mathbf{n}^-)$, at which time $\mathbf{n}$ indeed becomes a set of good quantum numbers.

The radiation field is now switched on, at a time most conveniently chosen to be $t = 0$. Once the field is switched on the eigenstates of $\hat{H}_M$ cease to be stationary and we can detect transitions between them. The exact dynamics, in the presence of the radiation field, is given by the (matter + radiation) Schrödinger equation:

$$\hat{H}\Psi(t) = \frac{i\hbar\, \partial \Psi(t)}{\partial t} \tag{13}$$

where $\hat{H}$ is the *total* Hamiltonian:

$$\hat{H} = \hat{H}_M + \hat{H}_R + \hat{H}_I \equiv \hat{H}_f + \hat{H}_I \tag{14}$$

where $\hat{H}_M$ is the matter Hamiltonian (Eq. 2) and $\hat{H}_R$ is the free-radiation Hamiltonian:

$$\hat{H}_R = \sum_k \hbar\omega_k \hat{a}_k^\dagger \hat{a}_k \tag{15}$$

where k is the radiation mode index, $\hat{a}_k^\dagger$ and $\hat{a}_k$ are, respectively, the creation and annihilation operators of photons in the k mode and ω_k is the photon frequency in the k mode. $\hat{H}_I$, the radiation–matter interaction, is written, in the electric dipole approximation,[24] as

$$\hat{H}_I = i\sum_k \left(\frac{\hbar\omega_k}{2\varepsilon_0 V}\right)^{1/2} \hat{\boldsymbol{\varepsilon}}_k \cdot \boldsymbol{\mu}(\hat{a}_k - \hat{a}_k^\dagger) \tag{16}$$

where V is the cavity volume, ε_0 is the electric permitivity in free space, $\hat{\boldsymbol{\varepsilon}}_k$ is the polarization direction and $\boldsymbol{\mu}$ is the electric dipole operator:

$$\boldsymbol{\mu} = e\sum_m \mathbf{r}_m \tag{17}$$

where m goes over both electrons and nuclei.

At $t = 0$ we assume that the state of our system is given by one of the eigenstates of $\hat{H}_f$, the non-interacting Hamiltonian (see Eq. 14):

$$(E_i + N_i\hbar\omega - \hat{H}_f)|\phi_M(E_i), N_i\rangle = 0 \tag{18}$$

We have implicitly assumed the free radiation to be in a well-defined photon number state $|N_i\rangle$. This does not present a serious limitation to treating more realistic situations, because we can, by a simple summation over such states, also consider a coherent state or a statistical mixture of coherent states.

As the system progresses in time we inquire about the photodissociation probability and rate. We define the photodissociation probability as that of eventually populating (i.e. as $t \to \infty$) a non-interacting continuum state. In order to calculate this probability we need not wait till infinite times. Instead, at any given time, it is simpler to calculate the transition probability to a state that is *guaranteed* to evolve to the non-interacting state of interest as

$t \to \infty$. In order to find this state we examine the non-interacting state of interest at asymptotic times and evolve it *back* in time.The resulting state is

$$|\psi(E, \mathbf{n}^-, \mathbf{N}^-, t)\rangle = \lim_{t_1 \to \infty} \exp\left[\frac{-i\hat{H}(t - t_1)}{\hbar}\right] |\phi_M(E, \mathbf{n}^-), \mathbf{N}, t_1\rangle \qquad (19)$$

where $\mathbf{N}$ is a collection of photon occupation numbers, $\mathbf{N} \equiv (N_{k_1}, N_{k_2}, \ldots, N_{k_m}, \ldots)$. This state can be shown to be an eigenstate of the total Hamiltonian:

$$\left(E + \sum_{N_k} N_k \hbar\omega_k - \hat{H}\right)\psi(E, \mathbf{n}^-, \mathbf{N}^-, t) = 0 \qquad (20)$$

provided that as $t \to \infty$ the matter–radiation interaction vanishes, i.e.

$$\hat{H} \underset{t \to \infty}{\to} \hat{H}_f = \hat{H}_M + \hat{H}_R \qquad (21)$$

This can be accomplished formally[24] by switching off the interaction at an infinitely slow rate (adiabatic switching), i.e. we set

$$\hat{H}_I(t) = \hat{H}_I \exp(-\varepsilon t) \qquad (22)$$

and let $\varepsilon \to +0$ after evaluating the (ε-dependent) transition rates. Notice that although in Eq. (20) the total energy of the system is written as $E + \sum_{N_k} N_k \hbar\omega_k$, neither E nor $\mathbf{N}$ are good quantum numbers at finite times. As before, when dealing with the pure matter states (Eq. 4), the $\mathbf{n}^-$, $\mathbf{N}^-$ notation simply serves as a reminder for the asymptotic state (given by $|\phi_M(E, \mathbf{n}^-), \mathbf{N}, t\rangle$).

The (photodissociation) transition amplitude can now be written as

$$\begin{aligned} A(\mathbf{n}, \mathbf{N} | i, N_i) &= \langle \psi(E, \mathbf{n}^-, \mathbf{N}^-, t) | \phi_M(E_i), N_i \rangle \\ &= \lim_{t_1 \to \infty} \left\langle \phi_M(E, \mathbf{n}^-), \mathbf{N}, t_1 \middle| \exp\left[\frac{i\hat{H}(t - t_1)}{\hbar}\right] \middle| \phi_M(E_i), N_i \right\rangle \\ &= \left\langle \phi_M(E, \mathbf{n}^-), \mathbf{N} \middle| \exp\left(\frac{i\hat{H}t}{\hbar}\right) \middle| \phi_M(E_i), N_i \right\rangle \end{aligned} \qquad (23)$$

where we have used Eq. (21) and the fact that

$$\exp\left(\frac{i\hat{H}_f t_1}{\hbar}\right) |\phi_M(E, \mathbf{n}^-), \mathbf{N}, t_1\rangle = |\phi_M(E, \mathbf{n}^-), \mathbf{N}\rangle \qquad (24)$$

In order to calculate the transition *rate* defined as

$$R(\mathbf{n}, \mathbf{N} | i, N_i) \equiv -\frac{\mathrm{d}}{\mathrm{d}t} |A(\mathbf{n}, \mathbf{N} | i, N_i)|^2 \qquad (25)$$

(where the $-$ sign is introduced because the transition probability *decreases* with time due to the switching *off* of the interaction) we make use of the integral equation for the evolution operator:[24]

$$\exp\left(\frac{i\hat{H}t}{\hbar}\right) = \lim_{\varepsilon\to+0, t_1\to\infty}\left[1 - \frac{i}{\hbar}\int_t^{t_1} dt' \exp\left(\frac{i\hat{H}t'}{\hbar}\right)\hat{H}_1 \right.$$
$$\left. \times \exp(-\varepsilon t')\exp\left(\frac{-i\hat{H}_f t'}{\hbar}\right)\right]\exp\left(\frac{i\hat{H}_f t}{\hbar}\right) \tag{26}$$

Inserting this expression in Eq. (23) and making use of Eq. (19) yields the following equation for the transition amplitude:

$A(\mathbf{n}, \mathbf{N}|i, N_i)$

$$= \lim_{\varepsilon\to+0, t_1\to\infty}\left(\frac{-i}{\hbar}\right)\left\langle \psi(E, \mathbf{n}^-, \mathbf{N}^-)\right|\int_t^{t_1} dt' \exp\frac{i\left(E + i\varepsilon\hbar + \sum_{N_k} N_k\, \hbar\omega_k\right)t'}{\hbar}$$
$$\times \hat{H}_I \exp\left[\frac{-i(E_i + N_i\hbar\omega)(t'-t)}{\hbar}\right]\left|\phi_M(E_i), N_i\right\rangle$$
$$= \lim_{\varepsilon\to+0}\frac{\langle\psi(E, \mathbf{n}^-, \mathbf{N}^-)|\hat{H}_I|\phi_M(E_i), N_i\rangle \exp\left[i(E + i\varepsilon\hbar + \sum_{N_k} N_k\hbar\omega_k)t/\hbar\right]}{E + i\varepsilon\hbar + \sum_{N_k} N_k\hbar\omega_k - E_i - N_i\hbar\omega} \tag{27}$$

where we have used the orthogonality between $\phi_M(E_i)$ and $\phi_M(E, \mathbf{n}^-)$.

With the use of Eq. (27) for the amplitude we immediately obtain for the transition rate (Eq. 25) that

$$R(\mathbf{n}, \mathbf{N}|i, N_i) = \lim_{\varepsilon\to+0}\frac{2\varepsilon\exp(-2\varepsilon t)|\langle\psi(E, \mathbf{n}^-, \mathbf{N}^-|\hat{H}_I|\phi_M(E_i), N_i\rangle|^2}{(E + \sum_{N_k} N_k\hbar\omega_k - E_i - N_i\hbar\omega)^2 + \varepsilon^2\hbar^2}$$
$$= \frac{2\pi}{\hbar}|\langle\psi(E, \mathbf{n}^-, \mathbf{N}^-)|\hat{H}_I|\phi_M(E_i), N_i\rangle|^2$$
$$\times\,\delta\left[E - E_i + \hbar\left(\sum_{N_k} N_k\,\omega_k - N_i\omega\right)\right] \tag{28}$$

The exact transition rate is independent of time. This result holds true regardless of the strength of the radiation field. It is a direct consequence of the fact that we probe the non-interacting final state in the *absence* of the radiation field (the adiabatic switching off). In contrast, the transition rate to a given non-interacting final state in the *presence* of the radiation field may be time dependent. Notice also that the transition rate is zero if total energy is not conserved, whereas the transition amplitude is non-zero even if energy is not conserved (see Eq. 27). This is surely a manifestation of the fact that the initial state is a wavepacket of different energy eigenstates of the total Hamiltonian.

In the dipole approximation (Eq. 16), the expression for the transition rate assumes the form

$$R(\mathbf{n}, \mathbf{N}|i, N_i) = 2\pi \sum_{k'} \left(\frac{\omega_{k'}}{2\varepsilon_0 V} \right) |\langle \psi(E, \mathbf{n}^-, \mathbf{N}^-)| \, \hat{\varepsilon}_{k'} \cdot \boldsymbol{\mu} (\hat{a}_{k'} - \hat{a}_{k'}^\dagger)$$

$$\times |\phi_M(E_i), N_i\rangle|^2 \delta \left[E - E_i + \hbar \left(\sum_{N_k} N_k \omega_k - N_i \omega \right) \right] \tag{29}$$

Since $\psi(E, \mathbf{n}^-, \mathbf{N}^-)$ contains, in principle, contributions from all modes and all occupation numbers we cannot eliminate the sum over k'. However, if N_i is not too large, we can use first-order perturbation theory in which we replace

$$|\psi(E, \mathbf{n}^-, \mathbf{N}^-)\rangle \simeq |\phi_M(E, \mathbf{n}^-), N_i - 1\rangle \tag{30}$$

The sum in Eq. (29) collapses to just one term:

$$R^{(1)}(\mathbf{n}, N_i - 1 | i, N_i) = 2\pi \left(\frac{\omega}{2\varepsilon_0 V} \right) N_i$$

$$\times |\langle \phi_M(E, \mathbf{n}^-) | \hat{\varepsilon}_i \cdot \boldsymbol{\mu} | \phi_M(E_i) \rangle|^2 \tag{31}$$

where we have integrated over ω to account for the non-zero frequency spread of the initial photon state. The cross-section for photodissociation is defined as the rate of energy removal from the radiation field, due to the transition of interest, divided by the intensity of the incident radiation. We therefore have

$$\sigma^{(1)}(E, \mathbf{n} | i, \omega) = \frac{\hbar\omega R^{(1)}(\mathbf{n}, N_i - 1 | i, N_i)}{N_i \hbar\omega c / V}$$

$$= 2\pi \left(\frac{\omega}{2c\varepsilon_0} \right) |\langle \phi_M(E, \mathbf{n}^-) | \hat{\varepsilon}_i \cdot \boldsymbol{\mu} | \phi_M(E_i) \rangle|^2 \tag{32}$$

if, as is customary, electrostatic units ($\varepsilon_0 = 1/4\pi$) are used and the frequency is expressed in cycles per seconds (by setting $\omega = 2\pi\nu$). The cross-section assumes the more familiar[25] form:

$$\sigma^{(1)}(E, \mathbf{n} | i, \nu) = \frac{8\pi^3 \nu}{c} |\langle \phi_M(E, \mathbf{n}^-) | \hat{\varepsilon}_i \cdot \boldsymbol{\mu} | \phi_M(E_i) \rangle|^2 \tag{33}$$

It is instructive to write a similar expression for the *exact* transition amplitude of Eq. (29) for $N_i \gg 1$ using the properties of the creation and annihilation operators:

$$\hat{a}_{k'}^\dagger |N_i\rangle = \begin{cases} (N_i + 1)^{1/2} |N_i + 1\rangle & k' = i \\ 2^{1/2} |1\rangle & k' \neq i \end{cases} \tag{34a}$$

$$\hat{a}_{k'} |N_i\rangle = \begin{cases} N_i^{1/2} |N_i - 1\rangle & k' = i \\ 0 & k' \neq i \end{cases} \tag{34b}$$

We obtain from Eq. (29), assuming $2^{1/2} \ll N_i^{1/2}$ and $(N_i + 1)^{1/2} \simeq N_i^{1/2}$ that

$$R(\mathbf{n}, \mathbf{N} | \mathbf{i}, N_i) = 2\pi \left(\frac{\omega}{2\varepsilon_0 V} \right) N_i | \langle \psi(E, \mathbf{n}^-, \mathbf{N}^-) | \hat{\boldsymbol{\varepsilon}} \cdot \boldsymbol{\mu} | (|N_i - 1\rangle - |N_i + 1\rangle) \\ \times | \phi_{\mathrm{M}}(E_i) \rangle |^2 \delta \left(E - E_i + \sum_{N_k} N_k \hbar\omega_k - N_i \hbar\omega \right) \tag{35}$$

The cross-section (on the energy shell) in electrostatic units is

$$\sigma(E, \mathbf{n}, \mathbf{N} | i, N_i) = \frac{8\pi^3 v}{c} | \langle \psi(E, \mathbf{n}^-, \mathbf{N}^-) | \hat{\boldsymbol{\varepsilon}}_i \cdot \boldsymbol{\mu} | (|N_i - 1\rangle - |N_i + 1\rangle) | \phi_{\mathrm{M}}(E_i) \rangle |^2 \tag{36}$$

where

$$E = E_i + \hbar \left(N_i \omega - \sum_{N_k} N_k \omega_k \right) \tag{37}$$

The expression for the exact cross-section is similar in form to the first-order result, except that the non-interacting state is replaced by a fully interacting stationary state. The formula applies to transitions to any final photon state and is correct to all orders of the radiation strength.

In order to compute the fully interacting state we must expand it in a complete set of photon occupation number states:

$$|\psi^-\rangle = \sum_{\mathbf{N}'} |\mathbf{N}'\rangle \langle \mathbf{N}' | \psi^- \rangle \tag{38}$$

where, as above, $|\mathbf{N}'\rangle \equiv |N_{k_1}, N_{k_2}, \ldots, N_{k_m}, \ldots\rangle$, and we use the shorthand notation of $|\psi^-\rangle$ for $|\psi(E, \mathbf{n}^-, \mathbf{N}^-)\rangle$. Since our main interest is in the single photon dissociation process we now treat specifically the case where $\mathbf{N} = (0, 0, \ldots, N_i - 1, 0, 0, \ldots)$, i.e. the final state corresponds to absorbing one photon from the incident radiation. This also means, by Eq. (37), that

$$E = E_i + \hbar\omega \tag{39}$$

Under these circumstances we expect the sum of Eq. (38) to be dominated by the projections on the $|N_i - 1\rangle$ photon state, $(\langle N_i - 1 | \psi^- \rangle)$ and 'neighbouring' photon number states. We therefore truncate the basis set used in Eq. (38) to be composed of a smaller number of states belonging to the i mode only:

$$|\psi^-\rangle \cong \sum_{N = N_i - m}^{N_i + m} |N\rangle \langle N | \psi^- \rangle \tag{40}$$

When substituting Eq. (40) in the Schrödinger equation (Eq. 20) we obtain

$$(E_i + (N_i - N)\hbar\omega - \hat{H}_{\mathrm{M}}) \langle N | \psi^- \rangle = i \left(\frac{\hbar\omega}{2\varepsilon_0 V} \right)^{1/2} \hat{\boldsymbol{\varepsilon}}_i \cdot \boldsymbol{\mu} \\ \times [(N+1)^{1/2} \langle N+1 | \psi^- \rangle - N^{1/2} \langle N-1 | \psi^- \rangle] \\ N = N_i - m, \ldots, N_i + m \tag{41}$$

If $N \gg 1$ we can equate $(N+1)^{1/2} \simeq N^{1/2} \simeq N_i^{1/2}$ and identify the light intensity as[24]

$$I = \left(\frac{c\hbar\omega}{V}\right) N_i \tag{42}$$

from which we obtain, in electrostatic units, that

$$\begin{aligned}[E_i + (N_i - N)\hbar\omega - \hat{H}_M]\langle N|\psi^-\rangle &= i(2\pi I/c)^{1/2} \\ &\times \hat{\boldsymbol{\varepsilon}}_i \cdot \boldsymbol{\mu}(\langle N+1|\psi^-\rangle - \langle N-1|\psi^-\rangle) \\ N &= N_i - m, \ldots, N_i + m \end{aligned} \tag{43}$$

We can further simplify the equations by considering only two electronic states, $|1\rangle$ and $|2\rangle$, defined as the eigenstates of H_{el} (Eq. 2):

$$\begin{aligned}[\hat{H}_{\text{el}}(\mathbf{q}|\mathbf{r}, \mathbf{R}) - V_1(\mathbf{r}, \mathbf{R})]\langle \mathbf{q}|1\rangle = 0 \\ [\hat{H}_{\text{el}}(\mathbf{q}|\mathbf{r}, \mathbf{R}) - V_2(\mathbf{r}, \mathbf{R})]\langle \mathbf{q}|2\rangle = 0 \end{aligned} \tag{44}$$

and confine the discussions to photodissociation due to a (visible or UV) photon which excites the system from a stable state $|1\rangle$ (the ground state) to a dissociating state $|2\rangle$ (the excited state). We therefore expand $\psi|E, \mathbf{n}^-, (N_i - 1)^-\rangle$ in the two electronic states

$$|\psi^-\rangle = (|1\rangle\langle 1| + |2\rangle\langle 2|)|\psi^-\rangle \tag{45}$$

and substitute the expansion in Eq. (43). We obtain the following equations:

$$\begin{aligned}[E_i + (N_i - N)\hbar\omega - \hat{K}(\mathbf{R}) - \hat{K}(\mathbf{r}) - V_1(\mathbf{r}, \mathbf{R})]\langle N, 1|\psi^-\rangle \\ = i\left(\frac{2\pi I}{c}\right)^{1/2} \langle 1|\hat{\boldsymbol{\varepsilon}}_i \cdot \boldsymbol{\mu}|2\rangle(\langle N+1, 2|\psi^-\rangle - \langle N-1, 2|\psi^-\rangle) \end{aligned} \tag{46a}$$

$$\begin{aligned}[E_i + (N_i - N)\hbar\omega - \hat{K}(\mathbf{R}) - \hat{K}(\mathbf{r}) - V_2(\mathbf{r}, \mathbf{R})]\langle N, 2|\psi^-\rangle \\ = i\left(\frac{2\pi I}{c}\right)^{1/2} \langle 2|\hat{\boldsymbol{\varepsilon}}_i \cdot \boldsymbol{\mu}|1\rangle(\langle N+1, 1|\psi^-\rangle - \langle N-1, 1|\psi^-\rangle) \end{aligned} \tag{46b}$$

$$N = N_i - m, \ldots, N_i + m$$

where we have assumed that

$$\langle 1|\hat{\boldsymbol{\varepsilon}}_i \cdot \boldsymbol{\mu}|1\rangle = \langle 2|\hat{\boldsymbol{\varepsilon}}_i \cdot \boldsymbol{\mu}|2\rangle = 0$$

and have neglected the non-Born–Oppenheimer coupling terms arising from the implicit dependence of states $|1\rangle$ and $|2\rangle$ on the nuclear coordinates **R** and **r**.

As pointed out above, given the initial photon number state $|N_i\rangle$, the states of greatest interest for a single-photon transition are $\langle N_i|\psi^-\rangle$ and $\langle N_i \pm 1|\psi^-\rangle$. The $\langle N_i + m|\psi^-\rangle$ components (even the projections on the ground electronic state) are closed, i.e. they must vanish asymptotically as $t \to \infty$, since their asymptotic energy is higher than $E_i + N_i\hbar\omega$—the total energy. At finite times, while the system is still interacting with the radiation, these components are in principle non-zero. However, because we are dealing with visible or UV photons, the closed components are so far removed in energy from the initial state that they hardly interact with it. The same may be said about the $\langle N_i - m, 1|\psi^-\rangle$ and $\langle N_i, 2|\psi^-\rangle$ components. Thus, given the initial photon occupancy number, N_i, and only two electronic states, we find that two photon states contribute strongly to $|\psi^-\rangle$, $\langle N_i, 1|\psi^-\rangle$ and $\langle N_i - 1, 2|\psi^-\rangle$. The resulting equation

$$[E_i + \hbar\omega - \hat{K}(\mathbf{R}) - V_1(\mathbf{r}, \mathbf{R})]\langle N_i, 1|\psi^-\rangle = -i\left(\frac{2\pi I}{c}\right)^{1/2}\langle 1|\hat{\boldsymbol{\varepsilon}}_i\cdot\boldsymbol{\mu}|2\rangle\langle N_i - 1, 2|\psi^-\rangle \tag{47a}$$

$$[E - \hat{K}(\mathbf{R}) - \hat{K}(\mathbf{r}) - V_2(\mathbf{r}, \mathbf{R})]\langle N_i - 1, 2|\psi^-\rangle = i\left(\frac{2\pi I}{c}\right)^{1/2}\langle 2|\hat{\boldsymbol{\varepsilon}}_i\cdot\boldsymbol{\mu}|1\rangle\langle N_i, 1|\psi^-\rangle \tag{47b}$$

where $E = E_i + \hbar\omega$ (Eq. 39), and we chose the zero-point energy equal to $(N_i - 1)\hbar\omega$.

Equations (47) are identical to those obtained in the rotating wave approximation[26] in the quasi-classical treatment of the radiation field. They are correct to all orders in the field but incorporate only a small number of photon states and neglect the contributions of all other modes. Thus spontaneous emission is manifestly excluded. In order to incorporate the spontaneous fluorescence from the dissociating molecule one must solve the more complete set of equations in which we expand, as in Eq. (38), in all the radiation modes.

In the weak field limit it is possible to solve directly for the transition amplitude of Eqs. (31) to (33) using first-order perturbation theory to solve Eq. (47). This is done by assuming that $\langle N_i, 1|\psi^-\rangle$ always remains larger than $\langle N_i - 1, 2|\psi^-\rangle$, which means that it is possible to neglect the right-hand side of Eq. (47a). Under these circumstances, Eq. (47a) becomes identical to Eq. (18) whose solution is $\langle 1|\phi_M(E_i)\rangle$. This solution can be substituted in Eq. (47b) which now becomes an inhomogeneous differential equation with a *known* source

$$[E - \hat{K}(\mathbf{R}) - \hat{K}(\mathbf{r}) - V_2(\mathbf{r}, \mathbf{R})]\psi_2^-(E) = i\left(\frac{2\pi I}{c}\right)^{1/2}\hat{\boldsymbol{\varepsilon}}_i\cdot\boldsymbol{\mu}_{21}\phi_1(E_i) \tag{48}$$

where $\psi_2^-(E)$ is a shorthand notation for $\langle N_i - 1, 2|\psi^-\rangle$ and μ_{21} stands for $\langle 2|\boldsymbol{\mu}|1\rangle$ and $\phi_1(E_i)$ for $\langle 1|\phi_m(E_i)\rangle$. The continuum component $\langle N_i - 1, 2|\psi^-\rangle$ is therefore 'driven' by the bound state times the coupling induced by the radiation. Driven equations of the type written here are implicitly used in most practical computational schemes developed for the molecular photodissociation problem.[18,27–29] These methods, discussed in detail in Section III, can also be used in treating the problem for strong fields in the rotating wave approximation (Eqs. 47) and beyond.

III. CALCULATION OF PHOTOFRAGMENATION CROSS-SECTIONS

In this section we discuss how photofragmentation cross-sections may be calculated. In particular we show how the photofragmentation cross-section may be written in terms of a photofragmentation amplitude and specially defined photofragmentation **T** matrix elements. In the first subsection A the general outline of the theory for the photodissociation of a molecule by a strong (laser) field is developed. The theory is based on that of Section II but is given in more explicit detail. In particular the partial wave expansion of the wavefunction in the presence of a strong laser field is developed, and the all-important differential equations and boundary conditions which determine the radial components of the wavefunction are discussed.

In the following subsection B the equations are specialized to consider the photodissociation of a triatomic molecule by a weak radiation field. We show how all the dynamics of the photodissociation process is contained in the photofragmentation **T** matrix elements. In Subsection C methods for the evaluation of the photofragmentation **T** matrix are discussed. In particular, the details of the artificial channel method for the calculation of the required bound-continuum integrals are described and compared with other ways of evaluating these integrals.

A. The Differential Equations, Boundary Conditions and Cross-sections for Photodissociation by a Strong Laser Field

We consider a photodissociation process of the form (see Eq. 1):

$$A - B + h\nu \rightarrow A + B$$

This process is illustrated schematically in Fig. 1. The product photofragments are produced in specific internal quantum states ($\mathbf{n}$) and they are scattered into direction $\hat{\mathbf{k}}$ with relative momentum $k_{\mathbf{n},m}\hbar$. We discuss the theory of this process within the framework of the dipole[30–33] and rotating wave[30,34–36] approximations. The general form of the expression for the photofragmentation cross-section is

$$\sigma(\hat{\mathbf{k}}E\mathbf{n}|i) = \left(\frac{8\pi^3\nu}{c}\right)|\langle\psi^{-(\hat{\mathbf{k}}\mathbf{n}N)}(\mathbf{R},\mathbf{r},E)| \\ \times \hat{\boldsymbol{\varepsilon}}\cdot\boldsymbol{\mu}|[|N_i - 1\rangle|\Psi_i(\mathbf{R},\mathbf{r},E_i)\rangle]|^2 \tag{49}$$

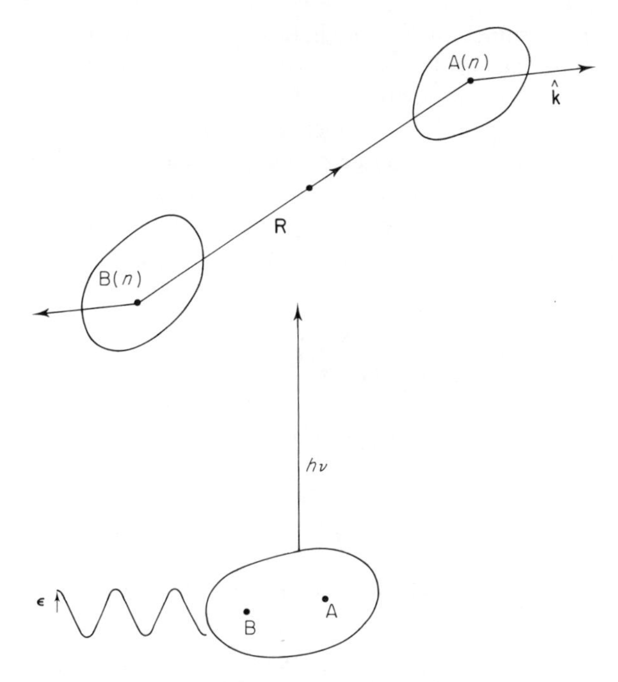

Fig. 1. Schematic representation of the photodissociation process. The molecule absorbs one or more photons of linearly polarised light and decomposes into two fragments, each with internal quantum states (**n**). Fragment A is scattered into direction $\hat{\mathbf{k}}$.

This form is similar to that of Eq. (36). As we are interested primarily in the *absorption* of photons, we have made the approximation of ignoring coupling to the closed $|N_i + 1\rangle$ photon states (see the discussion below Eq. 46b). This, and other approximations made in the theory discussed below, may easily be removed if desired. The wavefunction $\psi^{-(\hat{\mathbf{k}}\mathbf{n}N)}$ corresponds to the $|\psi^-\rangle$ wavefunction of the preceding section ($\psi^{-(\hat{\mathbf{k}}\mathbf{n}N)} \equiv |\psi^-\rangle \equiv |\psi(E, \mathbf{n}^-, \mathbf{N}^-)\rangle$). It is evaluated within the Born–Oppenheimer approximation, and therefore depends on the electronic coordinates only implicity through the specification of the electronic states of the fragments contained in the set of quantum numbers **n**. The boundary conditions to be imposed on this wavefunction are that asymptotically, at large **R** (and t), they correspond to a continuum state propagating in direction $\hat{\mathbf{k}}$ with the fragments in internal states **n**, and have a total energy

$$E = E_i + (N_i - N)h\nu = E_i + mh\nu$$

where E_i is the initial energy of the system and $m = (N_i - N)$ is the number of photons of the ith mode absorbed from the radiation field. We have furthermore invoked the approximation of Eq. (40), and have expanded the full scattering wavefunction ψ^- only in terms of photon states corresponding to absorption of photons from the ith mode. Ψ_i is the initial wavefunction for for nuclear motion in the absence of the radiation field. $\hat{\boldsymbol{\varepsilon}}$ is the direction of polarization of the radiation, which we assume to be linearly polarized. $\boldsymbol{\mu}(R)$ is the transition dipole moment function, calculated within the Born–Oppenheimer approximation, which couples the initial and final electronic states. $\mathbf{R}$ is the vector connecting the centres of mass of the fragments while $\mathbf{r}$ denotes collectively all the internal coordinates of fragments. The asymptotic form of the continuum wavefunction may be written as[37]

$$\psi^{-(\hat{\mathbf{k}}\mathbf{n}N)}(\mathbf{R},\mathbf{r},E) \underset{R\to\infty}{\sim} \left(\frac{\mu k_{\mathbf{n},m}}{h^2 2\pi}\right)^{1/2} \times \left[e^{i\mathbf{k}_{\mathbf{n},m}\cdot\mathbf{R}} \phi^{\mathrm{A}}_{\mathbf{n}}(\mathbf{r}^{\mathrm{A}})\phi^{\mathrm{B}}_{\mathbf{n}}(\mathbf{r}^{\mathrm{B}})|N\rangle + \sum_{\mathbf{n}',m'} f_{\mathbf{n}m,\mathbf{n}'m'}(\hat{\mathbf{k}},\hat{\mathbf{R}}) \frac{e^{-ik_{\mathbf{n}',m'}R}}{R} \phi^{\mathrm{A}}_{\mathbf{n}'}(\mathbf{r}^{\mathrm{A}})\phi^{\mathrm{B}}_{\mathbf{n}'}(\mathbf{r}^{\mathrm{B}})|N_i - m'\rangle \right] \tag{50}$$

where the wavefunction has been normalized on the energy scale.[21] The wavevector $k_{\mathbf{n},m}$ is defined by

$$k^2_{\mathbf{n},m} = \frac{[E_i + m\hbar\nu - \varepsilon_{\mathbf{n}}]2\mu}{\hbar^2} \tag{51}$$

where $\varepsilon_{\mathbf{n}}$ is the internal energy of the two fragments in their $\mathbf{n}^{\text{th}}$ quantum states. We now expand the plane wave in the first part of Eq. (50) in the form:

$$e^{i\mathbf{k}_{\mathbf{n},m}\cdot\mathbf{R}} = 4\pi \sum_{lm_l} i^l Y^*_{lm_l}(\hat{\mathbf{k}}) Y_{lm_l}(\hat{\mathbf{R}}) j_l(k_{\mathbf{n},m}R) \tag{52}$$

The scattering amplitude $f_{\mathbf{n}m,\mathbf{n}'m'}(\hat{\mathbf{k}},\mathbf{R})$ may also be expanded in the form:

$$f_{\mathbf{n}m,\mathbf{n}'m'}(\hat{\mathbf{k}},\hat{\mathbf{R}}) = 4\pi \sum_{lm_l}\sum_{l'm'_l} f_{lm_l\mathbf{n}m,l'm'_l\mathbf{n}'m'} Y^*_{lm_l}(\hat{\mathbf{k}}) Y_{l'm'_l}(\hat{\mathbf{R}}) \tag{53}$$

Substituting Eqs. (52) and (53) into (50) the asymptotic form of the wavefunction becomes

$$\psi^{-(\hat{\mathbf{k}}\mathbf{n}N)}(\mathbf{R},\mathbf{r},E) \underset{R\to\infty}{\sim} 4\pi\left(\frac{\mu k_{\mathbf{n},m}}{h^2 \times 2\pi}\right)^{1/2} \sum_{lm_l}\sum_{l'm'_l}\sum_{\mathbf{n}'m'} i^l Y^*_{lm_l}(\hat{\mathbf{k}}) Y_{l'm'_l}(\hat{\mathbf{R}}) \times \left[j_l(k_{\mathbf{n},m}R)\delta_{\mathbf{n},\mathbf{n}'}\delta_{m,m'}\delta_{l,l'}\delta_{m_l,m'_l} + i^{-l} f_{lm_l\mathbf{n}m,l'm'_l\mathbf{n}'m'} \frac{e^{-ik_{\mathbf{n}',m'}R}}{R}\right] \times \phi^{\mathrm{A}}_{\mathbf{n}'}(\mathbf{r}^{\mathrm{A}})\phi^{\mathrm{B}}_{\mathbf{n}'}(\mathbf{r}^{\mathrm{B}})|N_i - m'\rangle \tag{54}$$

We now use the asymptotic form of the spherical Bessel functions:

$$j_l(kR) \underset{R\to\infty}{\sim} \frac{1}{2ikR}\left(e^{i(kR - l\pi/2)} - e^{-i(kR - l\pi/2)}\right) \tag{55}$$

to substitute into Eq. (54) and obtain, for the asymptotic form of the wavefunction:

$$\psi^{-(\hat{\mathbf{k}}\mathbf{n}N)}(\mathbf{R},\mathbf{r},E)\underset{R\to\infty}{\sim}4\pi\left(\frac{\mu k_{\mathbf{n},m}}{h^2\times 2\pi}\right)^{1/2}\sum_{l,m_l}\sum_{l',m_l'}\sum_{\mathbf{n}'m'}\frac{i^{l-1}}{2}Y^*_{lm_l}(\hat{\mathbf{k}})Y_{l'm_l'}(\hat{R})$$

$$\times\left[\frac{e^{i(k_{\mathbf{n},m}R-l\pi/2)}}{k_{\mathbf{n},m}R}\delta_{\mathbf{n},\mathbf{n}'}\delta_{m,m'}\delta_{l,l'}\delta_{m_l,m_l'}\right.$$

$$-\left(\frac{\delta_{\mathbf{n},n'}\delta_{m,m'}\delta_{l,l'}\delta_{m_l,m_l'}}{k_{\mathbf{n},m}}-2i^{1-l-l'}f_{lm_l\mathbf{n}m,l'm_l'\mathbf{n}'m'}\right)$$

$$\left.\times\frac{e^{-i(k_{\mathbf{n}',m'}R-l'\pi/2)}}{R}\right]\phi^{\mathrm{A}}_{\mathbf{n}'}(\mathbf{r}^{\mathrm{A}})\phi^{\mathrm{B}}_{\mathbf{n}'}(\mathbf{r}^{\mathrm{B}})|N_i-m'\rangle \tag{56}$$

We have perservered through the analysis of the asymptotic form of ψ^- (i.e. Eqs. 50 to 56) so as to establish the form of the boundary conditions to be applied to the radial part of the wavefunction.

The total scattering wavefunction with the required boundary conditions may now be expanded in the form:

$$\psi^{-(\hat{\mathbf{k}}\mathbf{n}N)}(\mathbf{R},\mathbf{r},E)=4\pi\left(\frac{\mu\mathbf{k}_{\mathbf{n},m}}{h^2\times 2\pi}\right)^{1/2}\sum_{lm_l}\sum_{l'm_l'}\sum_{\mathbf{n}'m'}i^lY^*_{lm_l}(\hat{\mathbf{k}})Y_{l'm_l'}(\hat{\mathbf{R}})$$

$$\times\frac{\psi^{-lm_l\mathbf{n}m}_{l'm_l'n'm'}(R)}{k_{\mathbf{n},m}R}\phi^{\mathrm{A}}_{\mathbf{n}'}(\mathbf{r}^{\mathrm{A}})\phi^{\mathrm{B}}_{\mathbf{n}'}(\mathbf{r}^{\mathrm{B}})|N_i-m'\rangle \tag{57}$$

By comparing with Eq. (56) we see that the boundary conditions to be imposed on the radial wavefunctions are

$$\psi^{-lm_l\mathbf{n}m}_{l'm_l'\mathbf{n}'m'}(R)\underset{R\to\infty}{\sim}\frac{1}{2i}\left[e^{i(k_{\mathbf{n},m}R-l\pi/2)}\delta_{\mathbf{n},\mathbf{n}'}\delta_{m,m'}\delta_{l,l'}\delta_{m_l,m_l'}\right.$$

$$\left.-S^*_{l'm_l'\mathbf{n}'m',lm_l\mathbf{n}m}\left(\frac{k_{\mathbf{n},m}}{k_{\mathbf{n}',m'}}\right)^{1/2}e^{-i(k_{\mathbf{n}'m'}R-l'\pi/2)}\right] \tag{58}$$

The subscripts on the radial component of the wavefunctions $l'm_l'$, $\mathbf{n}'m'$ are the so-called channel indices in the present problem. Each channel represents one member of a set of coupled second-order differential equations. The index $\mathbf{n}'$ labels both the electronic state of the system and the internal vibrational–rotational quantum numbers of the fragments. The label m' indicates the number of photons which have been absorbed from the radiation field. The channels correspond asymptotically to pure states of the fragments 'dressed' by m' photons. The scattering wavefunction of Eq. (57) is analogous to that of Eq. (40) in the preceding section, where the wavefunction is expanded in terms of photon occupation number states of the ith mode only. By substituting the expansion of Eq. (57) into the full Schrödinger equation, multiplying from the left by the channel basis functions $Y_{l'm_l'}(\hat{\mathbf{R}})\phi^{A}_{\mathbf{n}'}(\mathbf{r}^{A})\phi^{B}_{\mathbf{n}'}(\mathbf{r}^{B})|N_i - m'\rangle$ and then integrating over all 'internal' coordinates we obtain the set of coupled differential equations for the radial wavefunctions $\psi^{-lm_l,\mathbf{n}m}_{l'm_l',n'm'}(R)$. These equations are

$$\left[-\frac{\mathrm{d}^2}{\mathrm{d}R^2} - k^2_{\mathbf{n}',m'} + \frac{l'(l'+1)}{R^2}\right]\psi^{-lm_l,\mathbf{n}m}_{l'm_l',n'm'}(R)$$

$$+\sum_{l''m_l''}\sum_{\mathbf{n}''}\frac{2\mu}{\hbar^2}V_{l'm_l'\mathbf{n}',l''m_l''\mathbf{n}''}(R)\,\psi^{-lm_l,\mathbf{n}m}_{l''m_l'',n''m'}(R)$$

$$+i\left(\frac{2\pi I}{c}\right)^{1/2}\frac{2\mu}{\hbar^2}\sum_{l''m_l''}\sum_{\mathbf{n}''}\langle \mathbf{n}'l'm_l'|\hat{\boldsymbol{\varepsilon}}\cdot\boldsymbol{\mu}|n''l''m_l''\rangle$$

$$\times\left[\psi^{-lm_l,\mathbf{n}m}_{l''m_l'',n''m'+1)}(R) - \psi^{-lm_l,\mathbf{n}m}_{l''m_l'',n''m''(m'-1)}(R)\right] = 0 \tag{59}$$

The above set of differential equations define the 'dressed molecule' picture of treating the molecule–radiation interaction. The *initial bound states* of the system exist *in the absence of the radiation field* and correspond to all those states with $E_i < 0$ which satisfy the bound state boundary conditions $\Psi_i(\mathbf{R},\mathbf{r},\mathbf{E}_i) \underset{R\to 0}{\sim} 0(1/R)$ and $\Psi_i(\mathbf{R},\mathbf{r},\mathbf{E}_i) \underset{R\to\infty}{\sim} 0$. These bound state wavefunctions may be expanded in terms of radial channel wavefunctions in a similar manner to the scattering wavefunction[39,40] (Eq. 57). The radial channel wavefunctions in this expansion are all associated with negative energies (i.e. $k^2_{\mathbf{n}m} < 0$). They obey the same set of coupled differential equations as the scattering states but the radial channel wavefunctions obey bound state boundary conditions (i.e. $\psi^{-lm_l,\mathbf{n}0}_{l'm_l',n'0}(R) \underset{R\to 0}{\sim} 0$ and $\psi^{-lm_l,\mathbf{n}0}_{l'm_l',n'0}(R) \underset{R\to\infty}{\sim} 0$).

In the presence of the radiation field these 'bound' radial channel wavefunctions are coupled to continuum dissociative channels by what may be viewed is a virtual two-step process (see Eq. 59). Firstly, the bound channels are coupled to the corresponding 'dressed' channels by the radiation field acting through the matrix elements of the dipole moment operator. Only dressed (or undressed) channels differing by a single photon are directly coupled in this way, but of course an entire 'ladder' of dressed channels originates from each initial set of quantum numbers **n**, and the channels in any given 'ladder' are coupled directly to each of their neighbours and therefore indirectly (through second- and higher-order effects) to each other. The second part of the coupling occurs through the agency of the potential coupling matrix elements $V_{l'm_l'\mathbf{n}',l''m_l''\mathbf{n}''}(R)$. These potential matrix elements couple only channels which are dressed by the same number of photons. It may happen that an undressed channel energy is approximately degenerate with a more highly dressed channel from a different 'ladder'. In such cases it is likely that the indirect 'two-step' coupling process will lead to a strong mixing of the two channels if the radiation field is sufficiently intense. The two-step coupling mechanism is illustrated in Fig. 2 (for just two 'ladders').

The set of coupled differential equations (Eq. 59) which describes the molecule and its interaction with a strong radiation field are of *exactly* the same form as those which arise in the quantum theory of inelastic (or reactive) molecular collisions.[41–43] Several different efficient techniques for solving these types of equations have been developed[44–51] and any one of these can be adapted to the present problem.

We have now completed the description of the derivation of the coupled differential equations (Eq. 59) and of the boundary conditions which their

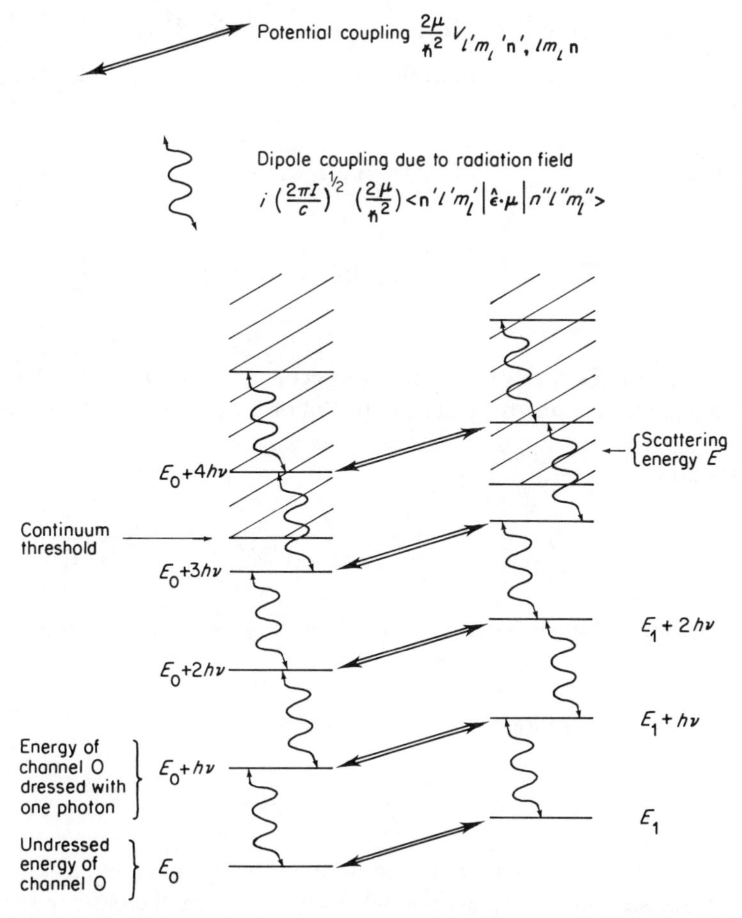

Fig. 2. Illustration of twofold mechanism coupling dressed channel wavefunctions in a strong laser field. The ladder of dressed channel energies originating in only two undressed channel states is shown. E_o corresponds to the energy associated with channel o at some fixed value of R (i.e. $E_o = E_o + \langle o|V|o\rangle$).

solutions are required to obey (Eq. 58). Using modern scattering theory techniques,[44-53] both the bound state wavefunctions[39,40] and the dissociative scattering wavefunctions (solutions of Eq. 59) may be evaluated, and the integral appearing in Eq. (49) may then be calculated so as to yield the photofragmentation cross-section. (See note added in proof at end of chapter.)

Techniques for the execution of such calculations will be discussed at greater length in subsection C. We now show how, through the expansion of the *scattering* wavefunction in Eq. 49, we may express the photofragmentation cross-section in terms of photofragmentation **T** matrices.

We define the photofragmentation amplitude as the integral on the right-hand side of Eq. (49).

$$f(\hat{\mathbf{k}}E\mathbf{n}|i) = \langle \psi^{-(\mathbf{k}\mathbf{n}N)}(\mathbf{R}, \mathbf{r}, E)|\hat{\boldsymbol{\varepsilon}}\cdot\boldsymbol{\mu}|[|N_i - 1\rangle|\Psi_i(\mathbf{R}, \mathbf{r}, E_i)]\rangle \tag{60}$$

We now expand the *scattering* wavefunction $\psi^{-(\hat{\mathbf{k}}\mathbf{n}N)}$ according to the expansion of Eq. (57) to obtain

$$f(\hat{\mathbf{k}}E\mathbf{n}|i) = 4\pi\left(\frac{\mu}{h^2 \times 2\pi k_{\mathbf{n},m}}\right)^{1/2} \sum_{lm_l} i^l Y^*_{lm_l}(\hat{\mathbf{k}})$$
$$\times\left[\sum_{l'm_l'}\sum_{\mathbf{n}'}\left\langle \psi^{-lm_l,\mathbf{n}m}_{l'm_l',\mathbf{n}'1}(R)\frac{1}{R}Y_{l'm_l'}(\hat{\mathbf{R}})\phi^{A}_{\mathbf{n}'}(\mathbf{r}^A)\phi^{B}_{\mathbf{n}'}(\mathbf{r}^B)|\hat{\boldsymbol{\varepsilon}}\cdot\boldsymbol{\mu}|\Psi_i(\mathbf{R}, \mathbf{r}, E_i)\right\rangle\right] \tag{61}$$

where we have integrated over the photon occupation number states. The final term in square brackets is defined as the 'photofragmentation **T** matrix':

$$t(lm_l\mathbf{n}mE|i) = 4\pi\left(\frac{\mu}{h^2 \times 2\pi k_{\mathbf{n},m}}\right)^{1/2}$$
$$\times\sum_{l'm_l'}\sum_{\mathbf{n}'}\left\langle \psi^{-lm_l,\mathbf{n}m}_{l'm_l',\mathbf{n}'1}(R)\frac{1}{R}Y_{l'm_l'}(\hat{\mathbf{R}})\phi^{A}_{\mathbf{n}'}(\mathbf{r}^A)\phi^{B}_{\mathbf{n}'}(\mathbf{r}^B)|\hat{\boldsymbol{\varepsilon}}\cdot\boldsymbol{\mu}|\Psi_i(\mathbf{R}, \mathbf{r}, E_i)\right\rangle \tag{62}$$

Thus the photofragmentation cross-sections may be written in either of the two forms:

$$\sigma(\hat{\mathbf{k}}E\mathbf{n}|i) = \frac{8\pi^3\nu}{c}|f(\hat{\mathbf{k}}E\mathbf{n}|i)|^2 \tag{63a}$$

$$= \frac{8\pi^3\nu}{c}|\sum_{lm_l} i^l Y^*_{lm_l}(\hat{\mathbf{k}})t(lm_l\mathbf{n}mE|i)|^2 \tag{63b}$$

The treatment above has set out to develop the theory of photofragmentation processes in the simplest possible form. We have, therefore, purposely not used eigenfunctions of the total angular momentum in the expansion of the *scattering* wavefunction; nor have we utilized the fact that the bound state wavefunction is itself an eigenfunction of the total angular momentum. The utilization of these facts considerably simplifies the computation of the photofragmentation **T** matrices, while simultaneously complicating the formal theory and obscuring the physical origins of the coupled differential equations and the coupling terms in them. In the following subsection we will consider in detail the theory of photofragmentation of a triatomic molecule in a weak field. The formulae derived there will appear to be more complicated than those considered above. This extra complication is due entirely to the detailed consideration of the angular momentum properties of the wavefunctions which we have largely omitted in the above treatment.

B. The Differential Equations and Photofragmentation Cross-sections for Photodissociation of a Triatomic Molecule by a Weak Radiation Field

In the weak-field limit the electric field strength is small and the matter–radiation coupling term may be treated as a perturbation. We must now solve the standard type of coupled differential equations of quantum mechanical scattering theory[37] for the scattering wavefunction. As the radiation field is weak we will only be interested in processes involving the absorption of a single photon from the radiation field. Our scattering wavefunctions are therefore eigenfunctions of a Hamiltonian dressed by a single photon and containing no matter–radiation interaction terms. In this subsection we consider in detail the application of this theory to the single-photon dissociation of a triatomic molecule.

The bound state vibrational–rotational eigenfunctions of a molecule are normally best expressed in terms of a body-fixed coordinate system.[39,40] Recent experience in molecular scattering theory has also demonstrated that, even for the calculation of scattering cross-sections, the use of a body-fixed coordinate system may be computationally beneficial.[41,54–57] These considerations motivate us to express the *scattering* wavefunction needed in the calculation of the photofragmentation cross-section in a body-fixed coordinate system. The coordinate system we use is the standard one adopted in the quantum theory of inelastic scattering cross-sections and is shown in Fig. 3. The body-fixed z axis is taken to lie along the line joining the centre of masses of the two dissociating fragments (i.e. the diatomic BC and the atom A).

The photofragmentation cross-section, in the weak-field limit, may be written[37] as (see Eq. 49):

$$\sigma(\hat{\mathbf{k}}Evjm_j|E_iJ_iM_i) = \frac{8\pi^3\nu}{c}|\langle {}^{\circ}\psi^{-(\hat{\mathbf{k}}vjm_j)}(\mathbf{R},\mathbf{r},E)|\,\hat{\varepsilon}\cdot\boldsymbol{\mu}|\Psi^{J_iM_i}(\mathbf{R},\mathbf{r},E_i)\rangle|^2 \quad (64)$$

where vjm_j are the vibrational and rotational quantum numbers of the fragment diatomic and J_iM_i are the quantum numbers for the total angular momentum of the original bound state and for its z component in the

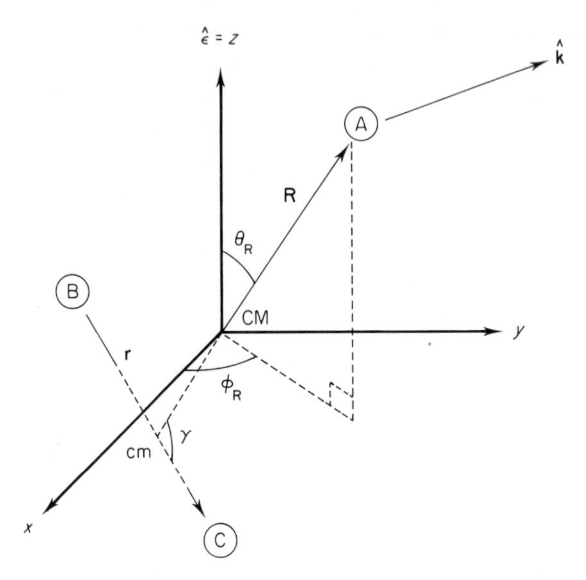

Fig. 3. Centre-of-mass space-fixed and body-fixed coordinate systems for photodissociation of a triatomic molecule. The direction of polarization of the light is $\hat{\boldsymbol{\varepsilon}}$ which defines the space-fixed z axis. $\hat{\mathbf{k}}$ is the direction of observation of the fragment A. CM is the location of the total centre of mass of the ABC system and cm is the BC centre of mass. The body-fixed z axis lies along the $\mathbf{R}$ vector (i.e. along the vector from the centre of mass of BC to A).

space-fixed axes. The superscript o has been added to the scattering wavefunction to indicate that it is now the solution of a field-free problem, corresponding to an energy of $E = E_i + h\nu$.

The total scattering wavefunction is now expanded in the form:

$$
\begin{aligned}
{}^{o}\psi^{-(\hat{\mathbf{k}}vjm_j)}(\mathbf{R}, \mathbf{r}, E) = 4\pi\left(\frac{\mu k_{vj}}{h^2 2\pi}\right)^{1/2} & \sum_{lm_l}\sum_{JM} Y^{*}_{lm_l}(\hat{\mathbf{k}})(jlJM\,|\,jm_j lm_l) \\
& \times \sum_{v'j'} \chi_{v'j'}(r) \sum_{\lambda} (Jjlo|J\lambda j-\lambda) \\
& \times \sum_{\lambda'} \Theta^{JMp}_{j'\lambda'}(\hat{\mathbf{R}}, \hat{\mathbf{r}}) \frac{\Phi^{-(Jvj\lambda p)}_{v'j'\lambda'}(R)}{k_{vj}R}
\end{aligned}
\tag{65}
$$

The angular functions[37,39] $\Theta^{JMp}_{j\lambda}(\hat{\mathbf{R}}, \hat{\mathbf{r}})$ are functions of the four angles which specify the directions of the vectors $\hat{\mathbf{R}}$ and $\hat{\mathbf{r}}$ (see Fig. 3). They are eigenfunctions of the total angular momentum (J) and of its space-fixed z component (M). The p quantum number is ± 1 and is related to the parity (parity $= (-1)^J p$). The quantum number λ is the absolute value of the quantum number corresponding to the component of the total angular momentum about the body-fixed z axis. The functions $\chi_{vj}(r)$ are the radial components of the bound state diatomic wavefunctions for the BC fragment. The increased complexity of the expansion of the wavefunction given in Eq. (65), as compared with previous expansions (e.g. Eq. 57), arises mainly from the fact that:

1. The angular expansion functions are eigenfunctions of the total angular momentum and
2. A body-fixed coordinate system is used.

The total angular momentum is conserved in the scattering process and consequently the use of eigenfunctions of the total angular momentum in the expansion of the scattering wavefunctions leads to sets of coupled differential equations with no coupling between differing values of the total angular momentum.

The coupled differential equations which determine the radial functions $\Phi_{v'j'\lambda'}^{-Jvj\lambda p}(R)$ may be written in the form:[37,39]

$$\left[-\frac{\mathrm{d}^2}{\mathrm{d}R^2} - k_{vj}^2 + \frac{J(J+1) + j'(j'+1) - 2\lambda^2}{R^2}\right]\Phi_{v'j'\lambda'}^{-(Jvj\lambda p)}(R) + C_{\lambda',\lambda'+1}^{Jp}\Phi_{v'j',\lambda'+1}^{-(Jvj\lambda p)}(R) + C_{\lambda',\lambda'-1}^{Jp}\Phi_{v'j',\lambda'-1}^{-(Jvj\lambda p)}(R) + \sum_{v''j''} U_{v'j',v''j''}^{J\lambda'}(R)\Phi_{v''j''\lambda'}^{-(Jvj\lambda p)}(R) = 0 \tag{66}$$

where

$$\left.\begin{aligned}
C_{\lambda,\lambda+1}^{Jp} &= C_{\lambda,\lambda+1}^{J} && \text{if } \lambda > 0\\
C_{\lambda-1,\lambda}^{Jp} &= C_{\lambda-1,\lambda}^{J} && \text{if } \lambda > 1\\
C_{0,1}^{Jp} &= \sqrt{2}C_{0,1}^{J} && \text{if } p = 1\\
C_{1,0}^{Jp} &= \sqrt{2}C_{1,0}^{J} && \text{if } p = 1\\
C_{0,1}^{Jp} &= C_{1,0}^{Jp} = 0 && \text{if } p = -1\\
C_{\lambda,\lambda\pm1}^{J} &= -[J(J+1) - \lambda(\lambda\pm1)]^{1/2}[j(j+1) - \lambda(\lambda\pm1)]^{1/2}\frac{1}{R^2}
\end{aligned}\right\} \tag{67}$$

$$U_{vj,v'j'}^{J\lambda}(R) = \left(\frac{2\mu}{\hbar^2}\right)\int \Theta_{j\lambda}^{JMp*}(\hat{\mathbf{R}},\hat{\mathbf{r}})\chi_{vj}(r)V(\mathbf{R},\mathbf{r}) \times \Theta_{j'\lambda}^{JMp}(\hat{\mathbf{R}},\hat{\mathbf{r}})\chi_{v'j'}(r)\,\mathrm{d}\mathbf{r}\,\mathrm{d}\mathbf{R} \tag{68}$$

$$k_{vj}^2 = \left(\frac{2\mu}{\hbar^2}\right)(E - \varepsilon_{vj}) = \left(\frac{2\mu}{\hbar^2}\right)(E_i + h\nu - \varepsilon_{vj}) \tag{69}$$

The functions $\chi_{vj}(r)$ are the radial part of the vibrational–rotational eigenfunctions of the diatomic fragment BC. ε_{vj} are its vibrational–rotational energy levels and the *scattering* energy is $E = E_i + hv$. $V(\mathbf{R}, \mathbf{r})$ is the full potential energy surface which governs the motions of the three nuclei.

The asymptotic boundary conditions which are required by the physics of the problem, are that the scattering wavefunction correspond to a pure plane wave with specific internal states and radial incoming waves in all other channels:

$$^{\circ}\psi^{-(\hat{\mathbf{k}}vjm_j)}(\mathbf{R}, \mathbf{r}, E) \underset{R\to\infty}{\sim} \left(\frac{\mu k_{vj}}{h^2 \times 2\pi}\right)^{1/2} \left[e^{i\mathbf{k}_{vj}\cdot\mathbf{R}} \chi_{vj}(r) Y_{jm_j}(\hat{\mathbf{r}}) + \sum_{v'j'm_j'} f_{vjm_j,v'j'm_j'}(\hat{\mathbf{k}}, \hat{\mathbf{R}}) \frac{e^{-ik_{v'j'}R}}{R} \chi_{v'j'}(r) Y_{j'm_j'}(\hat{\mathbf{r}}) \right] \quad (70)$$

This boundary condition implies a corresponding boundary condition on the radial wavefunctions:

$$\Phi_{v'j'\lambda'}^{-(Jvj\lambda p)}(R) \underset{R\to\infty}{\sim} \frac{1}{2i}\left(\frac{4t_\lambda}{1+\delta_{\lambda o}}\right)(-1)^{j+\lambda} \times \left[e^{ik_{v'j'}R}\delta_{vv'}\delta_{jj'}\delta_{\lambda\lambda'} - S^{Jp*}_{v'j'\lambda'jvj\lambda}\left(\frac{k_{vj}}{k_{v'j'}}\right)^{1/2} e^{-ik_{v'j'}R} \right] \quad (71)$$

The photofragmentation $\mathbf{T}$ matrix elements are written in terms of the radial wavefunction as

$$t(EJvj\lambda p|E_iJ_ip_i) = 4\pi\left(\frac{\mu k_{vj}}{h^2 \times 2\pi}\right)^{1/2}\left[\begin{pmatrix} J & 1 & J_i \\ -M_i & 0 & M_i \end{pmatrix}\right]^{-1} \times (-1)^{M_i} \sum_{v'j'\lambda'} \left\langle \frac{\Phi_{v'j'\lambda'}^{-(Jvj\lambda p)}(R)}{k_{vj}R} \chi_{v'j'}(R)\Theta_{j'\lambda'}^{JMp}(\hat{\mathbf{R}}, \hat{\mathbf{r}}) | \hat{\varepsilon}\cdot\boldsymbol{\mu} | \Psi^{J_iM_i}(\mathbf{R}, \mathbf{r}, E_i) \right\rangle \quad (72)$$

These photofragmentation $\mathbf{T}$ *matrix elements contain all the dynamical information which may be extracted from a photofragmentation process.* The term in square brackets on the right-hand side of Eq. (72) is a $3j$ symbol.[58] It has been factored out at this stage to remove from the $\mathbf{T}$ matrix elements their 'geometric' dependence on the M_i quantum number.

The photofragmentation cross-section may now be expressed in terms of the photofragmentation $\mathbf{T}$ matrix (Eq. 72):[37]

$$\sigma(\hat{\mathbf{k}}Evjm_j|E_iJ_iM_ip_i) = \frac{8\pi^3 v}{c}\left| \sum_{J\lambda}(2J+1)^{1/2} \times \begin{pmatrix} J & 1 & J_i \\ -M_i & 0 & M_i \end{pmatrix} D^J_{\lambda M_i}(\phi, \theta, 0) D^j_{-\lambda -m_j}(\phi, \theta, 0) \times t(EJvj\lambda p|E_iJ_ip_i) \right|^2 \quad (73)$$

The angles θ, ϕ are the polar and azimuthal angles, in the space-fixed coordinate system, of the direction into which the fragments are scattered. We recall that the z axis of this coordinate system is taken to lie along the direction of polarization of the incident, linearly polarized light. The cross-section of Eq. (73) is the most detailed type of photofragmentation cross-section which can be measured. The probability of finding the dissociated fragments in the internal state vjm_j scattered into a solid angle $\mathrm{d}\Omega$ about the direction $\hat{\mathbf{k}}$ and possessing a total energy of between E and $E + h\,\mathrm{d}\nu$ may be expressed in terms of the cross-section of Eq. (73) as

$$\sigma(\hat{\mathbf{k}}Evjm_j|E_iJ_iM_i)\rho(\hat{\varepsilon})ch\,\mathrm{d}\Omega\,\mathrm{d}\nu \tag{74}$$

where $\rho(\hat{\varepsilon})ch\,\mathrm{d}\nu$ is the energy flux of linearly polarized radiation of frequency ν to $\nu + \mathrm{d}\nu$. The two angular functions of the type $D^J_{\lambda M_i}(\phi, \theta, 0)$ appearing in Eq. (73) are standard angular functions.[58,59]

In Eq. (73) for the photofragmentation cross-section the intrinsic or irreducible dynamics of the photofragmentation processes is contained entirely within the photofragmentation **T** matrix element. The other terms in the expression have essentially geometric origins. The differential cross-section is in fact completely independent of the azimuthal angle ϕ, and ϕ may for convenience be put equal to zero without altering the cross-section. The summation over J is limited by the requirement $J = J_i$, $J_i \pm 1$. If we are, therefore, able to measure the differential photofragmentation cross-section for selected initial states J_i, M_i and selected final diatomic fragment states jm_j, then there are only $3(J_i + 1)$ or $3(j + 1)$ terms in the summation, whichever is the smaller.

C. The Artificial Channel Method and the Evaluation of the Photofragmentation T Matrix

The dynamics of the photodissociation process are contained in the photofragmentation **T** matrix elements (Eqs. 62 and 72). In this section we concentrate on describing how this centrally important **T** matrix element may be evaluated. In particular, we will discuss the calculation of the **T** matrix elements for the photodissociation of a triatomic molecule in the weak-field limit. There are basically three different methods for calculating these quantities:

1. The artificial channel method,
2. The driven equation method, and
3. The direct method.

The direct method[28,60,61] simply involves the separate computation of the radial components of the scattering, or continuum, wavefunction and of the bound state wavefunction. These components are then used to directly compute the bound-continuum wavefunction. The efficiency of the computation is, of course, greatest if the radial integrals are accumulated at the same time as the radial components of the wavefunction are computed.[28,60,61] The driven equations method[29,62] essentially solves a set of differential equations containing an inhomogeneous 'source' term (see Eq. 48). This source term is proportional to $\hat{\varepsilon}\cdot\boldsymbol{\mu}\Psi_i(\mathbf{R}, \mathbf{r}, E_i)$ or to its relevant channel component. Clearly in order to use this method the bound state wavefunction must be known before the calculation of the scattering wavefunction.

1. *The Artificial Channel Method*

In the present article we concentrate mainly on the description of the artificial channel method.[27] The close relationship between this method and the driven equation approach will be readily apparent from the derivation. The advantages of the artificial channel method are twofold. Firstly, the dynamically significant photofragmentation **T** matrix elements are directly computed, without the need to explicitly evaluate the bound-continuum integrals, whose direct evaluation is often very difficult. Secondly, the bound state wavefunctions may be computed to any desired degree of accuracy, rather than assumed to be simple harmonic normal mode wavefunctions.

The formalism of the artificial channel method requires that the coupled differential equations of the bound state problem be coupled, via the transition dipole moment function, to those associated with the scattering wavefunction. This *apparent* increase in the size of the set of coupled differential equations need not lead to increased computational expense. It may be shown[63] that, if consideration is given to the special (asymmetric) nature of the coupling, all the standard methods for the solution of coupled differential equations may be adapted to the artificial channel method. After such adaptation the work required to solve the artificial channel equations is essentially equal to the sum of the work required to solve the scattering and bound state problems separately. Furthermore, as the bound state part of the calculation is the same for different photon energies, it is possible to perform this part of the calculation only once and to save the results for subsequent use at different energies. In practice the artificial channel method has proved to be a highly efficient method for the computation of photodissociation cross-sections.[18,64–71]

In the artificial channel method[18,39] a set of coupled differential equations is solved which encompasses both the bound state and continuum state differential equations and furthermore includes an artificial channel. The bound state subspace is coupled to the continuum state subspace by the matrix elements of the space-fixed z component of the transition dipole function. The artificial channel is coupled to the channels of the bound state manifold by an artificial coupling. This permits particle flux to flow from the artificial channel into the bound state. The flux then flows into the channels of the continuum (or photodissociative) state manifold 'via' the dipole moment matrix elements which couple these two manifolds. The set of coupled differential equations which must be solved are explicitly:

Coupled equations for bound channels

$$\left[-\frac{d^2}{dR^2}+\frac{2\mu}{\hbar^2}(\varepsilon_{v'j'}-E_i)+W^{BJ_i}_{v'j'\lambda',v'j'\lambda'}(R)\right]\psi^{J_ip_i\gamma}_{v'j'\lambda'}(R)$$
$$+\sum_{v''j''\lambda''}W^{BJ_i}_{v'j'\lambda',v''j''\lambda''}(R)\psi^{J_ip_i\gamma}_{v''j''\lambda''}(R)=-U_{v'j'\lambda',\gamma}(R)\psi^{\gamma}_{\gamma}(R) \tag{75}$$

Coupled equations for continuum channels

$$\left[-\frac{d^2}{dR^2}+\frac{2\mu}{\hbar^2}(\varepsilon_{v'j'}-E_i-h\nu)+W^{cJ}_{v'j'\lambda',v'j'\lambda'}(R)\right]\psi^{-(Jvj\lambda p\gamma)}_{v'j'\lambda'}(R)$$
$$+\sum_{v''j''\lambda''}W^{cJ}_{v'j'\lambda',v''j''\lambda''}(R)\psi^{-(Jvj\lambda p\gamma)}_{v''j''\lambda''}(R)$$
$$=-\sum_{v''j''\lambda''}\frac{2\mu}{\hbar^2}U^{JJ_i}_{v'j'\lambda',v''j''\lambda''}(R)\psi^{J_ip_i\gamma}_{v''j''\lambda''}(R) \tag{76}$$

Equation for artificial channel

$$\left[-\frac{d^2}{dR^2}+\frac{2\mu}{\hbar^2}(\varepsilon_{\gamma}-E_i)+U_{\gamma,\gamma}(R)\right]\psi^{\gamma}_{\gamma}(R)=0 \tag{77}$$

Except for the presence of the inhomogeneous terms on the right-hand side of Eqs. (75) and (76), these two sets of coupled differential equations are identical to those obeyed by the radial components of the bound state[39] and continuum state[37] (Eq. 66) wavefunctions respectively.

In deriving the set of coupled differential equations for the bound state manifold the bound state wavefunction is expanded in the form:

$$\Psi^{J_iM_i}(\mathbf{R},\mathbf{r},E_i)=\sum_{vj\lambda}\frac{\Phi^{J_ip_i}_{vj\lambda}(R)}{R}\chi_{vj}(r)\Theta^{J_iM_ip_i}_{j\lambda}(\hat{\mathbf{R}},\hat{\mathbf{r}}) \tag{78}$$

In Eq. (75) the radial wavefunctions $\psi^{J_ip_i\gamma}_{v'j'\lambda'}(R)$ are solutions of an inhomogeneous differential equation with a source term. The radial wavefunctions $\Phi^{J_ip_i}_{vj\lambda}(R)$ (Eq. 78) are solutions of the equivalent set of homogeneous equations.[37,39] The coupling matrix elements in Eq. (75) are given by

$$W^{BJ_i}_{vj\lambda,v'j'\lambda'}(R)=$$
$$\left\langle\chi_{vj}\Theta^{J_iM_ip_i}_{j\lambda}\left|\frac{(\hat{J}-\hat{j})^2}{\hbar^2R^2}+\frac{2\mu}{\hbar^2}V^B(\mathbf{R},\mathbf{r})\right|\chi_{v'j'}\Theta^{J_iM_ip_i}_{j'\lambda'}\right\rangle \tag{79}$$

$V^B(\mathbf{R},\mathbf{r})$ is the potential energy surface governing the motion of the bound state. The matrix elements $W^{cJ}_{vj\lambda,v'j'\lambda'}(R)$ are defined in an analogous manner but using the potential surface $V^c(\mathbf{R},\mathbf{r})$ which governs the motion of the nuclei in the continuum or predissociative state (see Eqs. 66 to 68). The potential matrix elements coupling the artificial channel with the other channels are taken to be simple exponentials. The energy ε_γ is chosen so that the artificial channel has a small kinetic energy. The results of the final calculations are independent of the precise choice of these quantities.

The matrix elements $U^{JJ_i}_{v'j'\lambda',v''j''\lambda''}(R)$ which couple the bound state to the photodissociative manifold are simply the matrix elements of the z component of the transition dipole moment function:

$$U^{JJ_i}_{v'j'\lambda',v''j''\lambda''}(R) = (-1)^{M_i}\begin{pmatrix} J & 1 & J_i \\ -M_i & 0 & M_i \end{pmatrix}^{-1} \times \left\langle \chi_{v'j'}\Theta^{JM_iP}_{j'\lambda'} \middle| \hat{\boldsymbol{\varepsilon}}\cdot\boldsymbol{\mu} \middle| \chi_{v''j''}\Theta^{J_iM_ip_i}_{j''\lambda''} \right\rangle \tag{80}$$

The set of coupled differential equations (Eqs. 75–77) are solved subject to the boundary conditions:

$$\psi^{J_ip_i\gamma}_{v'j'\lambda'}(R) \underset{R\to\infty}{\sim} 0 \tag{81a}$$

$$\psi^{-(Jvj\lambda p\gamma)}_{v'j'\lambda'}(R) \underset{R\to\infty}{\sim} -T_{v'j'\lambda',\gamma}\left(\frac{k_\gamma}{k_{v'j'}}\right)^{1/2} e^{ik_{v'j'}R} \tag{81b}$$

$$\psi^{\gamma}_{\gamma}(R) \underset{R\to\infty}{\sim} -2ie^{i\delta_\gamma}\sin(k_\gamma R + \delta_\gamma) \tag{81c}$$

In Fig. 4 we show, schematically the diagonal 'channel potentials' (i.e. $W^{BJ_i}_{v'j'\lambda',v'j'\lambda'}(R) + 2\mu\varepsilon_{v'j'}/\hbar^2$, $W^{cJ}_{v'j'\lambda',v'j'\lambda'}(R) + 2\mu\varepsilon_{v'j'}/\hbar^2$ and $U_{\gamma,\gamma}(R) + 2\mu\varepsilon_\gamma/\hbar^2$). In order to clarify the physical situation only 'one channel' from each of the bound state and photodissociative manifolds is shown. We set out now to calculate the **T** matrix elements $T_{v'j'\lambda',\gamma}$ which gives the probability of flux initially incident in the artificial channel being 'scattered' into the photodissociative channel $v'j'\lambda'$. As we shall see from the analysis below, this **T** matrix element contains within itself the photofragmentation **T** matrix element (Eq. 72) which we seek to evaluate.

The analysis follows along the lines of that given by Shapiro and Balint-Kurti.[39] While the derivation does not contain anything fundamentally new beyond that originally given by Shapiro,[27] the alternative approach to the problem presented here may assist in clarifying the method. We first construct a Green's function using the regular and irregular solutions obtained from the homogeneous equation corresponding to the left-hand side of Eq. (76) (i.e. the continuum state coupled differential equations). These solutions have boundary conditions of the form:

Irregular solution:

$$\phi^{i,vj\lambda}_{v'j'\lambda'}(R) \underset{R\to\infty}{\sim} e^{ik_{v'j'}R} \tag{82a}$$

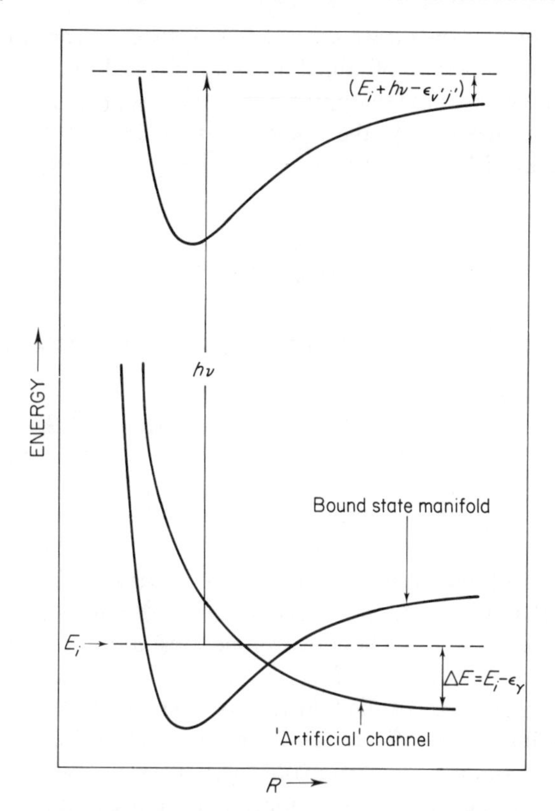

Fig. 4. Form of diagonal channel potentials for artificial channels method. Only one diagonal channel potential is shown for each of the bound state and continuum state manifolds.

Regular solution:

$$\phi^{r,vj\lambda}_{v'j'\lambda'}(R) \underset{R\to\infty}{\sim} \sin(k_{v'j'}R)\delta_{vv'}\delta_{jj'}\delta_{\lambda\lambda'} + T^*_{v'j'\lambda',vj\lambda}\frac{1}{2i}\left(\frac{k_{vj}}{k_{v'j'}}\right)^{1/2} e^{-ik_{v'j'}R} \quad (82b)$$

Using the Green's function constructed from these functions yields the following expression for the desired **T** matrix element:

$$T_{vj\lambda,\gamma} = \sum_{v'j'\lambda'}\sum_{v''j''\lambda''}\frac{1}{2i(k_{vj}k_\gamma)^{1/2}} \times \int_0^\infty \phi^{r,vj\lambda*}_{v'j'\lambda'}(R)U^{JJ_i}_{v'j'\lambda',v''j''\lambda''}(R)\psi^{J_ip_i\gamma}_{v''j''\lambda''}(R)\,dR \quad (83)$$

The functions $\psi^{J_ip_i\gamma}_{v''j''\lambda''}(R)$ are the solutions of Eq. (75). These functions can be expressed in terms of the Green's function constructed from the solutions of the homogeneous equation obtained by setting the right-hand side of Eq. (75) to zero (see ref. 39). We note that these homogeneous equations are exactly the same as those obeyed by the radial components of the bound state wavefunction (Eq. 78). In this way we obtain the following expression for the **T** matrix element:

$$
\begin{aligned}
T_{vj\lambda,\gamma} = {} & \left(\frac{2\mu}{\hbar^2}\right)\frac{1}{2i(k_{vj}k_\gamma)^{1/2}}\sum_n \sum_{v'j'\lambda'} \\
& \times \left[\sum_{v''j''\lambda''}\int_0^\infty \phi^{r,vj\lambda^*}_{v'j'\lambda'}(R)\,U^{JJ_i}_{v'j'\lambda',v''j''\lambda''}(R)\,\Phi^{J_np_n}_{v''j''\lambda''}(R)\,\mathrm{d}R\right] \\
& \times \left[\sum_{v'''j'''\lambda'''}\int_0^\infty \Phi^{J_np_n^*}_{v'''j'''\lambda'''}(R)\,U_{v'''j'''\lambda'',\gamma}(R)\,\psi^\gamma_\gamma(R)\,\mathrm{d}R\right] \\
& \times \frac{1}{E_n - E'} + \text{continuum term}
\end{aligned}
\tag{84}
$$

where E' is an arbitrary energy which has been substituted for E_i in Eqs. 75–77 and $\Phi^{J_np_n}_{v'''j'''\lambda'''}(R)$ is the radial components of the bound state wavefunction (Eq. 78) with energy E_n. These wavefunctions are the solutions of the homogeneous bound state wavefunction (Eq. 75) (i.e. the left-hand side). The continuum term, whose explicit form has been omitted from Eq. 84 (see Shapiro and Balint-Kurti[39]) arises from the contribution to the Green's function of the continuum solutions of the homogeneous part of Eq. (75).

The energy dependence of the **T** matrix element connecting the artificial channel γ to the continuum photodissociative channel $v'j'\lambda'$ (Eq. 84) has a first-order pole at the bound state energies E_n. The residue of this pole (i.e. the terms multiplying $1/(E_n - E')$) contains two bound-continuum integrals. One of these integrals involves exactly the same wavefunctions as those which arise in the photofragmentation **T** matrix element of Eq. (72), which is the quantity we seek to evaluate. We now analyse this photofragmentation **T** matrix element and express it in a form more easily related to Eq. (84). Substituting the expansion of the bound state wavefunction (Eq. 78) into Eq. (72) and using the definition of the transition dipole moment matrix elements (Eq. 80) we obtain

$$
\begin{aligned}
t(EJvj\lambda p\,|\,E_iJ_ip_i) = {} & 4\pi(\mu k_{vj}/h^2 2\pi)^{1/2}\sum_{v'j'\lambda'}\sum_{v''j''\lambda''}\int_0^\infty \Phi^{-(Jvj\lambda p)^*}_{v'j'\lambda'}(R) \\
& \times U^{JJ_i}_{v'j'\lambda',v''j''\lambda''}(R)\,\Phi^{J_ip_i}_{v''j''\lambda''}(R)\,\mathrm{d}R
\end{aligned}
\tag{85}
$$

By comparing Eqs. (71) and (82b) and taking account of the manner in which the total scattering wavefunction has been expanded, we note the relationship:

$$
\Phi^{-(Jvj\lambda p)}_{v'j'\lambda'}(R) = (-1)^{j+\lambda}\left(\frac{4t_\lambda}{1+\delta_{\lambda o}}\right)\phi^{r,vj\lambda}_{v'j'\lambda'}(R) \tag{86}
$$

Using Eqs. (85) and (86) we may now rewrite Eq. (84) as

$$T_{vj\lambda,\gamma}=\frac{(-1)^{j+\lambda}}{k_{vj}}\frac{(2\pi\mu)^{1/2}}{2i\hbar}\left(\frac{1+\delta_{\lambda o}}{4t_\lambda}\right)\sum_n\left\{t(EJvj\lambda p|E_nJ_np_n)\right.$$
$$\times\left[\left(\frac{1}{k_\gamma}\right)^{1/2}\sum_{v'''j'''\lambda'''}\int_0^\infty\Phi^{J_np_n*}_{v'''j'''\lambda'''}(R)U_{v'''j'''\lambda''',\gamma}(R)\right.$$
$$\left.\left.\times\,\psi^\gamma_\gamma(R)\,\mathrm{d}R\right]\right\}\frac{1}{E_n-E'}+\text{continuum term} \tag{87}$$

The computation of the photofragmentation **T** matrix element now proceeds as follows. We first locate the bound state energy E_i corresponding to the initial bound state of the system. This is done (see Shapiro and Balint-Kurti[39] and Kidd, Balint-Kurti and Shapiro[40]) by replacing the photodissociative manifold of channels in Eq. (76) by a single artificial channel β which is taken to have the same diagonal channel potential as channel γ (i.e. $U_{\beta,\beta}=U_{\gamma,\gamma}$). The coupling between the artificial channel β and the bound state manifold is also taken to be the same as that between γ and the bound state manifold (i.e. $U_{\beta,v'j'\lambda'}=U_{v'j'\lambda',\gamma}$). The **T** matrix elements connecting the two continuum artificial channels then take the form:

$$T_{\beta,\gamma}=\frac{1}{2i}\left[\left(\frac{1}{k_\beta}\right)^{1/2}\sum_{v''j''\lambda''}\int_0^\infty\psi^{\beta*}_\beta(R)U_{\beta,v''j''\lambda''}(R)\Phi^{J_ip_i}_{v''j''\lambda''}(R)\mathrm{d}R\right]$$
$$\times\left[\left(\frac{1}{k_\gamma}\right)^{1/2}\sum_{v'''j'''\lambda'''}\int_0^\infty\Phi^{J_ip_i*}_{v'''j'''\lambda'''}(R)U_{v'''J'''\lambda''',\gamma}(R)\psi^\gamma_\gamma(R)\mathrm{d}R\right]\frac{1}{E_i-E'} \tag{88}$$

at energies close to the bound state energy E_i.

We locate the required pole of this **T** matrix (i.e. the energy level E_i) and evaluate the residue at this pole. Near the pole we may ignore all contributions to the **T** matrix element except those due to the pole in question. The **T** matrix element may then be represented as

$$T_{\beta,\gamma}=\frac{\mathrm{Res}(i)}{E_i-E} \tag{89}$$

The integral involving the artificial channel γ which occurs in Eq. (87) may now be evaluated in terms of this residue:

$$\left(\frac{1}{k_\gamma}\right)^{1/2}\sum_{v'''j'''\lambda'''}\int_0^\infty\Phi^{J_ip_i*}_{v'''j'''\lambda'''}(R)U_{v'''j'''\lambda''',\gamma}(R)\psi^\gamma_\gamma(R)\mathrm{d}R=[2i\,\mathrm{Res}(i)]^{1/2} \tag{90}$$

The set of coupled differential equations (Eqs. 75 to 77) are now solved at an energy close to E_i. Under these conditions, only one term in the summation over n in the expression for the **T** matrix element (Eq. 87) is important and the continuum contribution may be ignored. The **T** matrix element of

Eq. (87) may now be written in the form:

$$T_{vj\lambda,\gamma} = \frac{(-1)^{j+\lambda}}{k_{vj}} \frac{(2\pi\mu)^{1/2}}{\hbar} \left(\frac{1+\delta_{\lambda o}}{4t_\lambda} \right) t(EJvj\lambda p | E_i J_i p_i) \left[\frac{\text{Res}(i)}{2i} \right]^{1/2} \frac{1}{E_i - E} \quad (91)$$

This equation may be reexpressed to provide an equation for the photofragmentation **T** matrix directly in terms of computed quantities as

$$t(EJvj\lambda p | E_i J_i p_i) = (-1)^{j+\lambda} \hbar k_{vj} (2\pi\mu)^{-1/2} \left(\frac{4t_\lambda}{1+\delta_{\lambda o}} \right) \times \left[\frac{\text{Res}(i)}{2i} \right]^{-1/2} T_{vj\lambda,\gamma}(E_i - E) \quad (92)$$

The above equation shows explicitly how standard scattering theory algorithms may be used to evaluate the photofragmentation **T** matrix elements directly without the need to explicitly perform the bound-continuum integration.

2. *The Driven Equation Method*

We see clearly from the above discussion of the artificial channel method that the bound-continuum integral needed in the evaluation of the photofragmentation cross-section (Eqs. 36 and 49) may be evaluated by introducing an inhomogeneous term into the differential equation for the continuum wavefunction (Eq. 76). The introduction of an artificial channel permits us to calculate the necessary integrals implicitly (rather than explicitly) by providing a channel through which incident particle flux can enter the system. Alternatively, this incident particle flux may be introduced into the system entirely through an inhomogeneous source term in the continuum equations.

Let us consider the inhomogeneous coupled differential equations:

$$\left[-\frac{d^2}{dR^2} + \frac{2\mu}{\hbar^2}(\varepsilon_{v'j'} - E_i - h\nu) + W^{cJ}_{v'j'\lambda',v'j'\lambda'}(R) \right] \Phi^{-J}_{v'j'\lambda'}(R) + \sum_{v''j''\lambda''} W^{cJ}_{v'j'\lambda',v''j''\lambda''}(R) \Phi^{-J}_{v''j''\lambda''}(R) = \langle \chi_{v'j'}(r) \Theta^{JMp}_{j'\lambda'}(\hat{\mathbf{R}}, \hat{\mathbf{r}}) | \hat{\varepsilon} \cdot \boldsymbol{\mu} | \Psi^{J_i M_i}(\mathbf{R}, \mathbf{r}, E_i) \rangle \quad (93)$$

These equations are almost identical to Eq. (76) used in the artificial channel method as the right-hand side of Eq. (94) is equal to the right-hand side of Eq. (76) multiplied by $(-1)^{M_i} \begin{pmatrix} J & 1 & J_i \\ -M_i & 0 & M_i \end{pmatrix}$. The radial equations must be solved subject to the boundary conditions:

$$\Phi^{-J}_{v'j'\lambda'}(R) \underset{R \to 0}{\sim} 0 \quad (94)$$

$$\Phi^{-J}_{v'j'\lambda'}(R) \underset{R\to\infty}{\sim} -T^{J}_{v'j'\lambda'}e^{ik_{v'j'}R} \tag{95}$$

Using the Green's function formed from the solutions of the homogeneous equation obtained by setting the right-hand side of Eq. (94) to zero (see Eqs. 82a and 82b), we may write:

$$T^{J}_{v'j'\lambda'} = \sum_{v''j''\lambda''}\sum_{v'''j'''\lambda'''}\frac{1}{2ik_{v'j'}}\left[\int_0^\infty \phi^{r,v'j'\lambda'}_{v''j''\lambda''}(R)U^{JJ_i}_{v''j''\lambda'',v'''j'''\lambda'''}(R)\Phi^{J_ip_i}_{v'''j'''\lambda'''}(R)\,\mathrm{d}R\right] \times(-1)^{M_i}\begin{pmatrix} J & 1 & J_i \\ -M_i & 0 & M_i\end{pmatrix} \tag{96}$$

Using Eqs. (85) and (86) we now obtain:

$$T^{J}_{v'j'\lambda'} = (-1)^{j'+\lambda'+M_i}\left(\frac{1+\delta_{\lambda o}}{4t_\lambda}\right)\begin{pmatrix} J & 1 & J_i \\ -M_i & 0 & M_i\end{pmatrix}\frac{h}{4ik^{3/2}_{v'j'}} \times\frac{1}{(2\pi\mu)^{1/2}}t(Ev'j'\lambda'p|E_iJ_ip_i) \tag{97}$$

We can see that the solution of the driven equations (Eq. 93) will also directly yield the photofragmentation **T** matrix elements. The solution of the inhomogeneous equations, however, in general requires the explicit evaluation of bound-continuum integrals[28,60] which is circumvented in the artificial channel approach.

IV. DIFFERENTIAL AND INTEGRAL PHOTOFRAGMENTATION CROSS-SECTIONS

The detailed photofragmentation cross-section for a triatomic molecule[37] is given, in terms of the photofragmentation **T** matrix elements, by Eq. (73). This equation may be rewritten in the slightly simplified form:

$$\sigma(\hat{\mathbf{k}}Evjm_j|E_iJ_iM_ip_i) = \frac{8\pi^3\nu}{c}\left|\sum_{J\lambda}(2J+1)^{1/2}\begin{pmatrix} J & 1 & J_i \\ -M_i & 0 & M_i\end{pmatrix}d^{J}_{\lambda M_i}(\theta) \times d^{j}_{-\lambda-m_j}(\theta)t(EJvj\lambda p|E_iJ_ip_i)\right|^2 \tag{98}$$

This cross-section gives the probability of a triatomic molecule, initially in a well-defined state (with energy E_i, total angular momentum quantum number J_i, quantum number for the z component of the total angular momentum M_i and parity p_i), being dissociated into fragments with internal quantum states vjm_j processing a total energy E and scattered into the direction $\hat{\mathbf{k}}$ corresponding to spherical polar angles (θ, ϕ) in the centre-of-mass reference frame.

A. Differential Photofragmentation Cross-sections

As a specific example of how dynamical information may be extracted from a measured differential photofragmentation cross-section let us suppose that we have successfully measured the cross-section for a molecule initially in a state with $J_i = 0$, $M_i = 0$ dissociating to a fragment in state $j = 1$, $m_j = 0, \pm 1$. On substitution into Eq. (98) we find:

$$\sigma_0(\theta) = \sigma(\hat{\mathbf{k}}Evj = 1, m_j = 0 | E_i J_i = 0, M_i = 0, p_i = 0)$$
$$= \frac{8\pi^3 v}{3c} \left| \cos^2(\theta) t_0 - \frac{\sin^2(\theta)}{2} t_1 \right|^2 \tag{99a}$$

$$\sigma_1(\theta) = \sigma(\hat{\mathbf{k}}Evj = 1, m_j = 1 | E_i J_i = 0, M_i = 0, p_i = 0)$$
$$= \frac{8\pi^3 v}{3c} \frac{1}{2} \left| \cos(\theta)\sin(\theta) t_0 + \sin(\theta) \frac{(1 + \cos(\theta))}{2} t_1 \right|^2 \tag{99b}$$

$$\sigma_{-1}(\theta) = \sigma(\hat{\mathbf{k}}Evj = 1, m_j = -1 | E_i J_i = 0, M_i = 0, p_i = 0)$$
$$= \frac{8\pi^3 v}{3c} \frac{1}{2} \left| \cos(\theta)\sin(\theta) t_0 - \sin(\theta) \frac{(1 - \cos(\theta))}{2} t_1 \right|^2 \tag{99c}$$

where

$$t_0 = t(EJ = 1, v, j = 1, \lambda = 0, p | E_i J_i = 0, p_i = 0)$$

and

$$t_1 = t(EJ = 1, v, j = 1, \lambda = 1, p | E_i J_i = 0, p_i = 0)$$

The dynamical information is contained entirely in the *complex* quantities t_0 and t_1. Clearly Eqs. (99) enable us to extract the **T** matrix elements t_0 and t_1 from the experimental measurements. As the elements are only measureable to within an arbitrary phase factor we may take t_0 to be real, but t_1 will have a real and a complex part.

The photofragmentation **T** matrix elements t_0 and t_1 depend upon the transition dipole moment function, and conversely experimentally determined information regarding these **T** matrix elements carries with it information about the transition dipole moment function. As an example of how the differential photofragmentation cross-section depends on the photofragmentation **T** matrix elements and on the nature of the transition dipole moment function we consider two extreme cases.

Example 1: $t_0 = 1, t_1 = 0$

This situation will occur whenever the dipole moment function $\boldsymbol{\mu}(\mathbf{R}, \mathbf{r})$ is an even function of the angle γ between $\mathbf{R}$ and $\mathbf{r}$. An example of this is the Ar—H_2 van der Waals molecule. The $\sigma_0(\theta)$ and $\sigma_1(\theta)$ differential cross-sections for this case are plotted in Fig. 5.

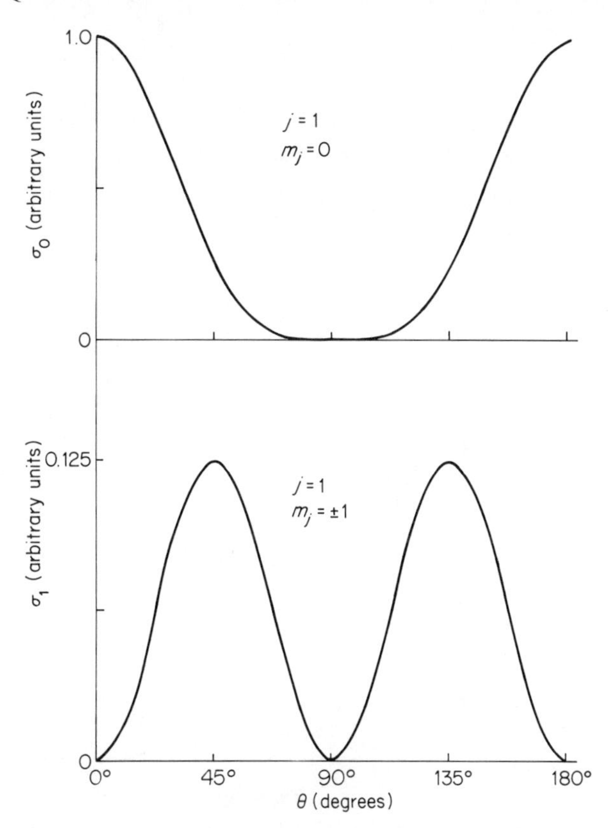

Fig. 5. Form of differential photofragmentation cross-sections for $J_i = 0 \rightarrow j = 1$, $m_j = 0$, 1 transition in a triatomic molecule. The transition dipole moment function is assumed to have an even dependence on γ. See Example 1 of text.

Example 2: $$t_0 = 0, t_1 = 1$$

This situation would pertain in the case of a photodissociation process occurring via a 'perpendicular' electronic transition where the transition dipole moment was oriented perpendicular to the molecular plane. An example would be the $\tilde{A}^1B_1 \leftarrow \tilde{X}^1A_1$ electronic transition in H_2O. The $\sigma_0(\theta)$ and $\sigma_1(\theta)$ differential cross-sections for this case are plotted in Fig. 6.

A comparison of Figs. 5 and 6 clearly shows a dramatic difference between the angular distributions in the two cases discussed and illustrates how the analysis of the form of such differential photofragmentation cross-sections leads to a knowledge of the form of the transition dipole moment function responsible for the transition.

No experiments have as yet been able to measure the detailed differential

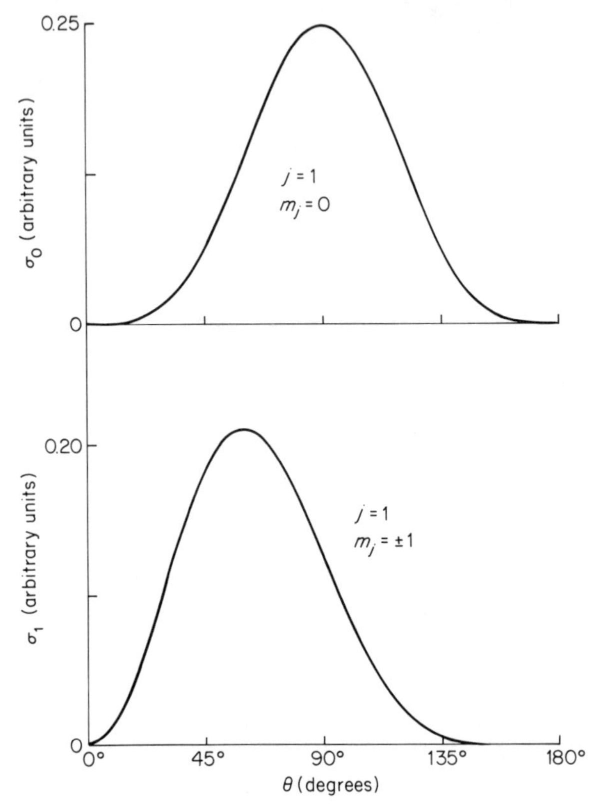

Fig. 6. Form of differential photofragmentation cross-section for $J_i = 0 \rightarrow j = 1$, $m_j = 0, 1$ transition in a triatomic molecule. The transition dipole moment function is assumed to lie perpendicular to the molecular plane. See Example 2 of text.

photofragmentation cross-sections of Eq. (98). Various averaged cross-sections are more accessible to experimental measurement. We now review the formulae for these cross-sections, in each case emphasizing the dynamical information which may be extracted from the experiment.

Let us consider an experiment in which there is no selection of initial M_i angular momentum components, but a differential cross-section is measured with sufficient energy resolution to resolve the vibrational (but not rotational) states of the fragments, and a polarizing field is used to select different m_j components of the products. The expression for the relevant cross-section is then

$$\bar{\sigma}(\hat{\mathbf{k}}Evjm_j|E_iJ_ip_i) = \frac{8\pi^3\nu}{c}\frac{(4\pi)^{1/2}}{(2J_i+1)}(-1)^{J_i}\sum_{\lambda\lambda'}\sum_{\mu=0,2}$$
$$\times \left[Y_{\mu,(\lambda-\lambda')}(\theta,0)d^j_{\lambda m_j}(\theta)d^j_{\lambda' m_j}(\theta)\right]$$

$$\times (-1)^{\lambda}(2\mu+1)^{1/2}\begin{pmatrix} 1 & 1 & \mu \\ 0 & 0 & 0 \end{pmatrix} \sum_{JJ'} (-1)^{J+J'}$$

$$\times [(2J+1)(2J'+1)]^{1/2}\begin{pmatrix} J & J' & \mu \\ \lambda & -\lambda' & \lambda'-\lambda \end{pmatrix}$$

$$\times \begin{Bmatrix} 1 & 1 & \mu \\ J & J' & J_i \end{Bmatrix} t(EJvj\lambda p|E_iJ_ip_i)$$

$$\times t^*(EJ'vj\lambda' p|E_iJ_ip_i) \tag{100}$$

From the formal expression for the differential cross-section we see that if we analyse the angular distribution of fragments in a particular m_j state in terms of the angular functions $[Y_{\mu,\lambda-\lambda'}(\theta,0)\, d^j_{\lambda m_j}(\theta)\, d^j_{\lambda' m_j}(\theta)]$ the coefficients of these angular functions tell us the contributions of the individual j rotational states to the scattering. Thus the differential cross-section in an m_j-resolved experiment may be analysed to provide the rotational state distribution of the product fragment *without* energy resolving the different fragment rotational j states.

In the experiment just considered we were unable, through the analysis of the photofragment angular distribution, to gain any information concerning the contributions of different J (and therefore also J_i) states to the photofragmentation dynamics. This situation arose because the experiment we were considering had not selected a definite initial M_i state but had averaged over all such states. By contrast, the angular distribution of this experiment carried with it information concerning the rotational energy distribution (j quantum numbers) of the photofragments. This arose because we had selected m_j states when analysing the product distribution. The converse situation is also found to hold, i.e. if we select the initial M_i state but average over all final m_j states then the angular distribution of the photofragments carries information about the contribution of different J's (total angular momenta) to the photofragmentation dynamics. The formal expression for a differential cross-section of this type is

$$\bar{\sigma}(\hat{\mathbf{k}}Evj|E_iJ_iM_ip_i) = \frac{8\pi^3 v}{c} \sum_{JJ'} \sum_{\lambda} [d^J_{\lambda M_i}(\theta) d^{J'}_{\lambda M_i}(\theta)]$$

$$\times \left\{ [(2J+1)(2J'+1)]^{1/2} \begin{pmatrix} J & 1 & J_i \\ -M_i & 0 & M_i \end{pmatrix} \right.$$

$$\left. \times \begin{pmatrix} J' & 1 & J_i \\ -M_i & 0 & M_i \end{pmatrix} t(EJvj\lambda p|E_iJ_ip_i) t^*(EJ'vj\lambda p|E_iJ_ip_i) \right\} \tag{101}$$

This type of cross-section should therefore be analysed in terms of the angular functions $[d^J_{\lambda M_i}(\theta)\, d^{J'}_{\lambda M_i}(\theta)]$. The coefficients of these angular functions carry information as to the contribution of different total angular momenta to the photofragmentation cross-section. If the initial molecules were not selected to be in a specified J_i state but only in specific M_i states, this information also permits an analysis of the extent to which different J_i states contribute to the photodissociation process. The same analysis of the angular behaviour of the differential cross-section provides a breakdown of the contributions to the cross-section of different λ quantum numbers (this quantum number is analogous to the K quantum number used to describe the rotational states of a symmetric top molecule). The contributions arising from different λ quantum numbers are highly sensitive to the functional form of the transition dipole moment governing the transition.

The differential cross-section averaged over initial M_i quantum numbers and summed over final m_j quantum numbers may be written in the well-known form:

$$\bar{\sigma}(\hat{\mathbf{k}}Evj|E_iJ_ip_i) = A(Evj|E_iJ_i)[1 + \beta(Evj|E_iJ_i)P_2(\cos\theta)] \tag{102}$$

The exact quantum mechanical form for the anisotropy parameter $\beta(Evj|E_iJ_i)$ has been derived in detail elsewhere.[37] In Eq. (102) we denote explicitly that the anisotropy parameter, and therefore the angular dependence of the differential cross-section, depends on the quantum numbers of both the initial and final states. The dynamical information contained in a differential photofragmentation cross-section from an experiment in which there is no selection of J_i, M_i, j or m_j is therefore very limited. Using polarizers to select either M_i or m_j (or both!) greatly increases the dynamical information which is available from the results.

B. Integral Photofragmentation Cross-sections

There have been very few measurements of differential photofragmentation cross-sections of the type discussed above.[2,8,12,72–79] A vastly greater number of experiments concentrate on the measurement of attributes which depend on the integral photofragmentation cross-sections. These include the quantum state distribution of the photofragmentation products[3–7,9–12,13,15,80,81] and measurements of the polarization of the photofragment distribution[19,82–84] either through the polarization of the light emitted by the excited products [19,82–84] or by laser-induced fluorescence measurements using polarized light.[14,85]

The integral cross-sections may be obtained from the corresponding differential cross-sections by integrating over the solid angles. Two cross-sections are of particular interest. Firstly, the integral photofragmentation cross-section averaged over the initial state magnetic quantum numbers[37] M_i:

$$\bar{\sigma}(Evjm_j|E_iJ_ip_i) = \frac{1}{2J_i+1}\sum_{M_i}\sigma(Evjm_j|E_iJ_iM_iP_i)$$

$$
\begin{aligned}
&= \frac{8\pi^3 \nu}{c} \frac{4\pi}{2J_i + 1} \sum_{J\lambda} \sum_{J'\lambda'} \\
&\times \Bigg\{ [(2J+1)(2J'+1)]^{1/2} (-1)^{m_j} \\
&\times \sum_{\mu=0,2} (-1)^{\lambda - \lambda' + J + J' + J_i} (2\mu + 1) \\
&\times \begin{pmatrix} J & J' & \mu \\ \lambda - \lambda' & \lambda' - \lambda \end{pmatrix} \begin{pmatrix} j & j & \mu \\ -\lambda & \lambda' & \lambda - \lambda' \end{pmatrix} \\
&\times \begin{pmatrix} j & j & \mu \\ 0 & 0 & 0 \end{pmatrix} \begin{Bmatrix} 1 & 1 & \mu \\ J & J' & J_i \end{Bmatrix} \Bigg\} \\
&\times t(EJvj\lambda p | E_i J_i p_i) \\
&\times t^*(EJ'vj\lambda' p | E_i J_i p_i)
\end{aligned} \tag{103}
$$

This cross-section gives the probability of producing diatomic fragments in a particular m_j state, from an initially unpolarized population. If the products are produced in an electronically excited state which can fluoresce, the polarization of the resulting fluorescence can be predicted by calculating the polarization of the fluorescence from each of the j, m_j states and summing this polarization over all j, m_j states, weighting each state by its cross-section.[86]

The second type of integral cross-section which is of particular interest is that corresponding to no selection of magnetic quantum states in either the original molecule or in the dissociation products. This requires the detailed integral photofragmentation cross-section to be averaged over the initial M_i quantum numbers and to be summed over the final m_j quantum numbers. The resulting expression is:[37]

$$
\begin{aligned}
\bar{\bar{\sigma}}(Evj | E_i J_i p_i) &= \frac{1}{2J_i + 1} \sum_{M_i} \sum_{m_j} \sigma(Evjm_j | E_i J_i M_i p_i) \\
&= \frac{8\pi^3 \nu}{c} \frac{4\pi}{3} \frac{1}{2J_i + 1} \sum_{J\lambda} \delta(JJ_i 1) \\
&\times |t(EJvj\lambda p | E_i J_i p_i)|^2
\end{aligned} \tag{104}
$$

where $\delta(JJ_i 1) = 1$ if J, J_i and 1 satisfy the 'triangular' condition and zero otherwise.

This cross-section gives the product quantum state distribution of the photofragments originating from a pure quantum state of the parent molecule. The cross-section of Eq. (104) is often referred to as a partial cross-section, the total cross-section being obtained by summation of the partial cross-sections over all possible final quantum states:

$$
\sigma^{\text{tot}}(E | E_i J_i p_i) = \sum_{vj} \bar{\bar{\sigma}}(Evj | E_i J_i p_i) \tag{105}
$$

In an absorption spectrum of a photodissociative or predissociative transition of a molecule, the absorption lineshape $I(\omega)$ is given by

$$I(\omega) = \sum_i f_i \sigma^{\text{tot}}(E|E_i J_i p_i) \tag{106}$$

where $\hbar\omega = E - E_i$ and f_i is the population of the ith quantum state of the parent molecule.

In our discussion of polarized photofragment fluorescence following on from Eq. (103) we treated the fluorescence and the photodissociation as two distinct and separate processes. This is only valid when the photofragment quantum states are long lived. In principle the entire process from the absorption of the initial photon to the emission of the fluorescent photon and the production of the stable (ground electronic state) photofragments should be treated as a single concerted process.[19] It is certainly possible to do this by an extension of the theory given in this review. This concerted treatment is only expected to yield significantly different results from the two-step treatment of the problem in cases where the (electronically) excited photofragments have lifetimes shorter than or of the order of their rotational periods.

V. APPLICATIONS

The theoretical treatment of photodissociation processes is still in its infancy. This is clearly demonstrated by the very small number of thorough quantum mechanical calculations which have so far been performed. Many approximate treatments of photofragmentation dynamics have been presented, and these have often led to invaluable insights as to the origins of experimentally observed features of both the absorption spectra [66–68,87–98] and the product state distribution.[15,64,65,67,68,99,100–125] The various methods of calculating photodissociation cross-sections and predissociative linewidths have been the subject of several recent reviews.[16–18,20,126,127]

One of the most extensively applied approximate methods for calculating final photofragment quantum state distributions has been the so-called 'generalized Franck–Condon theory'.[7,15,108–120,127,128] This theory treats the ground state wavefunction of the parent molecule as a product of simple harmonic normal-mode wavefunctions. No account is taken of the dynamics of the dissociation process on the upper final state electronic surface. The distribution of the products among their quantum states is calculated by evaluating the overlap of the asymptotic, non-interacting fragment wavefunctions with the ground state wavefunction. Such a method of calculating the photofragment quantum state distribution provides in general only a crude approximation. The method has proved very useful because of the problems encountered in the application of more exact theories.

More sophisticated methods attempt to treat the dynamics of the 'half-collision' on the final dissociative potential energy surface in an approximate way. Within the quantum mechanical approach to photodissociation, the approximations which may be used to simplify the problem of evaluating the continuum scattering wavefunction derive directly from those which have been developed and extensively tested in the field of inelastic molecular scattering theory. These include the distorted wave approach, which may be applied either within a framework of an adiabatic[99,94,129] or a diabatic[91-93,103,121] treatment of the internal fragment wavefunctions. Other standard scattering theory approximations, in particular the infinite order sudden[29,88,130] (IOS) and the *p*-helicity decoupling[67,68] (or coupled states) approximations have also been used to great effect to make otherwise intractable problems amenable to solution. Classical[131,132] and semi-classical[122-125,133,134] methods of treating the dynamics of the half-collisions problem have also been developed and shown to be extremely valuable.

In the process of developing a complete theory of photodissociation processes, many model problems in which the scattering dynamics on the final dissociative surface are treated exactly have been solved. These model problems have, in general, been firmly based on physical reality, but various simplifying assumptions have been made. These assumptions are generally chosen so as to reduce the mathematical dimension of the problem to two independent variables, one 'internal' variable and the scattering coordinate. Model calculations of this type have been performed for the systems ICN,[99,103,135,136] HCN,[103,136,137] CO_2,[28,87] N_2O[64,89], which have all been approximated as collinear processes occurring along a fixed direction in space, and for CH_3I,[65] where only the CH_3 'umbrella' motion and the C—I stretching motion were treated.

Another type of simplifying approximation, which is particularly appropriate in the case of van der Waals molecules, is to freeze the fragment vibrational coordinates. This type of approximation has been quite common both in the early calculations of scattering resonances[138-142] which were the precursors of the present photodissociation theory and in current realistic calculations of photodissociation cross-sections.[66] For dissociation processes, when the intermediate metastable state to which the molecule is excited by the absorption of the photon is long lived, the analysis of the **S** matrices in the vicinity of the scattering resonances on the final dissociative potential energy surface[143] often yields reliable estimates of the linewidths and frequencies of the spectral transitions to the predissociative final states.[144,145] Another popular method of calculating the properties of these metastable resonant states is the complex coordinate method, in which the Hamiltonian is extended into the complex coordinate plane and the homogeneous differential equations are solved subject to purely outgoing Siegert boundary conditions.[20,145-152]

There are three essential types of molecular information which must be properly taken into account to correctly describe a photodissociation process. These are: (1) the initial bound state wavefunction, (2) the transition dipole moment function and (3) the scattering on the upper dissociative state. There are only a very small number of theoretical calculations which have attempted to address all three aspects of the problem within a realistic three-dimensional treatment.[60,66–71,88] One of the obstacles militating against such calculations is the almost complete lack of calculated dipole moment functions for systems larger than diatomic molecules. For the Ar—H_2 van der Waals molecule, both a dipole moment function[153] and a potential energy surface[126] have been

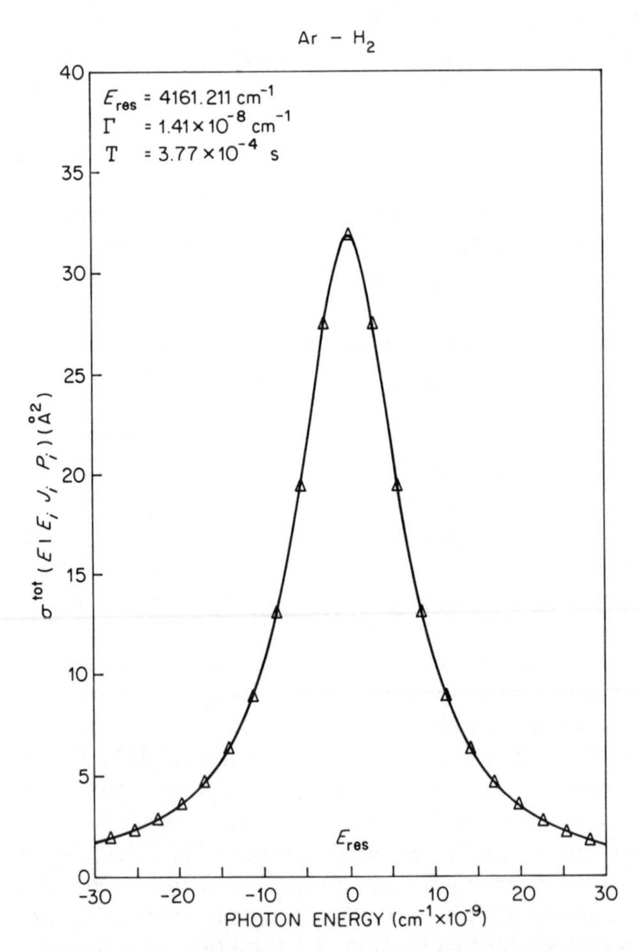

Fig. 7. Total photodissociation cross-section for the Ar—H_2 ($v = 0$, $j = 0, J = 0, l = 0, n = 0$) $\rightarrow$ [Ar—H_2^* ($v = 1, j = 0, J = 1, l = 1, n = 0$)] $\rightarrow$ Ar + H_2 ($v = 0$) transition.

obtained in an empirical manner by fitting the positions and intensities of some of the lines observed in the infrared spectrum of the system.[154] For this system, therefore, the necessary prerequisites for an accurate photodissociation calculation are available.

Figure 7 shows the total photofragmentation cross-section calculated using the theory of Section III.C for a spectroscopic transition from the ground state of the system to the lowest metastable state with H_2 in its ($v = 1, j = 0$) vibrational–rotational state.[69,71] The total cross-section is calculated at a series of energies in the vicinity of the 'line centre', thus mapping out the lineshape for this particular transition. The metastable intermediate state of the transition can only dissociate by the transfer of one quantum of H_2 vibrational energy

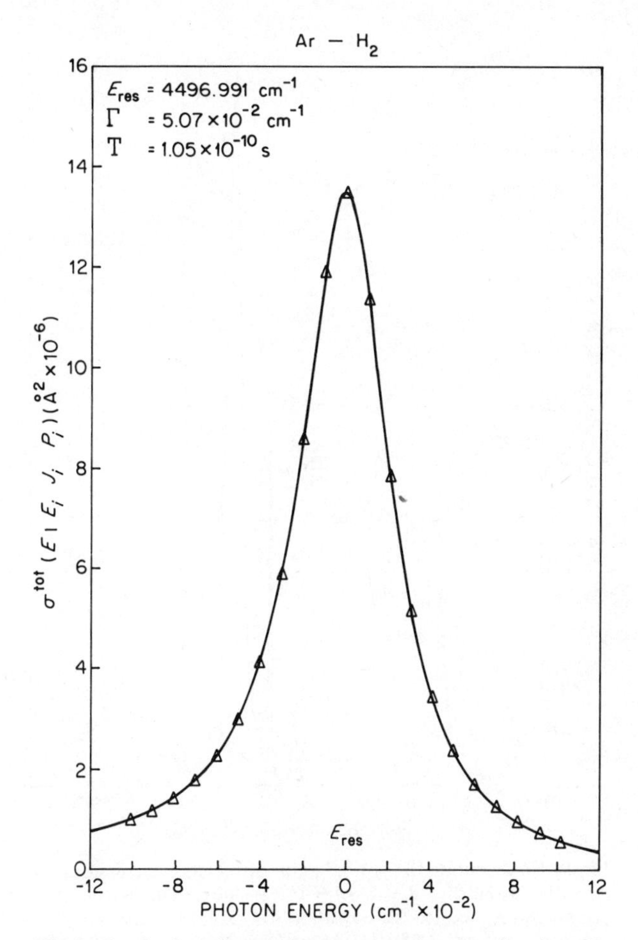

Fig. 8. Total photodissociation cross-section for the Ar—H_2 ($v = 0, j = 0, J = 0, l = 0, n = 0$) → [Ar—$H_2^*$ ($v = 1, j = 2, J = 1, l = 1, n = 0$)] → Ar + H_2 transition.

from the internal H_2 motion to the Ar—H_2 relative translational motion. The figure demonstrates that the lifetime of the metastable state is very long, which reflects the small probability for this intramolecular T–V transition process.

Figure 8 shows the total photodissociation cross-section for a spectroscopic transition from the ground state of the Ar—H_2 van der Waals molecule to the lowest metastable state of odd parity in which the H_2 is in its ($v = 1, j = 2$) internal state.[57] This metastable state can decompose by release of H_2

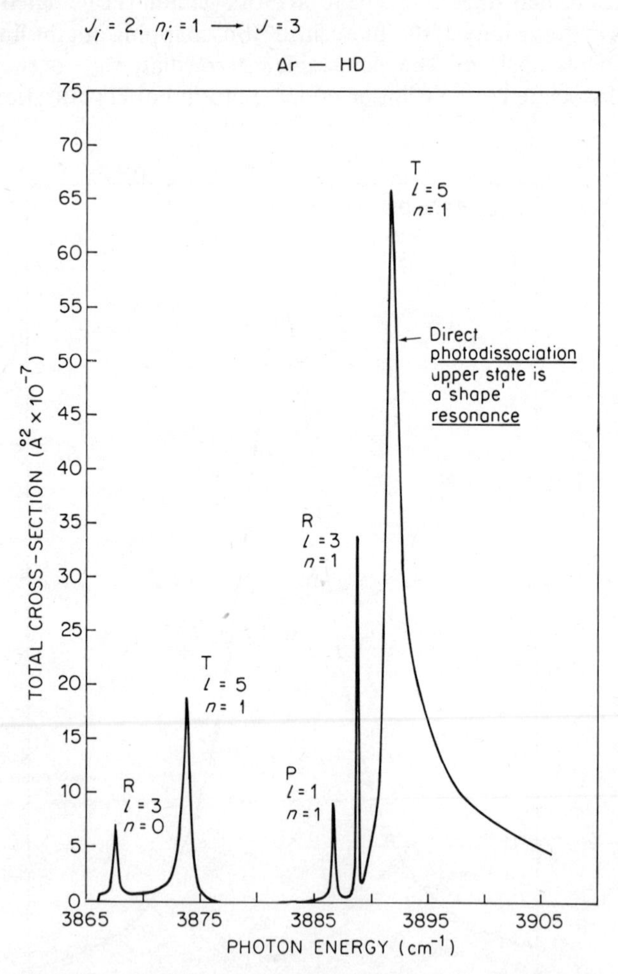

Fig. 9. Contribution to the total photodissociation cross-section of Ar—HD from the $J_i = 2 \rightarrow J = 3$ transition. The cross-section shown corresponds to the transition Ar—HD ($J_i = 2$, $v_i = 0$, $j_i = 0$, $n_i = 1$, $l_i = 2$) + $h\nu \rightarrow$ [Ar—HD* ($J = 3$, $v = 1$, $j = 2$, n, l)] $\rightarrow$ Ar + HD ($v = 1$, $j = 1$ or 0).

fragment rotational energy. As this rotational predissociation process occurs much more readily than the vibrational predissociation depicted in Fig. 1, we see that the lifetime of the metastable intermediate state is over five orders of magnitude shorter. The cross-sections depicted in Figs. 7 and 8 are computed using the full artificial channel formalism (Section III.C) with a converged basis set. They therefore constitute benchmark calculations against which more approximate approaches may be validated.

Similar calculations have been carried out for the Ar—HD van der Waals molecule,[70] and have been used to simulate part of its observed infrared spectrum at 77 K. Very many different transitions contribute to the spectrum. In Fig. 9 we show the $J_i = 2 \rightarrow J = 3$ contribution to the total cross-section for the process

$$\begin{aligned} &\mathrm{Ar{-}HD}(J_i = 2, v_i = 0, j_i = 0, n_i = 1, l_i = 2) + h\nu \\ &\quad \rightarrow [\mathrm{Ar{-}HD}^*(J = 3, v = 1, j = 2, n, l)] \\ &\quad \rightarrow \mathrm{Ar} + \mathrm{HD}(v = 1, j = 1 \text{ or } 0) \end{aligned}$$

where the quantum numbers n label the Ar—HD stretching vibration and l corresponds to the Ar—HD relative orbital angular momentum quantum number. The aspect of greatest interest in the figure is the resonance at around 3892 cm^{-1}. The lineshape of this resonance is clearly very non-Lorentzian. This is because the upper metastable state of the resonance has an energy above its channel asymptote. The scattering resonance corresponding to this line would be a mixture of a shape resonance (i.e. a resonance trapped by a centrifugal barrier) and a Feschbach resonance (i.e. one arising out of internal fragment excitation). The non-Lorentzian lineshape arises from the strong contribution of the continuum (shape resonance) wavefunction to the metastable state.

NOTE ADDED IN PROOF

The method presented in section III-A is closely related to previous work on photodissociation[155–160] and predissociation[161] in strong laser fields, as well as to work on other aspects of molecular dynamics in such fields. This latter type of work has been particularly concerned with laser-assisted reactions[36,162–166], laser-assisted inelastic collision processes[167–170], multi-photon absorption[171–175] and resonance Raman scattering[176,177].

Acknowledgement

We thank Mr. I.F. Kidd for valuable discussions and for performing the calculations depicted in Figs. 7, 8 and 9.

References

1. Ashfold, N. M. R., Macpherson, M. T., and Simons, J. P., *Top. Current Chem.*, **86**, 1 (1979).
2. Welge, K. H., and Schmiedl, R., *Adv. Chem. Phys.*, **47**, (Pt. 2), 133 (1981).
3. Leone, S. R., *Adv. Chem. Phys.*, **50**, 255 (1982).
4. Simons, J. P., *J. Phys. Chem.* (in press).
5. Zacharias, H., Meier, K., and Welge, K. H., in *Energy Storage and Redistribution in Molecules* (Ed. J. Hinze), Plenum Press, 1983.
6. Donovan, R. J., Fotakis, C., Hopkirk, A., McKendrick, C. B., and Torre, A., *Can. J. Chem.* (in press).
7. Bower, R. D., Hawkins, W. G., Houston, P. L., Jones, R. W., Kim, H-R., Marinelli, W. J., and Sivakumar, N., in *Proceedings of the Fourth Symposium on Recent Advances in Laser Spectroscopy*, Polytechnic Institute of New York, John Wiley (in press).
8. Sparks, R. K., Shobatake, K., Carlson, L. R., and Lee, Y. T., *J. Chem Phys.*, **75**, 3838 (1981).
9. Halpern, J. B. and Jackson, W. M., *J. Chem. Phys.*, **86**, 973 (1982).
10. Lahmani, F., Lardeux, C., and Solgadi, D., *J. Chem. Phys.*, **77**, 275 (1982).
11. Pfabb, J. Häger, J., and Krieger, W., *J. Chem. Phys.*, **78**, 266 (1983).
12. Ondrey, G. S., Kanfer, S., and Bersohn, R., *J. Chem. Phys.*, **79**, 179 (1983).
13. Ondrey, G., van Veen, N., and Bersohn, R., *J. Chem. Phys.*, **78**, 3732 (1983)
14. Andersen, P., and Rothe, E. W., *J. Chem. Phys.*, **78**, 898 (1983).
15. Vasudev, R., Zare, R. N., and Dixon, R. N., *J. Chem. Phys.* (in press).
16. Gelbert, W. M., *Ann. Rev. Phys. Chem.*, **28**, 323 (1977).
17. Beswick, J. A., and Jortner, J., *Adv. Chem. Phys.*, **47**, (Pt. I), 363 (1981).
18. Shapiro, M., and Bersohn, R., *Ann. Rev. Phys. Chem.*, **33**, 409 (1982).
19. Greene, C. H., and Zare, R. N., *Ann. Rev. Phys. Chem.*, **33**, 119 (1982).
20. Reinhardt, W. P., *Ann. Rev. Phys. Chem.*, **33**, 223 (1982).
21. Levine, R. D., *Quantum Mechanics of Molecular Rate Processes*, Clarendon Press, Oxford, 1969.
22. Newton, R. G., *Scattering Theory of Waves and Particles*, McGraw-Hill, 1966.
23. Zare, R. N., *Mol. Photochem.*, **4**, 1 (1972).
24. Loudon, R., *The Quantum Theory of Light*, Clarendon Press, Oxford, 1973.
25. Herzberg, G., *Molecular Spectra and Molecular Structure, Vols. I to III*, Van Nostrand, Princeton, N. J., 1966.
26. Macomber, J. D., *The Dynamics of Spectroscopic Transitions*, John Wiley, New York, 1976.
27. Shapiro, M., *J. Chem. Phys.*, **56**, 2582 (1972).
28. Kulander, K. C., and Light, J. C., *J. Chem. Phys.*, **73**, 4337 (1980).
29. Band, Y. B., Freed, K. F., and Kouri, D. J., *J. Chem, Phys.*, **74**, 4380 (1981).
30. Dupont-Roc, J., and *Proceedings of the Summer School on Chemical Photophysics*, June 1979, Editions du Centre National de La Recherche Scientifique, 15 quai Anatole-France, 75700 Paris, France.
31. Power, E. A., and Thirunamachandran, T., *Am. J. Phys.*, **46**, 370 (1978).
32. Babiker, M., Power, E. A., Thirunamachandron, T., *Proc. Roy. Soc. London*, **A338**, 235 (1974).
33. Woolley, R. G., *Proc. Roy. Soc. London*, **A321**, 557 (1971).
34. Cohen-Tannoundji, C., Diu, B., and Laloë, F., *Mécaneque Quantique, Complements*, F_{IV} *and* C_{XIII}, Herman, 1973.
35. Series, G. W., *Phys. Rep.*, **43**, 1 (1978).

36. George, T. F., Yuan, J.-M., Zimmerman, I. H., and Laing, J. R., *Faraday Disc. Chem. Soc.*, **62**, 246 (1977).
37. Balint-Kurti, G. G., and Shapiro, M., *Chem. Phys.*, **61**, 137 (1981), erratum **72**, 456 (1982).
38. Kanfer, S. and Shapiro, M. *Chem. Phys. Lett*, (1983).
39. Shapiro, M. and Balint-Kurti, G. G., *J. Chem. Phys.*, **71**, 1461 (1979).
40. Kidd, I. F., Balint-Kurti, G. G., and Shapiro, *Faraday Disc. Chem. Soc.*, **71**, 287 (1981).
41. Balint-Kurti, G. G., in *International Reviews of Science, Phys. Chem. Ser. 2* (Eds. A. D. Bukingham and C. A. Coulson), Vol.1, Butterworths, London, 1975.
42. Child, M. S., *Molecular Collision Theory*, Academic Press, New York, 1974.
43. Lester, W. A., Jr., *Meth. Comp. Phys.*, **10**, 211 (1971).
44. *Algorithms and Computer Codes for Atomic and Molecular Quantum Scattering Theory* (Ed. L. D. Thomas), NRCC Proceedings No. 5, available from National Technical Information Service, U.S. Dept. of Commerce, 5285, Port Royal Road, Springfield, VA 22161, USA.
45. *Meth. Comp. Phys.*, **10**.
46. Secrest, D., and Johnson, B. R., *J. Chem. Phys.*, **45**, 4556 (1966).
47. Johnson, B. R., *J. Comp. Phys.*, **13**, 445 (1973).
48. Mrugaka, F. and Secrest, D., *J. Chem. Phys.*, **78**, 5754 (1983).
49. Gordon, R. G., *J. Chem. Phys.*, **51**, 14 (1969).
50. Parker, G. A., Schmalz, T. G., and Light, J. C., *J. Chem. Phys.*, **73**, 1757 (1980).
51. Stechel, E. B., Walker, R. B., and Light, J. C., *J. Chem. Phys.*, **69**, 3518 (1978).
52. Mies, F. H., and Julienne, P. J., *J. Chem. Phys.*, **80**, 2526 (1984).
53. Mies, F. H., *J. Chem. Phys.*, **80**, 2514 (1984).
54. Pack, R. T., *J. Chem. Phys.*, **60**, 633 (1974).
55. McGuire, P., and Kouri, D., *J. Chem. Phys.*, **60**, 2488 (1974).
56. Shapiro, M., and Tamir, *Chem. Phys.*, **13**, 215 (1976).
57. Launay, J. M., *J. Phys.*, **B9**, 1823 (1976).
58. Edmonds, A. R., *Angular Momentum in Quantum Mechanics*, 2nd ed., Princeton University Press, Princeton, 1960.
59. Rose, M. E., *Elementary Theory of Angular Momentum*, John Wiley., New York, 1957. We use the $D^J_{MM'}$ as defined by Rose in this review.
60. Heather, R. W., and Light, J. C., *J. Chem. Phys.*, **78**, 5513 (1983).
61. Engel, V., and Schinke, R. (to be published).
62. Singer, S., Freed, K. F., and Band, Y. B., *J. Chem. Phys.*, **77**, 1942 (1982).
63. Balint-Kurti, G. G., unpublished work.
64. Shapiro, M., *Chem. Phys. Lett.*, **46**, 442 (1977).
65. Shapiro, M., and Bersohn, R., *J. Chem. Phys.*, **73**, 381 (1980).
66. Beswick, J. A., and Shapiro, M., *Chem. Phys.*, **64**, 333 (1982).
67. Segev, E., and Shapiro, M., *J. Chem. Phys.*, **73**, 2001 (1980).
68. Segev, E. and Shapiro, M., *J. Chem. Phys.*, **77**, 5604 (1982).
69. Kidd, I. F., and Balint-Kurti, G. G., *Chem. Phys. Lett.*, **101**, 419 (1983).
70. Kidd, I. F., and Balint-Kurti, G. G., *Chem. Phys. Lett.*, **105**, 91 (1984).
71. Kidd, I. F., and Balint-Kurti, G. G., *J. Chem. Phys.* (in press).
72. Busch, G. E., and Wilson, K. R., *J. Chem., Phys.*, **56**, 3638 (1972).
73. Busch, G. E., Cornelius, J. R., Mahoney, R. T., Morse, R. I., Schlosser, D. W., and Wilson, K. R., *Rev. Sci. Instr.*, **41**, 1066 (1970).
74. Wilson, K. R., in *Chemistry of the Excited State* (Ed. J. N. Pitts, Jr.), Gordon and Breach, New York, 1970.
75. Dzvonik, M. J., Yong, S. C., and Bersohn, R., *J. Chem Phys.*, **61**, 4408 (1974).

76. Yong, S. C., Freedman, A., Kawasuki, M., and Bersohn, R., *J. Chem. Phys.* **72**, 4058 (1980).
77. Diesen, R. W., Wahr, J. and Adler, S. E., *J. Phys. Chem.*, **55**, 2812 (1971).
78. Gilpin R., and Welge, K. H. *J. Chem. Phys.*, **54**, 975, 4224 (1971).
79. Schmiedl, R., Bottner, R., Zackarias, H., Meirer, U., and Welge, K. H., *Opt. Commun.*, **31**, 329 (1979).
80. Andresen, P., Ondrey, G. S., and Titze, B., *Phys. Rev. Lett.* (in press).
81. Vasudev, R., Zare, R. N., and Dixon, R. N., *Chem. Phys. Lett.*, **96**, 399 (1983).
82. Chamberlain, G. A., and Simons, J. P., *J. Chem. Soc. Faraday II*, **71**, 2043 (1975).
83. Macpherson, M. T., and Simons, J. P., *Chem. Phys. Lett.*, **51**, 261 (1977).
84. Macpherson, M. T., Simons, J. P., and Zare, R. N., *Mol. Phys.*, **38**, 2049 (1979).
85. Greene, C. H., and Zare, R. N., *J. Chem. Phys.*, **78**, 6741 (1983).
86. Segev, E., Ph.D. Thesis, Weizmann Institute, 1982; also Segev, E., and Shapiro, M., unpublished work.
87. Pack, R. T., *J. Chem. Phys.*, **65**, 4765 (1976).
88. Segev, E., and Shapiro, M., *J. Chem. Phys.*, **78**, 4969 (1983).
89. Shapiro, M., *Chem. Phys. Lett.*, **46**, 442 (1977).
90. Abgrall, H. and Figuet-Fayard, F., *J. Chem. Phys.*, **60**, 4497 (1974).
91. Beswick, J. A., and Requena, A., *J. Chem. Phys.*, **72**, 3018 (1980).
92. Beswick, J. A., and Delgado-Barrio, G., *J. Chem. Phys.*, **73**, 3653 (1980).
93. Beswick, J. A., and Requena, A., *J. Chem. Phys.*, **73**, 4347 (1980).
94. Halberstadt, N., and Beswick, J.A., *Faraday Dis. Chem. Soc.*, **73**, 357 (1983).
95. Miret-Artes, S., Delgado-Barrio, G., Atabek, O., and Beswick, J. A., *Chem. Phys. Lett.*, (in press).
96. Beswick, J. A., and Jortner, J., *Chem. Phys.*, **24**, 1 (1977).
97. Beswick, J. A., and Jortner, J., *J. Chem. Phys.*, **69**, 512 (1978).
98. Beswick, J. A., Delgado-Barrio, G., and Jortner, J., *J. Chem. Phys.*, **70**, 3895 (1979).
99. Halavee, U., and Shapiro, M., *Chem. Phys.*, **21**, 105 (1977).
100. Mukamel, S., and Jortner, J., *J. Chem. Phys.*, **60**, 4760 (1974).
101. Shapiro, M., *Chem. Phys. Lett.*, **81**, 52 (1981).
102. Mukamel, S., and Jortner, J., *J. Chem, Phys.*, **65**, 3735 (1976).
103. Atabek, O., Beswick, J. A., Lefebvre, R., Mukamel, S., and Jortner, J. A., *J. Chem. Phys.*, **65**, 4035 (1976).
104. Atabek, O., Beswick, J. A., Lefebvre, R., Mukamel, S., and Jortner, J., *Chem. Phys. Lett.*, **45**, 211 (1977).
105. Beswick, J. A., and Jortner, J., *Chem. Phys. Lett.*, **49**, 13 (1977).
106. Beswick, J. A., and Jortner, J., *J. Chem. Phys.*, **68**, 2277 (1978).
107. Beswick, J. A., and Jortner, J., *J. Chem. Phys.*, **74**, 6725 (1981).
108. Band, Y.B., and Freed, K. F., *Chem. Phys. Lett.*, **28**, 328 (1974).
109. Band, Y. B., and Freed, K. F., *J. Chem. Phys.*, **63**, 3382 (1975).
110. Band, Y. B., and Freed, K. F., *J. Chem. Phys.*, **63**, 4479 (1975).
111. Band, Y. B., and Freed, K. F., *J. Chem. Phys.*, **64**, 4329 (1976).
112. Morse, M. D., Freed, K. F., and Band, Y. B., *Chem. Phys.*, **44**, 125 (1976).
113. Band, Y. B., and Freed, K. F., *J. Chem. Phys.*, **67**, 1462 (1977).
114. Morse, M. D., Freed, K. F., and Band, Y. B., *Chem. Phys. Lett.*, **49**, 399 (1977).
115. Morse, M. D., Freed, K. F., and Band, Y. B., *J. Chem. Phys.*, **70**, 3620 (1979).
116. Band, Y. B., Morse, M. D., and Freed, K. F., *Chem. Phys. Lett.*, **67**, 294 (1979).
117. Morse, M. D., and Freed, K. F., *Chem. Phys. Lett.*, **74**, 49 (1980).
118. Morse, M. D., and Freed, K. F., *J. Chem. Phys.*, **74**, 4395 (1981).
119. Morse, M. D., and Freed, K. F., *J. Chem. Phys.*, **78**, 6045 (1983).
120. Morse, M. D., Band, Y. B., and Freed, K. F., *J. Chem. Phys.*, **78**, 6066 (1983).

121. Morse, M. D., Freed, K. F., and Band, Y. B., *J. Chem. Phys.*, **70**, 3604 (1979).
122. Heller, E. J., *J. Chem. Phys.*, **68**, 2066 (1978).
123. Heller, E. J., *J. Chem. Phys.*, **58**, 3891 (1978).
124. Heller, E. J., *Acc. Chem. Res.*, **14**, 368 (1981).
125. Lee, S. Y., and Heller, E. J., *J. Chem. Phys.*, **76**, 3035 (1982).
126. LeRoy, R. J., and Carley, J. S., *Adv. Chem. Phys.*, **42**, 353 (1980).
127. Freed, K. F., and Band, Y. B., in *Excited States*, (Ed. E. Lin), Vol. 3, p. 109, Academic Press, New York, 1978.
128. Beswick, J. A., and Gelbart, W. M., *J. Phys. Chem.*, **84**, 3148 (1980).
129. Kresin, V. Z., and Lester, W. A., Jr., *J. Chem. Phys.*, **86**, 2182 (1982).
130. Kouri, D. J., in *Atom Molucule Collision Theory* (Ed. R. B. Bernstein), Plenum Press, New York, 1979.
131. Chuljian, D. T., Ozment, J., and Simons, J., *Int. J. Quant. Chem. Symp.*, **16**, 435 (1982).
132. Holdy, K. E., Klotz, L. C., and Wilson, K. R., *J. Chem. Phys.*, **52**, 4588 (1970); Hose, W. L., Duchovic, R. J., Swamy, K. N., and Wolf, R. J., *J. Chem. Phys.*, **80**, 714 (1984).
133. Swammathan, P. K., and Micha, D. A., *Int. J. Quant. Chem.*, **S16**, 377 (1982).
134. Gray, S. K., and Child, M. S. (in press).
135. Shapiro, M., and Levine, R. D., *Chem. Phys. Lett.*, **5**, 449 (1970).
136. Atabek, O., and Lefebvre, R., *J. Chem. Phys.*, **67**, 4983 (1977).
137. Beswick, J. A., Shapiro, M., and Sharon, R., *J. Chem. Phys.*, **67**, 4045 (1977).
138. Micha, D. A., *Chem. Phys. Lett.*, **1**, 139 (1967).
139. Micha, D. A., *Phys. Rev.*, **162**, 88 (1967).
140. Micha, D. A., *Accts. Chem. Res.*, **6**, 138 (1973).
141. Redmon, M. J., and Micha, D. A., *Chem. Phys. Lett.*, **28**, 341 (1974).
142. Levine, R. D., Johnson, B. R., Muckerman, T., and Bernstein, R. B., *J. Chem. Phys.*, **49**, 56 (1968).
143. Ashton, C. J., Child, M.S., and Hutson, J. M., *J. Chem. Phys.*, **78**, 4025 (1983).
144. Hutson, J. M., Ashton, C. J., and LeRoy, R. J., *J. Phys. Chem.*, **87**, 2713 (1983).
145. Hutson, J. M., and LeRoy, R. J., *J. Chem. Phys.*, **78**, 4040 (1983).
146. Atabek, O., and Lefebvre, R., *Chem. Phys.*, **52**, 199 (1980).
147. Atabek, O., and Lefebvre, R., *Chem,. Phys.*, **55**, 395 (1981).
148. Atabek, O., and Lefebvre, R., *Chem, Phys.*, **56**, 195 (1981).
149. Special issue of *Intern. J. Quant. Chem.*, **14** (6) (1978).
150. Chu, S., *J. Chem. Phys.*, **72**, 4772 (1980).
151. Datta, K. K., and Chu, S.-I., *Chem. Phys. Lett.*, **95**, 38 (1983).
152. Moiseyev, N., *Chem. Phys. Lett.*, **99**, 364 (1983).
153. Dunker, A. M., and Gordon, R. G., *J. Chem. Phys.*, **68**, 700 (1978).
154. McKellar, A. R. W., *Faraday, Disc. Chem. Soc.*, **73**, 89 (1982).
155. Whaley, K. B., and Light, J. C., *J. Chem. Phys.*, **77**, 1818 (1982); *J. Chem. Phys.*, **79**, 3604 (1983).
156. Mukamel, S., and Jortner, J., *Chem. Phys. Lett.*, **40**, 150 (1976).
157. Mukamel, S., and Jortner, J., *J. Phys.*, **B10**, 2583 (1977).
158. Brandrauk, A., and Sink, M. L., *J. Chem. Phys.*, **74**, 1110 (1981).
159. Leforestier, C., and Wyatt, R. E., *Phys. Rev.*, **A25**, 1250 (1982).
160. Leforestier, C., and Wyatt, R. E., *J. Chem. Phys.*, **78**, 2334 (1983).
161. Lan, A. M. F., *Phys. Rev.*, **A18**, 172 (1978).
162. Light, J. C., and Altenberger–Siczek, A., *J. Chem. Phys.*, **70**, 4108 (1979).
163. George, T. F., Zimmerman, I. H., Yuan, J. M., Laing, J. R., and De-Vries, P. L., *Accts. Chem. Research*, **10**, 449 (1977).

164. George, T. F., *J. Phys. Chem.*, **86**, 10 (1982).
165. Kulander, K. C., and Orel, A. E., *J. Chem. Phys.*, **74**, 6529 (1981).
166. Kulander, K. C., and Orel, A. E., *J. Chem. Phys.*, **75**, 675 (1981).
167. Lam, K. S., Zimmerman, I. H., Yuan, J. M., Laing, J. R., and George, T. F., *Chem. Phys.*, **26**, 455 (1977).
168. Light, J. C., and Szoke, A., *Phys. Rev.*, **A18**, 1363 (1978).
169. Mies, F. H. in *Theoretical Chemistry: Advances and Perspectives, Vol. 6B*, Academic Press, New York, 1981.
170. Kulander, K. C., and Rebentrost, F., *J. Chem. Phys.*, **80**, 5623 (1984).
171. Leasure, S. C., Milfeld, K. F., and Wyatt, R. E., *J. Chem. Phys.*, **74**, 6197 (1981).
172. Schek, I., and Jortner, J., *Chem. Phys. Lett.*, **63**, 5 (1979).
173. Schek, I., Jortner, J., and Sage, M. L., *Chem. Phys.*, **59**, 11 (1981).
174. Schek, I., Jortner, J., and Sage, M. L., *Chem. Phys. Lett.*, **64**, 209 (1979).
175. Schek, I., Sage, M. L., and Jortner, J., *Chem. Phys. Lett.*, **63**, 230 (1979).
176. Brandrauk, A. D., Turcotte, G., and Lefebure, R., *J. Chem. Phys.*, **76**, 225 (1982).
177. Brandrauk, A. D., and Turcotte, G., *J. Chem. Phys.*, **77**, 3867 (1982).

AUTHOR INDEX

SUBJECT INDEX

The following abbreviations are used in the index: LIF = laser induced fluorescence, LIS = laser isotope separation, MPD = multiphoton dissociation, MPI = multiphoton ionization, PES = photoelectron spectroscopy, PD = photodissociation, PI = photoionization.

Substances listed by formula precede the main entries under each letter.